ECOLOGY AND EVOLUTION OF CANCER

ECOLOGY AND EVOLUTION OF CANCER

---

*Centre for Integrative Ecology, School of Life and Environmental Sciences,
Deakin University, Geelong, VIC, Australia*

BENJAMIN ROCHE
*UMMISCO (International Center for Mathematical and Computational Modeling
of Complex Systems), UMI IRD/UPMC UMMISCO, Bondy, France;
MIVEGEC (Infectious Diseases and Vectors: Ecology, Genetics, Evolution
and Control), UMR IRD/CNRS/UM 5290, Montpellier, France;
CREEC (Centre for Ecological and Evolutionary Research on Cancer),
Montpellier, France*

FRÉDÉRIC THOMAS
*MIVEGEC (Infectious Diseases and Vectors: Ecology, Genetics, Evolution
and Control), UMR IRD/CNRS/UM 5290, Montpellier, France;
CREEC (Centre for Ecological and Evolutionary Research
on Cancer), Montpellier, France*

ELSEVIER

ACADEMIC PRESS
An imprint of Elsevier
elsevier.com

Academic Press is an imprint of Elsevier
125 London Wall, London EC2Y 5AS, United Kingdom
525 B Street, Suite 1800, San Diego, CA 92101-4495, United States
50 Hampshire Street, 5th Floor, Cambridge, MA 02139, United States
The Boulevard, Langford Lane, Kidlington, Oxford OX5 1GB, United Kingdom

Notices
Knowledge and best practice in this field are constantly changing. As new research and experience broaden
our understanding, changes in research methods, professional practices, or medical treatment may become
necessary.

Practitioners and researchers must always rely on their own experience and knowledge in evaluating and
using any information, methods, compounds, or experiments described herein. In using such information
or methods they should be mindful of their own safety and the safety of others, including parties for whom
they have a professional responsibility.

To the fullest extent of the law, neither the Publisher nor the authors, contributors, or editors, assume any
liability for any injury and/or damage to persons or property as a matter of products liability, negligence
or otherwise, or from any use or operation of any methods, products, instructions, or ideas contained in the
material herein.

Library of Congress Cataloging-in-Publication Data
A catalog record for this book is available from the Library of Congress

British Library Cataloguing-in-Publication Data
A catalogue record for this book is available from the British Library

ISBN: 978-0-12-804310-3

For information on all Academic Press publications visit our website at
https://www.elsevier.com/books-and-journals

Working together
to grow libraries in
developing countries

www.elsevier.com • www.bookaid.org

Publisher: Sara Tenney
Acquisition Editor: Kristi Gomez
Editorial Project Manager: Pat Gonzalez
Production Project Manager: Julia Haynes
Designer: Victoria Pearson

Typeset by Thomson Digital

Contents

List of Contributors

Jerôme Abadie   LUNAM University, Oniris, AMaROC, Nantes, France

Ole Ammerpohl   Institute of Human Genetics, Christian-Albrechts-University Kiel and University Hospital Schleswig-Holstein, Campus Kiel, Kiel, Germany

Audrey Arnal   MIVEGEC (Infectious Diseases and Vectors: Ecology, Genetics, Evolution and Control), UMR IRD/CNRS/UM 5290; CREEC (Centre for Ecological and Evolutionary Research on Cancer), Montpellier, France

Maureen Banach   Department of Microbiology and Immunology, University of Rochester Medical Center, Rochester, NY, United States

Christa Beckmann   Centre for Integrative Ecology, School of Life and Environmental Sciences, Deakin University, Geelong, VIC, Australia

Florence Bernex   CREEC (Centre for Ecological and Evolutionary Research on Cancer); Montpellier University; RHEM, IRCM, Institute of Cancer Research Montpellier, INSERM; ICM Regional Cancer Institute of Montpellier, Montpellier, France

Amy M. Boddy   Department of Psychology, Arizona State University, Tempe, AZ; Center for Evolution & Cancer, University of California San Francisco, San Francisco, CA, United States

Joel S. Brown   Cancer Biology and Evolution Program, H. Lee Moffitt Cancer Center and Research Institute, Tampa, FL; Department of Biological Sciences, University of Illinois at Chicago; Department of Evolutionary Biology, University of Illinois, Chicago, IL, United States

Sam P. Brown   School of Biological Sciences, Georgia Institute of Technology, Atlanta, GA, United States

Francisco De Jesús Andino   Department of Microbiology and Immunology, University of Rochester Medical Center, Rochester, NY, United States

James DeGregori   Department of Biochemistry and Molecular Genetics, University of Colorado School of Medicine, Aurora, CO, United States

Eva-Stina Edholm   Department of Microbiology and Immunology, University of Rochester Medical Center, Rochester, NY, United States

Pedro M. Enriquez-Navas   H. Lee Moffitt Cancer Center and Research Institute, Tampa, FL, United States

Paul W. Ewald   Department of Biology; Department of Biology and the Program on Disease Evolution, University of Louisville, Louisville, KY, United States

Dominique Faugère   MIVEGEC (Infectious Diseases and Vectors: Ecology, Genetics, Evolution and Control), UMR IRD/CNRS/UM 5290; CREEC (Centre for Ecological and Evolutionary Research on Cancer), Montpellier, France

Jasmine Foo   School of Mathematics, University of Minnesota, Minneapolis, MN, United States

Déborah Garcia   MIVEGEC (Infectious Diseases and Vectors: Ecology, Genetics, Evolution and Control), UMR IRD/CNRS/UM 5290; CREEC (Centre for Ecological and Evolutionary Research on Cancer), Montpellier, France

Colleen M. Garvey   Lawrence J. Ellison Institute for Transformative Medicine, University of Southern California, Los Angeles, CA, United States

Robert A. Gatenby   Department of Integrated Mathematical Oncology; Department of Radiology; Cancer Biology and Evolution Program, H. Lee Moffitt Cancer Center and Research Institute, Tampa, FL, United States

Cindy Gidoin   MIVEGEC (Infectious Diseases and Vectors: Ecology, Genetics, Evolution and Control), UMR IRD/CNRS/UM 5290, Montpellier, France

Robert J. Gillies   Department of Cancer Imaging and Metabolism; Department of Radiology, H. Lee Moffitt Cancer Center and Research Institute, Tampa, FL, United States

Christoph Grunau   University of Perpignan Via Domitia, IHPE UMR 5244, CNRS, IFREMER, University of Montpellier, Perpignan, France

Valerie K. Harris   Cancer and Evolution Lab, Arizona State University; School of Biological and Health Systems Engineering, Arizona State University, Tempe, AZ, United States

Kirsten Hattermann   Institute of Anatomy, Christian-Albrechts-University Kiel, Kiel, Germany

Janka Held-Feindt   Department of Neurosurgery, University Hospital Schleswig-Holstein Campus Kiel, Kiel, Germany

Henry H. Heng   Center for Molecular Medicine and Genetics, Wayne State University School of Medicine; Department of Pathology, Wayne State University School of Medicine; Karmanos Cancer Institute, Detroit, MI, United States

Arig Ibrahim-Hashim   Department of Cancer Imaging and Metabolism, H. Lee Moffitt Cancer Center and Research Institute, Tampa, FL, United States

Irina Kareva   Simon A. Levin Mathematical, Computational and Modeling Sciences Center (SAL MCMSC), Arizona State University, Tempe, AZ, United States

Hanna Kokko   Department of Evolutionary Biology and Environmental Studies, University of Zurich, Winterthurerstrasse, Zurich, Switzerland

Sophie Labrut   LUNAM University, Oniris, AMaROC, Nantes, France

Karin Lemberger   Vet Diagnostics, Lyon, France

Danika Lindsay   School of Mathematics, University of Minnesota, Minneapolis, MN, United States

Mark C. Lloyd   Inspirata, Inc., Tampa, FL; Department of Biological Sciences, University of Illinois at Chicago, Chicago, IL, United States

Anders Pape Møller Ecology Systematic Evolution, CNRS UMR 8079, University Paris-Sud, Orsay, France

Thomas Madsen School of Biological Sciences, University of Wollongong, Wollongong, NSW; Centre for Integrative Ecology, School of Life and Environmental Sciences, Deakin University, Geelong, VIC, Australia

Andriy Marusyk Department of Cancer Imaging and Metabolism, H Lee Moffitt Cancer Center and Research Institute, Tampa, FL, United States

John F. McDonald School of Biological Sciences; Integrated Cancer Research Center; Parker H. Petit Institute for Bioengineering and Bioscience, Georgia Institute of Technology, Atlanta, GA, United States

Shannon M. Mumenthaler Lawrence J. Ellison Institute for Transformative Medicine, University of Southern California, Los Angeles, CA, United States

Aurora M. Nedelcu Department of Biology, University of New Brunswick, Fredericton, NB, Canada

Randolph M. Nesse The Center for Evolution and Medicine, Arizona State University, Tempe, AZ, United States

Leonard Nunney Department of Biology, University of California Riverside, Riverside, CA, United States

Andrew F. Read Center for Infectious Disease Dynamics, Pennsylvania State University, University Park, PA, United States

Kun Hyoe Rhoo Department of Microbiology and Immunology, University of Rochester Medical Center, Rochester, NY, United States

Christoph Röcken Department of Pathology, Christian-Albrechts-University Kiel and University Hospital Schleswig-Holstein, Campus Kiel, Kiel, Germany

Jacques Robert Department of Microbiology and Immunology, University of Rochester Medical Center, Rochester, NY, United States

Benjamin Roche UMMISCO (International Center for Mathematical and Computational Modeling of Complex Systems), UMI IRD/UPMC UMMISCO, Bondy; MIVEGEC (Infectious Diseases and Vectors: Ecology, Genetics, Evolution and Control), UMR IRD/CNRS/ UM 5290, Montpellier; CREEC (Centre for Ecological and Evolutionary Research on Cancer), Montpellier, France

Shonagh Russell Department of Cancer Biology Ph.D. Program, University of South Florida, Tampa, FL; Department of Cancer Imaging and Metabolism, H. Lee Moffitt Cancer Center and Research Institute, Tampa, FL, United States

Heiner Schäfer Institute for Experimental Cancer Research, Christian-Albrechts-University Kiel and University Hospital Schleswig-Holstein, Campus Kiel, Kiel, Germany

Christian Schem Department of Obstetrics and Gynecology, University Hospital Schleswig-Holstein, Campus Kiel, Kiel, Germany

Denis Schewe Department of Pediatrics, University Hospital Schleswig-Holstein, Campus Kiel, Kiel, Germany

Joshua D. Schiffman Departments of Pediatrics and Oncological Sciences, Huntsman Cancer Institute, University of Utah, Salt Lake City, UT, United States

Susanne Schindler Department of Evolutionary Biology and Environmental Studies, University of Zurich, Winterthurerstrasse, Zurich, Switzerland

Hinrich Schulenburg Department of Evolutionary Ecology Genetics, Zoological Institute, Christian-Albrechts-University Kiel, Kiel, Germany

Susanne Sebens Institute for Experimental Cancer Research, Christian-Albrechts-University Kiel and University Hospital Schleswig-Holstein, Campus Kiel, Kiel, Germany

Kathleen Sprouffske Department of Evolutionary Biology and Environmental Studies, University of Zurich, Winterthurerstrasse, Zurich, Switzerland

Holly A. Swain Ewald Department of Biology, University of Louisville, Louisville, KY, United States

Michael Synowitz Department of Neurosurgery, University Hospital Schleswig-Holstein Campus Kiel, Kiel, Germany

Aurélie Tasiemski Lille1 University, UMR CNRS 8198, Evolution, Ecology and Paleontology Unit, Villeneuve d'Ascq, France

Frédéric Thomas MIVEGEC (Infectious Diseases and Vectors: Ecology, Genetics, Evolution and Control), UMR IRD/CNRS/UM 5290, Montpellier; CREEC (Centre for Ecological and Evolutionary Research on Cancer), Montpellier, France

Sanjay Tiwari Department of Diagnostic Radiology and Neuroradiology, University Hospital Schleswig-Holstein Campus Kiel, Kiel, Germany

Arne Traulsen Department of Evolutionary Theory, Max Planck Institute for Evolutionary Biology, Plön, Germany

Anna Trauzold Institute for Experimental Cancer Research, Christian-Albrechts-University Kiel and University Hospital Schleswig-Holstein, Campus Kiel, Kiel, Germany

Beata Ujvari Centre for Integrative Ecology, School of Life and Environmental Sciences, Deakin University, Geelong, VIC, Australia

Thomas Valerius Department of Internal Medicine II, Section for Stem cell transplantation and Immunotherapy, University Hospital Schleswig-Holstein, Campus Kiel, Kiel, Germany

Mark Vincent Department of Oncology, University of Western Ontario, London, ON, Canada

Marion Vittecoq MIVEGEC (Infectious Diseases and Vectors: Ecology, Genetics, Evolution and Control), UMR IRD/CNRS/UM 5290, Montpellier; CREEC (Centre for Ecological and Evolutionary Research on Cancer), Montpellier; Research Center of the Tour du Valat, Arles, France

Wollein Waldetoft School of Biological Sciences, Georgia Institute of Technology, Atlanta, GA, United States

Daniela Wesch Institute of Immunology, Christian-Albrechts-University Kiel and University Hospital Schleswig-Holstein, Campus Kiel, Kiel, Germany

Preface

Cancer is not only a major cause of human death worldwide that touches nearly every family on the planet, but also a disease that affects all other multicellular organisms. Despite this, oncology and other biological sciences, such as ecology and evolution have until very recently developed in relative isolation. Although the first synergistic approaches to understand cancer were proposed in the mid-1970s by J. Cairns and P.C. Nowell, it is only during the last decade that the scientific community started to fully realize that adopting evolutionary principles and ecological approaches to cancer could greatly enhance our understanding of neoplastic progression, improve cancer prevention and therapies.

This view has transformed our understanding of cancer.

Today, most scientists accepts that cancer is a disease associated with clonal evolution and cell competition within the body, that appeared with the transition to multicellularity more than half a billion years ago. Specifically, somatic selection and adaptation to local and distant microenvironments are fundamental processes leading to malignancy, and its manifestations, that is, neoangiogenesis, immune system evasion, metastasis, resistance to therapies, and even contagion.

Although the congruence between the theory of cancer initiation, progression, and evolutionary and ecological concepts are increasingly accepted, this area of research is still in its infancy and a considerably enhanced research effort is urgently needed.

Applying evolutionary ecology to oncology is particularly crucial because treatment strategies have so far not lived up to expectations. Consequently, cancer research is now at a crossroad, needing novel ideas, major innovation, and new and unprecedented transdisciplinary approaches. The traditional separation between disciplines is more than ever a fundamental limitation that needs to be conquered if complex processes, such as oncogenesis, are to be understood.

Auspiciously, an increasing number of scientists and clinicians are now actively involved in pursuing interdisciplinary research and apply an evolutionary ecology view to cancer emergence and progression. This scientific community capitalizes on a panel of specialists using different, yet complementary approaches to cancer: mathematics, ecology, cell and evolutionary biology, and clinical research. This interdisciplinary field of study is also now moving beyond its descriptive phase and into the new dimensions of applying the theoretical understanding of cancer adaptations to treatment and prevention.

Apart from cancer being a problem of humanity, it also has, a so-far largely underestimated but significant impact on ecosystem functioning. Similar to humans, benign and malignant tumors are frequent in animals and prior to eventually causing death, cancer is likely to influence the organisms' fitness by reducing competitive abilities, increasing susceptibility to pathogens, and vulnerability to predation.

Despite the potential importance of these ecological impacts, oncogenic phenomena are rarely incorporated into ecosystem modeling. Acknowledging that cancer incidence in animal species may experience an upsurge due to the cumulative effect of ever-increasing pollution of our ecosystems and the detrimental impact of ongoing climate change, it is essential to improve our knowledge of the reciprocal actions between oncogenic processes, intra-/interspecific interactions and animal behavior.

For all these reasons we felt that it was timely to provide an up-to-date, authoritative and challenging synopsis on the topic of ecology, evolution, and cancer that assesses the current state of developments in the field and importantly lays down a framework for future research.

We hope that this book presents materials that will be useful to a broad audience with a wide range of interest and expertise, from oncologists to ecologists. To achieve this goal, we assembled a team of experts to provide an overview of this engrossing topic, with the aim, above all, to provide an integrated understanding of cancer in *Life*.

Beata Ujvari, Benjamin Roche,
Frédéric Thomas.

Acknowledgments

While compiling the book we lost dear friends and family members to cancer. The book grew out of desperation to unable to help and feeling impotent when facing the suffering of our loved ones. We sincerely hope that by laying down novel trajectories to understand the life history and evolution of cancer, we will eventually make an impact on tackling this devastating disease. We are grateful to our families and colleagues who patiently stood by us, provided endless support, and cheered us along the months of editing the book.

This work was supported by the *Agence Nationale de la Recherche* (Blanc project EVOCAN), the Centre National de la Recherche Scientifique (CNRS, Institute of Ecology and Evolution, INEE), the French National Research Institute for Sustainable Development (IRD), the University of Montpellier, the Labex CEMEB, the Australian Academy of Science's French-Australian Science Innovation Collaboration Program Early Career Fellowship, André HOFFMANN and la Fondation MAVA, and an International Associated Laboratory Project France/Australia.

Introduction: Five Evolutionary Principles for Understanding Cancer

Randolph M. Nesse

The Center for Evolution and Medicine, Arizona State University, Tempe, AZ, United States

Evolutionary perspectives are transforming our understanding of cancer. The plural is appropriate because evolution offers five somewhat separate principles useful for understanding cancer. Each is a landmark on a map of a new scientific territory. Descriptions of each principle and their relationships can help maintain orientation in an exponentially growing new field.

Three of the perspectives focus on how malignancies grow within the body. The greatest excitement is coming from recognition that cancers evolve within the body by somatic selection, by which the fastest reproducing malignant cells increase, their prevalence. In retrospect, it is amazing that this was not always obvious. The second perspective uses ecological principles to analyze how cancers create and interact with microenvironments that speed or slow tumor growth. The third uses principles from behavioral ecology to analyze how cancer clones compete and cooperate with each other. The final two perspectives focus on natural selection at the organismal level. The fourth uses principles of phylogeny and natural selection to understand why cancer is so rare. The ultimate perspective uses principles from evolutionary medicine to understand why we remain so vulnerable to cancer.

> **Five evolutionary principles for understanding cancer**
> 1. Somatic selection shapes malignancies.
> 2. Ecological principles explain how cancers interact with microenvironments.
> 3. Behavioral ecological principles explain competition and cooperation among cancer clones.
> 4. Natural selection explains why cancer is rare.
> 5. Evolutionary medicine explains why cancer is common.

All of these principles have been well described in recent articles and books (Aktipis and Nesse, 2013; Greaves, 2000, 2007, 2015; Hochberg et al., 2012; Merlo et al., 2006; Nunney, 2013). Each of the chapters in this book elaborates details of one or another aspect. The bird's eye overview of the landscape in this chapter cannot begin to summarize those details, it can only provide a rough map of this new exciting scientific territory.

SOMATIC SELECTION AND HOW CANCERS EVOLVE

An older view envisioned cancer as resulting from a defective cell that replicated identical copies of itself out of control. Remnants of this preevolutionary view of cancer persist, but recognition that cancer is an evolutionary process (Nowell, 1976) is growing fast, thanks in large part to new data showing massive heterogeneity among the cells in a single malignancy, a theme in many chapters in this book. That heterogeneity turns out to be important not only theoretically, but also as a predictor of the future trajectory of premalignant and malignant cell lines (Andor et al., 2016; Maley et al., 2006). As would be expected by Fisher's fundamental theorem, the rate of fitness change is proportional to the amount of variation, in this case, massive variation among individual cells in a tumor promotes rapid changes by somatic selection.

This revised view transforms our view of cancer (Aktipis and Nesse, 2013; Crespi and Summers, 2005; Greaves, 2015; Nunney, 2013; and many chapters in this book). The "war on cancer" encouraged thinking of it as one disease with one cause, but every decade of subsequent research has revealed new layers of complexity among related disorders that turn out to be diverse (Aktipis et al., 2011). Early studies looked for abnormalities in cancer cell lines, with little attention to selection that had taken place over the course of generations of replication in vitro. People still tend to think of cancers from each tissue as specific diseases: lung cancer, breast cancer, prostate cancer and so on. It's now clear, however, that all cancers share deep similarities in the ways cell cycle control is disrupted, and that apparently similar cancers from the same tissue can be very different. There is continuing hope that distinct genotypes will define specific subtypes of cancer. Identifying such variations is certainly useful, especially for making decisions about chemotherapy, but an evolutionary perspective suggests it is unrealistic to expect to find a few genetically uniform specific subtypes. Instead, we should expect that almost all cancer clones, even those from different location in the same malignancy, will be to some extent distinct. Heterogeneity is intrinsic to cancer (see especially Chapters 5, 10, and 17).

This new perspective has, in combination with new sequencing technologies and phylogenetic methods, encouraged investigations into the sequence by which individual cancers develops. The order in which various driver and passenger mutations arise is important in changing the selection forces that shape subsequent tumor development. Mutations that disrupt DNA replication and repair have special significance because they vastly increase the variation on which somatic selection can act. Variation among cells in different parts of a tumor is the norm, and that variation is an important variable that predicts progression (Andor et al., 2016). For instance, the heterogeneity of cells in Barrett's esophagus is a strong predictor of progression to cancer independent of information about individual cells (Maley et al., 2006).

The process of somatic selection acting on cells is, however, somewhat different from natural selection shaping species. The generic principle of selection explains changes in a group that inevitably result when variations between individuals influence their future prevalence in a group. For instance, the collection of items in a cottage cupboard will tend to shift over the years toward robust items because fragile glassware is more likely to break and be discarded. Natural selection is the same process, with the addition of reproduction and heritability. When variations in heritable factors influence the number of offspring who make it to maturity, the average characteristics of a group will change over time to be more like those of the individuals who had more offspring than average.

This process of natural selection shapes mechanisms that maintain and transmit the information code with extraordinary fidelity. Almost all new mutations are harmful, so selection shapes mechanisms that minimize them, repair them, and compensate for them. The process is different in somatic selection. Cancers arise from driver mutations that give a cell line a selective advantage. Some of them initiate mutations that cause genome chaos beyond the mutation rate that would maximize tumor growth. Somatic selection can be for or against

cancer, and the direction of selection can shift as a malignancy develops in different loci (Michor et al., 2003).

Natural selection shapes phenotypes that maximize transmission of genes to future generations. For selection within the body, there is no surviving generation (with the informative exception of transmissible cancers (see Chapter 12). Whether somatic selection shapes traits that benefit the tumor as a whole at the expense of individual cells is a fascinating question, considered later.

Convergent evolution is prevalent in somatic selection, as it is in natural selection (see also Chapter 17). All cancers face the same set of constraints, and mutations that allow a tumor to escape from those constraints get a selective advantage and become more common. Thus, somatic selection increases the prevalence of mutations that disrupt apoptosis, preserve telomeres, stimulate angiogenesis, interfere with recognition by the immune system, and foster plasticity that enables coping with rapidly changing microenvironments (see Chapter 8). These ecological challenges faced by every cancer provide an evolutionary framework for understanding cancer that complements the well-known hallmarks of cancer (Hanahan and Weinberg, 2011).

Somatic selection and natural selection also differ in the role of epigenetics. In naturally selected species, epigenetic marks that inhibit, or stimulate gene expression can transmit information between one or a few subsequent generations; the extent to which this reflects adaptations or epiphenomena is an area of active study. Epigenetic changes that arise in the course of normal phenotype development influence gene expression patterns that tend to persist across the lifespan. They account for the differences between normal cell types in an individual, including the differences between germ cells and somatic cells, so it should not be surprising that they also account for many aspects of malignant transformation. Epigenetic changes are increasingly recognized as important and perhaps essential mediators that can prevent or speed transformation to cancer (see also Chapters 5 and 6).

ECOLOGICAL ENVIRONMENTS INFLUENCE CANCER GROWTH

There is no such thing as fitness for a gene or individual except in relation to a specific ecological environment. Fitness is a characteristic not of genotypes, but of phenotypes interacting with environments. The microenvironments inhabited by cancers influence their growth, as much as their genotypes. Furthermore, the growth of a tumor creates microenvironments that can speed or slow subsequent growth.

Of particular interest, and emphasized by several chapters in this book, is the hypoxia that results when a tumor has an inadequate blood supply (see especially Chapters 8 and 19). Even in tumors that stimulate angiogenesis, unregulated growth is likely to block arterial routes providing oxygen and nutrients to small or large parts of the tumor. This can severely restrain growth or even eliminate a tumor. More often, however, the resulting hypoxia creates a microenvironment that compromises the efficiency of the usual immunologic tumor control mechanisms. This may have major implications for treatment strategies (see also pseudohypoxia in Chapter 4).

The rapid changes in microenvironments also may select for plasticity in tumors. Clones gain an advantage if they have the capacity to adapt their metabolism to sudden changes in oxygen tension or nutrient availability without changes in their genomes (see also Chapter 6). When nutrient supply is limited, cells that can enter a dormant state get an advantage, with major implications for how best to use chemotherapy agents that preferentially attack rapidly dividing cells (see also Chapter 20). An ecological approach offers the foundation for adaptive therapy that adjusts chemotherapy doses to minimize tumor growth, instead of to attempt to kill every malignant cell (see Chapter 14).

Many other ecological principles are useful for understanding the ecological setting in which tumors grow (Daoust et al., 2013; Ducasse et al., 2015; Thomas et al., 2013). The limitations of blood supply and its availability only along widely separated routes creates riparian environments with selection forces that differ depending on the distance of a cell from the blood supply. Principles of island ecology are relevant to understanding isolated tumors and metastasis. Thinking of islands of cancer cells as if they were an endangered species provides suggestions for how to speed their extinction (Korolev et al., 2014). Chapters 8 and 11 provide more about ecological applications in cancer, as do several articles in a special issue of Evolutionary Applications on cancer (Thomas et al., 2013). Of particular interest is the possibility that tumors can create their own niches, perhaps even with different clones cooperating to construct and expand the niche (Barcellos-Hoff et al., 2013).

BEHAVIORAL ECOLOGY AND COOPERATION AMONG CANCER CLONES

Heterogeneity in cancer is now well recognized, along with increasing understanding of its several possible origins. On the cutting edge is consideration of the possibility that different clones can provide resources that promote the survival and replication of other clones in a process that is something like cooperation (Aktipis et al., 2015, see also Chapter 17). There is no higher-level enforcement of cooperation in a tumor, and selection cannot shape persisting sophisticated coordinated systems like those that account for termite societies or the cooperation among cells in metazoan bodies. Nonetheless, it is important to consider how different clones may cooperate and compete with each other, and studies of such systems may yield new therapeutic strategies.

A simple example is when some malignant cells secrete substances that disrupt host cells and expand the ecological space while other cells manufacture angiogenesis factors that vascularize the space. Such interactions bring the principles from behavioral ecology into play. Somewhat separate groups of cells may well both benefit from mutualistic interactions. All clones are inevitably shaped, however, to maximize growth in whatever environment they exist in.

Hypoxia induces the transition of malignant cells from attached epithelial cells into mesenchymal forms that are free to circulate and initiate metastases. Many higher organisms have evolved behavior regulation mechanisms that monitor the environment and initiate movement elsewhere when conditions deteriorate. The analogy is attractive despite the major differences. It is also of special importance because of the possibility that treatments causing hypoxia might stimulate this transition in ways that influence metastasis.

EVOLUTIONARY EXPLANATIONS FOR WHY CANCER IS RARE

From an evolutionary perspective, cancer is astoundingly rare. Controlling unrestrained cell growth is the original giant problem that multicellular life had to overcome before large complex organisms could evolve (Smith and Szathmáry, 1995). Attachment between cells is easy to explain; individuals with tendencies to attach to each other can increase their nutrition and safety in ways that increase their own replication. Shaping mechanisms that reduce the fitness of individual cells in ways that benefit a completely larger organism is a challenge more difficult by orders of magnitude. The process took billions of years for good reasons, a main one being the challenge of controlling the unregulated cell divisions that are the hallmark of cancer (see Chapter 1).

The earth is about 4.5 billion years old. Life emerged, in the form of bacteria that leave their traces in stone about 3.7 billion years ago, about within the first 20% of time from earth's origin. Multicellular organisms didn't appear until over 3 billion years later. Complex multicellular organisms have existed for only the most recent 10% of the history of the planet. The fast emergence of life, and the extended delay until the emergence of complex multicellular organisms, suggests that simple life forms may exist on many other planets, but complex multicellular organisms are likely to be rare.

Why did it take about 75% of the time of earth's existence for simple life to evolve into complex multicellular life (see also Chapter 1)? Among several obstacles overcome by major transitions, difficulty of evolving mechanisms to enforce cell cooperation was crucial. As cells united to begin to form early multicellular organisms, their generation times increased, so faster-evolving smaller organisms could overcome their defenses. However, being larger also offers protection against predators, and ecosystems had open niches for larger organisms (Bonner, 2011). The related challenge was the need to inhibit cell division, foster division of labor, and create ways to purge cells that nonetheless replicated at the expense of the whole organism (Smith and Szathmáry, 1995).

The Cambrian explosion, in which small simple organisms evolved into large complex ones, was but an eye blink in the history of life (Chapter 16). It started 542 million years ago, and gave rise to the main phyla of metazoans in as little as 25 million years. This speed has been attributed to the emergence of predators whose presence created strong selection for larger prey in a rapid process of coevolution that also rapidly made predators larger (Niklas, 2014). These larger bodies required new mechanisms to enforce cooperation. It is fascinating to consider that external threats may have been a crucial factor that shaped cells go sacrifice their own reproduction for the benefit of larger bodies, and that there is still much debate about this (Erwin and Valentine, 2013).

Multicellular species in which all cells are capable of reproduction cannot become complex with division of labor because there is no way to enforce cooperation. The differentiation of a sterile line of somatic cells was the key innovation that made the transition to complex multicellular life possible (Buss, 1987; Smith and Szathmáry, 1995). Sterile cell lines can advance their genetic interests only by advancing the interests of the whole organism, or rather, the organism's genes. Once a sequestered germ line is separated from a sterile somatic cell line, selection can shape specialized kinds of somatic cells with separate functions, including enforcing cooperation among other cells. The dynamics of this transition have been the focus of intense study (Queller, 2000).

Any cell potentially capable of independent reproduction poses a threat that must be carefully monitored and controlled. For organisms as complex as vertebrates, somatic cells cannot establish their own continuing separate lineages, with the dramatic exception of transmissible cancers (see Chapter 12). Even stem cells cannot get out of the body and reproduce on their own. However, because they lack the controls that strictly limit replication of all normal somatic cells, stem cells are capable of indefinite replication, and thus cancer.

This helps to explain many curious aspects of stem cells (Greaves and Maley, 2012). They are relatively rare, for the good reason that they are vulnerable to malignant transformation. They tend to be sequestered in locations away from toxins and damage. The sequence with which they differentiate into descendent cells is strictly controlled. This means that mutations that immortalize cells are essential to malignant transformation.

This perspective distinguishes the several factors necessary to control aberrant cell division (Aktipis et al., 2015). First, there must be a restraint on division. Second, there must be policing to eliminate or dissociate cells that try to divide at a cost to the group. Third, there must be policing to prevent individual cells from taking advantage of group resources for their own benefit. Finally, there must be a mechanism to eliminate cells that go rogue. Chapter 7 provides comprehensive details in a comparative perspective.

These different factors have varied requirements and impacts in different species, making comparative studies valuable. The idea that cancer is only a disease of humans is contradicted by evidence for cancer in animals in the wild (see also Chapter 2) and an overview of cancer across the breadth of the living world (Aktipis et al., 2015). The similarity in rates of cancer in organisms with orders of magnitude different numbers of cells is Peto's paradox (Nunney et al., 2015; Peto, 1977). The solution to the paradox is that natural selection shaped mechanisms that provide protection against cancer that are adequate to however, many cells a phenotype has; the effectiveness of these mechanisms is limited not by the number of cells, but mainly by tradeoffs with their costs. This perspective has inspired research that has predicted and confirmed increased number and expression of tumor suppression genes in large organisms (Caulin and Maley, 2011; Nagy et al., 2007; Sulak et al., 2016).

The application of biological theories of sociality to cancer is opening up whole new realms of cancer biology. Who could have anticipated that deep thinking about altruistic behavior in bees (Hamilton, 1964) would develop into an elaborated and still-developing body of theory crucial for understanding cancer (Aktipis et al., 2015; Frank, 2007; Nunney, 2013)?

WHY IS CANCER SO COMMON? EVOLUTIONARY MEDICINE

Even as we marvel at the ability of natural selection to shape mechanisms to prevent cancer, the reality is that about one half of us will have cancer at sometime during our lives. Explaining such vulnerabilities to disease is a primary focus for evolutionary medicine (Nesse and Williams, 1994). A list of possible kinds of evolutionary explanations for apparent maladaptations helps to avoid the tendency to overemphasize one explanation

(Crespi, 2000; Nesse, 2005). For instance, a kerfuffle was aroused by a recent publication showing that variation in vulnerability to cancer across species is directly proportional to the number of stem cells and the number of their divisions (Tomasetti and Vogelstein, 2015). Journalists took the implication that most cancer is a result of "just bad luck," arousing rebuttals that emphasized the role of environmental factors, and the difference between explaining variations in cancer rates and causes of cancer (Couzin-Frankel, 2015).

Two kinds of questions need to be distinguished. First is the usual question about why some individuals get a disease and others do not. Such questions are answered by descriptions of individual differences in mechanisms at a reductionist level, and factors that influence those mechanisms. Evolutionary medicine asks a different question about why all members of a species are the same in ways that leave them vulnerable to disease. Such questions have six possible kinds of answers, each of which is summarized briefly later.

Constraints

The traditional explanation for traits that leave bodies vulnerable to disease is that natural selection has limits. Mutations happen. Natural selection tends to eliminate deleterious ones, but the process takes time and is subject to the vagaries of any stochastic process. Useful mutations tend to increase, but genetic drift makes the process uncertain. The limits of selection offer a powerful and correct explanation that is only part of the picture. It is often framed using an outmoded model of the body as a designed machine whose robustness depends only on redundancies, instead of the complex networks that make bodies stable (Nesse, 2016). Explanations that refer to mere chance are also unwelcome. People who have cancer want to know what caused it; understanding a specific cause gives hope of a specific cure, or at least better prevention. Few people are willing to accept the reality that natural selection is incapable of shaping mechanisms to prevent all cancers. However, there are five other good reasons why selection has not shaped better protections against cancer.

Tradeoffs

A second closely related explanation is that tradeoffs limit the perfection of all traits shaped by natural selection. In the case of cancer, tradeoffs are present in abundance. Tighter controls on cell division would decrease risk of cancer but would also decrease the ability of wounds to heal, and of cells in organs, such as the liver to divide in order to replace losses. Shorter telomeres, and increased sensitivity of apoptosis regulation

mechanisms, would decrease rates of cancer, but they would increase rates of aging. More aggressive immune systems would likewise decrease cancer, at the cost of increased tissue damage and possible autoimmune disease. On a macro level, tall people get many advantages in life, especially in mating, but every 10 cm increase in height increases the cancer risk by at least 10% because of increased cell number and the associated extra growth factor stimulation (Green et al., 2011). Light skin increases Vitamin D synthesis, but at the cost of increased vulnerability to skin cancer (Greaves, 2014). There is no free lunch; trade-offs are intrinsic to all cancer prevention mechanisms (Aktipis et al., 2013; Boddy et al., 2015, see also Chapter 1).

Mismatch

Natural selection can change a species only slowly, so many aspects of bodies become vulnerable to disease when environments change rapidly, as they have dramatically for humans in recent generations. The American Cancer Society estimates that more than half of all human cancers could be prevented by modifying environmental factors (Fontham et al., 2009). The big one is tobacco use, which accounts for a third of all cancers. Other environmental factors, including sun exposure, modern diets, and obesity also increase risks substantially. Much research looks at specific factors, such as the multiple aspects of modern environments that can increase rates of breast cancer. High levels of nutrition and leisure induce high levels of reproductive hormones that are correlated with breast cancer rates (Jasieńska and Thune, 2001). They also lead to early menarche, which combines with delayed first pregnancy to create an interval in which breast cells are particularly vulnerable to metaplastic transformation (Russo et al., 2005). Furthermore, in ancestral environments the average woman had only about 100 menstrual cycles because of lactation induced amenorrhea while in modern societies the number is over 400 per lifetime (Strassmann, 1997), driving the system with endocrine stimulation unprecedented in the history of our species. Finally, exposure to light at night is also associated with higher rates of breast cancer (Stevens, 2009). Combinations of aspects of modern environments help to explain increased vulnerability to other kinds of cancer, and offer a potent prescription for prevention (Hochberg et al., 2013).

These environmental factors interact with genes to influence cancer risk. Common genetic variations found to strongly influence risk are likely to be those with few deleterious effects in ancestral environments, otherwise, they would have been selected out. Such variations associated with cancer should not be assumed to be deleterious; they may be "quirks" that were harmless in the environments in which we evolved.

Benefits to Genes at the Expense of Health

Natural selection shapes maximum health and longevity only to the extent that they contribute to increased reproduction. An allele that increases reproduction will tend to spread, even if it increases the risk of cancer. Such variations that have gone to fixation are very hard to identify, because they would need to be compared to some alternative that would be hard or impossible to recognize. However, some of the increased vulnerability of males to cancer may be explained, because they get a relatively greater reproductive payoff for investments in competition, while females get a relatively greater payoff for tissue maintenance and repair (Kruger and Nesse, 2006). Also, specific alleles have been suggested to perhaps increase reproduction at the cost of cancer; for instance, BRCA1, seems to be correlated with increased fecundity in some studies (Smith et al., 2012), although it is hard to control for confounding factors that could potentially explain this finding. Transmission of cancers between individuals is a dramatic example that is rare for good evolutionary reasons.

Coevolution

That infectious agents cause cancer has long been recognized, but the phenomenon is increasingly studied in evolutionary perspective. The role of papilloma virus in causing cervical cancer is well-established, as is the efficacy of vaccine prevention. An evolutionary view of why infection causes cancer begins with the general observation that tissues subject to chronic inflammation are also subject to genetic damage that can increase the risk of cancer. However, Ewald and others have suggested, and shown, that certain pathogens get reproductive advantages in the body by disrupting cell junctions and other manipulations that allow them to persist within hosts (Ewald, 2009; Ewald and Swain Ewald, 2012, see also Chapter 3). This is a classic example of coevolution, in which every improvement in the body's ability to resist infection is countered by the rapid evolution in pathogens of mechanisms to get around the defenses. The resulting arms races are extremely expensive, leaving hosts with defenses that decrease the risk of infection but increase the risk of cancer.

Defenses

Many defenses are facultative adaptations shaped by natural selection that are aroused when needed, such as fever, pain, vomiting, and inflammation. Defenses against cancer provide similar protection, but only some are aroused by specific threats, more are continuously preventing malignant transformation and eliminating rouge clades. The long-recognized special role of the immune system in controlling cancer has led to dramatic new therapeutic options now under intense study (see also Chapter 9).

CONCLUSIONS

All five evolutionary perspectives offer useful insights about cancer. Keeping them separate encourages clear thinking. They are, however, intimately related. For instance, comparative studies looking at different rates of cancer among species need to be informed by both consideration of the role of chance mutations resulting from stem cell divisions, and also different life history patterns, and body designs with different compartments. Behavioral ecological approaches that analyze why cooperation sometimes breaks down among metazoan cells are closely connected with analyses of tradeoffs, and the costs and benefits of different strategies. These behavioral and ecological principles are also essential for understanding how cancers may shape their own ecological environments, and be shaped by those environments.

The transition to a fully evolutionary view of cancer offers huge promise, and many challenges (see Chapter 18). A description of five relevant principles provides useful landmarks as the territory continues to expand.

Acknowledgments

Thanks to Karla Moeller, Manfred Laubichler, Frédéric Thomas, and Beata Ujvari.

References

Aktipis, C.A., Boddy, A.M., Gatenby, R.A., Brown, J.S., Maley, C.C., 2013. Life history trade-offs in cancer evolution. Nat. Rev. Cancer 13 (12), 883–892.

Aktipis, C.A., Boddy, A.M., Jansen, G., Hibner, U., Hochberg, M.E., Maley, C.C., Wilkinson, G.S., 2015. Cancer across the tree of life: cooperation and cheating in multicellularity. Phil. Trans. R. Soc. B 370: 20140219. http://dx.doi.org/10.1098/rstb.2014.0219.

Aktipis, C.A., Nesse, R.M., 2013. Evolutionary foundations for cancer biology. Evol. Appl. 6 (1), 144–159.

Aktipis, C.A., Kwan, V.S., Johnson, K.A., Neuberg, S.L., Maley, C.C., 2011. Overlooking evolution: a systematic analysis of cancer relapse and therapeutic resistance research. PloS One 6 (11), e26100.

Andor, N., Graham, T.A., Jansen, M., Xia, L.C., Aktipis, C.A., Petritsch, C., et al., 2016. Pan-cancer analysis of the extent and consequences of intratumor heterogeneity. Nat. Med. 22 (1), 105–113.

Barcellos-Hoff, M.H., Lyden, D., Wang, T.C., 2013. The evolution of the cancer niche during multistage carcinogenesis. Nat. Rev. Cancer 13 (7), 511–518.

Boddy, A.M., Kokko, H., Breden, F., Wilkinson, G.S., Aktipis, C.A., 2015. Cancer susceptibility and reproductive trade-offs: a model of the evolution of cancer defences. Phil. Trans. R. Soc. B 370: 20140220 http://dx.doi.org/10.1098/rstb.2014.0220.

Bonner, J.T., 2011. Why size matters: from bacteria to blue whales. Princeton University Press, Princeton, Available from: https://

books.google.com/books?hl=en&lr=&id=Ad9hxKVg_xEC&oi=fnd&pg=PP2&dq=why+size+matters+bonner&ots=8QoLvuukFN&sig=cFHSa3gBs0cOUbDJ4f2-tfdbbBg

Buss, L.W., 1987. The Evolution of Individuality. Princeton University Press, Princeton.

Caulin, A.F., Maley, C.C., 2011. Peto's Paradox: evolution's prescription for cancer prevention. Trends Ecol. Evol. 26 (4), 175–182.

Couzin-Frankel, J., 2015. Backlash greets "bad luck" cancer study and coverage. Science 347 (6219), 224–1224.

Crespi, B.J., 2000. The evolution of maladaptation. Heredity 84 (Pt 6), 623–629.

Crespi, B.J., Summers, K., 2005. Evolutionary biology of cancer. Trends Ecol. Evol. 20 (10), 545–552.

Daoust, S.P., Fahrig, L., Martin, A.E., Thomas, F., 2013. From forest and agro-ecosystems to the microecosystems of the human body: what can landscape ecology tell us about tumor growth, metastasis, and treatment options? Evol. Appl. 6 (1), 82–91.

Ducasse, H., Arnal, A., Vittecoq, M., Daoust, S.P., Ujvari, B., Jacqueline, C., et al., 2015. Cancer: an emergent property of disturbed resource-rich environments? Ecology meets personalized medicine. Evol. Appl. 8 (6), 527–540.

Erwin, D.H., Valentine, J.W., 2013. The Cambrian Explosion: the Construction of Animal Biodiversity. Roberts and Company, Greenwood Village, Colorado.

Ewald, P.W., 2009. An evolutionary perspective on parasitism as a cause of cancer. Adv. Parasitol. 68, 21–43.

Ewald, P.W., Swain Ewald, H.A., 2012. Infection, mutation, and cancer evolution. J. Mol. Med. 90 (5), 535–541.

Fontham, E.T.H., Thun, M.J., Ward, E., Balch, A.J., Delancey, J.O.L., Samet, J.M., ACS Cancer and the Environment Subcommittee, 2009. American cancer society perspectives on environmental factors and cancer. Cancer J. Clin. 59 (6), 343–351.

Frank, S.A., 2007. Dynamics of Cancer: Incidence, Inheritance, and Evolution. Princeton University Press, Princeton, NJ.

Greaves, M., 2000. Cancer: The Evolutionary Legacy. New York: Oxford University Press, Oxford.

Greaves, M., 2007. Darwinian medicine: a case for cancer. Nat. Rev. Cancer 7 (3), 213–221.

Greaves, M., 2014. Was skin cancer a selective force for black pigmentation in early hominin evolution? Proc. R. Soc. B 281: 20132955 http://dx.doi.org/10.1098/rspb.2013.2955.

Greaves, M., 2015. Evolutionary Determinants of Cancer Mel Greaves. Cancer Discov August 1 2015 (5) (8), 806–820 DOI: 10.1158/2159-8290.CD-15-0439.

Greaves, M., Maley, C.C., 2012. Clonal evolution in cancer. Nature 481 (7381), 306–313.

Green, J., Cairns, B.J., Casabonne, D., Wright, F.L., Reeves, G., Beral, V., 2011. Height and cancer incidence in the Million Women Study: prospective cohort, and meta-analysis of prospective studies of height and total cancer risk. Lancet Oncol. 12 (8), 785–794.

Hamilton, W.D., 1964. The genetical evolution of social behavior I, and II. J. Theoretical Biol. 7, 1–52.

Hanahan, D., Weinberg, R.A., 2011. Hallmarks of cancer: the next generation. Cell 144 (5), 646–674.

Hochberg, M.E., Thomas, F., Assenat, E., Hibner, U., 2012. Preventive evolutionary medicine of cancers. Evol. Appl. 6, 134–143.

Hochberg, M.E., Thomas, F., Assenat, E., Hibner, U., 2013. Preventive evolutionary medicine of cancers. Evol. Appl. 6 (1), 134–143.

Jasieńska, G., Thune, I., 2001. Lifestyle, hormones, and risk of breast cancer. BMJ 322 (7286), 586–587.

Korolev, K.S., Xavier, J.B., Gore, J., 2014. Turning ecology and evolution against cancer. Nat. Rev. Cancer 14 (5), 371–380.

Kruger, D.J., Nesse, R.M., 2006. An evolutionary life-history framework for understanding sex differences in human mortality rates. Human Nat. 17 (1), 74–97.

Maley, C.C., Galipeau, P.C., Finley, J.C., Wongsurawat, V.J., Li, X., Sanchez, C.A., et al., 2006. Genetic clonal diversity predicts progression to esophageal adenocarcinoma. Nat. Genet. 38 (4), 468–473.

Merlo, L.M., Pepper, J.W., Reid, B.J., Maley, C.C., 2006. Cancer as an evolutionary and ecological process. Nat. Rev. Cancer 6 (12), 924–935.

Michor, F., Frank, S.A., May, R.M., Iwasa, Y., Nowak, M.A., 2003. Somatic selection for and against cancer. J. Theor. Biol. 225 (3), 377–382.

Nagy, J.D., Victor, E.M., Cropper, J.H., 2007. Why don't all whales have cancer? A novel hypothesis resolving Peto's paradox. Integr. Compar. Biol. 47 (2), 317–328.

Nesse, R.M., 2005. Maladaptation and natural selection. Q. Rev. Biol. 80 (1), 62–70.

Nesse, R.M., 2016. The body is not a machine. Available from: https://evmed.asu.edu/blog/body-not-machine

Nesse, R.M., Williams, G.C., 1994. Why We Get Sick: The New Science of Darwinian Medicine. Vintage Books, New York.

Niklas, K.J., 2014. The evolutionary-developmental origins of multicellularity. Am. J. Bo. 101 (1), 6–25.

Nowell, P.C., 1976. The clonal evolution of tumor cell populations. Science 194 (4260), 23–28.

Nunney, L., 2013. The real war on cancer: the evolutionary dynamics of cancer suppression. Evol. Appl. 6 (1), 11–19.

Nunney, L., Maley, C.C., Breen, M., Hochberg, M.E., Schiffman, J.D., 2015. Peto's paradox and the promise of comparative oncology. Phil. Trans. R. Soc. B 370: 20140177. http://dx.doi.org/10.1098/rstb.2014.0177.

Peto's paradox and the promise of comparative oncology Leonard Nunney, Carlo C. Maley, Matthew Breen, Michael E. Hochberg, Joshua D. Schiffman Phil. Trans. R. Soc. B 2015 370 20140177; DOI: 10.1098/rstb.2014.0177. Published 8 June 2015

Peto, R., 1977. Epidemiology, multistage models, and short-term mutagenicity tests. Orig. Hum. Cancer 4, 1403–1428.

Queller, D.C., 2000. Relatedness and the fraternal major transitions. Phil. Transact. R. Soc. B 355 (1403), 1647–1655.

Russo, J., Mailo, D., Hu, Y.-F., Balogh, G., Sheriff, F., Russo, I.H., 2005. Breast differentiation and its implication in cancer prevention. Clin. Cancer Res. 11 (2), 931s–936s.

Smith, J.M., Szathmáry, E., 1995. The Major Transitions of Evolution. W.H. Freeman, New York.

Smith, K.R., Hanson, H.A., Mineau, G.P., Buys, S.S., 2012. Effects of BRCA1 and BRCA2 mutations on female fertility. Proc. R. Soc. Lond. B 279 (1732), 1389–1395.

Stevens, R.G., 2009. Light-at-night, circadian disruption and breast cancer: assessment of existing evidence. Int. J. Epidemiol. 38 (4), 963–970.

Strassmann, B.I., 1997. The biology of menstruation in *Homo sapiens*: total lifetime menses, fecundity and nonsynchrony in a natural fertility population. Curr. Anthropol. 38 (1), 123–129.

Sulak, M., Fong, L., Mika, K., Chigurupati, S., Yon, L., Mongan, N.P., et al., 2016. *TP53* copy number expansion is associated with the evolution of increased body size and an enhanced DNA damage response in elephants. eLife 5.

Thomas, F., Fisher, D., Fort, P., Marie, J.-P., Daoust, S., Roche, B., et al., 2013. Applying ecological and evolutionary theory to cancer: a long and winding road. Evol. Appl. 6 (1), 1–10.

Tomasetti, C., Vogelstein, B., 2015. Variation in cancer risk among tissues can be explained by the number of stem cell divisions. Science 347 (6217), 78–81.

1

The Evolutionary Origins of Cancer and of Its Control by Immune Policing and Genetic Suppression

Leonard Nunney

Department of Biology, University of California Riverside, Riverside, CA, United States

Cancer is a malignant growth of cells capable of invading other tissues. As such, cancer is a disease of multicellular organisms and disrupts organismal function so severely that it is generally fatal. Based on this definition, cancer is restricted to animals because the cell walls of plants, algae, and fungi prevent cellular invasion (Doonan and Sablowski, 2010). In contrast, "benign" tumors, which do not invade other tissues, are relatively common across multicellular life and are generally not life-threatening, although by virtue of their location or abundance, their growth can cause serious problems, including death. However, sometimes it can be difficult to distinguish malignant and nonmalignant forms, leading to a category of "cancer-like" tumors (Aktipis et al., 2015). For example, crown galls, which are bacteria-induced benign tumors of plants, can produce secondary tumors nearby, but these additional tumors are due to a diffusible factor rather than cell movement (White, 1951; White and Braun, 1942). Despite these distinctions, an understanding of such noncancerous tumors can provide valuable insight into the unregulated cell division that is a prerequisite of all cancers.

Multicellularity is one of the major transitions of life (Maynard Smith and Szathmary, 1995), creating individual organisms, each made up of a "society" of cells that persists for a significant period of time and within which there is a division of labor between cells that are reproductive (in the sense that they contribute directly to the next generation of individuals) and nonreproductive cells. In animals, this division corresponds to the germline versus the soma. A critical component in the origin of such a society is the evolution of reproductive altruism in the cells that are nonreproductive. These cells can transmit only copies of their genes to future generations indirectly via the success of the reproductive cells.

At first sight, this division of labor does not seem to be a significant evolutionary barrier: the cells of a multicellular organism are clonal derivatives of a single fertilized egg, so discounting somatic mutation, the germline and somatic cells are 100% related. The theory of kin selection (Hamilton, 1964) predicts that, given a minimal benefit to multicellularity, such high relatedness would make the barrier to the evolution of reproductive altruists almost nonexistent. So why did the transition to multicellularity not occur until about 1 billion years after the first appearance of eukaryotes? One of the factors that may have contributed to this delay is the problem of "cheating."

All societies are vulnerable to antisocial cheats that fail to conform to their assigned role, and societies will fail to persist if they lack mechanisms for limiting the effect of cheating (Maynard Smith, 1964; Nunney, 1985). In the case of multicellularity, these cheats are tumor cells that, in the worst-case scenario, become cancerous. Cells can become antisocial in two ways: their behavior can be disrupted through infection by a pathogen (see Ewald and Swain Ewald, 2015, and Chapter 3) or via somatic mutation. Furthermore, these possibilities are nonexclusive in driving the occurrence of cancer. For example, infection with certain types of human papillomavirus (HPV) does not lead directly to cervical cancer, but it reduces the number of mutational changes needed to induce it (McLaughlin-Drubin and Münger, 2009).

There appear to have been five independent origins of multicellular life (Knoll, 2011). Each origin has required the evolution of reproductive altruism, which, once evolved, raises the problem of cheating. In plants, algae, and fungi

Ecology and Evolution of Cancer. http://dx.doi.org/10.1016/B978-0-12-804310-3.00001-6

an incidental benefit of a structural cell wall is that cancer is completely suppressed, although cheats that form non-invasive tumors often arise (Aktipis et al., 2015; Doonan and Sablowski, 2010). In large long-lived animals, cancer is a never-ending problem with the incidence of a given cancer resulting from a balance between the fitness advantage gained by preventing that cancer and its fitness cost (Nunney, 1999a, 2003). I will now consider how that balance is achieved.

LINEAGE SELECTION

Natural selection can occur whenever there exist reproduction and heritability, and can act at any level of biological organization. Viewed in this way, cancer is the result of natural selection favoring proliferating cells within an individual, selection that has the incidental effect of lowering fitness at the next organizational level, the level of the individual. Thus a conflict is created between natural selection acting at the cell and individual levels. The success of cancer occurs because evolution is blind to the future, and, if all else is equal, selection acting at the shortest timescale is the most effective. As we know to our cost, once established, a cancer is very hard to eliminate, and cancer cells succeed even though they jeopardize their own evolutionary future by causing the premature death of the individual.

But why should such a conflict arise among the cells of an individual given that these cells are 100% related? Unfortunately, the theory of kin selection is an equilibrium ideal that does not incorporate the problem of differences between the timescales of a kin-selected unit and its component parts. It is these timescale differences that lead to the problem of cheating, with the result that the occurrence of selfish cheats exhibiting antisocial characteristics is to be expected. However, despite the short-term advantage of cancer cells, the low incidence of cancer prior to old age demonstrates that this timescale problem has been overcome, allowing multicellular animals to thrive. The intrinsic advantage of cancer cells cannot be changed; however, a form of natural selection called lineage selection can be expected to act at the individual level to minimize the occurrence of cancer (Nunney, 1999a,b).

Lineage selection acts whenever there is a conflict within a group between the short-term success of the components of the group and the longer-term success of the group, and it favors longer-term success of the group. It does so provided that the members of the group form a single lineage (i.e., they are related), and it favors lineages that are best able to control cheats. Thus lineage selection reduces the likelihood of cheating because lineages that control cheating most effectively are the most successful over the long term (Fig. 1.1).

One mechanism favored by this form of selection is policing, which is the detection and removal of cheating elements. Policing is advantageous whenever conflicts arise within a biological society (Frank, 1995; Michod, 1996). For example, the removal of worker-produced eggs in honeybee colonies by workers is a form of policing (Ratnieks and Visscher, 1989). The second mechanism favored by lineage selection is suppression, by which the occurrence of cheats is inhibited. A classic example is the suppression of worker ovary development by the queen in honeybee colonies (Hoover et al., 2003).

Policing and suppression are fundamentally different ways of controlling antisocial behavior, but we can expect lineage selection to have favored both mechanisms in the "war on cancer," the phrase introduced by Richard Nixon in 1971 to describe a goal of medical research in the United States, but which perhaps applies more appropriately to the situation faced by all animals that are vulnerable to cancer (Nunney, 2013).

IMMUNE SYSTEM POLICING

In the case of cancer, a priori it seems plausible that the immune system would act as a policing agent; however, establishing this point has not been easy. More than 100 years ago, Coley (1893) was convinced that streptococcal infection following surgery could sometimes stimulate the immune system and result in tumor regression; however, his therapy of provoking an immune response by injecting bacteria (live or killed) was not well received, and other therapeutic approaches (primarily X-rays) shifted attention away from the immune system. This trend was not significantly reversed until the late 1950s, when Burnet (1957) suggested that cancer cells might provoke an immunological reaction because of their unique antigenic properties and Thomas (1959) took this possibility a step further by proposing that the response to tumor antigens represented an important natural defense against cancer. Based on these ideas, Burnet (1970, p. 3) defined the immunological surveillance hypothesis: "In large long-lived animals, like most of the warm-blooded vertebrates, inheritable genetic changes must be common in somatic cells and a proportion of these changes will represent a step toward malignancy. It is an evolutionary necessity that there should be some mechanism for eliminating or inactivating such potentially dangerous mutant cells and it is postulated that this mechanism is of immunological character." Furthermore, he concluded that the evidence pointed to the specific involvement of T cells in this process. In support of the immunological surveillance hypothesis, Burnet (1970) cited a range of observations. This included the extensive literature citing examples of spontaneous regression of malignant tumors, and related evidence from

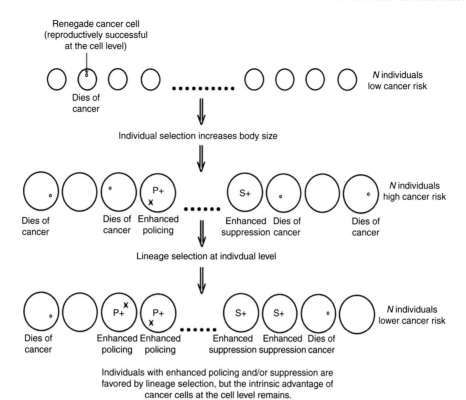

FIGURE 1.1 **The evolution of increased cancer protection by lineage selection.** The top row in the diagram shows a population with cancer controlled to a very low level. Only the occasional individual dies of cancer due to the origination and spread of a single "renegade" cell. Next it is assumed that conditions change and the population is subject to individual selection for increased body size. As a result, individuals become larger due to increased cell number, which in turn increases cancer risk (see Eq. (1.2)). The significant loss of fitness due to cancer now favors genotypes with enhanced policing of cancer cells ($P+$) or with enhanced cancer suppression ($S+$). As a result, these genotypes increase in frequency in the population. Note that some individuals are lucky and avoid cancer, but this "luck" is not heritable.

adrenal neuroblastomas in children, where it had been observed that there was an unusually high frequency of the disease in autopsies of newborns (who died of unrelated causes) compared with the frequency of clinical manifestation (Beckwith and Perrin, 1963). Similarly, evidence of regression was seen in the adrenals of older children (Mortensen et al., 1955). Further evidence was provided by the existence of tumor-specific antigens, which had been established through the use of tumor transplants among inbred mice (Klein, 1966).

In response to Burnet's proposal, a number of experimental tests were carried out (see Dunn et al., 2002), but, in general, these did not provide significant support. As Dunn et al. (2002, p. 992) noted, "the cancer immunosurveillance concept was considered dead by 1978." However, over the past 15 years or so, new data have resulted in a reversal of this view (reviewed by Dunn et al., 2002; Swann and Smyth, 2007; Vesely et al., 2011). Specifically, it became clear that endogenous interferon and lymphocytes were involved in process of cancer immunosurveillance (Shankaran et al., 2001). Thus, while the "hallmarks of cancer" listed by Hanahan and Weinberg (2000) did not include evading the immune system as a

feature of cancer, their updated version (Hanahan and Weinberg, 2011) included it as an "emerging hallmark," although perhaps it should be seen as a reemerging hallmark. The ambiguity was understandable because the immune system can have opposing effects on tumor growth. Prehn (1971) had speculated that the immune system could stimulate tumor growth, and it is now clear that the involvement of the immune system in chronic inflammation has important tumor-promoting effects (Grivennikov et al., 2010).

If the immune system is important in reducing the incidence of cancer, then a clear prediction is that immunosuppressed individuals should be at a significantly higher risk of cancer. It is certainly the case that immunosuppression, either induced following an organ transplant or resulting from AIDS, does increase the incidence of certain cancers; however, many of these cancers have a known viral involvement (Boshoff and Weiss, 2002), so their increased frequency is probably due to reduced resistance to viral infection. However, an increase in cancers with no known viral etiology has been consistently reported in immunosuppressed individuals (see Vesely et al., 2011). For example, an increase in lung cancer has been observed in individuals with AIDS (Kirk et al., 2007) and in heart and kidney transplant patients (Pham et al., 1995; Vajdic et al., 2006). Similarly, significant increases in both nonmelanoma (which may have a viral component; Boshoff and Weiss, 2002) and melanoma skin cancers have been reported (Aberg et al., 2008; Brewer et al., 2009; Loeffelbein et al., 2009; Moloney et al., 2006). While an increase in nonviral cancers of about two- to fourfold is general across transplant patients, it appears that not all cancers are equally affected, with the risk of some, such as breast and prostate cancers, appearing unaffected (Adami et al., 2003; Vesely et al., 2011).

There is increasing insight into the nature of the immune response to tumors (see also Chapter 9). For example, it is now known that the elevated expression of p53 associated with many tumors can lead to the formation of p53-specific antibodies (Soussi, 2000), suggesting that an immune response may be mounted against a range of changes associated with cell transformation. This view is supported by data from immunotherapy (see subsequent text). More direct evidence has been obtained from studies of melanoma. Individuals developing melanoma typically show evidence of mounting a T-cell immune response against their tumor (Boon et al., 2006); however, presumably due to successful selection for immune-system avoidance or inhibition, in almost all cases the tumor ultimately spreads. Only rarely does a T-cell response specific to a particular tumor result in spontaneous regression (Zorn and Hercend, 1999).

The clear evidence demonstrating a role for the immune system in regulating the growth of a tumor has led to a recent explosive growth of interest in immunotherapy (see Mellman et al., 2011). Much of this growth has centered around the finding that the tumor-specific T-cell response typically present in cancer patients is inhibited by "checkpoint" receptors such as cytotoxic T lymphocyte–associated protein 4 (CTLA-4) and programmed death 1 (PD-1) (see Ai and Curran, 2015; Pardoll, 2012). Once this inhibition is removed, the immune response is temporarily reestablished, at least in some patients, and research using mice suggests that much of the response is to tumor-specific mutant antigens (Gubin et al., 2014).

It is clear that immune policing has a significant effect on reducing the incidence of cancer. However, viewed from an evolutionary perspective, its effectiveness is limited by the strong selection acting on cancer cells favoring those that are capable of local inhibition of the immune attack (Mellman et al., 2011) or that lack the tumor-specific antigens targeted by the immune response. Unfortunately this problem remains when immunotherapy is used, predicting that some cells will be resistant and drive remission unless the cancer cell population is much reduced prior to therapy (thus limiting the opportunity for selection) or if one or more of the targeted antigens are essential to the success of the cells. The observation of spontaneous remission shows that the immune response

can sometimes be fully effective in eliminating a tumor; however, the extent to which this is a rare tumor-specific vulnerability or a property that can be exploited in most tumors remains to be seen. The success of checkpoint inhibition with melanoma, with CTLA-4 inhibition increasing the percentage of advanced melanoma patients surviving more than 5 years by more than 10% (Maio et al., 2015), gives cause for optimism.

The immune surveillance hypothesis proposed that controlling cancer in large long-lived animals was a primary function of the immune system (Burnet, 1970). Based on the data from immunosuppressed individuals, it is clear that the reduction in cancer risk due to the immune system, while significant, is generally modest. However, this does not undermine the possibility that cancer prevention was an important factor in the early evolution of the immune system. Burnet (1970) considered that, at least from the perspective of a mammal, the adaptive immune system is important in providing "a partial protection of the individual against cancer and a complete protection of the species against epidemic spread of any form of malignant disease" (p. 15). This statement is particularly resonant in the context of the current epidemic spread of devil facial tumor disease that threatens to cause the extinction of the Tasmanian devil (McCallum, 2008) (see also Chapter 12).

Burnet (1970) speculated that adaptive immunity may have evolved as a response to the parasitism by early cyclostomes (the jawless fish: lampreys and hagfish) on closely related host species, based on the advantage to hosts of recognizing foreign (but similar) cells. He argued that from this origin, the recognition of neoplastic cells could easily follow. Almost 50 years later, it is still considered that the adaptive immune system originated with the common ancestor of the jawed and jawless vertebrates (Flajnik and Kasahara, 2010). However, there is still no generally accepted hypothesis regarding the selection that favored its origination, and it may be time to reconsider Burnet's (1970) original proposal.

CANCER SUPPRESSION

The most important mechanism limiting the incidence of cancer is suppression, and the best-known examples are the tumor suppressor genes (TSGs) (Knudson, 1993; Weinberg, 1991). However, the number and nature of the genes involved in different cancers varies. To account for this pattern, Nunney (1999a,b) argued that cancer suppression is an evolving trait, such that over evolutionary time whenever individual fitness within a species is significantly reduced by some form of early onset cancer, lineage selection will favor any mechanism that reduces the risk of that specific cancer being initiated. It will do so by favoring individuals exhibiting novel mechanisms that increase control over unregulated cell proliferation in that tissue, such as the tissue-specific expression of an additional TSG. This proposition raises two important questions: first, what drives the need for increased cancer suppression; and second, what is the evidence that cancer suppression is an evolving trait?

Cancer originates following the accumulation of a series of mutations in a single cell, a process known as multistage carcinogenesis. This model was first proposed in the 1950s (Nordling, 1953) based on the pattern of the increasing incidence of cancer with age, and Armitage and Doll (1954) showed that this pattern was found across a range of different cancers. These data were consistent with a multistage model in which seven mutational steps were required in a single cell to initiate cancer. However, it is now known that other cancers initiate after fewer mutational steps, the classic example being retinoblastoma, a cancer of the embryonic retina, which is initiated following only two mutational "hits" disabling both copies of the *Rb* gene (Knudson, 1971). While the loss of Rb function is sufficient to cause retinoblastoma, it is also implicated in more complex sets of gene loss leading to other cancers including osteosarcomas, and small cell lung carcinomas (Classon and Harlow, 2002); however, loss of Rb function is not implicated in all cancers. These observations illustrate that cancer suppression is at least somewhat tissue specific in both the number and the nature of the genes involved.

Given multistage carcinogenesis, the hypothesis of lineage selection driving adaptation through increased cancer suppression can be modeled (Nunney, 1999a, 2003). Considering the simplest case, assume that, in a tissue with C stem cells, unregulated cell division (leading to cancer) is prevented by a set of n identical TSGs. Each gene copy can be disabled by somatic mutation occurring at a rate u per cell division, and the stem cells undergo K postgrowth divisions (again for simplicity, the early growth phase is ignored). Then the expected incidence of cancer (p) is as follows (Nunney, 1999a):

$$p = 1 - [1 - \{1 - \exp(-uK)\}^{2n}]^C \qquad (1.1)$$

Given that p is small (i.e., the incidence of that specific cancer is low), then to a good approximation, we have

$$p \approx C(uK)^{2n} \qquad (1.2)$$

This very simple formula provides the basis for a number of predictions concerning differences among tissues. For example, it predicts that while cancer incidence is expected to be related to the total number of (stem) cell divisions in a tissue (CK), this is a poor predictor since the risk of cancer increases linearly with tissue size (C) but as a power function with the number of cell divisions (K). These relationships undermine recent analyses based on CK (Tomasetti and Vogelstein, 2015; Wu et al., 2016), as noted by Nunney and Muir (2015) and Noble et al. (2015, 2016).

An important prediction of Eq. (1.2) is that the number of TSGs (n), or other controls, should be greater in a large and/or rapidly dividing tissue compared to that in a small and/or slowly dividing tissue. As noted previously, such tissue differences are general. For example, in humans, while the embryonic retina is protected from retinoblastoma by a single TSG (Rb), transforming fibroblasts requires perturbing six different pathways (Rangarajan et al., 2004). Moreover, it is to be expected that in some cases different tissues will be protected from cancer by mechanisms that involve different genes. This observation runs counter to the plausible expectation that all cancers would be suppressed by a single mechanism; however, it is consistent with layers of suppression being added to different tissues at different evolutionary times exploiting the genetic variation available in the population at that moment. This view is supported by the results of a study of the expression in normal human tissue of 15 different TSGs, each linked to a specific familial cancer (Muir and Nunney, 2015). They tested the null hypothesis that these genes would be expressed approximately equally across different tissues, but instead found that in the majority of cases the level of expression of these genes in the at-risk tissue was higher than their expression in other tissues. These results support the idea that tissue-specific expression has evolved to suppress specific cancers.

Another prediction is that any increase in the somatic mutation rate (u) will increase cancer risk, but with the greatest effect in tissues with the most tumor suppressors, that is, the largest and/or most rapidly dividing tissues. It is notable that the examples of inherited DNA-repair abnormalities listed by Hall et al. (1995) increase cancers such as lymphoma, leukemia, skin cancer, and colon cancer, which are derived from such tissues.

Viewing the model from a population genetic perspective yields additional insight (Nunney, 2003). Given that cancer suppression is an evolving trait, natural selection is expected to reduce the incidence of all early onset cancers to every low levels; however, the weakening effectiveness of natural selection as organisms age means that this pattern is not necessarily true in old age. A strong prediction of the evolutionary model is that the cancers expected to predominate in old age will be derived from tissues that are large and rapidly dividing (i.e., the ones expected to have evolved the most layers of suppression). This prediction arises because natural selection will act to minimize the occurrence of cancer prior to the cessation of reproduction, but not beyond. As a result, we can expect an increasing number of cells to be one mutational step away from being transformed as reproductive activity ceases. More of these at-risk cells are likely to be found in large tissues, and the transforming mutation is most likely to occur if the division rate of those cells is high, both of which contribute to a rapidly increasing incidence of cancer. This effect is consistent with the marked prevalence of epithelial carcinomas over sarcomas in old age (DePinho, 2000).

A central feature of multistage models is that increasing the number of cells [C in Eq. (1.2)] increases the risk of cancer. In humans, it is assumed that tall individuals have more cells than shorter ones, based on interspecific patterns (Savage et al., 2007), predicting that tall individuals should have a higher incidence of cancer. Albanes and Winick (1988) reviewed data that supported this view, but it took results from large-scale cancer studies, notably the Million Women Study, to firmly establish the positive association between height and overall cancer incidence, and between height and most forms of cancer (Green et al., 2011). Cancer risk increases by about 10–15% for every 10-cm increase in height. A similar pattern is found when large and small breeds of domestic dog are compared (see Nunney, 2013).

EVOLUTION AND PETO'S PARADOX

Although the multistage model predicted that tall humans should have an increased cancer risk, in the late 1980s there were only limited data supporting this prediction. However, Albanes and Winick (1988) recognized its importance, and proposed that within any species cancer risk increased with cell number. However, they did not speculate as to why this pattern would not be (and is not) found between species. Peto (1977) had previously pointed out that this prediction of the multistage model did not hold up in the comparison of large and small species. Specifically, he noted that humans are much larger than mice and live for much longer, and yet humans do not suffer the expected orders-of-magnitude greater incidence of cancer. Nunney (1999a) called this Peto's paradox and argued that the solution was adaptation through the evolution of increased cancer suppression in large long-lived animals.

It is indeed clear that humans and mice experience very comparable levels of cancer in their lifetime (Rangarajan and Weinberg, 2003). This suggests that humans are more effective at suppressing and/or policing cancer, as predicted by the evolutionary model. But is there any direct support for this thesis? One line of evidence concerns telomerase suppression. It is known that cultured human fibroblasts will stop dividing after about 40–60 divisions (the Hayflick limit) due to the deterioration of chromosomal telomeres resulting from a suppression of telomerase, and that cancer cells upregulate telomerase (Hayflick, 2000). In contrast, mouse fibroblasts do not exhibit telomerase suppression (see Wright and Shay, 2000). One interpretation of these observations is that telomerase suppression may have evolved in some ancestors of humans as an anticancer mechanism; however, these two data points from unrelated species do not provide very compelling support for this hypothesis. On the other hand, the comparative study of Seluanov et al. (2007) did. This study demonstrated a strong correlation between telomerase suppression in cultured fibroblasts and body size across 15 species of rodents (Fig. 1.2). The

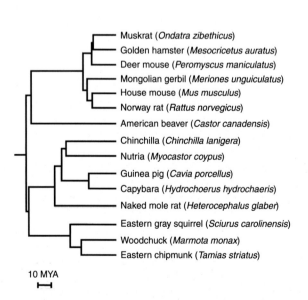
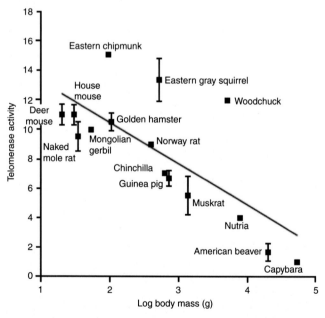

FIGURE 1.2   **The negative relationship between body mass and telomerase activity across different species of rodents using cultured fibroblasts in the context of the rodent phylogeny.** If all else is equal, large rodents will be more prone to cancer; hence, the observation of increasing telomerase suppression with increasing body mass is consistent with the hypothesis that this suppression is an anticancer mechanism, limiting the total number of cell divisions in the fibroblasts of larger rodents. Note that none of the four species with the greatest telomerase suppression are closely related. *Source: Redrawn from Gorbunova, V., Seluanov, A., Zhang, Z., Gladyshev, V.N., Vijg, J., 2014. Comparative genetics of longevity and cancer: insights from long-lived rodents. Nat. Rev. Genet. 15, 531–540.*

beaver and the capybara are the largest rodents and showed the strongest telomerase suppression, demonstrating that a high degree of suppression has evolved independently in two different rodent clades.

In their study, Seluanov et al. (2007) failed to find any association between telomerase suppression and the other factor expected to increase cancer risk, life span. In particular the small but very long-lived naked mole rat showed levels of telomerase comparable to those of the similarly sized but short-lived house mouse. Further study revealed that the cells of the naked mole rat showed a unique form of early contact inhibition, mediated through a high-molecular-mass hyaluronan that appears to be another independent mechanism of cancer suppression (Gorbunova et al., 2014).

The success of this study across rodents will hopefully stimulate more comparative research. Certainly the availability of an increasing number of genomes will help in identifying evolutionary shifts that have occurred in large organisms (see Caulin et al., 2015). For example, one potential response to the evolutionary pressure of increasing cancer risk with increasing body size is enhanced suppression through the duplication of existing TSGs. This possibility can be investigated directly with genomic data.

No evidence for the duplication of TSGs has yet been found in the largest mammals, the whales, but it has been shown that the African elephant has multiple retrogene copies of the tumor suppressor $TP53$ (Abegglen et al., 2015). The role of these copies is uncertain given that they are all shortened by premature stop codons, but it is clear that elephant cells are unusually sensitive to double-stranded breaks, consistent with enhanced activity of the $TP53$ gene leading to greater cancer suppression (Abegglen et al., 2015) (see also Chapter 7).

WHY DOES CANCER PERSIST?

From time to time there is speculation that since cancer persists it must provide some long-term species-level benefit, such as the purging of deleterious mutations (e.g., Lichtenstein, 2005). Such group selection arguments are both unsupportable theoretically (Maynard Smith, 1964; Nunney, 1985) and unnecessary. Cancer persists because of the interaction between natural selection and germline mutation that affects all fitness-related traits. Germline mutations that undermine evolved suppression mechanisms and increase the risk of early onset cancer continuously enter the population, while the reproductive fitness loss due to such mutations acts to remove them from the population. This dynamic leads to an equilibrium mutation–selection balance. Using a multistage model, the expected equilibrium frequency of such mutations due to mutation–selection balance

can be calculated (Nunney, 2003), and, not surprisingly, because of their devastating effect, early onset cancers are expected to be rare. The same is not true of late onset (i.e., postreproductive) cancers. Since these cause no significant fitness loss, selection is ineffective at enhancing or even maintaining cancer suppression (Nunney, 1999a, 2003).

The inefficiency of selection is not the only reason for cancer to persist. It is possible for cancer to occur at a higher frequency than that predicted by mutation–selection balance if there is some beneficial trade-off that outweighs the negative fitness effect of cancer (Boddy et al., 2015). In particular, it is possible that the allocation of energy to early life reproductive activity (e.g., costly sexual ornaments or fighting in males) may trade off with the allocation of energy to activities that serve to suppress cancer (e.g., DNA repair). At present this possibility is hard to evaluate because we have no indication of the costs that might be associated with maintaining cancer suppression.

Smith et al. (2012) presented evidence of a trade-off in humans linking BRCA1/2 mutations with increased early fecundity. Whether or not this is a genetic trade-off remains to be established, since there would appear to be a very real possibility that the women who are carriers of these mutations may modify their reproductive behavior in response to the early mortality of female relatives. Another potential trade-off is between immune suppression and cancer risk. Pregnant female mammals experience pregnancy-associated immune suppression to protect the fetus; however, a trade-off may be increased cancer risk (Donegan, 1983). Related to this reduced immune surveillance is the direct risk of gestational choriocarcinoma (Bardawil and Toy, 1959), which results from the invasion of placental trophoblasts, and, more speculatively, the pregnancy-related increase in breast cancers having a high risk of metastasis (Schedin, 2006).

CONCLUSIONS

The recognition that Peto's paradox can be resolved by adaptive evolution (Nunney, 1999a,b, 2003) is gradually changing the prevailing view from one in which baseline cancer incidence and cancer control are invariant features of animals to one in which the varying incidence of cancer is controlled by a dynamic adaptive process. It follows that among the diversity of animal species, we can expect to see a great diversity of mechanisms that have evolved to control the deadly effects of cancer. As noted earlier, the comparative study of rodents, using cell culture, proved to be remarkably productive in identifying mechanisms of cancer suppression relating to body size and to longevity (Gorbunova et al., 2014; Seluanov et al., 2007). A recent collection of papers focused on the evolution of cancer suppression (see Nunney et al., 2015) included compelling evidence for the value of domestic dogs in the study of cancer, given the wealth of breed-associated data available (Schiffman and Breen, 2015). Similarly, the wealth of genomic data that are becoming available for an increasingly wide range of animal species will enable genomic searches of the type initiated by Caulin et al. (2015); however, while we can be confident that large and/or long-lived animals have evolved added mechanisms for cancer prevention, finding them may prove difficult.

Peto (1977) posed the question of why there is not a vastly greater incidence of cancer in humans compared to that in mice and, given the importance of the mouse model in understanding human cancers, concern has been raised over the known differences in tumorigenesis between these two species (Rangarajan and Weinberg, 2003). Based on an evolutionary model, we can predict that mice and humans will share some features of cancer control based on their common ancestry more than 65 million years ago, but since that time the independent life-history evolution in the rodent and primate clades will have resulted in the addition and potentially the loss of some aspects of cancer suppression. Notably the increase in size and longevity that has occurred in the lineage leading to humans will have required the addition of novel mechanisms at least in some tissues. The time has come for more well-designed comparative studies based on life-history divergence of close relatives, and we can expect them to be productive in resolving the question of how species adapt to a changing risk of cancer.

References

Abegglen, L.M., Caulin, A.F., Chan, A., Lee, K., et al., 2015. Potential mechanisms for cancer resistance in elephants and comparative cellular response to DNA damage in humans. JAMA 314, 1850–1860.

Aberg, F., Pukkala, E., Hockerstedt, K., Sankila, R., Isoniemi, H., 2008. Risk of malignant neoplasms after liver transplantation: a population-based study. Liver Transpl. 14, 1428–1436.

Adami, J., Gabel, H., Lindelof, B., Ekstrom, K., Rydh, B., et al., 2003. Cancer risk following organ transplantation: a nationwide cohort study in Sweden. Br. J. Cancer 89, 1221–1227.

Ai, M., Curran, M.A., 2015. Immune checkpoint combinations from mouse to man. Cancer Immunol. Immunother. 64, 885–892.

Aktipis, C.A., Boddy, A.M., Jansen, G., Hibner, U., Hochberg, M.E., Maley, C.C., Wilkinson, G.S., 2015. Cancer across the tree of life: cooperation and cheating in multicellularity. Philos. Trans. R. Soc. B 370, 20140220 (21 pp.).

Albanes, D., Winick, M., 1988. Are cell number and cell proliferation risk factors for cancer? J. Natl. Cancer Inst. 80, 772–775.

Armitage, P., Doll, R., 1954. The age distribution of cancer and a multi-stage theory of carcinogenesis. Br. J. Cancer 8, 1–12.

Bardawil, W.A., Toy, B.L., 1959. The natural history of choriocarcinoma: problems of immunity and spontaneous regression. Ann. NY Acad. Sci. 80, 197–261.

Beckwith, J.B., Perrin, E.V., 1963. In situ neuroblastomas: a contribution to the natural history of neural crest tumors. Am. J. Pathol. 43, 1089–1104.

Boddy, A.M., Kokko, H., Breden, F., Wilkinson, G.S., Aktipis, C.A., 2015. Cancer susceptibility and reproductive trade-offs: a model of the evolution of cancer defences. Philos. Trans. R. Soc. B 370, 20140219 (12 pp.).

Boon, T., Coulie, P.G., Van den Eynde, B.J., van der Bruggen, P., 2006. Human T cell responses against melanoma. Annu. Rev. Immunol. 24, 175–208.

Boshoff, C., Weiss, R., 2002. AIDS-related malignancies. Nat. Rev. Cancer 2, 373–382.

Brewer, J.D., Colegio, O.R., Phillips, P.K., Roenigk, R.K., Jacobs, M.A., et al., 2009. Incidence of and risk factors for skin cancer after heart transplant. Arch. Dermatol. 145, 1391–1396.

Burnet, F.M., 1957. Cancer: a biological approach. Br. Med. J. 1, 779–786, 841–847.

Burnet, F.M., 1970. The concept of immunological surveillance. Prog. Exp. Tumor Res. 13, 1–27.

Caulin, A.F., Graham, T.A., Wang, L.-S., Maley, C.C., 2015. Solutions to Peto's paradox revealed by mathematical modelling and cross-species cancer gene analysis. Philos. Trans. R. Soc. B 370, 20140222 (12 pp.).

Classon, M., Harlow, E., 2002. The retinoblastoma tumour suppressor in development and cancer. Nat. Rev. Cancer 2, 910–917.

Coley, W.B., 1893. The treatment of malignant tumors by repeated innoculations of erysipelas: with a report of ten original cases. Am. J. Med. Sci. 10, 487–511.

DePinho, R.A., 2000. The age of cancer. Nature 408, 248–254.

Donegan, W.L., 1983. Cancer and pregnancy. CA Cancer J. Clin. 33, 194–214.

Doonan, J., Sablowski, R., 2010. Walls around tumours—why plants do not develop cancer. Nat. Rev. Cancer 10, 794–802.

Dunn, G.P., Bruce, A.T., Ikeda, H., Old, L.J., Schreiber, R.D., 2002. Cancer immunoediting: from immunosurveillance to tumor escape. Nat. Immunol. 3, 991–998.

Ewald, P.W., Swain Ewald, H.A., 2015. Infection and cancer in multicellular organisms. Philos. Trans. R. Soc. B 370, 20140224 (11 pp.).

Flajnik, M.F., Kasahara, M., 2010. Origin and evolution of the adaptive immune system: genetic events and selective pressures. Nat. Rev. Genet. 11, 47–59.

Frank, S.A., 1995. Mutual policing and repression of competition in the evolution of cooperative groups. Nature 377, 520–522.

Gorbunova, V., Seluanov, A., Zhang, Z., Gladyshev, V.N., Vijg, J., 2014. Comparative genetics of longevity and cancer: insights from long-lived rodents. Nat. Rev. Genet. 15, 531–540.

Green, J., Cairns, B.J., Casabonne, D., Wright, F.L., et al., 2011. Height and cancer incidence in the Million Women Study: prospective cohort, and metaanalysis of prospective studies of height and total cancer risk. Lancet Oncol. 12, 785–794.

Grivennikov, S.I., Greten, F.R., Karin, M., 2010. Immunity, inflammation, and cancer. Cell 140, 883–899.

Gubin, M.M., Zhang, X., Schuster, H., Caron, E., Ward, JP., 2014. Checkpoint blockade cancer immunotherapy targets tumour-specific mutant antigens. Nature 515, 577–581.

Hall, M., Norris, P.G., Johnson, R.T., 1995. Human repair deficiencies and predisposition to cancer. In: Ponder, B.A.J., Waring, M.J. (Eds.), The Genetics of Cancer. Kluwer Academic, Dordrecht, The Netherlands, pp. 123–157.

Hamilton, W.D., 1964. The genetical evolution of social behavior. I and II. J. Theor. Biol. 7, 1–52.

Hanahan, D., Weinberg, R.A., 2000. The hallmarks of cancer. Cell 100, 57–70.

Hanahan, D., Weinberg, R.A., 2011. Hallmarks of cancer: the next generation. Cell 144, 646–674.

Hayflick, L., 2000. The illusion of cell immortality. Br. J. Cancer 83, 841–846.

Hoover, S.E.R., Keeling, C.I., Winston, M.L., Slessor, K.N., 2003. The effect of queen pheromones on worker honey bee ovary development. Naturwissenschaften 90, 477–480.

Kirk, G.D., Merlo, C., O'Driscoll, P., Mehta, S.H., Galai, N., et al., 2007. HIV infection is associated with an increased risk for lung cancer, independent of smoking. Clin. Infect. Dis. 45, 103–110.

Klein, G., 1966. Tumor antigens. Annu. Rev. Microbiol. 20, 223–252.

Knoll, A.H., 2011. The multiple origins of complex multicellularity. Annu. Rev. Earth Planet. Sci. 39, 217–239.

Knudson, A.G., 1971. Mutation and cancer: statistical study of retinoblastoma. Proc. Natl. Acad. Sci. USA 68, 820–823.

Knudson, A.G., 1993. Antioncogenes and human cancer. Proc. Natl. Acad. Sci. USA 90, 10914–10921.

Lichtenstein, A.V., 2005. On evolutionary origin of cancer. Cancer Cell Int. 5, 5, (9 pp.).

Loeffelbein, D.J., Szilinski, K., Holzle, F., 2009. Immunosuppressive regimen influences incidence of skin cancer in renal and pancreatic transplant recipients. Transplantation 88, 1398–1399.

Maio, M., Grob, J.-J., Aamdal, S., et al., 2015. Five-year survival rates for treatment-naive patients with advanced melanoma who received ipilimumab plus dacarbazine in a phase III trial. J. Clin. Oncol. 33, 1191–1196.

Maynard Smith, J., 1964. Group selection and kin selection: a rejoinder. Nature 201, 1145–1147.

Maynard Smith, J., Szathmary, E., 1995. The Major Transitions in Evolution. Freeman, Oxford.

McCallum, H., 2008. Tasmanian devil facial tumour disease: lessons for conservation biology. Trends Ecol. Evol. 23, 631–637.

McLaughlin-Drubin, M.E., Münger, K., 2009. Oncogenic activities of human papillomaviruses. Virus Res. 143, 195–208.

Mellman, I., Coukos, G., Dranoff, G., 2011. Cancer immunotherapy comes of age. Nature 480, 480–489.

Michod, R.E., 1996. Cooperation and conflict in the evolution of individuality. II. Conflict mediation. Proc. R. Soc. B Biol. Sci. 263, 813–822.

Moloney, F.J., Comber, H., O'Lorcain, P., O'Kelly, P., Conlon, P.J., Murphy, G.M., 2006. A population-based study of skin cancer incidence and prevalence in renal transplant recipients. Br. J. Dermatol. 154, 498–504.

Mortensen, J.D., Wollne, L.B., Bennett, W.A., 1955. Gross and microscopic findings in clinically normal thyroid glands. J. Clin. Endocrinol. Metab. 15, 1270–1280.

Muir, B., Nunney, L., 2015. The expression of tumor suppressors and proto-oncogenes in tissues susceptible to their hereditary cancers. Br. J. Cancer 113, 345–353.

Noble, R., Kaltz, O., Hochberg, M.E., 2015. Peto's paradox and human cancers. Philos. Trans. R. Soc. B 370, 20150104 (9pp).

Noble, R., Kaltz, O., Nunney, L., Hochberg, M.E., 2016. Overestimating the role of environment in cancers. Cancer Prev. Res. 9, 773–776.

Nordling, C.O., 1953. A new theory on the cancer inducing mechanism. Br. J. Cancer 7, 68–72.

Nunney, L., 1985. Group selection, altruism, and structured-deme models. Am. Nat. 126, 212–230.

Nunney, L., 1999a. Lineage selection and the evolution of multistep carcinogenesis. Proc. R. Soc. B Biol. Sci. 266, 493–498.

Nunney, L., 1999b. Lineage selection: natural selection for long-term benefit. In: Keller, L. (Ed.), Levels of Selection in Evolution. Princeton University Press, Princeton, pp. 238–252.

Nunney, L., 2003. The population genetics of multistage carcinogenesis. Proc. R. Soc. B Biol. Sci. 270, 1183–1191.

Nunney, L., 2013. The real war on cancer: the evolutionary dynamics of cancer suppression. Evol. Appl. 6, 11–19.

Nunney, L., Breen, M., Maley, C., Hochberg, M., Schiffman, J., 2015. Peto's paradox and the promise of comparative oncology. Philos. Trans. R. Soc. B 370, 20140177 (8 pp.).

Nunney, L., Muir, B., 2015. Peto's paradox and the hallmarks of cancer: constructing an evolutionary framework for understanding the incidence of cancer. Philos. Trans. R. Soc. B 370, 2015161 (7pp).

Pardoll, D.M., 2012. The blockade of immune checkpoints in cancer immunotherapy. Nat. Rev. Cancer 12, 252–264.

Peto, R., 1977. Epidemiology, multistage models, and short-term mutagenicity tests. In: Hiatt, H.H., Watson, J.D., Winsten, J.A., (Eds.), The Origins of Human Cancer, pp. 1403-1428.(Cold Spring Harbor Conferences on Cell Proliferation, vol. 4, Cold Spring Harbor Laboratory, NY).

Pham, S.M., Kormos, R.L., Landreneau, R.J., Kawai, A., Gonzalez-Cancel, I., et al., 1995. Solid tumors after heart transplantation: lethality of lung cancer. Ann. Thorac. Surg. 60, 1623–1626.

Prehn, R.T., 1971. Perspectives on oncogenesis—does immunity stimulate or inhibit neoplasia. J. Reticuloendothelial Soc. 10, 1–16.

Rangarajan, A., Weinberg, R.A., 2003. Comparative biology of mouse versus human cells: modelling human cancer in mice. Nat. Rev. Cancer 3, 952–959.

Rangarajan, A., Hong, S.J., Gifford, A., Weinberg, R.A., 2004. Species- and cell type-specific requirements for cellular transformation. Cancer Cell 6, 171–183.

Ratnieks, F.L.W., Visscher, P.K., 1989. Worker policing in the honeybee. Nature 342, 796–797.

Savage, V.M., Allen, A.P., Brown, J.H., Gillooly, J.F., et al., 2007. Scaling of number, size, and metabolic rate of cells with body size in mammals. Proc. Natl. Acad. Sci. USA 104, 4718–4723.

Schedin, P., 2006. Pregnancy-associated breast cancer and metastasis. Nat. Rev. Cancer 6, 281–291.

Schiffman, J.D., Breen, M., 2015. Comparative oncology: what dogs and other species can teach us about humans with cancer. Philos. Trans. R. Soc. B 370, 20140231 (13 pp.).

Seluanov, A., Chen, Z., Hine, C., et al., 2007. Telomerase activity coevolves with body mass, not lifespan. Aging Cell 6, 45–52.

Shankaran, V., Ikeda, H., Bruce, A.T., White, J.M., Swanson, P.E., et al., 2001. IFNγ and lymphocytes prevent primary tumour development and shape tumour immunogenicity. Nature 410, 1107–1111.

Smith, K.R., Hanson, H.A., Mineau, G.P., Buys, S.S., 2012. Effects of BRCA1 and BRCA2 mutations on female fertility. Proc. R. Soc. Lond. B 279, 1389–1395.

Soussi, T., 2000. p53 antibodies in the sera of patients with various types of cancer: a review. Cancer Res. 60, 1777–1788.

Swann, J.B., Smyth, M.J., 2007. Immune surveillance of tumors. J. Clin. Investig. 117, 1137–1146.

Thomas, L., 1959. Discussion. In: Lawrence, H.S. (Ed.), Cellular and Humoral Aspects of the Hypersensitive State. Roeber, New York, p. 529.

Tomasetti, C., Vogelstein, B., 2015. Variation in cancer risk among tissues can be explained by the number of stem cell divisions. Science 347, 78–81.

Vajdic, C.M., McDonald, S.P., McCredie, M.R., van Leeuwen, M.T., Stewart, J.H., et al., 2006. Cancer incidence before and after kidney transplantation. JAMA 296, 2823–2831.

Vesely, M.D., Kershaw, M.H., Schreiber, R.D., Smyth, M.J., 2011. Natural innate and adaptive immunity to cancer. Annu. Rev. Immunol. 29, 235–271.

Weinberg, R.A., 1991. Tumor suppressor genes. Science 254, 1138–1146.

White, P.R., 1951. Neoplastic growth in plants. Q. Rev. Biol. 26, 1–16.

White, P.R., Braun, A.C., 1942. A cancerous neoplasm of plants. Autonomous bacteria-free crown-gall tissue. Cancer Res. 2, 597–617.

Wright, W.E., Shay, J.W., 2000. Telomere dynamics in cancer progression and prevention: fundamental differences in human and mouse telomere biology. Nat. Med. 6, 849–851.

Wu, S., Powers, S., Zhu, W., Hannun, Y.A., 2016. Substantial contribution of extrinsic risk factors to cancer development. Nature 529, 43–47.

Zorn, E., Hercend, T., 1999. A natural cytotoxic T cell response in a spontaneously regressing human melanoma targets a neoantigen resulting from a somatic point mutation. Eur. J. Immunol. 29, 592–601.

2

Cancer Prevalence and Etiology in Wild and Captive Animals

Thomas Madsen,**, Audrey Arnal[†,‡], Marion Vittecoq[†,‡,§],
Florence Bernex[‡,¶,††,‡‡], Jerôme Abadie[§§], Sophie Labrut[§§],
Déborah Garcia[†,‡], Dominique Faugère[†,‡], Karin Lemberger***,
Christa Beckmann**, Benjamin Roche[†,‡,†††], Frédéric Thomas[†,‡],
Beata Ujvari***

*School of Biological Sciences, University of Wollongong, Wollongong, NSW, Australia
**Centre for Integrative Ecology, School of Life and Environmental Sciences,
Deakin University, Geelong, VIC, Australia
[†]MIVEGEC (Infectious Diseases and Vectors: Ecology, Genetics, Evolution and Control),
UMR IRD/CNRS/UM 5290, Montpellier, France
[‡]CREEC (Centre for Ecological and Evolutionary Research on Cancer), Montpellier, France
[§]Research Center of the Tour du Valat, Arles, France
[¶]Montpellier University, Montpellier, France
[††]RHEM, IRCM, Institute of Cancer Research Montpellier, INSERM, Montpellier, France
[‡‡]ICM Regional Cancer Institute of Montpellier, Montpellier, France
[§§]LUNAM University, Oniris, AMaROC, Nantes, France
***Vet Diagnostics, Lyon, France
[†††]UMMISCO (International Center for Mathematical and Computational Modeling
of Complex Systems), UMI IRD/UPMC UMMISCO, Bondy, France

INTRODUCTION

Despite the evolution of numerous natural cancer suppressor mechanisms (DeGregori, 2011), neoplasia has been recorded in most metazoans (Leroi et al., 2003). Although, a few exceptional species, such as the naked mole-rat (*Heterocephalus glaber*) and sharks have been claimed to be resistant to cancer (Finkelstein, 2005; Tian et al., 2013). Recent studies have, however, shown that even these species may develop cancer (Delaney et al., 2016; Finkelstein, 2005) strongly suggesting that the vast majority of multicellular organisms are indeed susceptible to cancer. The frequent occurrence of cancer in metazoans suggests that neoplasia, similar to pathogens/parasites, may have a significant negative impact on host fitness in the wild (Vittecoq et al., 2013).

This is supported by a recent review of wildlife cancer by McAloose and Newton (2009) demonstrating that high prevalence of cancer in, for example, Tasmanian devils (*Sarcophilus harrisii*) and belugas (*Delphinapterus leucas*) resulted in concomitant significant increase in levels of mortality and reduction in fitness.

Wildlife cancer statistics are, however, highly scattered in the scientific literature and hence challenging to access. Moreover, tumors in wildlife are most commonly detected during postmortem examination and therefore hard to confirm without histopathological examinations. However, even such analyses can be inaccurate because of high levels of autolysis (organ disintegration) (McAloose and Newton, 2009). In addition, individuals harboring tumors often display a decrease in body condition frequently resulting in higher levels of

Ecology and Evolution of Cancer. http://dx.doi.org/10.1016/B978-0-12-804310-3.00002-8

parasite/pathogen infections and concomitant increased levels in morbidity and mortality (Vittecoq et al., 2013) further impeding a correct analysis of the ultimate cause of death. The combination of the negative effects of cancer and/or pathogen/parasite infections has also been shown to result in increased levels of predation (Vittecoq et al., 2013). In our view, the combination of the problems involved in accurately recording wildlife cancer, the increased risk of succumbing to pathogens/parasites, and/or predation has often led to a somewhat erroneous assumption that although cancer is common in domestic animals, it remains rare in the wild. If, as we suggest, cancer may be a significant determinant of animal fitness it is therefore crucial to determine cancer prevalence in the wild.

The etiology and prevalence of transmissible cancers are presented and discussed in Chapter 12; this chapter will therefore focus on the prevalence and etiology of nontransmissible cancers.

Via thorough searches of the available literature we provide a comprehensive and an updated list of cancer prevalence in wild animals ranging from fish to whales. We also provide a list of cancer recorded in captive animals from French zoological parks and compare our findings to that recorded at other zoological parks. Finally we provide an updated list of cancers recorded as single cases in the wild, as well as in captive animals demonstrating that cancer occurs in nearly every taxonomic order of the animal kingdom.

CANCER PREVALENCE AND ETIOLOGY IN WILD VERTEBRATES

Although cancers are frequently encountered in wild animals (see Table 2.1 starting on page 13), we were only able to retrieve robust data on cancer prevalence in 31 wild vertebrate species (Table 2.2). We were unable to find information on nontransmissible cancer prevalence in wild invertebrates and consequently this chapter focuses on cancer in wild vertebrates ranging from fish to mammals. In following sections, we provide a summary of cancer etiology and prevalence in each of the five vertebrate groups (Table 2.2).

Fish

FAO (2010) fisheries and aquaculture department published a report showing that the mean contribution of fish to global diets was 17 kg per person/year, supplying over three billion people with 15% of their animal protein intake. About 45% of the fish consumed were farmed but the remaining 55% of fish were caught in the wild clearly demonstrating the importance of wild fish in the human diet. In spite of their importance to

humans we have only been able to find information on cancer prevalence in 12 wild fish taxa.

In walleye (*Sander vitreus*) and Atlantic salmon (*Salmo salar*) retroviruses have been found to initiate cancer development (Coffee et al., 2013). In bicolor damselfish (*Stegastes partitus*) neurofibromatosis-like tumors are most likely caused by an "extrachromosomal DNA virus-like agent" (Coffee et al., 2013) whereas in European smelt (*Osmerus eperlanus*) cancer development have been suggested to be caused by a "herpesvirus-like agent" (Coffee et al., 2013). In northern pike (*Esox lucius*) a corona virus has been suggested to be the cause for the development of lymphosarcoma (Papas et al., 1976). This species also shows substantial seasonal variation in lymphosarcoma prevalence but the underlying etiology is unknown (Papas et al., 1976). In brown bullhead (*Ameiurus nebulosus*), however, the higher levels of liver neoplasms (15%) recorded on one of the lakes investigated have been suggested to be caused by pollution (Baumann et al., 2008). Similarly, a study of English sole (*Parophrys vetulus*) revealed that up to 24% of the fish had developed liver neoplasms of which etiology could be traced to have been caused by pollution/chemical carcinogens (Malins et al., 1987).

Cancer prevalence as high as 20% have been observed in several species, such as gizzard shad (*Dorosoma cepedianum*), northern pike, walleye, bicolor damselfish and in white sucker (*Catostomus commersoni*) cancer may affect up to 59% of the fish (Coffee et al., 2013). However, the epidermal papilloma recorded in the latter taxon appears to result in low mortality (Coffee et al., 2013). In contrast, cancers, such as plasmacytoid leukemia have been shown to result in up to 50% mortality in commercially important taxa, such as Chinook salmon (*Oncorhynchus tshawytscha*; Eaton et al., 1994) and neurofibromatosis-like tumors have been shown to result in 100% mortality in bicolor damselfish (Coffee et al., 2013). Apart from the latter two studies, the remaining studies do not provide any data on the effect of cancer on fish mortality. In spite of this we find it reasonable to suggest that the high tumor frequency observed in several species may have a significant negative impact on fish fitness. Given the importance of fish in the human diet the high cancer prevalence and associated mortality recorded in some fish taxa, clearly demonstrate the need of a substantial increase in research on the effect of cancer on both marine and freshwater fish.

Amphibians

Although cancer has been reported in numerous amphibians (Balls and Clothier, 1974) we have only been able to find three studies that incorporated data on cancer prevalence in the wild. In the North American leopard frog (*Rana pipiens*) McKinnel (1965) found that

TABLE 2.1   Examples of Neoplasia Across the Animal Kingdom[a]

Latin name	Common name	Neoplasia (including benign and malignant abnormal cell growths)	References
INVERTEBRATES			
Hydrozoa			
Pelmatohydra robusta	Hydra	Undetermined neoplasia	Domazet-Lošo et al. (2014)
Mollusca			
Arctica islandica	Ocean quahog	Germinoma	Peters et al. (1994)
Argopecten irradians	Atlantic bay scallop	Gonadal neoplasia or germinoma	Peters et al. (1994); Carballal et al. (2015)
Cerastoderma edule	Common cockle	Disseminated neoplasia of unknown origin, gonadal neoplasia, or germinoma	Peters et al. (1994); Barber (2004); Carballal et al. (2015)
Crassostrea gigas	Pacific oyster	Fibroma or myofibroma, gonadal neoplasia, gonadoblastoma, disseminated neoplasia of unknown origin	Peters et al. (1994); Carballal et al. (2015)
Crassostrea virginica	Eastern oyster	Germinoma, gonadoblastoma, disseminated neoplasia of unknown origin	Peters et al. (1994); Carballal et al. (2015)
Ensis magnus (=arcuatus)	Razor clam	Gonadal neoplasia or germinoma	Carballal et al. (2015)
Ensis siliqua	Pod razor	Gonadal neoplasia or germinoma	Carballal et al. (2015)
Macoma balthica	Baltic macoma	Disseminated neoplasia of unknown origin	Carballal et al. (2015)
Macoma calcarea	Chalky macoma	Germinoma, hemic neoplasia	Peters et al. (1994)
Mercenaria campechiensis	Southern quahog	Gonadal neoplasia or germinoma	Carballal et al. (2015)
Mercenaria campechiensis × Mercenaria mercenaria hybrid	Quahog hybrid	Gonadal neoplasia or germinoma	Peters et al. (1994); Carballal et al. (2015)
Mercenaria mercenaria	Hard-shell clam, quahog	Germinoma	Peters et al. (1994); Carballal et al. (2015)
Mya arenaria	Soft-shell clam	Disseminated neoplasia of unknown origin, gonadal neoplasia or germinoma, hemic neoplasia	Carballal et al. (2015); Metzger et al. (2015)
Mytilus edulis	Blue mussel	Focal polypoid hyperplasia of germinal epithelium, germinoma, disseminated neoplasia of unknown origin	Peters et al. (1994); Carballal et al. (2015)
Mytilus edulis (trossulus/ galloprovincialis hybrid)	Blue mussel hybrid	Myxomas in vesicular connective tissue	Peters et al. (1994)
Mytilus galloprovincialis	Mediterranean mussel	Gonadal neoplasia or germinoma	Peters et al. (1994); Carballal et al. (2015)
Mytilus trossulus	Bay mussel	Disseminated neoplasia of unknown origin	Peters et al. (1994); Ciocan and Sunila (2005); Ciocan et al. (2006); Carballal et al. (2015)
Ostrea edulis	European flat oyster	Disseminated neoplasia of unknown origin	Barber (2004)
Tiostrea chilensis	Dredge oyster	Germinoma, hemic neoplasia	Peters et al. (1994)
Venerupis aurea	Golden carpet shell	Disseminated neoplasia of unknown origin	Carballal et al. (2015)
Xenostrobus securis	Small brown mussel	Gonadal neoplasia or germinoma	Carballal et al. (2015)
Crustacea			
Lithodes aequispinus	Golden king crab	Probable tegmental gland adenocarcinoma	Morado et al. (2014)
Paralithodes camtschaticus	Red king crab	Midgut tumor, probable tegmental gland adenocarcinoma	Morado et al. (2014)

(Continued)

TABLE 2.1 Examples of Neoplasia Across the Animal Kingdom (*cont.*)

Latin name	Common name	Neoplasia (including benign and malignant abnormal cell growths)	References
Paralithodes platypus	Blue king crab	Anaplastic cells on the surface of the antennal gland, probable tegmental gland adenocarcinoma	Morado et al. (2014)
Insecta			
Drosophila melanogaster	Fruit fly	Gut and testis tumors	Salomon and Jackson (2008)
VERTEBRATE			
Fish			
Agonus cataphractus	Armed bullhead	Dermal fibromas, fibrosarcomas	Groff (2004)
Amia calva	Bowfin	Granuloplastic leukemia	Groff (2004)
Anguilla japonica	Japanese eel	Nephroblastoma	Groff (2004)
Astronotus ocellatus	Oscar	Adenocarcinomas	Groff (2004)
Barbus barbus plebejus	Italian barbel	Osteoblastic osteosarcoma	Groff (2004)
Carassius auratus	Goldfish	Fibrosarcoma, pigment cell neoplasm, neurofibromas, schwannomas, focal or multifocal cutaneous erythrophoromas	Groff (2004)
Carassius auratus × *Cyprinus carpio*	Goldfish hybrid	Gonadal neoplasms	Groff (2004)
Carcharhinus brachyurus	Bronze whaler shark	Proliferative, possibly neoplastic, lesions	Robbins et al. (2014)
Carcharhinus leucas	Bull shark	Cutaneous neoplasms	Robbins et al. (2014)
Carcharias taurus rafinesque	Gray nurse shark	Odontogenic, oral, and gingival neoplasms	Robbins et al. (2014)
Carcharodon carcharias	Great white shark	Proliferative, possibly neoplastic, lesions	Robbins et al. (2014)
Catostomus commersoni	White sucker	Cutaneous papillomas	Groff (2004)
Chaetodon multicinctus and *C. miliaris*	Butterflyfish hybrids	Pigment cell neoplasms	Groff (2004)
Chologaster agassizi	Spring cavefish	Spontaneous retinoblastomas	Groff (2004)
Corydoras spp.	Cory catfish	Pigment cell neoplasms	Groff (2004)
Cyprinus carpio	Common carp	Gonadal neoplasms, erythrophoromas	Groff (2004)
Danio rerio	Zebrafish	Malignant neoplasms of the intestine	Groff (2004)
Esox lucius	Northern pike	Lymphomas, undifferentiated sarcoma of the integument	Groff (2004)
Esox masquinongy	Muskellunge	Lymphoma	Groff (2004)
Fundulus heteroclitus	Mummichog	Hepatoblastoma	Groff (2004)
Gadus spp.	Alaska pollock	Pseudobranchial adenomas	Groff (2004)
Galeocerdo cuvier	Tiger shark	Cutaneous neoplasms	Robbins et al. (2014)
Ginglymostoma cirratum	Nurse shark	Melanoma	Robbins et al. (2014)
Hemichromis bimaculatus	African jewelfish	Osteochondroma	Groff (2004)
Hippocampus abdominalis	Pot-bellied sea horse	Reticuloendothelial hyperplasia	LePage et al. (2012)
Hippocampus erectus	Lined sea horse	Fibrosarcoma of the brood pouch	LePage et al. (2012)
Hippocampus kuda	Yellow sea horse	Renal adenoma, renal round cell tumor, exocrine pancreatic carcinoma, intestinal carcinoma	LePage et al. (2012)

2. CANCER PREVALENCE AND ETIOLOGY IN WILD AND CAPTIVE ANIMALS

TABLE 2.1 Examples of Neoplasia Across the Animal Kingdom (*cont.*)

Latin name	Common name	Neoplasia (including benign and malignant abnormal cell growths)	References
Hippocampus kuda and *Phyllopteryx taeniolatus*	Sea horse hybrids	Cardiac rhabdomyosarcoma, renal adenocarcinoma, renal adenoma, lymphomas, exocrine pancreatic carcinoma, intestinal carcinoma	LePage et al. (2012)
Ictalurus nebulosus	Brown bullhead	Hepatobiliary neoplasms	Groff (2004)
Ictalurus punctatus	Channel catfish	Osteosarcoma	Groff (2004)
Kryptolebias marmoratus	Mangrove rivulus	Chondrosarcomas, hemangiomas, hemangioendotheliomas, hemangioendotheliosarcomas	Groff (2004)
Lepomis sp.	Sunfish	Cutaneous carcinoma	Groff (2004)
Limanda limanda	Common dab	Papillomas	Groff (2004)
Microgadus tomcod	Atlantic tomcod	Hepatic neoplasm	Groff (2004)
Morone saxatilis	Striped bass	Nephroblastomas	Groff (2004)
Mustelus canis	Smooth dogfish	Cutaneous neoplasms	Groff (2004)
Nebrius ferrugineus	Tawny nurse shark	Cutaneous osteoma	Groff (2004)
Oncorhynchus kisutch	Coho salmon	Plasmacytoid leukemia (marine anemia), lymphomas	Groff (2004)
Oncorhynchus mykiss	Rainbow trout	Hepatobiliary neoplasms, nephroblastoma, adenopapillomas, lymphomas	Groff (2004)
Oncorhynchus tshawytscha	Chinook salmon	Plasmacytoid leukemia (marine anemia)	Groff (2004)
Oryzias latipes	Medaka	Lymphohematopoietic neoplasms, cutaneous lymphoma, adenomas, adenocarcinomas, retinoblastomas, teratoid medulloepitheliomas, neoplasms of embryonal origin, or teratomas	Groff (2004)
Osmerus eperlanus	European smelt	Papillomas and squamous cell carcinomas	Groff (2004)
Osmerus mordax	Rainbow smelt	Papillomas and squamous cell carcinomas	Groff (2004)
Pagrus major	Japanese seabream	Leukemia	Groff (2004)
Perca flavescens	Yellow perch	Ovarian and testicular leiomyomas and fibroleiomyomas	Groff (2004)
Phyllopteryx taeniolatus	Weedy sea dragon	Rhabdomyosarcoma	LePage et al. (2012)
Plecoglossus altivelis	Ayu	Rhabdomyoma	Groff (2004)
Plectropomus leopardus	Coral trout	Melanomas	Sweet et al. (2012)
Poecilia formosa	Amazon molly	Pigment cell neoplasms (or chromatophoromas)	Groff (2004)
Poecilia reticulata	Guppy	Epidermal cystadenoma, adenomas, adenocarcinomas, neoplasms of embryonal origin, or teratomas	Groff (2004)
Pomacentrus partitus	Bicolor damselfish	Neurofibromas, schwannomas	Groff (2004)
Prionace glauca	Blue shark	Cholangiocarcinoma, testicular mesothelioma, odontogenic, oral, and gingival neoplasms	Groff (2004); Robbins et al. (2014)
Pseudopleuronectes obscurus	Flatfish	Papillomas (wild)	Groff (2004)
Pterophyllum scalare	Angelfish	Labial fibromas (odontomas)	Groff (2004)

(Continued)

TABLE 2.1   Examples of Neoplasia Across the Animal Kingdom (*cont.*)

Latin name	Common name	Neoplasia (including benign and malignant abnormal cell growths)	References
Salmo salar	Atlantic salmon	Fibrosarcomas of the swimbladder, cutaneous papillomas, sarcomas	Groff (2004)
Sparus aurata	Gilt-head bream	Osteochondroma	Groff (2004)
Stizostedion vitreum	Walleye	Dermal sarcomas	Groff (2004)
Tilapia spp.	Tilapia	Adenocarcinomas, lymphomas	Groff (2004)
Xiphophorus maculatus	Southern platyfish	Melanoma, neoplasms of embryonal origin, or teratomas	Groff (2004)
Xiphophorus maculatus and *X. helleri*	Platyfish and swordtail hybrid	Pigment cell neoplasms	Groff (2004)
Amphibians			
Ambystoma tigrinum	Tiger salamander	Tumorous growths, type not specified	Rose (1976); Rose and Harshbarger (1977)
Bufo japonicus × *Bufo raddei*	Toad hybrids	Renal cell carcinomas	Masahito et al. (2003)
Calotriton arnoldi	Montseny brook newt	Pigmented skin tumors, melanophoroma, chromatophoromas	Martinez-Silvestre et al. (2011)
Litoria aurea	Green and golden bell frog	Nephroblastoma, carcinoma	Ladds (2009)
Litoria caerulea	Green tree frog	Renal adenocarcinoma, cutaneous papilloma and fibropapilloma of the maxillary region and upper lip, hepatoma, metastatic pancreatic adenocarcinoma, coelomic adenoma	Ladds (2009)
Litoria infrafrenata	Giant (white-lipped) tree frog	Lymphoma, renal tubular adenoma, squamous cell carcinoma, papilloma, sebaceous gland carcinoma	Ladds (2009)
Litoria lesueurii	Lesueur's frog	Melanoma	Ladds (2009)
Paramesotriton hongkongensis	Hong Kong warty newt	Seminoma	Chu et al. (2012)
Xenopus laevis	African clawed frog	Various types, the most common being hepatomas, ovarian tumors, and teratomas	Balls and Clothier (1974); Robert et al. (2009); Hardwick and Philpott (2015)
Reptiles			
Acanthophis antarcticus	Death adder	Leukemic lymphoma, melanoma	Mader (1996); Ladds (2009)
Acrantophis madagascariensis	Madagascar boa	Squamous cell carcinoma, biphasic neoplasm	Bera et al. (2008); Steeil et al. (2013)
Acrochordus javanicus	Elephant trunk snake	Fibroma	Mader (1996)
Agkistrodon contortrix	Southern copperhead	Myeloid leukemia, cholangiocarcinoma, hemangiosarcoma	Catão-Dias and Nichols (1999)
Agkistrodon halys brevicaudus	Korean mamushi	Adenocarcinoma, neurofibrosarcoma	Mader (1996)
Agkistrodon piscivorus	Cottonmouth	Squamous cell carcinoma, sarcoma, fibroma	Mader (1996)
Alligator mississippiensis	American alligator	Papilloma, seminoma, fibrosarcoma	Mader (1996); Elsey et al. (2013)
Anolis carolinensis	Carolina anole	Reticulum sarcoma	Hernandez-Divers and Garner (2003)
Apalone ferox	Florida softshell turtle	Lymphoreticular neoplasia	Hernandez-Divers and Garner (2003)

TABLE 2.1 Examples of Neoplasia Across the Animal Kingdom (*cont.*)

Latin name	Common name	Neoplasia (including benign and malignant abnormal cell growths)	References
Arizona elegans occidentalis	California glossy snake	Pheochromocytoma	Mader (1996)
Aspidites melanocephalus	Black-headed python	Gastric adenocarcinoma, angiolipoma	Ladds (2009); Dietz et al. (2016)
Aspidites ramsayi	Woma	Lymphoma, colonic adenocarcinoma	Ladds (2009)
Basiliscus plumifrons	Green basilisk	Fibrosarcoma	Hernandez-Divers and Garner (2003)
Bitis arietans	Puff adder	Leukemic lymphoma, adenoma	Mader (1996)
Bitis gabonica	Gaboon viper	Transitional cell carcinoma, carcinoma, adenocarcinoma, fibrosarcoma, lymphoma, squamous cell carcinoma	Mader (1996) Catão-Dias and Nichols (1999)
Bitis nasicornis	Rhinoceros viper	Lymphoma, leukemic lymphoma, leukemia	Mader (1996)
Boa constrictor	Boa constrictor	Fibrosarcoma, malignant peripheral nerve sheath tumor, malignant perivascular wall tumor, squamous cell carcinoma, fibrosarcoma, melanoma, hemangiosarcoma, lipoma, leukemia, adenocarcinoma, carcinoma, rhabdomyosarcoma	Mader (1996); Dietz et al. (2016)
Boa cookii	Cook's tree boa	Hemangiosarcoma	Mader (1996)
Boiga dendrophila	Mangrove snake	Fibrosarcoma	Mader (1996)
Bothrops atrox	Common lancehead	Adenocarcinoma	Mader (1996)
Caretta caretta	Loggerhead	Fibropapilloma, lymphoblastic lymphoma	Ladds (2009)
Chalcides ocellatus	Ocellated skink	Lymphoma	Chu et al. (2012)
Chamaeleo dilepis	Flap-necked chameleon	Hepatoma	Hernandez-Divers and Garner (2003)
Chelonia mydas	Green sea turtle	Papillomas, fibromas, fibropapillomas, fibroadenoma, carcinoma, myxofibroma, leiomyoma, papilloma of the gall bladder	Reichenbach-Klinke (1963); Brill et al. (1995); Mader (1996); Ladds (2009)
Chilabothrus inornatus	Yellow tree boa	Squamous cell carcinoma, hepatoma, leiomyosarcoma	Mader (1996)
Chondropython viridis	Green tree python	Lymphoid leukemia, fibrosarcoma, chromatophoroma (small intestine), thymoma, myeloid leukemia, lymphoma	Catão-Dias and Nichols (1999)
Clelia clelia	Mussurana	Hepatoma	Mader (1996)
Cnemidophorus uniparens	Desert grassland whiptail lizard	Teratoma	Hernandez-Divers and Garner (2003)
Coleonyx mitratus	Central American banded gecko	Coelom	Hernandez-Divers and Garner (2003)
Corallus caninus	Emerald tree boa	Leiomyosarcoma, lymphoma, adenocarcinoma, malignant peripheral nerve sheath tumor	Catão-Dias and Nichols (1999); Dietz et al. (2016)
Cordylus polyzonus	Karoo girdled lizard	Adenoma	Hernandez-Divers and Garner (2003)
Crocodylus acutus	American crocodile	Lipoma	Mader (1996)
Crocodylus porosus	Saltwater crocodile	Lymphoma, papilloma, cancer of the cerebellum, squamous cell carcinoma	Reichenbach-Klinke (1963); Hill et al. (2016)

(Continued)

TABLE 2.1   Examples of Neoplasia Across the Animal Kingdom (*cont.*)

Latin name	Common name	Neoplasia (including benign and malignant abnormal cell growths)	References
Crocodylus siamensis	Siamese crocodile	Fibrosarcoma	Hernandez-Divers and Garner (2003)
Crotalus atrox	Western diamondback rattlesnake	Fibrosarcoma	Mader (1996)
Crotalus horridus	Timber rattlesnake	Adenoma, adenocarcinoma, fibrosarcoma, leukemia, mesothelioma, hemangioma	Mader (1996)
Crotalus mitchellii pyrrhus	Southwestern speckled rattlesnake	Adenocarcinoma	Mader (1996)
Crotalus ruber	Red diamond rattlesnake	Sarcoma	Mader (1996)
Crotalus viridis helleri	Prairie rattlesnake	Hemangioma	Mader (1996)
Crotalus viridis viridis	Prairie rattlesnake	Fibrosarcoma	Mader (1996)
Cyclura cornuta	Rhinoceros iguana	Chondro-osteofibroma	Hernandez-Divers and Garner (2003)
Cyclura ricordi	Hispaniolan ground iguana	Biliary adenoma	Hernandez-Divers and Garner (2003)
Dipsosaurus dorsalis	Desert iguana	Adenoma, adenocarcinoma	Hernandez-Divers and Garner (2003)
Dispholidus typus	Boomslang	Adenoma	Mader (1996)
Drymarchon corais	Eastern indigo snake	Melanophoroma	Mader (1996)
Drymarchon couperi	Eastern indigo snake	Adenocarcinoma	Mader (1996)
Drymarchon melanurus erebennus	Texas indigo snake	Leiomyosarcoma	Mader (1996)
Echis carinatus	Saw-scaled viper	Hepatocarcinoma	Catão-Dias and Nichols (1999)
Elaphe guttata guttata	Corn snake	Lymphoma, carcinoma, chondrosarcoma, renal cell carcinoma, adenocarcinoma, myeloid leukemia, leiomyosarcoma, lipoma, fibrosarcoma, malignant peripheral nerve sheath tumor, rhabdomyosarcoma	Mader (1996); Catão-Dias and Nichols (1999); Dietz et al. (2016)
Elaphe obsoleta	Western rat snake	Adenocarcinoma, adenoma, fibrosarcoma, rhabdomyosarcoma	Mader (1996); Catão-Dias and Nichols (1999)
Elaphe obsoleta rossalleni	Everglades rat snake	Melanoma	Mader (1996)
Elaphe obsoleta quadrivittata	Yellow rat snake	Transitional cell carcinoma	Mader (1996)
Elaphe taeniura	Beauty snake	Hepatocarcinoma	Catão-Dias and Nichols (1999)
Elaphe taeniura friesei	Taiwan beauty rat snake	Malignant chromatophoroma	Chu et al. (2012)
Elaphe vulpina	Fox snake	Adenocarcinoma	Mader (1996)
Emys orbicularis	European pond turtle	Squamous cell carcinoma, fibroadenoma	Mader (1996)
Epicrates cenchria	Rainbow boa	Histiocytoma, lymphoma, adenoma, myelomonocytic leukemia, squamous cell carcinoma	Catão-Dias and Nichols (1999)
Epicrates subflavus	Jamaican boa	Malignant peripheral nerve sheath tumor, malignant perivascular wall tumor	Dietz et al. (2016)
Eryx conicus	Common sand boa	Squamous cell carcinoma, mixed cell tumor	Mader (1996)
Eublepharis macularis	Leopard gecko	Cholangiocarcinoma	Hernandez-Divers and Garner (2003)

TABLE 2.1 Examples of Neoplasia Across the Animal Kingdom (*cont.*)

Latin name	Common name	Neoplasia (including benign and malignant abnormal cell growths)	References
Eumeces fasciatus	Five-lined skink	Hepatocarcinoma	Hernandez-Divers and Garner (2003)
Eunectes murinus	Green anaconda	Lymphoma, fibrosarcoma, granulosa cell tumor	Mader (1996)
Eunectes notaeus	Yellow anaconda	Cystadenoma	Catão-Dias and Nichols (1999)
Geochelone carbonaria	Redfoot tortoise	Adenoma	Mader (1996)
Gopherus agassizii	Mojave desert tortoise	Adenoma, interstitial tumor	Mader (1996)
Gopherus trijuga	Ceylon terrapin	Carcinoma, squamous cell carcinoma	Mader (1996)
Heloderma suspectum	Gila monster	Squamous cell carcinoma, melanoma	Hernandez-Divers and Garner (2003)
Heterodon nasicus	Western hognose snake	Sarcoma, lymphoma	Mader (1996)
Hydrosaurus amboinensis	Amboina sailfin lizard	Lymphoma, plasma cell tumor	Hernandez-Divers and Garner (2003)
Iguana iguana	Green iguana	Lymphoma, hepatoma, cholangioma, adenocarcinoma, ovarian teratoma, adenoma	Hernandez-Divers and Garner (2003)
Indotestudo elongata	Yellow-headed tortoise	Leukemia	Chu et al. (2012)
Lacerta agilis	Sand lizard	Papilloma, squamous cell carcinoma	Reichenbach-Klinke (1963); Hernandez-Divers and Garner (2003)
Lacerta lepida	Ocellated lizard	Papilloma	Reichenbach-Klinke (1963)
Lacerta viridis	Green lizard	Papilloma, osteosarcoma	Reichenbach-Klinke (1963); Hernandez-Divers and Garner (2003)
Lamprophis fuliginosus	African house snake	Malignant peripheral nerve sheath tumor	Dietz et al. (2016)
Lampropeltis getula californiae	Eastern kingsnake	Adenoma, carcinoma, lymphoma, squamous cell carcinoma, cholangiocarcinoma, melanoma, malignant peripheral nerve sheath tumor	Mader (1996); Dietz et al. (2016)
Lampropeltis getula getula	Eastern kingsnake	Tubular adenoma	Catão-Dias and Nichols (1999)
Lampropeltis getula holbrooki	Speckled kingsnake	Adenoma	Mader (1996)
Lampropeltis triangulum annulata	Mexican milk snake	Sarcoma	Mader (1996)
Lampropeltis triangulum sinaloae	Sinaloan milk snake	Myxosarcoma, sarcoma, hepatoma	Mader (1996); Catão-Dias and Nichols (1999)
Lampropeltis triangulum triangulum	Eastern milk snake	Adenocarcinoma, adenoma	Catão-Dias and Nichols (1999)
Morelia spilota	Carpet python	Multicentric lymphoma, soft tissue sarcoma, fibrosarcoma, cholangiocarcinoma, coelomic carcinoma	Ladds (2009)
Morelia spilota spilota	Diamond python	Myxosarcoma, monocytic leukemia of azurophilic type, lymphoid leukemia	Ladds (2009)
Morelia spilota variegata	Darwin carpet python	Cholangiocarcinoma	Mader (1996)
Morelia viridis	Green tree python	Ossifying fibrosarcoma	Mader (1996); Ladds (2009)
Naja naja	Indian cobra	Leiomyosarcoma, adenocarcinoma, adenoma, lymphoma, hepatocarcinoma	Mader (1996); Catão-Dias and Nichols (1999)

(Continued)

TABLE 2.1 Examples of Neoplasia Across the Animal Kingdom (*cont.*)

Latin name	Common name	Neoplasia (including benign and malignant abnormal cell growths)	References
Naja nigricollis	Black-necked spitting cobra	Adenoma, lymphoma	Mader (1996)
Naja nivea	Cape cobra	Adenocarcinoma	Mader (1996)
Natrix natrix	Grass snake	Pancreatic adenocarcinoma, malignant peripheral nerve sheath tumor	Reichenbach-Klinke (1963); Dietz et al. (2016)
Ophiophagus hannah	King cobra	Tubular adenoma	Catão-Dias and Nichols (1999)
Pantherophis alleghaniensis	Black rat snake	Ameloblastoma	Comolli et al. (2015)
Pelomedusa subrufa	African helmeted turtle	Leukemia	Mader (1996)
Pelusios subniger	East African black mud turtle	Carcinoma	Mader (1996)
Pituophis melanoleucus	Pine snake	Adenocarcinoma, malignant chromatophoroma, carcinoma, adenoma	Mader (1996)
Pituophis melanoleucus mugitus	Florida pine snake	Rhabdomyosarcoma, adenoma, adenocarcinoma, melanoma	Mader (1996)
Pituophis melanoleucus sayi	Bullsnake	Papilloma, adenocarcinoma, malignant melanoma	Reichenbach-Klinke (1963); Mader (1996)
Podarcis muralis	Common wall lizard	Papilloma	Hernandez-Divers and Garner (2003)
Pogona vitticeps	Bearded dragon	Adenocarcinoma of the liver, disseminated myelogenous leukemia, monocytic leukemia, malignant nerve sheath tumor	Hernandez-Divers and Garner (2003); Ladds (2009)
Podarcis sicula	Italian wall lizard	Lymphoma, fibrosarcoma, undifferentiated mesenchymal tumor	Hernandez-Divers and Garner (2003)
Pseudechis porphyriacus	Red-bellied black snake	Cutaneous papillomas, adenomatous proliferation, adenoma of the bile duct	Mader (1996); Ladds (2009)
Pseudonaja affinis	Dugite	Melanoma	Ladds (2009)
Pseudonaja nuchalis	Western brown snake	Leukemic lymphoma	Ladds (2009)
Python molurus	Indian rock python	Ameloblastoma, fibroma	Mader (1996)
Python molurus bivittatus	Burmese python	Carcinoma, adenocarcinoma, interstitial cell tumor, osteosarcoma	Mader (1996)
Python molurus molurus	Indian python	Sarcoma, lymphoma, leukemia	Mader (1996)
Python regius	Ball python	Fibrosarcoma	Mader (1996)
Python reticulatus	Reticulated python	Carcinoma, melanoma, lymphoma	Mader (1996)
Python sebae	African rock python	Adenoma	Mader (1996)
Rhamphiophis oxyrhynchus	Rufous beaked snake	Hemangiosarcoma, lymphoma, fibrosarcoma	Catão-Dias and Nichols (1999)
Sistrurus catenatus	Massasauga	Adenoma, hemangioma, carcinoma	Mader (1996)
Spilotes pullatus	Yellow rat snake	Adenocarcinoma	Mader (1996)
Strophurus spinigerus	Spiny-tailed gecko	Neuroblastoma	Ladds (2009)
Terrapene carolina	Common box turtle	Adenocarcinoma	Mader (1996)
Testudo graeca	Spur-thighed tortoise	Adenoma	Mader (1996)
Testudo hermanni	Hermann's tortoise	Lymphoma, neurilemmoma	Mader (1996)
Testudo horsfieldii	Afghan tortoise	Fibroma, fibroadenoma	Mader (1996)
Thamnophis sauritus	Ribbon snake	Lipoma	Dietz et al. (2016)

TABLE 2.1 Examples of Neoplasia Across the Animal Kingdom (*cont.*)

Latin name	Common name	Neoplasia (including benign and malignant abnormal cell growths)	References
Thamnophis sirtalis	Common garter snake	Squamous cell carcinoma, cholangioma, granulosa cell tumor, Sertoli cell tumor, malignant perivascular wall tumor, malignant peripheral nerve sheath tumor	Mader (1996); Dietz et al. (2016)
Thamnophis elegans terrestris	Coast garter snake	Malignant chromatophoroma	Mader (1996)
Tiliqua rugosa	Shingle-back lizard	Subcutaneous osteoma, liposarcoma	Ladds (2009), Hernandez-Divers and Garner (2003)
Trachemys scripta elegans	Red-eared slider	Carcinoma, leukemia	Mader (1996)
Tupinambis nigropunctatus	Tegu	Squamous cell carcinoma	Hernandez-Divers and Garner (2003)
Tupinambis rufescens	Argentine red tegu	Hepatoma	Hernandez-Divers and Garner (2003)
Tupinambis teguixin	Golden tegu	Squamous cell carcinoma	Hernandez-Divers and Garner (2003)
Uromastyx acanthinura	Bell's dabb lizard	Lymphoma	Hernandez-Divers and Garner (2003)
Uromastyx aegyptia	Egyptian mastigure	Lymphoid neoplasia	Gyimesi et al. (2005)
Varanus bengalensis	Bengal monitor	Leukemia, osteochrondroma, enchondroma	Hernandez-Divers and Garner (2003)
Varanus exanthematicus	Savannah monitor	Lymphoma	Hernandez-Divers and Garner (2003)
Varanus komodoensis	Komodo dragon	Carcinoma, adenoma, islet cell tumor, pheochromocytoma, interstitial cell tumor	Hernandez-Divers and Garner (2003)
Varanus niloticus	Nile monitor	Plasma cell tumor	Hernandez-Divers and Garner (2003)
Varanus salvator	Water monitor	Lymphoma	Hernandez-Divers and Garner (2003)
Vipera ammodytes	Horned viper	Adenocarcinoma	Mader (1996)
Vipera palestine	Palestine viper	Adenocarcinoma	Mader (1996)
Vipera russelli	Russell's viper	Fibrosarcoma, leukemia, myofibroma	Mader (1996)
Walterinnesia aegyptia	Desert cobra	Pheochromocytoma	Mader (1996)
Birds			
Acanthagenys rufogularis	Spiny-cheeked honeyeater	Nephroblastoma	Ladds (2009)
Agapornis lilianae	Nyasa lovebird	Fibromas and fibrosarcomas of integument and/or skeletal muscle, Sertoli cell tumors	Reece (1992); Ladds (2009)
Agapornis roseicollis	Peach-faced lovebird	Fibromas and fibrosarcomas of the integument and/or skeletal muscle, visceral fibromas and fibrosarcomas, subcutaneous lipomas, intraabdominal lipomas, lymphoblastic lymphomas, lymphocytic lymphomas, and mixed-cell lymphomas, hepatocarcinomas, neurilemmoma	Reece (1992); Ladds (2009)
Ailuroedus crassirostris	Green catbird	Myelocytomas	Reece (1992); Ladds (2009)
Aix sponsa	North American wood duck	Malignant melanoma	Chu et al. (2012)

(Continued)

TABLE 2.1   Examples of Neoplasia Across the Animal Kingdom (*cont.*)

Latin name	Common name	Neoplasia (including benign and malignant abnormal cell growths)	References
Alectoris graeca	Chukar partridge	Liposarcomas, cholangiomas	Reece (1992); Ladds (2009)
Alisterus scapularis	Australian king parrot	Plasma cell tumors	Reece (1992); Ladds (2009)
Anas castanea	Chestnut teal	Lymphocytic lymphomas and mixed-cell lymphomas, metastatic abdominal adenocarcinomas	Reece (1992); Ladds (2009)
Anas cyanoptera	Cinnamon teal	Adenocarcinoma	Snyder and Ratcliffe (1966)
Anas novaehollandiae	New Zealand scaup	Dermal squamous cell carcinomas	Reece (1992); Ladds (2009)
Anas platyrhynchos	Mallard, domestic duck	Intraabdominal lipomas, chondromas, osteomas, lymphoblastic lymphomas, seminomas, metastatic abdominal adenocarcinomas, astrocytoma	Reece (1992); Ladds (2009); Chu et al. (2012)
Anas superciliosa	Pacific black duck	Myxomas and myxofibromas, malignant melanomas	Reece (1992); Ladds (2009)
Anhinga novaehollandiae	Darter	Hemangiomas	Reece (1992); Ladds (2009)
Anser anser	Graylag goose	Chondromas	Reece (1992); Ladds (2009)
Anser domesticus	Domestic goose	Fibrosarcoma	Ratcliffe (1933)
Aprosmictus scapularis	King parrot	Fibrosarcoma	Ratcliffe (1933)
Ara militaris	Military macaw	Squamous cell carcinoma	Ratcliffe (1933)
Barnardius barnardii	Mallee ring-neck	Lymphoblastic lymphomas	Reece (1992); Ladds (2009)
Barnardius zonarius	Port Lincoln parrot	Plasma cell tumors	Reece (1992); Ladds (2009)
Barnardius zonarius semitorquatus	Twenty eight parrot	Plasma cell tumors	Ladds (2009)
Brotogeris tirica	Plain parakeet	Fibrosarcoma	Ratcliffe (1933)
Bubo virginianus	Great horned owl	Myelogenous leukemia	Wiley et al. (2009)
Buteo jamaicensis	Red-tailed hawk	Cholangiocarcinoma	Hartup et al. (1995)
Cacatua galerita	Sulfur-crested cockatoo	Visceral fibromas and fibrosarcomas, subcutaneous lipomas, intraabdominal lipomas, osteosarcomas, lymphoblastic lymphomas, lymphocytic lymphomas, and mixed-cell lymphomas, granulosa cell tumors, squamous cell carcinoma, adenocarcinoma	Reece (1992); Ladds (2009); Ratcliffe (1933)
Cacatua leadbeateri	Pink cockatoo	Dermal squamous cell carcinomas	Reece (1992); Ladds (2009)
Cacatua moluccensis	Gang-gang cockatoo	Lymphomas	Ratcliffe (1933)
Cacatua roseicapilla	Galah	Subcutaneous lipomas, intraabdominal lipomas, granulosa cell tumors	Reece (1992); Ladds (2009)
Cacatua sanguinea	Little corella	Intraabdominal lipomas, lymphoblastic lymphomas	Reece (1992); Ladds (2009)
Callocephalon fimbriatum	Gang-gang cockatoo	Visceral fibromas and fibrosarcomas, osteosarcomas, plasma cell tumors	Reece (1992); Ladds (2009)
Calyptorhynchus baudinii	White-tailed black cockatoo	Myeloblastomas	Reece (1992); Ladds (2009)
Casuarius casarius johnsonii	Southern cassowary	Papilliform mesotheliomas, gastrointestinal adenocarcinomas	Reece (1992); Ladds (2009)
Centropus phasianinus	Pheasant coucal	Hepatoma, hepatocarcinoma, cholangioma	Ladds (2009)
Cereopsis novaehollandiae	Cape Barren goose	Plasma cell tumors	Reece (1992); Ladds (2009)

TABLE 2.1   Examples of Neoplasia Across the Animal Kingdom (*cont.*)

Latin name	Common name	Neoplasia (including benign and malignant abnormal cell growths)	References
Chalcophaps indica	Emerald dove	Pinealoma	Reece (1992); Ladds (2009)
Chloephaga leucoptera	Upland goose	Adenocarcinoma	Snyder and Ratcliffe (1966)
Chrysolophus pictus	Golden pheasant	Adenocarcinoma, adenoma	Ratcliffe (1933)
Columba livia	Domestic pigeon	Fibromas and fibrosarcomas of integument and/or skeletal muscle, visceral fibromas and fibrosarcomas, subcutaneous lipomas, intraabdominal lipomas, liposarcomas, rhabdomyoma, leiomyomas and leiomyofibromas, myelocytomas, lymphoblastic lymphomas, lymphocytic lymphomas and mixed-cell lymphomas, plasma cell tumors, basal cell tumors, crop carcinoma, cholangiomas, renal adenocarcinomas, metastatic abdominal adenocarcinomas, seminomas, thyroid adenomas	Reece (1992); Ladds (2009); Shimonohara et al. (2013)
Columba pulchricollis	Ashy wood pigeon	Cholangioma	Chu et al. (2012)
Conurus holochlorus	Green parakeet	Carcinomatoid embryoma	Ratcliffe (1933)
Coscoroba coscoroba	Coscoroba swan	Cholangiocarcinoma, renal cell carcinoma	Chu et al. (2012)
Coturnix australis	Brown quail	Subcutaneous lipomas	Reece (1992); Ladds (2009)
Coturnix chinensis	King quail	Fibromas and fibrosarcomas of integument and/or skeletal muscle, hepatocarcinomas, seminomas, metastatic abdominal adenocarcinomas	Reece (1992); Ladds (2009)
Coturnix coturnix japonica	Japanese quail	Fibromas and fibrosarcomas of integument and/or skeletal muscle, visceral fibromas and fibrosarcomas, osteosarcomas, hemangiomas, lymphocytic lymphomas and mixed-cell lymphomas, cholangiomas	Reece (1992); Ladds (2009)
Cyanoramphus novaezelandia	Red-fronted parakeet	Intraabdominal lipomas	Reece (1992); Ladds (2009)
Cygnus atratus	Black swan	Myxomas and myxofibromas, osteosarcoma	Reece (1992); Ladds (2009); Chu et al. (2012)
Cygnus olor	Mute swan	Myxomas and myxofibromas	Reece (1992); Ladds (2009)
Dacelo novaeguineae	Laughing kookaburra	Intraabdominal lipomas, cholangiomas	Reece (1992); Ladds (2009)
Dendrocygna autumnalis	Black-bellied whistling duck	Adenocarcinoma	Snyder and Ratcliffe (1966)
Dromaius novaehollandiae	Emu	Pancreatic adenocarcinomas	Reece (1992); Ladds (2009)
Dryonastes berthemyi	Buffy laughingthrush	Fibrosarcoma	Ratcliffe (1933)
Egretta novaehollandiae	White-faced heron	Lymphoma	Ladds (2009)
Emberiza icterica	Red-headed bunting	Lipoma	Ratcliffe (1933)
Emblema temporalis	Red-browed firetail	Esophageal papilloma	Reece (1992); Ladds (2009)
Eolophus roseicapilla	Galah	Lipomas	Ratcliffe (1933)
Erythrura gouldiae	Gouldian finch	Adrenocortical adenomas	Reece (1992); Ladds (2009)
Erythrura trichroa	Blue-faced parrot finch	Renal adenoma, adenocarcinoma	Ladds (2009)
Eudyptula minor	Little penguin	Lymphocytic lymphomas and mixed-cell lymphomas, fibroma, fibrosarcoma, cutaneous papillomas	Reece (1992); Ladds (2009)

(*Continued*)

TABLE 2.1   Examples of Neoplasia Across the Animal Kingdom (*cont.*)

Latin name	Common name	Neoplasia (including benign and malignant abnormal cell growths)	References
Falco naumanni	Lesser kestrel	Malignant intracranial teratoma	Lopez and Murcia (2008)
Gennaeus nycthemerus	Silver pheasant	Adenocarcinoma	Ratcliffe (1933); Snyder and Ratcliffe (1966)
Geopelia cuneata	Diamond dove	Leiomyomas and leiomyofibromas	Reece (1992); Ladds (2009)
Geopelia humeralis	Bar-shouldered dove	Osteosarcomas	Reece (1992); Ladds (2009)
Geopelia placida	Peaceful dove	Leiomyomas and leiomyofibromas	Reece (1992); Ladds (2009)
Ginnaeus swinhoii	Swinhoe's pheasant	Visceral fibromas and fibrosarcomas	Reece (1992); Ladds (2009)
Gracula religiosa	Greater hill mynah	Chondrosarcoma	Chu et al. (2012)
Gymnorhina tibicen	Australian magpie	Fibromas and fibrosarcomas of the integument and/or skeletal muscle, myeloblastomas, lymphocytic lymphomas, and mixed-cell lymphomas	Reece (1992); Ladds (2009)
Larus novaehollandiae	Silver gull	Lymphocytic lymphomas and mixed-cell lymphomas	Reece (1992); Ladds (2009)
Larus pacificus	Pacific gull	Chondromas, myelocytomas	Reece (1992); Ladds (2009)
Leipoa ocellata	Malleefowl	Lymphomas	Ladds (2009)
Leptolophus hollandicus	Cockatiel	Lipomas	Ratcliffe (1933)
Lonchura castaneothorax	Chestnut-breasted mannikin	Lymphocytic lymphomas and mixed-cell lymphomas	Reece (1992); Ladds (2009)
Lopholaimus antarcticus	Topknot pigeon	Lymphomas	Ladds (2009)
Macropygia amboinensis	Cuckoo-dove	Gastrointestinal adenocarcinomas	Ladds (2009)
Malurus cyaneus	Superb fairy-wren	Lymphocytic lymphomas and mixed-cell lymphomas	Reece (1992); Ladds (2009)
Mareca sibilatrix	Chiloe wigeon	Adenocarcinoma	Snyder and Ratcliffe (1966)
Megaquiscalus major	Boat-tailed grackle	Adenocarcinoma	Ratcliffe (1933)
Meleagris gallopavo	Wild turkey	Adenocarcinoma	Ratcliffe (1933)
Melopsittacus undulatus	Budgerigar	Adenoma, adenocarcinoma, carcinomas, fibromas and fibrosarcomas of integument and/or skeletal muscle, visceral fibromas and fibrosarcomas, myxomas and myxofibromas, subcutaneous lipomas, intraabdominal lipomas, osteomas, leiomyomas and leiomyofibromas, hemangiomas, myelocytomas, reticulum cell sarcoma, lymphoblastic lymphomas, lymphocytic lymphomas and mixed-cell lymphomas, plasma cell tumors, dermal squamous cell carcinomas, feather folliculomas, uropygial adenomas, proventricular adenocarcinomas, cholangiomas, renal adenocarcinomas, seminomas, Sertoli cell tumors, Leydig cell tumor, ovarian adenocarcinoma, granulosa cell tumors, oviduct adenomas, metastatic abdominal adenocarcinomas, adrenocortical adenomas, thyroid adenomas, thyroid mixed-cell tumor, neurofibroma, nephroblastoma, lipomas, glioma, lymphoma, teratoma	Ratcliffe (1933); Reece (1992); Ladds (2009)

TABLE 2.1 Examples of Neoplasia Across the Animal Kingdom (*cont.*)

Latin name	Common name	Neoplasia (including benign and malignant abnormal cell growths)	References
Neochmia ruficauda	Star finch	Myxomas and myxofibromas	Reece (1992); Ladds (2009)
Neophema pulchella	Turquoise parrot	Lymphocytic lymphomas and mixed-cell lymphomas	Reece (1992); Ladds (2009)
Neopsephotus bourkii	Bourke's parrot	Plasma cell tumors	Reece (1992); Ladds (2009)
Northiella haematogaster	Blue bonnet	Plasma cell tumors	Reece (1992); Ladds (2009)
Nycticorax caledonicus	Rufous night heron	Myelocytomas	Reece (1992); Ladds (2009)
Nymphicus hollandicus	Cockatiel	Liposarcomas, renal adenocarcinomas, fibroma, fibrosarcoma	Reece (1992); Ladds (2009)
Nyroca americana	Redhead duck	Adenocarcinoma	Snyder and Ratcliffe (1966)
Oxyura australis	Blue-billed duck	Cholangiomas, hepatoma, hepatocarcinoma	Reece (1992); Ladds (2009)
Padda oryzivora	Java sparrow	Lymphocytic lymphomas and mixed-cell lymphomas, metastatic abdominal adenocarcinomas	Reece (1992); Ladds (2009)
Palaeornis cyanocephala	Burmese parrakeet	Adenoma	Ratcliffe (1933)
Palaeornis eupatrius	Alexandrine parrakeet	Teratoma	Ratcliffe (1933)
Paroaria cucullata	Red crested cardinal	Myxosarcoma	Ratcliffe (1933)
Passer domesticus	House sparrow	Lymphocytic lymphomas and mixed-cell lymphomas	Reece (1992); Ladds (2009)
Pavo cristatus	Common peafowl	Esophageal papilloma	Chu et al. (2012)
Plegadis falcinellus	Glossy ibis	Intracutaneous keratoacanthomas	Reece (1992); Ladds (2009)
Phalacrocorax carbo	Great cormorant	Melanoma	Kusewitt and Ley (1996)
Phaps chalcoptera	Common bronze-wing	Dermal squamous cell carcinomas	Reece (1992); Ladds (2009)
Phasianus colchicus	Ring-necked pheasant	Lymphoblastic lymphomas, lymphocytic lymphomas and mixed-cell lymphomas, renal adenocarcinomas, cholangioma, pulmonary carcinoma, renal cell carcinoma, thyroid adenoma, fibroma	Reece (1992); Ladds (2009); Chu et al. (2012)
Phasianus versicolor	Green pheasant	Fibrosarcoma, lymphomas	Ratcliffe (1933)
Phylidonyris novaehollandiae	New Holland honeyeater	Cutaneous papillomas	Ladds (2009)
Planestictus m. migratorius	Three-legged robin	Nephroblastoma or renal carcinoma	Ratcliffe (1933)
Platycercus elegans	Crimson rosella	Lymphocytic lymphomas and mixed-cell lymphomas, plasma cell tumors	Reece (1992); Ladds (2009)
Platycercus eximius	Eastern rosella	Lymphocytic lymphomas and mixed-cell lymphomas	Reece (1992); Ladds (2009)
Plectorhyncha lanceolata	Striped honeyeater	Nephroblastoma	Ladds (2009)
Plectropterus gambensis	Spur-winged goose	Fibrosarcoma	Ratcliffe (1933)
Podargus strigoides	Tawny frogmouth	Intraabdominal lipomas	Reece (1992); Ladds (2009)
Polytelis swainsonii	Superb parrot	Lymphoma, plasma cell tumors	Reece (1992); Ladds (2009)
Prunella collaris	Alpine accentor	Hepatoma	Chu et al. (2012)
Psephotus dissimilis	Hooded parrot	Fibroma, fibrosarcoma	Ladds (2009)
Psephotus varius	Mulga parrot	Fibromas and fibrosarcomas of integument and/or skeletal muscle	Reece (1992); Ladds (2009)

(Continued)

TABLE 2.1   Examples of Neoplasia Across the Animal Kingdom (*cont.*)

Latin name	Common name	Neoplasia (including benign and malignant abnormal cell growths)	References
Quelea quelea	Red-billed quelea	Teratoma	Ratcliffe (1933)
Serinus canaria	Canary	Fibromas and fibrosarcomas of integument and/or skeletal muscle, visceral fibromas and fibrosarcomas, myxomas and myxofibromas, chondromas, leiomyomas and leiomyofibromas, lymphoblastic lymphomas, lymphocytic lymphomas and mixed-cell lymphomas, plasma cell tumors, dermal squamous cell carcinomas, intracutaneous keratoacanthomas, feather folliculomas, uropygial adenomas, proventricular adenocarcinomas, hepatocarcinomas, adrenocortical adenomas, pituitary adenoma	Ratcliffe (1933); Reece (1992); Ladds (2009)
Sicalis flaveola	Saffron finch	Adenocarcinoma	Ratcliffe (1933)
Spatula clypeata	Shoveler duck	Adenocarcinoma	Snyder and Ratcliffe (1966)
Strepera spp.	Currawong	Lymphocytic lymphomas and mixed-cell lymphomas	Reece (1992); Ladds (2009)
Struthidea cinerea	Apostlebird	Plasma cell tumors	Reece (1992); Ladds (2009)
Struthio camelus	Ostrich	Papilliform mesotheliomas	Reece (1992); Ladds (2009)
Tadoma radjah	Radjah shelduck	Oviduct adenomas, nephroblastoma	Reece (1992); Ladds (2009)
Tadoma variegata	Paradise shelduck	Plasma cell tumors	Reece (1992); Ladds (2009)
Taeniopygia bichenovii	Double-barred finch	Fibroma, fibrosarcoma	Ladds (2009)
Taeniopygia castanotis	Zebra finch	Adenocarcinoma, teratoma	Ratcliffe (1933)
Thraupis palmarum	Palm tanager	Lipoma	Ratcliffe (1933)
Torgos tracheliotus	African eared vulture	Adenocarcinoma	Snyder and Ratcliffe (1966)
Trichoglossus chloroepidotus	Scaly-breasted lorikeet	Pancreatic adenocarcinomas	Reece (1992); Ladds (2009)
Trichoglossus rubritorquis	Red-collared lorikeet	Intraabdominal lipomas, hepatocarcinomas, metastatic abdominal adenocarcinomas	Reece (1992); Ladds (2009)
Turdoides terricolor	Jungle babbler	Adenoma	Ratcliffe (1933)
Turdus merula	Blakbird	Nephroblastoma or renal carcinoma (hypernephroma)	Ratcliffe (1933)
Turnix melanogaster	Black-breasted button-quail	Metastatic abdominal adenocarcinomas	Reece (1992); Ladds (2009)
Vanellus miles	Masked lapwing	Lymphoma	Ladds (2009)
Mammals			
Acinonyx jubatus	Cheetah	Myometrial leiomyomas, uterine fibroleiomyoma	Munson et al. (1999); Walzer et al. (2003)
Acrobates pygmaeus	Feathertail glider	Biliary adenocarcinoma	Ladds (2009)
Addax nasomaculatus	Addax	Intestinal tubulopapillary carcinoma	Chu et al. (2012)
Aepyprymnus rufescens	Rufous rat-kangaroo	Thyroid adenoma, lymphoma, hemangiomas, carcinoma	Ladds (2009)
Ammotragus lervia	Barbary sheep	Lymphoma	Chu et al. (2012)
Antechinus stuartii	Brown antechinus	Squamous cell carcinoma, trichoepithelioma	Canfield et al. (1990)
Antechinomys laniger spenceri	Kultarr	Pulmonary adenomatosis	Attwood and Woolley (1973)

TABLE 2.1   Examples of Neoplasia Across the Animal Kingdom (*cont.*)

Latin name	Common name	Neoplasia (including benign and malignant abnormal cell growths)	References
Antechinus minimus	Swamp antechinus	Renal pelvic transitional cell proliferation	Canfield et al. (1990)
Arctictis binturong	Binturong	Hepatocarcinoma	Chu et al. (2012)
Arctocephalus forsteri	New Zealand fur seal	Renal adenocarcinoma, papilloma, basal cell carcinoma, osteosarcoma, anaplastic renal adenocarcinoma, neuroblastoma	Ladds (2009)
Arctocephalus pusillus	Afro-Australian Fur Seal	Hepatoma, hepatocarcinoma, uterine and intestinal leiomyomas, thyroid adenoma, lymphoma, ovarian granulosa cell tumor, adenocarcinoma, malignant melanoma	Newman and Smith (2006); Ladds (2009)
Atelerix albiventris	Four-toed hedgehog	Epithelial tumors, round cell tumors, mesenchymal or spindle cell tumors, endometrial stromal sarcomas, leiomyosarcoma, adenoleiomyoma, adenocarcinoma, lymphoma, oral squamous cell carcinoma, schwannoma or neurofibrosarcoma, plasma cell tumor, hemangiosarcoma, fibrosarcoma, osteosarcoma, undifferentiated or poorly differentiated sarcomas, mammary gland tumors, mast cell tumors, sebaceous carcinoma, lipoma	Mikaelian et al. (2004); Heatley et al. (2005)
Atherurus macrourus	Brush-tailed porcupine	Inflammatory myofibroblastic tumor	Chu et al. (2012)
Balaena mysticetus	Bowhead whale	Lipoma	Newman and Smith (2006)
Balaenoptera borealis	Sei whale	Melanocytoma (possibly hamartoma)	Newman and Smith (2006)
Balaenoptera musculus	Blue whale	Mediastinal ganglioneuroma, mucinous cystadenoma, granulosa cell tumor, gastric lipoma, fibroma of the pleura	Newman and Smith (2006); Ladds (2009)
Balaenoptera physalus	Fin whale	Neurofibroma of the cerebellum, Hodgkin's-like lymphoma, fibromas of the tongue, of the pleura, of the subcutis and skin, granulosa cell tumor, ovarian carcinoma, osteoma, lipoma	Newman and Smith (2006); Ladds (2009)
Bassariscus astutus	Ringtail cat	Basal cell carcinoma	Ratcliffe (1933)
Bison bison	Bison	Adenocarcinoma	Ratcliffe (1933)
Bos bubalis	Buffalo	Adenoma	Ratcliffe (1933)
Bos taurus	Domestic cattle	Esophageal papilloma, cutaneous squamous cell carcinoma	Chu et al. (2012)
Boselaphus tragocamelus	Nilgai	Fibroma	Ratcliffe (1933)
Callimico goeldii	Goeldi's marmoset	Myelolipoma	Porter et al. (2004)
Callithrix jacchus	Common marmoset	Myelolipoma	Porter et al. (2004)
Callorhinus ursinus	Northern fur seal	Granulosa cell tumor, lymphoma, lipoma, fibrosarcoma, squamous cell carcinoma, ganglioneuroblastoma, rhabdomyosarcoma	Newman and Smith (2006)
Camelus bactrianus	Bactrian camel	Hemangioma	Ratcliffe (1933)
Canis anthus	Senegalese wolf	Medullary carcinoma	Ratcliffe (1933)
Canis latrans	Coyote	Chondrosarcoma	Ratcliffe (1933)
Canis lupus baileyi	Mexican wolf	Basal cell carcinoma, squamous cell carcinoma, nephroblastoma, adenocarcinoma	Ratcliffe (1933)

(Continued)

TABLE 2.1   Examples of Neoplasia Across the Animal Kingdom (*cont.*)

Latin name	Common name	Neoplasia (including benign and malignant abnormal cell growths)	References
Canis lupus dingo	Dingo	Lymphoma, thymoma, lipoma of subcutis, fibromatous epulis, perianal adenoma, sebaceous adenoma, squamous cell carcinoma, bronchial adenoma	Ladds (2009)
Canis mesomelas	Black-backed jackal	Osteoma, hemangiosarcoma	Chu et al. (2012)
Canis rufus	Red wolf	Adenocarcinoma, carcinomas	Snyder and Ratcliffe (1966); Seeley et al. (2016)
Capra hircus	Domestic goat	Lymphoma	Ratcliffe (1933)
Caracal caracal	Caracal	Osteochondroma	Ratcliffe (1933)
Cebus albifrons	White-fronted capuchin	Cholangiocarcinoma	Porter et al. (2004)
Cebus apella fatuellus	Tufted capuchin	Adenoma	Ratcliffe (1933)
Cercocebus atys	Sooty mangabey	Hepatocarcinoma	Porter et al. (2004)
Cercocebus atys lunulatus	White-naped mangabey	Hepatocarcinoma	Porter et al. (2004)
Cercopithecus aethiops	African green monkey	Hepatoma, mixed hepatocellular and cholangiocellular carcinoma, uterine leiomyoma	Porter et al. (2004); Chu et al. (2012)
Cercopithecus diana	Diana monkey	Cholangiocarcinoma	Porter et al. (2004)
Cercopithecus mitis	Blue monkey	Biliary adenoma/cystadenoma	Porter et al. (2004)
Cercopithecus mitis ssp. *albogularis*	White-throated guenon	Biliary adenoma/cystadenoma	Porter et al. (2004)
Cercopithecus mona	Mona monkey	Biliary adenoma/cystadenoma	Porter et al. (2004)
Chalinolobus gouldii	Gould's wattled bat	Cutaneous papilloma of the wing	Ladds (2009)
Chlorocebus sabaeus	Green monkey	Adenocarcinoma	Ratcliffe (1933)
Connochaetes gnou	Black wildebeest	Squamous cell carcinoma	Ratcliffe (1933)
Cuniculus paca	Lowland paca	Lymphoma	Ratcliffe (1933)
Cynomys ludovicianus	Black-tailed prairie dogs	Hepatocarcinoma, hepatoma, biliary cystadenoma, cholangiocarcinoma, odontoma (elodontoma), lingual squamous cell carcinoma, salivary gland adenocarcinoma, gingival squamous cell carcinoma, intestinal leiomyoma, multicentric lymphoma, malignant round cell tumor, high grade lymphoma of liver and gall bladder, cutaneous lymphoma, malignant thymoma, atrial hemangiosarcoma, splenic hemangioma, thoracic lipoma, thyroid adenocarcinoma, pancreatic adenocarcinoma, cystadenocarcinoma, adenocarcinoma, probably mammary, basal cell tumor, squamous cell carcinoma, bronchioloalveolar carcinoma	Thas and Garner (2012)
Dasycercus cristicauda	Mulgara	Prostatic carcinoma	Canfield et al. (1990)
Dasykaluta rosamondae	Little red kaluta	Splenic myeloid hyperplasia	Canfield et al. (1990)
Dasyprocta albida	Agouti	Uterine fibroleiomyoma	Chu et al. (2012)
Dasyprocta azarae	Azara's agouti	Squamous cell carcinoma	Ratcliffe (1933)

TABLE 2.1 Examples of Neoplasia Across the Animal Kingdom (*cont.*)

Latin name	Common name	Neoplasia (including benign and malignant abnormal cell growths)	References
Dasyuroides byrnei	Kowari	Pulmonary adenoma, splenic hematopoietic hyperplasia, trichoepithelioma, dermal mastocytoma metastatic to spleen, squamous cell carcinoma, metastatic adenocarcinoma of unknown origin, spindle cell tumor of scapula, multiple hepatomas, schwannoma, apocrine gland cystadenoma, splenic and thoracic fibrosarcoma, squamous cell carcinoma, cerebellar medulloblastoma, schwannoma	Attwood and Woolley (1973); Canfield et al. (1990); Ladds (2009)
Dasyurus geoffroii	Western quoll	Metastatic facial fibrosarcoma	Canfield et al. (1990)
Dasyurus hallucatus	Northern quoll	Lymphoma, squamous cell carcinoma of teat, lymphocytic leukemia, histiocytoma	Canfield et al. (1990); Ladds (2009)
Dasyurus maculatus	Tiger quoll	Pulmonary carcinoma, mesothelioma of peritoneum, squamous cell carcinoma, renal adenoma, abdominal lipoma, splenic hemangiosarcoma, adrenal adenocarcinoma, adenocarcinoma of the small gut, ovarian hemangioma, cutaneous lipoma	Attwood and Woolley (1973); Canfield et al. (1990); Ratcliffe (1933); Ladds (2009); Chu et al. (2012)
Dasyurus viverrinus	Eastern quoll	Adrenal cortical nodular hyperplasia, multiple hepatomas, papillomas, metastatic squamous cell carcinoma to lung, trichoepithelioma, splenic leiomyosarcoma, mammary adenocarcinoma, ganglioneuroma of liver, metastatic mammary adenocarcinoma, splenic hemangioma, ovarian adenocarcinoma, dermal spindle cell tumor, sebaceous hyperplasia, papillomas of head and feet, carcinoma of the rectum, medullary carcinoma	Ratcliffe (1933); Attwood and Woolley (1973); Canfield et al. (1990); Ladds (2009)
Delphinus delphis ponticus	Short-beaked common dolphin	Fibroma of the epididymis, Leydig cell tumor, testicular neoplasia	Newman and Smith (2006); Ladds (2009); Diaz-Delgado et al. (2012)
Dendrolagus bennettianus	Bennett's tree kangaroo	Generalized sarcoma	Ladds (2009)
Didelphis marsupialis	Common opossum	Adenocarcinoma, squamous cell carcinoma	Ratcliffe (1933); Snyder and Ratcliffe (1966)
Didelphis virginiana	Virginia opossum	Transitional cell carcinoma of the bladder, pulmonary adenomatosis, lymphoma	Attwood and Woolley (1973); Canfield et al. (1990); Marrow et al. (2010); Higbie et al. (2015)
Dorcopsis muelleri	Brown forest wallaby	Pulmonary metastasis of carcinoma	Ladds (2009)
Elaphurus davidianus	Père David's deer	Cutaneous squamous cell carcinoma	Chu et al. (2012)
Elephas maximus	Asian elephant	Cutaneous fibrosarcoma, uterine leiomyoma	Chu et al. (2012)
Enhydra lutris nereis	Southern sea otter	Osteosarcoma, osteoma	Rodriguez-Ramos Fernandez et al. (2012)
Equus asinus	Donkey	Renal hemangiosarcoma	Chu et al. (2012)
Equus ferus przewalski	Przewalski's wild horse	Uterine adenocarcinoma	Thompson et al. (2014)
Equus quagga	Common zebra	Fibrosarcoma	Ratcliffe (1933)
Equus zebra zebra	Mountain zebra	Sarcoid tumors	Sasidharan (2006); Sasidharan et al. (2011)

(Continued)

TABLE 2.1 Examples of Neoplasia Across the Animal Kingdom (*cont.*)

Latin name	Common name	Neoplasia (including benign and malignant abnormal cell growths)	References
Erethizon dorsatum	North American porcupine	Chorion epithelioma	Ratcliffe (1933)
Erinaceus europaeus	Hedgehog	Adenocarcinoma, uterine leiomyoma	Chu et al. (2012)
Eulemur fulvus	Common brown lemur	Hepatocarcinoma	Porter et al. (2004)
Eulemur macaco	Black lemur	Biliary adenoma/cystadenoma, hepatocarcinoma	Porter et al. (2004)
Eulemur mongoz	Mongoose lemur	Adenoma	Ratcliffe (1933)
Eumetopias jubatus	Steller's sea lion	Fibroleiomyoma, adenocarcinoma	Newman and Smith (2006)
Galago crassicaudatus	Greater galago	Hepatocarcinoma, uterine leiomyoma	Chu et al. (2012)
Galagoides demidoff	Demidoff's dwarf galago	Cholangiocarcinoma	Porter et al. (2004)
Gazella dorcas	Dorcas gazella	Osteoma	Ratcliffe (1933)
Gazella thomsonii	Thomson's gazelle	Hepatocarcinoma	Chu et al. (2012)
Genetta genetta	Common genet	Basal cell carcinoma	Ratcliffe (1933)
Gerbilliscus robustus	Fringe-tailed gerbil	Fibrosarcoma, squamous cell carcinoma	Ratcliffe (1933)
Globicephala macrorhynchus	Short-finned pilot whale	Granulosa cell tumor	Newman and Smith (2006)
Globicephala melaena	Long-finned pilot whale	Fibroleiomyomas, leiomyoma	Newman and Smith (2006)
Gorilla gorilla gorilla	Lowland gorilla	Uterine adenocarcinoma, squamous cell carcinoma of vulva, cervix, and uterus	Stringer et al. (2010)
Herpestes urva	Carb-eating mongoose	Uterine leiomyoma	Chu et al. (2012)
Heterocephalus glaber	Naked mole-rat	Adenocarcinoma possibly of mammary or salivary origin, neuroendocrine carcinoma	Delaney et al. (2016)
Hyaena brunnea	Brown hyena	Mammary gland adenocarcinoma	Chu et al. (2012)
Hyaena hyaena	Striped hyena	Lymphoma, bronchioloalveolar carcinoma	Chu et al. (2012)
Hydromys chrysogaster	Water-rat	Mediastinal lymphoma, adenoma, pheochromocytoma	Ladds (2009)
Hydrurga leptonyx	Leopard seal	Fibromatous epulis	Ladds (2009)
Hystrix brachyura longicauda	Malayan porcupine	Scirrhous carcinoma	Ratcliffe (1933)
Hystrix cristata	Crested porcupine	Uterine leiomyosarcoma	Chu et al. (2012)
Inia geoffrensis	Amazon river dolphin	Squamous cell carcinoma	Newman and Smith (2006)
Isoodon auratus	Golden bandicoot	Unidentified cloacal neoplasia	Canfield et al. (1990); Marrow et al. (2010)
Jaculus jaculus	Lesser Egyptian jerboa	Angiolipoma	Ratcliffe (1933)
Lagenorhynchus obliquidens	Pacific white-sided dolphin	Squamous cell carcinoma, eosinophilic leukemia, lymphoma, teratoma, fibroma	Newman and Smith (2006)
Lagenorhynchus obscurus	Dusky dolphin	Dysgerminoma, uterine leiomyomas, fibroleiomyomas	Newman and Smith (2006); Ladds (2009)
Lagenorhynchus acutus	Atlantic white-sided dolphin	Fibropapilloma, adenoma, leiomyoma	Newman and Smith (2006)
Lama glama	Llama	Gastric squamous cell carcinoma	Chu et al. (2012)
Lemur catta	Ring-tailed lemur	Cholangiocarcinoma, biliary adenoma/cystadenoma, mammary gland Adenoma	Porter et al. (2004); Chu et al. (2012)

TABLE 2.1 Examples of Neoplasia Across the Animal Kingdom (*cont.*)

Latin name	Common name	Neoplasia (including benign and malignant abnormal cell growths)	References
Leopardus pardalis	Ocelot	Hepatocarcinoma	Miranda et al. (2015)
Leopardus wiedii	Margay	Cholangiocarcinoma, vaginal leiomyoma	McClure et al. (1977)
Leporillus conditor	Greater stick-nest rat	Sarcoma, mediastinal thymoma, adenocarcinoma	Ladds (2009)
Lutra canadensis	North American otter	Squamous cell carcinoma	Snyder and Ratcliffe (1966)
Lycaon pictus	African wild dog	Hemangioma	Ratcliffe (1933)
Macaca fascicularis	Crab-eating macaque	Hepatocarcioma, mixed hepatocellular and cholangiocellular carcinoma	Porter et al. (2004)
Macaca fuscata	Japanese macaque	Biliary adenoma/cystadenoma, hepatocarcinoma, squamous cell carcinoma, adenoma	Ratcliffe (1933); Porter et al. (2004)
Macaca sinica	Toque macaque	Papilloma	Ratcliffe (1933)
Macropus agilis	Agile wallaby	Focal hepatobiliary proliferation, biliary adenoma	Ladds (2009)
Macropus giganteus	Eastern gray kangaroo	Dermal lymphoma, metastatic hemangiosarcoma, hepatoma, trichoepithelioma, bronchioloalveolar carcinoma, bronchial carcinoma	Ladds (2009); Chu et al. (2012)
Macropus parma	Parma wallaby	Osteochondromatous proliferation, squamous cell carcinoma of the cervix and vagina	Canfield et al. (1990); Ladds (2009); Marrow et al. (2010)
Macropus parryi	Whiptail wallaby	Lymphoblastic lymphoma	Ladds (2009)
Macropus robustus	Common wallaroo	Hepatic vascular proliferation, hamartoma, biliary adenoma	Ladds (2009)
Macropus rufogriseus	Bennett's wallaby	Oral melanoma, oral adenocarcinoma, lymphoma, bile duct proliferation	Brust (2013); Ladds (2009)
Macropus rufus	Red kangaroo	Adenocarcinoma, lymphoma, squamous cell carcinoma, basal cell carcinoma of the pouch, squamous cell carcinoma of the oral cavity, gastric carcinoma, pulmonary carcinoma	Brust (2013); Ratcliffe (1933); Ladds (2009)
Macrotis lagotis	Greater bilby	Fibrosarcoma of skin and lung, osteosarcoma, hemangioma of pancreas, hemangiosarcoma, histiocytoma, basal cell carcinoma, pulmonary sclerosing squamous cell carcinoma and adenocarcinoma, pulmonary adenomatosis, lymphoma	Ratcliffe (1933); Ladds (2009)
Manis pentadactyla	Pangolin	Hepatocarcinoma, hepatoma	Chu et al. (2012)
Marmota monax	Groundhog	Adenoma	Ratcliffe (1933)
Megaptera novaeangliae	Humpback whale	Lipoma, fibroma	Newman and Smith (2006)
Melomys burtoni	Grassland mosaic-tailed rat	Fibrosarcoma, hepatoma, adenocarcinoma, carcinoma	Ladds (2009)
Mephitis mephitis	Striped skunk	Biliary cystadenoma, renal cell carcinoma, adenocarcinoma	Snyder and Ratcliffe (1966); Chu et al. (2012)
Mesembriomys gouldii	Black-footed tree-rat	Thymic lymphoma, hepatoma	Ladds (2009)
Mesoplodon densirostris	Blainville's beaked whale	Vaginal fibromas	Newman and Smith (2006); Ladds (2009)

(Continued)

TABLE 2.1 Examples of Neoplasia Across the Animal Kingdom (*cont.*)

Latin name	Common name	Neoplasia (including benign and malignant abnormal cell growths)	References
Microcebus murinus	Gray mouse lemur	Hepatocarcinoma	Porter et al. (2004)
Mirounga leonina	Southern elephant seals	Adrenocortical adenoma, malignant granulosa cell tumor	Ladds (2009)
Mus musculus molossinus	Japanese waltzing mice	Adenocarcinoma, fibroadenoma	Ratcliffe (1933)
Mustela putorius furo	Ferret	Sebaceous carcinoma, adrenocortical carcinoma	Chu et al. (2012)
Mustela vison	American mink	Cutaneous squamous cell carcinoma, hepatocarcinoma, adrenocortical carcinoma, lymphoma, hemangiosarcoma, hepatoma	Chu et al. (2012)
Myocastor coypus	Nutria	Adenocarcinoma, fibroma	Ratcliffe (1933)
Myrmecophaga tridactyla	Giant anteater	Multicentric lymphoma	Sanches et al. (2013)
Nasua nasua	South American coati	Squamous cell carcinoma	Ratcliffe (1933)
Neofelis nebulosa	Clouded leopard	Pheochromocytoma, uterine leiomyoma, mesothelioma, hemangioma	Snyder and Ratcliffe (1966); Chu et al. (2012)
Neophocaena phocaenoides	Indo-Pacific finless porpoise	Fibroma	Newman and Smith (2006)
Notomys alexis	Spinifex hopping mouse	Fibroma, lipoma, rhabdomyosarcoma, cavernous hemangioma, thymic lymphoma, multicentric lymphoma, melanoma	Ladds (2009); Old and Price (2016)
Nyctereutes procyonoides	Raccoon dog	Adenocarcinoma	Ratcliffe (1933)
Nycticebus coucang	Slow loris	Cholangioma, adrenocortical adenoma, myeloid leukemia	Chu et al. (2012)
Nyctophilus geoffroyi	Lesser long-eared bat	Fibrosarcoma of the abdomen	Ladds (2009)
Odocoileus hemionus	Mule deer	Intracerebral malignant plasma cell tumor	Clancy et al. (2016)
Odocoileus virginianus	White-tailed deer	Fibroadenoma, oligodendrogliomas	Ratcliffe (1933); Gottdenker et al. (2012)
Orcinus orca	Killer whale	Hodgkins-like lymphoma, papilloma	Newman and Smith (2006); Ladds (2009)
Ornithorhynchus anatinus	Duck-billed platypus	Papilloma, hepatoma, adrenocortical adenoma	Ladds (2009)
Oryx gazella gazelle	Gemsbok	Adrenocortical adenoma	Chu et al. (2012)
Otolemur crassicaudatus	Brown greater galago	Cholangiocarcinoma, hepatoma	Porter et al. (2004)
Otospermophilus beecheyi	California ground squirrel	Osteoma	Ratcliffe (1933)
Pan troglodytes	Chimpanzee	Hepatocarcinoma, hepatoma, reproductive neoplasia, uterine leiomyomas	Stringer et al. (2010)
Panthera leo	Lion	Gallbladder adenocarcinomas, mammary gland adenocarcinoma, uterine leiomyoma, hepatocarcinoma, biliary cystadenoma, malignant histiocytosis, scirrhous carcinoma	Ratcliffe (1933); Sakai et al. (2003); Chu et al. (2012)
Panthera onca	Jaguar	Adrenocortical carcinoma, pancreatic islet cell carcinoma, metastatic leiomyosarcoma, leiomyoma, mammary fibroadenoma, lymphangioma	Port et al. (1981); Chu et al. (2012); Ratcliffe (1933)
Panthera pardus	Leopard	Hepatoma, parathyroid carcinoma, mammary gland adenocarcinoma, cholangiocarcinoma, lymphangioma	Sakai et al. (2003); Chu et al. (2012); Ratcliffe (1933)

TABLE 2.1 Examples of Neoplasia Across the Animal Kingdom (*cont.*)

Latin name	Common name	Neoplasia (including benign and malignant abnormal cell growths)	References
Panthera tigris	Tiger	Mammary gland adenocarcinoma, adenomatous polyps, squamous cell carcinoma	Chu et al. (2012); Ratcliffe (1933)
Panthera tigris bengalensis	Bengal tiger	Endometrial adenocarcinoma	Linnehan and Edwards (1991)
Papio cynocephalus	Yellow baboon	Adenocarcinoma, fibroadenoma	Ratcliffe (1933)
Papio cynocephalus anubis	Anubis baboon	Cutaneous squamous cell carcinoma, trichofolliculoma	Chu et al. (2012)
Papio hamadryas	Hamadryas baboon	Gall bladder adenocarcinoma, biliary adenoma/cystadenoma	Ratcliffe (1933); Porter et al. (2004)
Papio papio	Guinea baboon	Gall bladder adenocarcinoma	Porter et al. (2004)
Papio sphinx	Mandrill	Uterine leiomyosarcoma, cutaneous lipoma	Chu et al. (2012)
Papio ursinus	Chacma baboon	Gall bladder cystadenocarcinoma, fibrosarcoma	Ratcliffe (1933); Porter et al. (2004)
Paradoxurus hermaphroditus	Asian palm civet	Adenocarcinoma	Ratcliffe (1933)
Parantechinus apicalis	Dibbler	Lymphoma (leukemic)	Canfield et al. (1990)
Perameles bougainville	Western barred bandicoot	Pulmonary carcinoma, prostatic carcinoma, cutaneous papillomatosis, and carcinomatosis	Ladds (2009)
Perameles gunnii	Eastern barred bandicoot	Colonic leiomyosarcoma and leiomyoma, leiomyosarcoma of skin and lymph nodes, fibriohistiocytoma, cutaneous histiocytoma, mast cell tumor, basal cell tumor of the larynx	Canfield et al. (1990); Ladds (2009); Marrow et al. (2010)
Perodicticus potto	Potto	Cholangiocarcinoma	Porter et al. (2004)
Perodipus richardsoni	Kangaroo rat	Fibrosarcoma, lymphoma	Ratcliffe (1933)
Peromyscus leucopus	White-footed mouse	Fibrosarcoma, adenocarcinoma	Ratcliffe (1933)
Petaurus breviceps	Sugar glider	Subcutaneous fibroma, histiocytoma, sebaceous carcinoma, lymphoma, leukemia, fibrosarcoma, myxosarcoma, adenocarcinomas and carcinomas of the adrenals, intestines, liver, and mammary glands, transitional cell carcinoma of the urinary bladder	Brust (2013); Ladds (2009); Marrow et al. (2010)
Phalanger gymnotis	Ground cuscus	Cutaneous lymphoma	Goodnight et al. (2008)
Phascogale tapoatafa	Brush-tailed phascogale	Hemangiopericytoma, trichoepithelioma, fibrosarcoma, lymphoma, hemangioma, hemangiosarcoma, basal cell tumor, squamous cell carcinoma, hepatocarcinoma, melanoma	Canfield et al. (1990); Ladds (2009)
Phascolarctos cinereus	Koala	Lymphoma, meothelioma, tumors of the carilaginous or osseous tissues of the craniofacial region, testicular teratoma, myeloid leukemia, rhabdomyosarcoma, myxofibroma of the subcutis, leiomyoma of the intestine, oral fibrosarcoma, biliary adenoma, hepatoma, ovarian tumor, adenoma of fimbria, mammary adenocarcinoma, cutaneous papilloma and squamous cell carcinoma, pilomatrixoma, chromophobe adenoma, adenoma of the frontal sinus, serosal adenocarcinoma	Ladds (2009)

(Continued)

TABLE 2.1   Examples of Neoplasia Across the Animal Kingdom (*cont.*)

Latin name	Common name	Neoplasia (including benign and malignant abnormal cell growths)	References
Phocoena phocoena	Harbor porpoise	Papilloma, adenocarcinoma	Newman and Smith (2006)
Physeter macrocephalus	Sperm whale	Uterine leiomyoma, fibroleiomyoma, fibroleiomyosarcoma, hemangioma, fibromas of the jaw and skin, penil papillomatosis	Newman and Smith (2006); Ladds (2009)
Planigale maculata	Common planigale	Dermal spindle cell tumor, uterine adenocarcinoma, squamous cell carcinoma	Canfield et al. (1990)
Pongo pygmaeus	Bornean orangutan	Malignant granulosa cell tumor	Stringer et al. (2010)
Presbytis entellus	Gray langur	Hepatoma	Porter et al. (2004)
Procyon cancrivorus	Crab-eating raccoon	Adenocarcinoma	Ratcliffe (1933)
Procyon lotor	Raccoon	Adenoma, adenocarcinoma, pancreatic exocrine adenocarcinoma, hepatocarcinoma, mammary gland adenocarcinoma, sweat gland adenocarcinoma	Ratcliffe (1933); Chu et al. (2012)
Proteles cristatus	Aardwolf	Peritoneal mesothelioma	Chu et al. (2012)
Pseudantechinus bilarni	Sandstone antechinus	Lymphoma or splenic erythroid hyperplasia, sebaceous adenoma	Canfield et al. (1990)
Pseudantechinus macdonellensis	False antechinus	Lymphoma (leukemic)	Canfield et al. (1990)
Pseudocheirus peregrinus	Common ringtail possum	Lymphoma, metastatic adenocarcinoma	Ladds (2009)
Pseudomys albocinereus	Ash-gray mouse	Liposarcoma, lymphoma	Ladds (2009)
Pseudomys australis	Plains rat	Hemangiosarcoma, adenocarcinoma, carcinoma	Ladds (2009)
Pteropus d. dasymallus	Flying fox	Hepatocarcinoma, chondrosarcoma, uterine adenocarcinoma	Chu et al. (2012)
Pteropus poliocephalus	Gray-headed flying fox	Metastatic carcinoma, fibropapilloma, subcutaneous fibrosarcoma, rhabdomyoma	Ladds (2009)
Puma concolor	Mountain lion	Fibrosarcoma, thyroid carcinoma	Chu et al. (2012)
Rangifer tarandus tarandus	Reindeer	Lymphoma	Jarplid and Rehbinder (1995)
Rattus norvegicus	Rat	Mammary gland fibroadenoma	Chu et al. (2012)
Rattus tunneyi	Pale field rat	Adenocarcinoma	Ladds (2009)
Saguinus oedipus	Cotton-top tamarin	Intestinal adenocarcinoma	Chu et al. (2012)
Saimiri boliviensis	Black-headed squirrel monkey	Hepatocarcinoma	Porter et al. (2004)
Saimiri sciureus	Squirrel monkey	Cutaneous lipoma, adenocarcinoma of vaginal wall, hepatocarcinoma	Porter et al. (2004); Chu et al. (2012)
Sarcophilus harrisii	Tasmanian devil	Papillomas, splenic erythroid hyperplasia, adrenocortical nodular hyperplasia, keratoacanthoma, mammary cystadenoma, metastatic squamous cell carcinoma of lung, trichoepithelioma, fibrosarcoma of lung, squamous cell carcinoma, sebaceous, and apocrine hyperplasia; adrenal dermal lymphosarcoma, hemangioma, smooth muscle hyperplasia of esophagus, sebaceous hyperplasia or adenoma of pouch, apocrine and mammary gland hyperplasia, pyloric leiomyoma, hepatoma, Tasmanian devil facial tumor disease	Ratcliffe (1933); Attwood and Woolley (1973); Canfield et al. (1990); Ladds (2009)

TABLE 2.1 Examples of Neoplasia Across the Animal Kingdom (*cont.*)

Latin name	Common name	Neoplasia (including benign and malignant abnormal cell growths)	References
Sciurus niger	Fox squirrel	Adenocarcinoma and adenoma of the kidney	Ratcliffe (1933)
Sciurus carolinensis pennsylvanicus	Northern gray squirrel	Hypernephroma	Ratcliffe (1933)
Setonix brachyurus	Quokka	Liposarcoma, papilloma	Ladds (2009)
Sminthopsis crassicaudata	Fat-tailed dunnart	Dermal spindle cell tumor, splenic lymphoma, squamous cell carcinoma, round cell sarcoma of the upper forelimb	Canfield et al. (1990)
Stenella coeruleoalba	Striped dolphin	Myelogenous leukemia, squamous cell carcinoma	Newman and Smith (2006)
Suricata suricatta	Meerkat	Rhabdomyosarcoma	Chu et al. (2012)
Sus barbatus	Black small-eared pig	Uterine adenocarcinoma, uterine leiomyoma	Chu et al. (2012)
Sus scrofa	Pig	Adenocarcinoma	Ratcliffe (1933)
Tachyglossus aculeatus	Short-beaked echidna	Lymphoma of spleen, fibroma of subcutis, leiomyoma of the cloaca, pericloacal leiomyosarcoma, fibroma of the beak, myocardial lymphom, lymphoma with leukemia, cystic adenoma of the thyroid	Ladds (2009)
Taurotragus oryx	Common eland	Cutanous lymphoma	Chu et al. (2012)
Taxidea taxus	American badger	Peritoneal epithelioid leiomyosarcoma, undetermined adenocarcinomas	Chu et al. (2012)
Thylacomys lagotis	Rabbit-eared bandicoot	Squamous cell carcinoma	Snyder and Ratcliffe (1966)
Thylogale billardierii	Tasmanian pademelon	Squamous tumor of the stomach, melanoma	Kusewitt and Ley (1996); Ladds (2009)
Tragelaphus eurycerus isaaci	Bongo	Uterine leiomyomas	Napier et al. (2005)
Tragelaphus strepsiceros	Greater kudu	Renal cell carcinoma	Chu et al. (2012)
Trichosurus vulpecula	Common brushtail possum	Thoracic chondrosarcoma	Ladds (2009)
Tupaia belangeri	Northern treeshrew	Hepatocarcinoma	Porter et al. (2004)
Tursiops truncatus	Common bottlenose dolphin	Lymphoma, myeloma, malignant seminoma, hepatic and thyroid adenoma, pancreatic carcinoma, reticuloenditheliosis of lung, liver, lymphoma of spleen, myelogenous leukemia, immunoblastic lymphoma, lymphadenopathy, splenomegaly, plasmacytoid neoplastic cells, sublingual squamous cell carcinoma, uterine adenocarcinoma, renal adenoma, teratoma	Newman and Smith (2006); Ladds (2009)
Urocyon cinereoargenteus	Gray fox	Adenoma, splenic myelolipoma, cutaneous squamous cell carcinoma, sweat gland adenoma of eyelid, lymphoma	Ratcliffe (1933); Chu et al. (2012)
Ursus americanus	American black bear	Medullary carcinoma, basal cell carcinoma	Ratcliffe (1933)
Ursus arctos	Brown bear	Bronchial adenoma, cholangiocarcinoma, hypernephroma	Ratcliffe (1933); Chu et al. (2012)
Ursus maritimus	Polar bear	Cutaneous lymphoma, adenocarcinoma	Ratcliffe (1933); Chu et al. (2012)
Ursus thibetanus	Asiatic black bear	Biliary cystadenoma	Chu et al. (2012)

(Continued)

TABLE 2.1 Examples of Neoplasia Across the Animal Kingdom (*cont.*)

Latin name	Common name	Neoplasia (including benign and malignant abnormal cell growths)	References
Ursus thibetanus formosanus	Formosan black bear	Bronchioloalveolar carcinoma	Chu et al. (2012)
Varecia variegata	Black-and-white ruffed lemur	Hepatocarcinoma, biliary adenoma/cystadenoma	Porter et al. (2004)
Vicugna pacos	Alpaca	Medullary carcinoma	Ratcliffe (1933)
Viverra tangalunga	Malayan civet	Adenocarcinoma, squamous cell carcinoma	Ratcliffe (1933); Snyder and Ratcliffe (1966)
Viverra zibetha	Large Indian civet	Adenocarcinoma	Ratcliffe (1933)
Vombatus ursinus	Coarse-haired wombat	Fibropapilloma, lymphoma, leukemia, adenocarcinoma	Ladds (2009)
Vulpes corsac	Corsac fox	Adenoma	Ratcliffe (1933)
Vulpes v. pennsylvanicus	American red fox	Adenoma	Ratcliffe (1933)
Zalophus californianus	California sea lion	Adenocarcinoma, hypernephroma, squamous-cell carcinoma, leiomyoma, fibroma, carcinoma, adenoma, ovarian granulosa cell tumor, lymphoma, islet cell adenoma or carcinoma, transitional cell carcinoma, adenoma, duct adenoma, hepatocarcinoma, lipoma, nephroblastoma, sarcoma, neuroendocrine tumor, fibrosarcoma, myosarcoma, melanoma, mesenchymoma, multicentric neurofibromatosis	Newman and Smith (2006); Rush et al. (2012)

[a]*We tried to provide a comprehensive list of examples of neoplasia in wild and captive animals, but understandably the list cannot be complete due to limited space. Review articles were used primarily due to restricted space for citations. We are well aware of that several taxonomic revisions have been undertaken since many of the listed references were published. We, have, however in virtually all cases followed the taxonomy used in the studies referred to in the table.*

TABLE 2.2 Cancer Prevalence in Wild Animals

Species	Neoplasia	Prevalence (%)	References
Fish			
Atlantic salmon (*Salmo salar*)	Leiomyosarcoma	4.60	Coffee et al. (2013)
Bicolor damselfish (*Stegastes partitus*)	Neurofibromatosis-like disease	23	Coffee et al. (2013)
Brown bullhead (*Ameiurus nebulosus*)	Liver neoplasms	5–15	Baumann et al. (2008)
Chinook salmon (*Oncorhynchus tshawytscha*)	Plasmacytoid leukemia	6	Eaton et al. (1994)
Dab (*Limanda limanda*)	Epidermal papilloma	1–7	Dethlefsen et al. (2000)
English sole (*Parophrys vetulus*)	Carcinomas, adenomas, hepatic mesenchymal neoplasms	up to 24	Malins et al. (1987)
European smelt (*Osmerus eperlanus*)	Spawning papillomatosis	5.50	Coffee et al. (2013)
Gizzard shad (*Dorosoma cepedianum*)	Spindle cell neoplasms	20	Geter et al. (1998)
Northern pike (*Esox lucius*)	Lymphosarcoma	21	Papas et al. (1976)
Roach (*Rutilus rutilus*)	Epidermal papillomatosis	3–31	Korkea-aho et al. (2006)
Walleye (*Sander vitreus*)	Dermal sarcoma	20–30	Coffee et al. (2013)
Walleye (*Sander vitreus*)	Epidermal hyperplasia	up to 20	Coffee et al. (2013)
White sucker (*Catostomus commersoni*)	Epidermal papilloma	59	Coffee et al. (2013)

TABLE 2.2   Cancer Prevalence in Wild Animals (*cont.*)

Species	Neoplasia	Prevalence (%)	References
Amphibians			
Japanese fire belly newt (*Cynops pyrrhogaster*)	Skin papilloma	5.50	Asashima et al. (1982)
Leopard frog (*Rana pipiens*)	Renal adenocarcinoma	9	McKinnel (1965)
Montseny brook newt (*Calotriton arnoldi*)	Chromatophoroma melanocytoma	27	Martinez-Silvestre et al. (2011)
Reptiles			
Green turtle (*Chelonia mydas*)	FP	23	Foley et al. (2005)
Green turtle (*Chelonia mydas*)	FP	22	Adnyana et al. (1997)
Green turtle (*Chelonia mydas*)	FP	[a]	Chaloupka et al. (2009)
Green turtle (*Chelonia mydas*)	FP	58	dos Santos et al. (2010)
Green turtle (*Chelonia mydas*)	FP	16	Aguirre et al. (1999b)
Loggerhead turtle (*Caretta caretta*)	FP	6	Aguirre et al. (1999b)
Birds			
Canada geese (*Branta canadensis*)	Spindle cell sarcomas	0.2	Gates et al. (1992)
White-fronted geese (*Anser albifrons*)	Multicentric intramuscular lipomatosis/fibromatosis	23	Daoust et al. (1991)
Mammals			
Baltic gray seal (*Halichoerus grypus*)	Uterine leiomyomas	64	Bäcklin et al. (2003)
Beluga (*Delphinapterus leucas*)	Adenocarcinoma, squamous cell carcinoma, dysgerminoma, lymphosarcoma	18	Martineau et al. (2002)
Brown hare (*Lepus europaeus occidentalis*)	Ovarian tumors	5.60	Flux (1965)
California sea lion (*Zalophus californianus*)	Metastatic carcinoma, spindle cell sarcoma, adenocarcinoma, adrenocortical adenoma	18–26	Gulland et al. (1996)
Cape mountain zebra (*Equus zebra zebra*)	Equine sarcoid	53	Marais et al. (2007)
Gray squirrel (*Sciurus carolinensis*)	Fibromatosis	[b]	Terrell et al. (2002)
Northern sea otter (*Enhydra lutris*)	Uterine leiomyomas	2	Williams and Pulley (1981)
Pacific walrus (*Odobenus rosmarus divergens*)	Uterine leiomyomas, ovarian leiomyoma, mesenteric leiomyoma, gastric gastrointestinal stromal tumors, ovarian dysgerminomas, intestinal hemangioma, hepatic hemangioma, mammary adenoma	17	Fleetwood et al. (2005)
Roe deer (*Capreolus capreolus*)	Fibropapillomas	33	Erdélyi et al. (2009)
Roe deer (*Capreolus capreolus*)	Adenoma, brain tumors, bile duct carcinoma, hemangiosarcoma, lymphoma, osteosarcoma, rhabdomyosarcoma	2	Aguirre et al. (1999a)
Santa Catalina Island fox (*Urocyon littoralis catalinae*)	Ceruminous gland tumors	52	Vickers et al. (2015)
Sea otter (*Enhydra lutris*)	Leiomyoma	1.80	Williams and Pulley (1981)
Western barred bandicoot (*Perameles bougainville*)	Cutaneous papillomatosis and carcinomatosis	[c]	Woolford et al. (2008)

FP, Fibropapillomatosis.

[a]*Significant temporal decrease in prevalence.*

[b]*Epizootic, no data provided on exact prevalence.*

[c]*High, no data provided on exact prevalence.*

up to 9% of the frogs were diagnosed with renal adenocarcinoma in 1965. However, no information about tumor etiology or its possible effects on the frogs was provided. Interestingly, in a later study McKinnell and Martin (1979) observed a gradual temporal decline in tumor prevalence and in 1978 no tumors were observed in 1216 dissected frogs. McKinnell and Martin (1979) suggested that the decline in cancer prevalence was caused by a significant reduction in frog numbers and a concomitant reduction in the release of oncogenic viruses into the breeding ponds. However, the authors could not rule out that a gradual reduction of carcinogenic pollutants into the breeding ponds could have caused the decline in tumor prevalence.

Asashima et al. (1982) studied the occurrence of spontaneous skin papillomas in Japanese newts (*Cynops pyrrhogaster*) in northern Japan. The prevalence of papillomas showed a seasonal variation, being highest in autumn, ranging from 1.93% to 5.45%, whereas during the rest of the year the prevalence ranged between 0.16% and 0.50%. A spatial difference in cancer prevalence was also recorded with newts collected from the northern, seaside prefectures having higher papilloma rates (1.00–5.45%) than newts from the southern, Pacific Ocean prefectures (0–0.27%). No intersexual differences in tumor prevalence were recorded. Virus-like bodies, resembling herpes-type virus, were found in the cytoplasm of the epitheliomas, suggesting that tumor may have been caused by a viral agent. Unfortunately the study does not provide any information of the underpinning(s) of the temporal and spatial variation in tumor prevalence or whether the tumors affected newt mortality.

In a recent study skin tumor prevalence was investigated in the Montseny brook newt (*Calotriton arnoldi*) in Spain (Martinez-Silvestre et al., 2011). The range of this taxon is restricted to a small geographic area <40 km^2 of the North Eastern Iberian Peninsula. Similar to the Japanese newt a profound spatial population difference in tumor prevalence was observed ranging from 0%, 2% to 29%. The tumors were only observed in adult newts, which led the authors to suggest that the tumors may be caused by increased UV-B exposure. Yet again no data are provided on whether the tumors may affect newt mortality.

Although our sample is small, it shows that cancer prevalence may affect a substantial proportion of wild amphibians. Considering the dramatic decline in amphibians caused by the chytrid fungus, *Batrachochytrium dendrobatidis* (Daszak et al., 1999) makes it even more important to further investigate the possible negative effects of cancer in this group of vertebrates.

Reptiles

The only reptile taxa for which we have been able to retrieve data on cancer prevalence in the wild are restricted to marine turtles. Although fibropapillomatosis (FP) mainly affects green turtles (*Chelonia mydas*) it has also been documented in loggerheads (*Caretta caretta*; Aguirre et al., 1999b). Green turtles have been subjected to numerous and extensive research projects and here we summarize the major findings from some of these studies. FP in green turtles results in tumor growth on eyes, oral cavity, skin, carapace, plastron, and/or internal organs (Santos et al., 2010). Consequently the disease may significantly reduce turtle foraging efficiency. FP shows significant geographic variation not only in prevalence (ranging from 0% to 92%) but also in severity (Santos et al., 2010). Moreover, in Brazil the disease is absent in juvenile green turtles but increases in prevalence in older turtles (Santos et al., 2010). In contrast in Hawaii, FP affects mainly juvenile turtles (Balazs and Pooley, 1990; Work and Balazs, 1999). However, the reason(s) for the age-specific increase in FP prevalence in Brazil and the age-specific difference in cancer development in Brazil and Hawaii is unknown. Interestingly, Chaloupka et al. (2009) reported on cases where FP had regressed and even completely disappeared in some individual green turtles in Hawaii, and that the diseases since the mid-1990s has showed a significant decline in prevalence.

Although we have not been able to find data on how FP affects green turtle mortality rates, the disease results in high parasite load, immune suppression, increased physiological cost (Work and Balazs, 1999; Work et al., 2001, 2005) and is the most common cause of green turtle stranding on Hawaii (Chaloupka et al., 2008, 2009). Consequently, we find it highly likely that FP may impose considerable mortality cost on green turtles in the wild. In spite of FP's high prevalence in some areas and its possible severe effects on green turtle fitness the etiology of FP is still not known. Some studies have found an association between herpesviruses and FP (Greenblatt et al., 2005), whereas others have implicated that pollution and habitat quality may be major factors explaining the presence of FP (Herbst and Klein, 1995).

The detrimental impact of cancer on marine turtles and the emergence of a novel fungal disease in squamate reptiles (Guthrie et al., 2015) warrant increased research efforts to investigate how cancer might affect the demography of reptiles in the wild.

Birds

We find it remarkable that although birds are often abundant in both urban and rural habitats we have only been able to find a handful of studies that have recorded cancer prevalence in wild birds. Jennings (1968) estimated the prevalence of neoplasia in wild birds in Great Britain to be between 0.1% and 1.0%. Similar low cancer prevalence was recorded by Gates et al. (1992) in

Canada geese (*Branta canadensis interior*; 2 out of 1272 birds, 0.2%). Both birds were young and emaciated and microscopical analyses suggested that the tumors "had the typical appearance of spindle cell sarcomas" (Gates et al., 1992). Similar results on low cancer prevalence in wild birds were published by Siegfried (1983) who found tumors in only 9 out of more than 18,000 birds examined (0.05%). Although based on a significantly smaller sample size, 3 out of 13 (23%) ruffed grouse (*Bonasa umbellus*) were diagnosed having tumors (Howerth et al., 1986). One bird was diagnosed with a lipoma, the second bird a fibroma, while the third bird had developed a renal carcinoma metastatic to the liver (Howerth et al., 1986). The high prevalence recorded in ruffed grouse should, however, be interpreted with caution as all three birds were delivered for examination because they all suffered from obvious lesions. Reece (1992) reported 383 cases of cancer from a collection of more than 10,000 birds (3.8%) submitted for necroscopy in Victoria, Australia from 1977 to 1987. As the birds examined included both wild and captive birds and no data are provided on the number of birds in each of the two groups, again the data on cancer prevalence should be interpreted with caution.

The only publication we have found showing that cancer prevalence in birds may reach similar levels as that found among other vertebrates is a study by Daoust et al. (1991) who reported that out of 30 wild white-fronted goose (*Anser albifrons*) killed by hunters 7 (23%) were diagnosed as having developed multicentric mesenchymal tumors. Daoust et al. (1991) suggested that the high prevalence could have been caused by "a genetically influenced susceptibility to the disease." Unfortunately, however, no data are provided to support this statement.

In their review of wildlife cancer McAloose and Newton (2009) listed the endangered North American Attwater's prairie chicken (*Tympanicus cupido attwateri*) as an example of a species being further threatened by extinction due to cancer. Although reticuloendotheliosis virus may infect up to 50% of the captive birds, we have not been able to find any publication that relate this high infection level to mortality in the wild.

The few publications that we have been able to retrieve suggest that cancer prevalence in birds in general appears to be low. Although the sample size in some of the studies were very high, they were often restricted to large-bodied and long-lived species, such as geese. In order to make any robust generalization of cancer prevalence among wild birds, future research should incorporate birds representing a significantly more diverse taxonomic range.

Mammals

Cancer prevalence and its effect on some wild mammal population, such as California sea lions (*Zalophus californianus*) and belugas (*Delphinapterus leucas*) have been subjected to intensive research (Gulland et al., 1996; Martineau et al., 2002). Between 1979 and 1994 the prevalence of a metastatic carcinoma of urogenital origin in stranded California sea lions was reported to be 18% (Gulland et al., 1996). However, between 1998 and 2012 the prevalence of this cancer increased to 26% (Browning et al., 2015). This metastatic carcinoma appears to result in 100% mortality as all animals died during rehabilitation (Gulland et al., 1996). The actual prevalence of this cancer is most likely lower as only sick animals are likely to strand, but despite this the cancer represents a significant cause of death (Browning et al., 2015). Recent studies have found that the etiology of the cancer is associated with individual genotype, persistent organic pollutants, and/or a herpesvirus (Browning et al., 2015). Similar high cancer prevalence has been recorded in an isolated beluga population living in the St. Lawrence estuary (Martineau et al., 2002). Although the primary causes of death were respiratory and gastrointestinal infections with metazoan parasites, observed in 22% of the belugas, cancer was the second most common cause of death across all age groups and observed in 18% of the stranded belugas (Martineau et al., 2002). Cancer prevalence in adults was even higher (27%) and Martineau et al. (2002) estimated the annual rate of all cancer types in belugas to 163 out of 100,000 animals, a rate significantly higher than that reported for any other cetacean populations and similar to that of recorded in humans. Beluga habitat in the St. Lawrence estuary is highly contaminated by polycyclic aromatic hydrocarbons produced by the local aluminum smelters, strongly suggesting that polycyclic aromatic hydrocarbons are a major cause of the high cancer prevalence recorded in this population (Martineau et al., 2002).

High cancer prevalence has also been recorded in other marine mammals, such as Pacific walrus (*Odobenus rosmarus divergens*) and Baltic gray seal (*Halichoerus grypus*). In the former, 18 neoplasms were found during examination of tissues collected from 107 carcasses (17%) from Alaskan subsistence hunting over a 10-year period (Fleetwood et al., 2005). However, no data regarding cancer etiology or pathogenesis of the walrus examined are presented. Between 1975 and 1997, 53 female Baltic gray seals aged between 15 and 40 years were found dead along the Baltic coast of Sweden, of which 34 (64%) where diagnosed having developed uterine leiomyomas (Bäcklin et al., 2003). Although little is known about the etiology and pathogenesis of leiomyoma in Baltic gray seals, Bäcklin et al. (2003) tentatively suggested an association between cancer prevalence and pollutants, such as organochlorines. However, as mentioned earlier, the actual prevalence of uterine leiomyomas in female Baltic gray seals is most likely lower as only sick animals are likely to strand. Regardless, similar to the California sea lion, this cancer may constitute a significant cause of mortality in this species.

In contrast to the four marine species mentioned previously, neoplasia in northern sea otters (*Enhydra lutris*) appears to be rare and Williams and Pulley (1981) only found tumors in 2 females out of 112 otters examined (1.8%). Similar to the female Baltic gray seals, tumors of the female otters were diagnosed as uterine leiomyomas (Williams and Pulley, 1981).

Our review of the literature also revealed significant geographical species-specific difference in cancer prevalence. For example, of 42 roe deer (*Capreolus capreolus*) carcasses examined in Hungary, 14 (33%) showed macrosopic lesions consistent with skin FP (Erdélyi et al., 2009), whereas out of 985 carcasses examined in Sweden only 19 were diagnosed having neoplasia (2%) and only 1 of the 19 having developed FP (Aguirre et al., 1999a). Moreover, within the Hungarian study neoplasia was confined to certain geographical areas (Erdélyi et al. 2009). As FP is caused by the infection of papillomavirus (CcPV1) this led Erdélyi et al. (2009) to suggest that genetic factors may underpin roe deer susceptibility to FP.

In humans increased cancer prevalence has shown to be associated with reduced genetic diversity (Assié et al., 2008; Rudan et al., 2003). It is therefore interesting to note that some of the highest cancer prevalence's observed in wild mammals (>50%) have been recorded in species/populations with low genetic diversity, such as the Santa Catalina Island foxes (*Urocyon littoralis catalinae*; Funk et al., 2016; Vickers et al., 2015) and the South African Cape mountain zebra (*Equus zebra zebra*; Marais and Page, 2011; Marais et al., 2007; Sasidharan et al., 2011). Vickers et al. (2015) suggested that the high prevalence of ceruminous gland tumors (carcinomas and adenomas) observed in the Santa Catalina Island foxes may have a genetic basis. Similarly, the high cancer prevalence observed in one of the South African Cape mountain zebra populations has also been suggested to be associated with concomitant low genetic diversity (Marais et al., 2007; Sasidharan et al., 2011). Interestingly, as mentioned earlier, the high prevalence of cancer in California sea lions may also, at least partly, have a genetic basis (Browning et al., 2015). The possible association between reduced genetic diversity and cancer prevalence is further supported by the high prevalence of cancer observed in both captive and wild western barred bandicoot (*Perameles bougainville*), a highly endangered Australian marsupial once widespread across western and southern Australia but now restricted to two small islands off the Western Australian coast (Woolford et al., 2008, 2009). Captive breeding of this species has been severely hampered by debilitating cutaneous and mucocutaneous papillomatosis and carcinomatosis, associated with infection of papillomatosis carcinomatosis virus type 1 (BPCV1) (Woolford et al., 2008, 2009).

Low genetic diversity in the wild has been found to result in increased risk of inbreeding depression and concomitant increased risk of extinctions (Madsen et al., 1996, 1999, 2004). If low genetic diversity results in an increased risk of cancer, as suggested by the examples mentioned previously, this may further imperil the long-term survival of the numerous wild organisms presently suffering from low genetic diversity.

CANCER ETIOLOGY AND PREVALENCE IN FRENCH ZOOLOGICAL PARKS

Although conditions (and hence associated cancer risks) in zoological parks are often significantly different from those experienced in nature (e.g., altered levels of activity and food and abnormal breeding frequency; Vittecoq et al., 2013), cancer studies in captive animals are facilitated by the absence of masking variables, such as predation. In addition, because of curative and preventive improvements in veterinary medicine, diseases of captive animals are closely monitored and routine necropsies are performed using microscopy analysis (Hubbard et al., 1983; Lombard and Witte, 1959).

Materials and Methods

The study was conducted from September 2013 to February 2015. Thirty zoological parks were contacted through a partnership with two French animal histopathology laboratories (ONIRIS in Nantes, VetDiagnostic in Lyon) and the French Association of Zoological Park Veterinarians (AFVPZ). Data collection consisted of (1) consultation of veterinarian archives in the zoological parks and (2) analysis of centralized data by veterinarian histopathology laboratories.

As accurate cancer diagnosis relies on histopathological examination of samples from biopsies, resection, or autopsy/necropsy (Martineau et al., 2002), we therefore only entered tumor type (benign or malignant) into our database when they had been confirmed by histological analyses. We also recorded the organs affected, and, if any, the presence and the location of metastases. In order to facilitate data presentation, we classified the tumors into 12 anatomical systems.

Results

The database consisted of 343 tumor references, including 271 cases of cancer in mammals, 46 in birds, and 26 in reptiles representing 27 different orders (Table 2.3).

TABLE 2.3 Number of Tumors Recorded Among 27 Vertebrate Orders in French Zoological Parks

Birds (*n* = 46)		Mammals (*n* = 271)	
Order	*n*	Order	*n*
Accipitriformes (e.g., birds of prey)	2	Afrosoricida (e.g., tenrecs and golden moles)	2
Anseriformes (e.g., ducks and geese)	5	Carnivora (e.g., cats and wolves)	114
Bucerotiformes (e.g., hornbills and hoopoes)	3	Cetartiodactyla (e.g., pigs and deer)	49
Ciconiiformes (e.g., storks)	1	Chiroptera (e.g., bats and flying foxes)	4
Columbiformes (e.g., pigeons and doves)	3	Cingulata (e.g., armadillos)	2
Galliformes (e.g., turkeys and chickens)	2	Diprotodontia (e.g., kangaroos and koalas)	7
Gruiformes (e.g., cranes, coots, and rails)	1	Lagomorpha (e.g., hares and rabbits)	1
Pelecaniformes (e.g., pelicans and cormorants)	8	Perissodactyla (e.g., zebras and rhinoceros)	14
Phoenicopteriformes (e.g., flamingos)	2	Pilosa (e.g., anteaters)	2
Psittaciformes (e.g., parrots and parakeets)	10	Primates (e.g., monkeys and apes)	70
Rheiformes (e.g., rhea)	2	Rodentia (e.g., rats and capybaras)	6
Sphenisciformes (e.g., penguins)	3		
Strigiformes (e.g., owls)	4		

Reptiles (*n* = 26)	
Order	*n*
Crocodilia (e.g., crocodiles and alligators)	1
Squamata (e.g., snakes and lizards)	23
Testudines (e.g., turtles and tortoises)	2

Anatomical Distribution by Class

The tumor frequencies observed in the three vertebrate classes revealed remarkable similarities. High frequencies of digestive (18.4–34.8%), hematopoietic (17.6–27.9%), and skin tumors (14.2–18.6%) were observed in mammals, birds, and reptiles whereas tumors in the remaining 9 anatomical systems occurred in similar low frequencies (0–9.5%; Table 2.4).

TABLE 2.4 Anatomical Percentage Distribution of Tumors in Three Vertebrate Classes in French Zoological Parks

	Birds	Mammals	Reptiles
	(*n* = 46)	(*n* = 271)	(*n* = 26)
Mammary	N/A	4.9	N/A
Cardiovascular	4.6	5.6	4.4
Digestive	20.9	18.4	34.8
Endocrine	2.3	7.9	4.4
Genital	4.6	7.1	4.4
Hematopoietic	27.9	17.6	21.7
Musculoskeletal	7.0	6.7	4.4
Neural	0.0	1.9	0.0
Oral cavity	0.0	3.4	4.4
Pulmonary	4.6	6.4	0.0
Skin	18.6	14.2	17.4
Urinary tract	9.3	6.0	4.4

Benign and Malignant Tumors Recorded in Mammals, Birds, and Reptiles

The prevalence of malignant tumors differed among taxa (χ^2 = 8.68, df = 2, P = 0.01; Table 2.5) and posthoc tests revealed that reptiles had a higher prevalence of malignant tumors than mammals (P = 0.018), while no significant difference in prevalence of malignant tumors were observed between mammals and birds and reptiles and birds (P > 0.17).

TABLE 2.5 Number of Benign and Malignant Tumors in Three Vertebrate Classes in French Zoological Parks

Class	Number of benign tumors	Number of malignant tumors
Aves	10	25
Mammalia	94	142
Reptilia	4	20

DISCUSSION

Comparison to Other Studies

Mammals

The results from the present study show that the highest tumor prevalence was observed in the carnivores (42.1%, 114 of 271). Similar high cancer prevalence in this group of mammals was observed by Lombard and Witte (1959) and Effron et al. (1977). Carnivores include both domestic cats and domestic dogs, of which both have been shown to be subjected to high prevalence of tumors (Merlo et al., 2008; Zambelli, 2015). However, we have not been able to find any information explaining the high prevalence in these two groups of mammals. Our results also revealed similar levels of cancer prevalence in mammal digestive system (18.4%) to that recorded by Lombard and Witte (1959) (20%). Moreover, the second most common tumors observed in mammals by Effron et al. (1977) was hematopoietic/lymphosarcoma (8.9%) followed by skin tumors (8.7%). Our results thus again show a remarkable similarity with the results obtained by Effron et al. (1977) as we also found hematopoietic and skin tumors being the second and third most common tumors recorded (17.6 and 14.2%, respectively).

In contrast Lombard and Witte (1959) found that the second most prevalent tumors were confined to the endocrine system (18.4%) whereas in our study endocrine tumors were only found in 7.9% of mammals. Both Effron et al. (1977) and Lombard and Witte (1959) found that pulmonary tumors were the most prevalent cancer recorded (14 and 16%, respectively) whereas in our study pulmonary tumors were only found in 6.4% of the animal investigated that is, the 7th of the 12 anatomical systems.

Birds

Comparing our results of tumor prevalence in birds with those obtained by Effron et al. (1977) again revealed some striking similarities. In both studies hematopoietic/lymphosarcoma were the most prevalent tumors recorded (27.9 and 32.4%, respectively). In the studies by both Effron et al. (1977) and Lombard and Witte (1959), as well as in our study the second most prevalent cancers were confined to the gastric/digestive system (20.9, 22.2, and 12.6%, respectively). In all three studies the third most prevalent tumors were fibrosarcoma/skin tumors (18.6, 11.1, and 9.9%, respectively). In contrast the second most common tumors recorded by Lombard and Witte (1959) were confined to genital system (20.2%) whereas these tumors were the fourth most common tumors recorded by Effron et al. (1977) (9.9%) and the sixth most common tumors in our study (4.6%).

Reptiles

The most common cancers recorded by Effron et al. (1977) in reptiles were lymphosarcoma (25%) followed by tumors in the intrahepatic biliary/digestive system (21%). Again our results are quite similar to that recorded by Effron et al. (1977) although the order of the two cancer types was reversed, that is, our results showed a highest prevalence in the digestive system (34.8%) followed by the hematopoietic system (21.7%).

CONCLUDING REMARKS

The high prevalence of cancers observed in our study affecting the digestive, hematopoietic, and skin systems recorded across the three vertebrate classes is remarkable and certainly warrants further studies to investigate whether these high prevalences also occur at other zoological parks. As these animals are kept under quite different conditions, that is, most mammals and birds are kept in outdoor cages whereas reptiles are mostly kept indoors we presently have no explanation for the similarities in tumor prevalences among these three groups of vertebrates.

The results from the present study suggest that malignant tumors in reptiles were more prevalent than that observed in mammals. However, this is not supported by the study of Effron et al. (1977) who did not find any significant difference in malignant tumors among the three vertebrate classes. As our data on tumor prevalence in reptiles are based on fairly small number of individuals we therefore suggest that our results should be interpreted with caution.

Although many of the results from the present study are similar to that observed by Effron et al. (1977) and Lombard and Witte (1959) we do emphasize that the cancer etiology and prevalence were all obtained from animals kept in captivity. Cancer in captive animals has been shown to develop predominantly in older age cohorts. For example, although cancer prevalence in captive black-footed ferrets (*Mustela nigripes*) has been shown to affect 55% of the ferrets, the cancer almost exclusively affected postreproductive animals (Lair et al., 2002). The age-specific increase in cancer prevalence recorded in captive animals suggests that the significance of cancers recorded, similar to that recorded in black-footed ferrets, may therefore have limited or in some cases even no fitness effect in the wild. Regardless, cancer statistics recorded in captive animals remain an important source of information for studies in comparative oncology, as well as providing data on cancer etiology.

Acknowledgments

We acknowledge the French zoological parks who welcomed us into their premises and provided data on their animals: Safari de Peaugres, Réserve

Africaine de Sigean, Zoo de la Barben, Zoo La Palmyre, Montpellier Parc Zoologique, Réserve de la Haute-Touche, ZooParc de Beauval, Zoo de La Boissière-du-Doré, Planète sauvage, Bioparc Zoo de Doué, Zoo de Lyon, La Ménagerie, le zoo du Jardin des Plantes, Zoo/Fauverie du Mont Faron.

This work was supported by the ANR (Blanc project EVOCAN), the CNRS (INEE), the Australian Academy of Science's French–Australian Science Innovation Collaboration Program Early Career Fellowship, and an International Associated Laboratory Project France/Australia.

References

Adnyana, W., Ladds, P.W., Blair, D., 1997. Observations of fibropapillomatosis in green turtles (*Chelonia mydas*) in Indonesia. Aust. Vet. J. 75, 736–742.

Aguirre, A.A., Bröjer, C., Mörner, T., 1999a. Descriptive epidemiology of roe deer mortality in Sweden. J. Wildl. Dis. 35, 753–762.

Aguirre, A.A., Limpus, C.J., Spraker, T.R., Balazs, G.H. (Eds.), 1999b. Proceedings of the 19th Annual Symposium on Sea Turtle Biology and Conservation. Texas.

Asashima, M., Komazaki, S., Satou, C., Oinuma, T., 1982. Seasonal and geographical changes of spontaneous skin papillomas in the Japanese newt *Cynops pyrrhogaster*. Cancer Res. 42, 3741–3746.

Assié, G., LaFramboise, T., Platzer, P., Eng, C., 2008. Frequency of germline genomic homozygosity associated with cancer cases. JAMA 299, 1437–1445.

Attwood, H.D., Woolley, P.A., 1973. Spontaneous malignant neoplasms in dasyurid marsupials. J. Comp. Pathol. 83, 569–581.

Bäcklin, B.-M., Eriksson, L., Olovsson, M., 2003. Histology of uterine leiomyoma and occurrence in relation to reproductive activity in the Baltic gray seal (*Halichoerus grypus*). Vet. Pathol. 40, 175–180.

Balazs, G.H., Pooley, S.G. (Eds.), 1990. Marine Turtle Fibropapilloma Disease Workshop, Honolulu Laboratory. Southwest Fisheries Science Centre, La Jolla, CA.

Balls, M., Clothier, R., 1974. Spontaneous tumors in amphibia. Oncology 29, 501–519.

Barber, B.J., 2004. Neoplastic diseases of commercially important marine bivalves. Aquat. Living Resour. 17, 449–466.

Baumann, P.C., LeBlanc, D.R., Blazer, V., Meier, J.R., Hurley, S.T., Kiryu, Y., 2008. Prevalence of Tumors in Brown Bullhead from Three Lakes in Southeastern Massachusetts, 2002. U.S. Geological Survey. In: Scientific Investigations Report.

Bera, M.M., Veeramachaneni, D.N.R., Pandher, K., 2008. Characterization of a biphasic neoplasm in a Madagascar tree boa (*Sanzinia madagascariensis*). Vet. Pathol. 45, 259–263.

Brill, R.W., Balazs, G.H., Holland, K.N., Chang, R.K.C., Sullivan, S., George, J.C., 1995. Daily movements, habitat use, and submergence intervals of normal and tumor-bearing juvenile green turtles (*Chelonia mydas* L.) within a foraging area in the Hawaiian Islands. J. Exp. Mar. Biol. Ecol. 185, 203–218.

Browning, H.M., Gulland, F.M.D., Hammond, J.A., Colegrove, K.M., Hall, A.J., 2015. Common cancer in a wild animal: the California sea lion (*Zalophus californianus*) as an emerging model for carcinogenesis. Philos. Trans. R. Soc. B 370, 20140228.

Brust, D.M., 2013. Gastrointestinal diseases of marsupials. J. Exot. Pet Med. 22, 132–140.

Canfield, P.J., Hartley, W.J., Reddacliff, G.L., 1990. Spontaneous proliferations in Australian marsupials—a survey and review. 1. Macropods, koalas, wombats, possums and gliders. J. Comp. Pathol. 103, 135–146.

Carballal, M.J., Barber, B.J., Iglesias, D., Villalba, A., 2015. Neoplastic diseases of marine bivalves. J. Invertebr. Pathol. 131, 83–106.

Catão-Dias, J.L., Nichols, D.K., 1999. Neoplasia in snakes at the National Zoological Park, Washington, DC (1978–1997). J. Comp. Pathol. 120, 89–95.

Chaloupka, M., Balazs, G.H., Work, T.M., 2009. Rise and fall over 26 years of a marine epizootic in Hawaiian green sea turtles. J. Wildl. Dis. 45, 1138–1142.

Chaloupka, M., Work, T.M., Balazs, G.H., Murakawa, S.K.K., Morris, R., 2008. Cause-specific temporal and spatial trends in green sea turtle strandings in the Hawaiian Archipelago (1982-2003). Mar. Biol. 154, 887–898.

Chu, P.-Y., Zhuo, Y.-X., Wang, F.-I., Jeng, C.-R., Pang, V.F., Chang, P.-H., Chin, S.-C., Liu, C.-H., 2012. Spontaneous neoplasms in zoo mammals, birds, and reptiles in Taiwan—a 10-year survey. Anim. Biol. 62, 95–110.

Ciocan, C.M., Moore, J.D., Rotchell, J.M., 2006. The role of ras gene in the development of haemic neoplasia in *Mytilus trossulus*. Mar. Environ. Res. 62 (Suppl. 1), S147–S150.

Ciocan, C., Sunila, I., 2005. Disseminated neoplasia in blue mussels, *Mytilus galloprovincialis*, from the Black Sea, Romania. Mar. Pollut. Bull. 50, 1335–1339.

Clancy, C.S., Roug, A., Armien, A.G., Van Wettere, A.J., 2016. Intracerebral malignant plasmacytoma in a mule deer (*Odocoileus hemionus*). J. Comp. Pathol. 154, 268–271.

Coffee, L.L., Casey, J.W., Bowser, P.R., 2013. Pathology of tumors in fish associated with retroviruses: a review. Vet. Pathol. 50, 390–403.

Comolli, J.R., Olsen, H.M., Seguel, M., Schnellbacher, R.W., Fox, A.J., Divers, S.J., Sakamoto, K., 2015. Ameloblastoma in a wild black rat snake (*Pantherophis alleghaniensis*). J. Vet. Diagn. Invest. 27, 536–539.

Daoust, P.-Y., Wobeser, G., Rainnie, D.J., Leighton, F.A., 1991. Multicentric intramuscular lipomatosis/fibromatosis in free-flying white-fronted and Canada geese. J. Wildl. Dis. 27, 135–139.

Daszak, P., Berger, L., Cunningham, A.A., Hyatt, A.D., Green, D.E., Speare, R., 1999. Emerging infectious diseases and amphibian population declines. Emerg. Infect. Dis. 5, 735–748.

DeGregori, J., 2011. Evolved tumor suppression: why are we so good at not getting cancer? Cancer Res. 71, 3739–3744.

Delaney, M.A., Ward, J.M., Walsh, T.F., Chinnadurai, S.K., Kerns, K., Kinsel, M.J., Treuting, P.M., 2016. Initial case reports of cancer in naked mole-rats (*Heterocephalus glaber*). Vet. Pathol. 53, 691–696.

Dethlefsen, V., Lang, T., Koves, P., 2000. Regional patterns in prevalence of principal external diseases of dab *Limanda limanda* in the North Sea and adjacent areas 1992–1997. Dis. Aquat. Organ. 42, 119–132.

Diaz-Delgado, J., Arbelo, M., Sacchini, S., Quesada-Canales, O., Andrada, M., Rivero, M., Fernandez, A., 2012. Pulmonary angiomatosis and hemangioma in common dolphins (*Delphinus delphis*) stranded in Canary Islands. J. Vet. Med. Sci. 74, 1063–1066.

Dietz, J., Heckers, K.O., Aupperle, H., Pees, M., 2016. Cutaneous and subcutaneous soft tissue tumours in snakes: a retrospective study of 33 cases. J. Comp. Pathol. 155, 76–87.

Domazet-Lošo, T., Klimovich, A., Anokhin, B., Anton-Erxleben, F., Hamm, M.J., Lange, C., Bosch, T.C.G., 2014. Naturally occurring tumours in the basal metazoan Hydra. Nat. Commun. 5, 4222.

dos Santos, R.G., Martins, A.S., Torezani, E., Baptistotte, C., da Nobrega, F.J., Horta, P.A., Work, T.M., Balazs, G.H., 2010. Relationship between fibropapillomatosis and environmental quality: a case study with Chelonia mydas off Brazil. Dis. Aquat. Organ. 89, 87–95.

Eaton, W.D., Folkins, B., Kent, M.L., 1994. Biochemical and histologic evidence of plasmacytoid leukemia and salmon leukemia virus (SLV) in wild-caught chinook salmon *Oncorhynchus tshawytscha* from British Columbia expressing plasmacytoid leukemia. Dis. Aquat. Organ. 19, 147–151.

Effron, M., Griner, L., Benirschke, K., 1977. Nature and rate of neoplasia found in captive wild mammals, birds, and reptiles at necropsy. J. Natl. Cancer Inst. 59, 185–198.

Elsey, R.M., Nevarez, J.G., Boundy, J., Bauer, R.W., 2013. Massive distal forelimb fibromyxoma in a free-ranging American alligator (*Alligator mississippiensis*). Southeast. Nat. 12, N31–N34.

Erdélyi, K., Dencső, L., Lehoczki, R., Heltai, M., Sonkoly, K., Csányi, S., Solymosi, N., 2009. Endemic papillomavirus infection of roe deer (*Capreolus capreolus*). Vet. Microbiol. 138, 20–26.

FAO, 2010. The State of the World Fisheries and Aqualculture. Food and Agriculture Organization of the United Nations, Rome.

Finkelstein, J.B., 2005. Sharks do get cancer: few surprises in cartilage research. J. Natl. Cancer Inst. 97, 1562–1563.

Fleetwood, M., Lipscomb, T.P., Garlich-Miller, J. (Eds.), 2005. Int. Assoc. Aquat. Anim. Med. Proc. 36, ps. 172.

Flux, J.E.C., 1965. Incidence of ovarian tumors in hares in New Zealand. J. Wildl. Manage. 29, 622–624.

Foley, A.M., Schroeder, B.A., Redlow, A.E., Fick-Child, K.J., Teas, W.G., 2005. Fibropapillomatosis in stranded green turtles (Chelonia mydas) from the eastern United States (1980–98): trends and associations with environmental factors. J. Wildl. Dis. 41, 29–41.

Funk, W.C., Lovich, R.E., Hohenlohe, P.A., Hofman, C.A., Morrison, S.A., Sillett, T.S., Ghalambor, C.K., Maldonado, J.E., Rick, T.C., Day, M.D., et al., 2016. Adaptive divergence despite strong genetic drift: genomic analysis of the evolutionary mechanisms causing genetic differentiation in the island fox (Urocyon littoralis). Mol. Ecol. 25, 2176–2194.

Gates, R.J., Woolf, A., Caithamer, D.F., Moritz, W.E., 1992. Prevalence of spindle cell sarcomas among wild Canada geese from Southern Illinois. J. Wildl. Dis. 28, 666–668.

Geter, D.R., Hawkins, W.E., Means, J.C., Ostrander, G.K., 1998. Pigmented skin tumors in gizzard shad (Dorosoma cepedianum) from the south-central United States: range extension and further etiological studies. Environ. Toxicol. Chem. 17, 2282–2287.

Goodnight, A.L., Couto, C.G., Green, E., Barrie, M., Myers, G., 2008. Chemotherapy and radiotherapy for treatment of cutaneous lymphoma in a ground cuscus (Phalanger gymnotis). J. Zoo Wildl. Med. 39, 472–475.

Gottdenker, N.L., Gerhold, R., Cartoceti, A., Keel, M.K., Goltz, J.P., Howerth, E., 2012. Reports of oligodendrogliomas in three white-tailed deer (Odocoileus virginianus). J. Vet. Diagn. Invest. 24, 202–206.

Greenblatt, R.J., Quackenbush, S.L., Casey, R.N., Rovnak, J., Balazs, G.H., Work, T.M., Casey, J.W., Sutton, C.A., 2005. Genomic variation of the fibropapilloma-associated marine turtle herpesvirus across seven geographic areas and three host species. J. Virol. 79, 1125–1132.

Groff, J.M., 2004. Neoplasia in fishes. Vet. Clin. North Am. Exot. Anim. Pract. 7, 705–756, vii.

Gulland, F.M.D., Trupkiewicz, J.G., Spraker, T.R., Lowenstine, L.J., 1996. Metastatic carcinoma of probable transitional cell origin in 66 free-living California sea lions (Zalophus californianus), 1979 to 1994. J. Wildl. Dis. 32, 250–258.

Guthrie, A.L., Knowles, S., Ballmann, A.E., Lorch, J.M., 2015. Detection of snake fungal disease due to Ophidiomyces ophiodiicola in Virginia, USA. J. Wildl. Dis. 52, 143–149.

Gyimesi, Z.S., Garner, M.M., Burns, III, R.B., Nichols, D.K., Brannian, R.E., Raymond, J.T., Poonacha, K.B., Kennedy, M., Wojcieszyn, J.W., Nordhausen, R., 2005. High Ophidiomyces ophiodiicola a colony of Egyptian spiny-tailed lizards (Uromastyx aegyptius). J. Zoo Wildl. Med. 36, 103–110.

Hardwick, L.J.A., Philpott, A., 2015. An oncologist's friend: how Xenopus contributes to cancer research. Dev. Biol. 408, 180–187.

Hartup, B.K., Steinberg, H., Forest, L.J., 1995. Cholangiocarcinoma in a red-tailed hawk (Buteo jamaicensis). In: Proceedings of a Joint Conference American Association of Zoo Veterinarians, Wildlife Disease Association, and American Association of Wildlife Veterinarians, p. 448.

Heatley, J.J., Mauldin, G.E., Cho, D.Y., 2005. A review of neoplasia in the captive African hedgehog (Atelerix albiventris). Semin. Avian Exot. Pet Med. 14, 182–192.

Herbst, L.H., Klein, P.A., 1995. Green turtle fibropapillomatosis: challenges to assessing the role of environmental cofactors. Environ. Health Perspect. 103 (Suppl. 4), 27–30.

Hernandez-Divers, S.M., Garner, M.M., 2003. Neoplasia of reptiles with an emphasis on lizards. Vet. Clin. North Am. Exot. Anim. Pract. 6, 251–273.

Higbie, C.T., Carpenter, J.W., Choudhary, S., DeBey, B., Bagladi-Swanson, M., Eshar, D., 2015. Cutaneous epitheliotropic T-cell lymphoma with metastases in a Virginia opossum (Didelphis virginiana). J. Zoo Wildl. Med. 46, 409–413.

Hill, A.G., Denis, M.M., Pyne, M., 2016. Squamous cell carcinoma with hepatic metastasis in a saltwater crocodile (Crocodylus porosus). Aust. Vet. J. 94, 83–86.

Howerth, E.W., Schorr, L.F., Nettles, V.F., 1986. Neoplasia in free-flying ruffed grouse (Bonasa umbellus). Avian Dis. 30, 238–240.

Hubbard, G., Schmidt, R., Fletcher, K., 1983. Neoplasia in zoo animals. J. Zoo Anim. 14, 33–40.

Jarplid, B., Rehbinder, C., 1995. Lymphoma in reindeer (Rangifer tarandus tarandus L.). Rangifer 15, 37–38.

Jennings, A.H., 1968. Tumours in free-living wild mammals and birds in Great Britain. Symp. Zool. Soc. Lond. 24, 273–287.

Korkea-aho, T., Vainikka, A., Taskinen, J., 2006. Factors affecting the intensity of epidermal papillomatosis in populations of roach, Rutilus rutilus (L.), estimated as scale coverage. J. Fish Dis. 29, 115–122.

Kusewitt, D.F., Ley, R.D., 1996. Animal models of melanoma. Cancer Surv. 26, 35–70.

Ladds, P., 2009. Pathology of Australian Native Wildlife. CSIRO Publishing, Collingwood, Australia.

Lair, S., Barker, I.K., Mehren, K.G., Williams, E.S., 2002. Epidemiology of neoplasia in captive black-footed ferrets (Mustela nigripes), 1986–1996. J. Zoo Wildl. Med. 33, 204–213.

LePage, V., Dutton, C.J., Kummrow, M., McLelland, D.J., Young, K., Lumsden, J.S., 2012. Neoplasia of captive yellow sea horses (Hippocampus kuda) and weedy sea dragons (Phyllopteryx taeniolatus). J. Zoo Wildl. Med. 43, 50–58.

Leroi, A.M., Koufopanou, V., Burt, A., 2003. Cancer selection. Nat. Rev. Cancer 3, 226–231.

Linnehan, R.M., Edwards, J.L., 1991. Endometrial adenocarcinoma in a Bengal tiger (Panthera tigris bengalensis) implanted with melengestrol acetate. J. Zoo Wildl. Med. 22, 130–134.

Lombard, L.S., Witte, E.J., 1959. Frequency and types of tumors in mammals and birds of the Philadelphia Zoological Garden. Cancer Res. 19, 127–141.

Lopez, R.M., Murcia, D.B., 2008. First description of malignant retrobulbar and intracranial teratoma in a lesser kestrel (Falco naumanni). Avian Pathol. 37, 413–414.

Mader, D.R., 1996. Reptile Medicine and Surgery. W.B. Saunders Company, Philadelphia, PA, USA.

Madsen, T., Shine, R., Olsson, M., Wittzell, H., 1999. Conservation biology: restoration of an inbred adder population. Nature 402, 34–35.

Madsen, T., Stille, B., Shine, R., 1996. Inbreeding depression in an isolated population of adders Vipera berus. Biol. Conserv. 75, 113–118.

Madsen, T., Ujvari, B., Olsson, M., 2004. Novel genes continue to enhance population growth in adders (Vipera berus). Biol. Conserv. 120, 145–147.

Malins, D.C., McCain, B.B., Myers, M.S., Brown, D.W., Krahn, M.M., Roubal, W.T., Schiewe, M.H., Landahl, J.T., Chan, S.L., 1987. Field and laboratory studies of the etiology of liver neoplasms in marine fish from Puget Sound. Environ. Health Perspect. 71, 5–16.

Marais, H.J., Nel, P., Bertschinger, H.J., Schoeman, J.P., Zimmerman, D., 2007. Prevalence and body distribution of sarcoids in South African Cape mountain zebra (Equus zebra zebra). J. S. Afr. Vet. Assoc. 78, 145–148.

Marais, H.J., Page, P.C., 2011. Treatment of equine sarcoid in seven Cape mountain zebra (Equus zebra zebra). J. Wildl. Dis. 47, 917–924.

Marrow, J.C., Carpenter, J.W., Lloyd, A., Bawa, B., 2010. A transitional cell carcinoma with squamous differentiation in a pericloacal mass in a sugar glider (Petaurus breviceps). J. Exot. Pet Med. 19, 92–95.

Martineau, D., Lemberger, K., Dallaire, A., Labelle, P., Lipscomb, T.P., Michel, P., Mikaelian, I., 2002. Cancer in wildlife, a case study: beluga from the St. Lawrence estuary, Québec, Canada. Environ. Health Perspect. 110, 285–292.

Martinez-Silvestre, A., Amat, F., Bargallo, F., Carranza, S., 2011. Incidence of pigmented skin tumors in a population of wild Montseny brook newt (Calotriton arnoldi). J. Wildl. Dis. 47, 410–414.

Masahito, P., Nishioka, M., Kondo, Y., Yamazaki, I., Nomura, K., Kato, Y., Sugano, H., Kitagawa, T., 2003. Polycystic kidney and renal cell carcinoma in Japanese and Chinese toad hybrids. Int. J. Cancer 103, 1–4.

McAloose, D., Newton, A.L., 2009. Wildlife cancer: a conservation perspective. Nat. Rev. Cancer 9, 517–526.

McClure, H.M., Chang, J., Golarz, M.N., 1977. Cholangiocarcinoma in a Margay (Felis wiedii). Vet. Pathol. 14, 510–512.

McKinnel, R.G., 1965. Incidence and histology of renale tumors of leopard frogs from the north central states. Ann. NY Acad. Sci. 126, 85–98.

McKinnell, R.G., Martin, F.B., 1979. Continued diminished prevalence of the Lucke renal adenocarcinoma in Minnesota leopard frogs. Am. Midl. Nat. 104, 402–404.

Merlo, D.F., Rossi, L., Pellegrino, C., Ceppi, M., Cardellino, U., Capurro, C., Ratto, A., Sambucco, P.L., Sestito, V., Tanara, G., et al., 2008. Cancer incidence in pet dogs: findings of the animal tumor registry of Genoa, Italy. J. Vet. Intern. Med. 22, 976–984.

Metzger, M.J., Reinisch, C., Sherry, J., Goff, S.P., 2015. Horizontal transmission of clonal cancer cells causes leukemia in soft-shell clams. Cell 161, 255–263.

Mikaelian, I., Reavill, D.R., Practice, A., 2004. Spontaneous proliferative lesions and tumors of the uterus of captive African hedgehogs (Atelerix albiventris). J. Zoo Wildl. Med. 35, 216–220.

Miranda, D.F.H., Souza, F.A.L., Fonseca, L.S., de Almeida, H.M., Braga, J.F.V., Costa, F.A.L., Silva, S.M.M.S., 2015. Metastatic hepatocellular carcinoma in ocelot (Leopardus pardalis). Pesq. Vet. Bras. 35, 913–918.

Morado, J.F., Shavey, C.A., Ryazanova, T., White, V.C., 2014. Diseases of king crabs and other anomalies. King Crabs of the World: Biology and Fisheries ManagementCRC Press, Boca Raton, FL.

Munson, L., Nesbit, J.W., Meltzer, D.G., Colly, L.P., Bolton, L., Kriek, N.P., 1999. Diseases of captive cheetahs (Acinonyx jubatus jubatus) in South Africa: a 20-year retrospective survey. J. Zoo Wildl. Med. 30, 342–347.

Napier, J.E., Murray, S., Garner, M.M., Viner, T., Murphy, H., 2005. Uterine leiomyomas in three captive eastern bongo (Tragelaphus eurycerus isaaci). J. Zoo Wildl. Med. 36, 709–711.

Newman, S.J., Smith, S.A., 2006. Marine mammal neoplasia: a review. Vet. Pathol. 43, 865–880.

Old, J.M., Price, M.D., 2016. A case of melanoma in a native Australian murid, the spinifex hopping-mouse (Notomys alexis). Aust. Mammal. 38, 117–119.

Papas, T.S., Dahlberg, J.E., Sonstegard, R.A., 1976. Type C virus in lymphosarcoma in northern pike (Esox lucius). Nature 261, 506–508.

Peters, E.C., Yevich, P.P., Harshbarger, J.C., Zaroogian, G.E., 1994. Comparative histopathology of gonadal neoplasms in marine bivalve mollusks. Dis. Aquat. Organ. 20, 59–76.

Port, C.D., Maschgan, E.R., Pond, J., Scarpelli, D.G., 1981. Multiple Neoplasia in a jaguar (Panthera onca). J. Comp. Pathol. 91, 115–122.

Porter, B.F., Goens, S.D., Brasky, K.M., Hubbard, G.B., 2004. A case report of hepatocellular carcinoma and focal nodular hyperplasia with a myelolipoma in two chimpanzees and a review of spontaneous hepatobiliary tumors in non-human primates. J. Med. Primatol. 33, 38–47.

Ratcliffe, H.L., 1933. Incidence and nature of tumors in captive wild mammals and birds. Am. J. Cancer 17, 116–135.

Reece, R.L., 1992. Observations on naturally occurring neoplasms in birds in the state of Victoria, Australia. Avian Pathol. 21, 3–32.

Reichenbach-Klinke, H.-H., 1963. Krankheiten der Reptilien. Gustav Fischer Verlag, Stuttgart.

Robbins, R., Bruce, B., Fox, A., 2014. First reports of proliferative lesions in the great white shark, Carcharodon carcharias L., and bronze whaler shark, Carcharhinus brachyurus Gunther. J. Fish Dis. 37, 997–1000.

Robert, J., Goyos, A., Nedelkovska, H., 2009. Xenopus, a unique comparative model to explore the role of certain heat shock proteins and non-classical MHC class Ib gene products in immune surveillance. Immunol. Res. 45, 114–122.

Rodriguez-Ramos Fernandez, J., Thomas, N.J., Dubielzig, R.R., Drees, R., 2012. Osteosarcoma of the maxilla with concurrent osteoma in a southern sea otter (Enhydra lutris nereis). J. Comp. Pathol. 147, 391–396.

Rose, F.L., 1976. Tumorous growths of the tiger salamander, Ambystoma tigrinum, associated with treated sewage effluent. Prog. Exp. Tumor Res. 20, 251–262.

Rose, F.L., Harshbarger, J.C., 1977. Neoplastic and possibly related skin lesions in neotenic tiger salamanders from a sewage lagoon. Science 196, 315–317.

Rudan, I., Rudan, D., Campbell, H., Carothers, A., Wright, A., Smolej-Narancic, N., Janicijevic, B., Jin, L., Chakraborty, R., Deka, R., et al., 2003. Inbreeding and risk of late onset complex disease. J. Med. Genet. 40, 925–932.

Rush, E.M., Ogburn, A.L., Garner, M.M., 2012. Multicentric neurofibromatosis with rectal prolapse in a California sea lion (Zalophus californianus). J. Zoo Wildl. Med. 43, 110–119.

Sakai, H., Yanai, T., Yonemaru, K., Hirata, A., Masegi, T., 2003. Gallbladder adenocarcinomas in two captive African lions (Panthera leo). J. Zoo Wildl. Med. 34, 302–306.

Salomon, R.N., Jackson, F.R., 2008. Tumors of testis and midgut in aging flies. Fly 2, 265–268.

Sanches, A.W.D., Werner, P.R., Margarido, T.C.C., Pachaly, J.R., 2013. Multicentric lymphoma in a giant anteater (Myrmecophaga tridactyla). J. Zoo Wildl. Med. 44, 186–188.

Santos, R.G., Martins, A.S., Torezani, E., Baptistotte, C., Farias, J.N., Horta, P.A., Work, T.M., Balazs, G.H., 2010. Relationship between fibropapillomatosis and environmental quality: a case study with Chelonia mydas off Brazil. Dis. Aquat. Organ. 89, 87–95.

Sasidharan, S.P., 2006. Sarcoid tumours in Cape mountain zebra (Equus zebra zebra) populations in South Africa: a review of associated epidemiology, virology and genetics. Trans. R. Soc. S. Afr. 61, 11–18.

Sasidharan, S.P., Ludwig, A., Harper, C., Moodley, Y., Bertschinger, H.J., Guthrie, A.J., 2011. Comparative genetics of sarcoid tumour-affected and non-affected mountain zebra (Equus zebra) populations. S. Afr. J. Wildl. Res. 41, 36–49.

Seeley, K.E., Garner, M.M., Waddell, W.T., Wolf, K.N., 2016. A survey of diseases in captive red wolves (Canis rufus), 1997–2012. J. Zoo Wildl. Med. 47, 83–90.

Shimonohara, N., Holland, C.H., Lin, T.L., Wigle, W.L., 2013. Naturally occurring neoplasms in pigeons in a research colony: a retrospective study. Avian Dis. 57, 133–139.

Siegfried, L.M., 1983. Neoplasms identified in free-flying birds. Avian Dis. 27, 86–99.

Snyder, R.L., Ratcliffe, H.L., 1966. Primary lung cancers in birds and mammals of the Philadelphia Zoo. Cancer Res. 26, 514–518.

Steeil, J.C., Schumacher, J., Hecht, S., Baine, K., Ramsay, E.C., Ferguson, S., Miller, D., Lee, N.D., 2013. Diagnosis and treatment of a pharyngeal squamous cell carcinoma in a Madagascar ground boa (Boa madagascariensis). J. Zoo Wildl. Med. 44, 144–151.

Stringer, E.M., De Voe, R.S., Valea, F., Toma, S., Mulvaney, G., Pruitt, A., Troan, B., Loomis, M.R., 2010. Medical and surgical management of reproductive neoplasia in two western lowland gorillas (Gorilla gorilla gorilla). J. Med. Primatol. 39, 328–335.

Sweet, M., Kirkham, N., Bendall, M., Currey, L., Bythell, J., Heupel, M., 2012. Evidence of melanoma in wild marine fish populations. PLoS One 7, e41989.

Terrell, S.P., Forrester, D.J., Mederer, H., Regan, T.W., 2002. An epizootic of fibromatosis in gray squirrels (Sciurus carolinensis) in Florida. J. Wildl. Dis. 38, 305–312.

Thas, I., Garner, M.M., 2012. A retrospective study of tumours in black-tailed prairie dogs (*Cynomys ludovicianus*) submitted to a zoological pathology service. J. Comp. Pathol. 147, 368–375.

Thompson, R., Armien, A.G., Rasmussen, J.M., Wolf, T.M., 2014. Uterine adenocarcinoma in a Przewalski's wild horse (*Equus ferus przewalskii*). J. Zoo Wildl. Med. 45, 441–445.

Tian, X., Azpurua, J., Hine, C., Vaidya, A., Myakishev-Rempel, M., Ablaeva, J., Mao, Z., Nevo, E., Gorbunova, V., Seluanov, A., 2013. High-molecular-mass hyaluronan mediates the cancer resistance of the naked mole rat. Nature 499, 346–349.

Vickers, T.W., Clifford, D.L., Garcelon, D.K., King, J.L., Duncan, C.L., Gaffney, P.M., Boyce, W.M., 2015. Pathology and epidemiology of ceruminous gland tumors among endangered Santa Catalina Island foxes (*Urocyon littoralis catalinae*) in the Channel Islands, USA. PLoS One 10, e0143211.

Vittecoq, M., Roche, B., Daoust, S.P., Ducasse, H., Misse, D., Abadie, J., Labrut, S., Renaud, F., Gauthier-Clerc, M., Thomas, F., 2013. Cancer: a missing link in ecosystem functioning? Trends Ecol. Evol. 28, 628–635.

Walzer, C., Kubber-Heiss, A., Bauder, B., 2003. Spontaneous uterine fibroleiomyoma in a captive cheetah. J. Vet. Med. A Physiol. Pathol. Clin. Med. 50, 363–365.

Wiley, J.L., Whittington, J.K., Wilmes, C.M., Messick, J.B., 2009. Chronic myelogenous leukemia in a great horned owl (*Bubo virginianus*). J. Avian Med. Surg. 23, 36–43.

Williams, T.D., Pulley, L.T., 1981. Leiomyomas in two sea otters, *Enhydra lutris*. J. Wildl. Dis. 17, 401–404.

Woolford, L., Bennett, M.D., Sims, C., Thomas, N., Friend, J.A., Nicholls, P.K., Warren, K.S., O'Hara, A.J., 2009. Prevalence, emergence, and factors associated with a viral papillomatosis and carcinomatosis syndrome in wild, reintroduced, and captive Western barred bandicoots (*Perameles bougainville*). EcoHealth 6, 414–425.

Woolford, L., O'Hara, A.J., Bennett, M.D., Slaven, M., Swan, R., Friend, J.A., Ducki, A., Sims, C., Hill, S., Nicholls, P.K., et al., 2008. Cutaneous papillomatosis and carcinomatosis in the Western barred bandicoot (*Perameles bougainville*). Vet. Pathol. 45, 95–103.

Work, T.M., Balazs, G.H., 1999. Relating tumor score to hematology in green turtles with fibropapillomatosis in Hawaii. J. Wildl. Dis. 35, 804–807.

Work, T.M., Balazs, G.H., Schumacher, J.L., Marie, A., 2005. Epizootiology of spirorchid infection in green turtles (*Chelonia mydas*) in Hawaii. J. Parasitol. 91, 871–876.

Work, T.M., Rameyer, R.A., Balazs, G.H., Cray, C., Chang, S.P., 2001. Immune status of free-ranging green turtles with fibropapillomatosis from Hawaii. J. Wildl. Dis. 37, 574–581.

Zambelli, A.B., 2015. Feline cancer prevalence in South Africa (1998–2005): contrasts with the rest of the world. J. Basic Appl. Sci. 11, 370–380.

3

Infection and Cancer in Nature

Paul W. Ewald, Holly A. Swain Ewald

Department of Biology, University of Louisville, Louisville, KY, United States

INTRODUCTION

Cancer is responsible for approximately 13% of human deaths (Ferlay et al., 2010). Estimates of cancer incidence in humans, dogs, cattle, and horses are roughly similar (Dorn and Priester, 1987). In contrast, cancers are not commonly reported from animals in nature (see also Chapter 2). Direct evidence is insufficient to determine whether this paucity of reports is due to a true rarity of cancer or from a reporting bias; however, evolutionary and ecological considerations can provide insight into the probable presence of cancer in nature and fruitful areas for research.

Cancer could be rare in nature if it is a manifestation of organismal senescence and animals rarely survive to old ages (Niccoli and Partridge, 2012). In this case, a low frequency of precancerous tumors might develop to a malignant state or precancerous tumors might be rare. The incidence of cancer in nature also might be rare relative to that in humans if wild animals are less exposed to man-made carcinogens.

Alternatively cancer could be common in nature but unreported because survival after malignancy is brief for ecological reasons (Roche et al., 2012; Stroud and Amundson, 1983). Cancer may reduce survival in nature by making afflicted animals more prone to death from predation, parasitism, or competition for resources (Vittecoq et al., 2013).

An evolutionary perspective emphasizes the possibility that prevalence of cancer in any species may be damped by anticancer adaptations. Natural selection should favor enhancement of these adaptations until the rarity of cancer makes the selection pressure against cancer too weak for further reductions. By this argument the prevalence of cancer could be similar among species in nature (Caulin and Maley, 2011; Roche et al., 2012) even though it may be lower than in humans and domestic animals because of increased longevity in nonnatural environments.

INFECTIOUS CAUSES OF CANCER

Cancers can spread in a population if a cancer-causing pathogen or the cancer cell itself is transmitted from one host to another. These two categories are referred to in this chapter as infectious cancers, because they are dependent on infectious processes: infection of a host by an oncogenic pathogen or by a cancer cell that was acquired from a different host.

The prevalence of infectious cancers in nature is certain to be underestimated relative to the prevalence of cancers, because cancers first need to be identified before the possibilities of infectious causation can be assessed. At present considerations of the scope for infectious cancers must therefore interpret available information in the context of what is known about infectious oncogenesis. This chapter will focus on the prevalence of parasite-induced cancers in nature, that is, cancers caused by subcellular, unicellular, or multicellular parasites (cancers that result from the transmission of cancer cells are discussed in Chapter 12).

In domestic animals as in humans, most of the identified oncogenic pathogens (i.e., cellular or subcellular parasites) are viruses (Ewald and Swain Ewald, 2016; Theilen and Madewell, 1987). Although the literature on cancer in wild vertebrates is less complete than the literature on cancer in domestic animals, the current state of knowledge implicates a disproportionate involvement of viruses relative to other pathogens (Table 3.1; McAloose and Newton, 2009; McCallum and Jones, 2012).

Most of the information on pathogen-induced cancers comes from the study of virally induced cancers of humans and domestic animals (Ewald and Swain Ewald, 2012). Viruses have been accepted as causes of cancer for about a century. Rous (1910, 1911) showed that a nonfilterable agent, now known as Rous sarcoma virus, caused a rapidly fulminant cancer of chickens. During the ensuing half-century, several viruses were accepted as causes of cancers: a papillomavirus in rabbits

Ecology and Evolution of Cancer. http://dx.doi.org/10.1016/B978-0-12-804310-3.00003-X

TABLE 3.1 Viruses That Are Oncogenic or Borderline Oncogenic in Wildlife

Virus	Host	Reference	Taxonomic group of viruses	Human oncogenic viruses in same taxonomic group
Woodchuck hepatitis virus	Woodchuck (*Marmota monax*)	Summers et al. (1978)	Orthohepadnavirus	HBV
Otarine herpesvirus-1	California sea lion (*Zalophus californianus*)	King et al. (2002)	Gamma herpes virus	EBV, KSHV
Bandicoot papillomatosis carcinomatosis virus-1	Western barred bandicoots (*Perameles bougainville*)	Woolford et al. (2008)	Papillomavirus–polyomavirus mosaic	HPV, MCPyV
Simian lymphotropic virus-1	Various primates (see text)	Homma et al. (1984), Lee et al. (1985), Blakeslee et al. (1987), Mone et al. (1992)	Deltaretrovirus	HTLV-1
Avian reticuloendotheliosis virus	Attwater prairie chicken (*Tympanuchus cupido attwateri*)	Barbosa et al. (2007)	Gammaretrovirus	EBV, KSHV
Fibropapillomatosis virus	Sea turtles	Aguirre and Lutz (2004)	Alpha herpes virus	HHSV
Rana virus-1	Leopard frog (*Rana pipiens*)	Granoff (1973)	Iridoviridae	None
Walleye dermal sarcoma virus	Walleye pike (*Sander vitreus*)	Martineau et al. (1992)	Epsilonretrovirus	None
Atlantic salmon swim bladder sarcoma virus	Atlantic Salmon (*Salmo salar*)	Paul et al. (2006)	Epsilonretrovirus/gammaretrovirus intermediate	None

EBV, Epstein–Barr virus; HBV, hepatitis B virus; HHSV, human herpes simplex virus (possibly oncogenic); HPV, human papillomavirus; HTLV-1, human T-lymphotropic virus type 1; KSHV, Kaposi sarcoma–associated herpes virus (=human herpes virus 8); MCPyV, Merkel cell polyomavirus.

(Shope and Hurst, 1933), mouse mammary tumor virus in mice (Bittner, 1942), murine leukemia virus in mice (Gross, 1951), and simian virus 40 in hamsters (Rabson and Kirschstein, 1962). Subsequently, many other tumor viruses have been associated with cancer in domestic animals, mostly in poultry, rodents, and cats (Leroux et al., 2007; Madewell and Theilen, 1987).

For most of the past century, researchers have presumed that the oncogenic mechanisms are indirect, involving increases in mutations or cellular proliferation arising largely from immunological responses to infection. Researchers have found evidence for indirect immune enhancement of oncogenesis (Mantovani et al., 2008; Moss and Blaser, 2005; Nath et al., 2010; Trinchieri, 2012) as well as for immune suppression that compromises control of oncogenesis, as in the case of increased vulnerability to cancer that can be associated with human immunodeficiency virus (HIV) infection (Grulich and Vajdic, 2015). However, the greatest clarity in understanding the role of pathogens in oncogenesis has come from the identification of mechanisms by which pathogens, particularly viruses, contribute by encoding proteins that compromise cellular barriers to oncogenesis. Prominent among these effects are the abrogation of the infected cell's cap on the total number of divisions it

can undergo, its ability to destroy itself, its adhesion to other cells, and its arrest of cellular division (Ewald and Swain Ewald, 2012, 2013). The compromising of these barriers involves precise interactions between viral and host cell molecules that abrogate the function of the host molecules.

These sophisticated mechanisms of oncogenicity have evolved independently in different viruses, with entirely different viral proteins being modified to negate the same cellular barriers to oncogenesis (Ewald and Swain Ewald, 2012). This convergent evolution accords with the idea that selection for persistence will often favor the evolution of oncogenic tendencies that result from the propagation of viral genomes through enhancement of host cell survival and proliferation (Ewald and Swain Ewald, 2013). More generally it suggests that pathogen-induced oncogenesis may be common in natural populations.

In humans, oncogenic viruses are disproportionately transmitted by routes that tend to have relatively long intervals between transmission opportunities: sexual contact, milk, and needles (Ewald and Swain Ewald, 2012, 2016). Pathogens evolve oncogenicity if the fitness benefits of the characteristics that contribute to oncogenicity (i.e., increased proliferation and survival of infected cells) outweigh the negative effects. When long intervals occur between transmission opportunities, persistence within hosts should be especially beneficial because it fosters transmission to new hosts. Cancer itself negatively affects the viral fitness because severe illness reduces and host death eliminates chances of transmission to new hosts. However, these fitness costs are incurred by only a small portion of oncogenic pathogens because the proportion of infected individuals that develop cancer is generally small.

Natural selection acting on pathogens will tend to curb but not entirely eliminate oncogenic effects for two reasons: (1) as oncogenicity evolves to lower levels, selection pressure to dampen it declines; and (2) the beneficial effects of infected cell proliferation and survival may always be linked to some extent to increased risk of oncogenesis. This evolutionary trade-off leads to the conclusion that natural selection could act on oncogenic pathogens of short-lived hosts to cause cancer in periods of time that are much shorter than the years to decades that are required for human oncogenic pathogens (Ewald and Swain Ewald, 2015). In support of this conclusion, time intervals between the onset of infection and the onset of cancer can range from a few weeks (e.g., in Rous sarcoma) to many decades (e.g., in adult T-cell leukemia), and can occur in both short-lived and long-lived organisms (Ewald and Swain Ewald, 2015). This range emphasizes that infectious cancer could be widespread but unreported across the spectrum of animals in nature.

In captive undomesticated animals, cancer has been implicated in about 1–5% of deaths (Effron et al., 1977; Hernandez-Divers and Garner, 2003; Hubbard et al., 1983a,b). This occurrence shows that undomesticated species are vulnerable to cancer and accords with the idea that longer life associated with captivity increases the incidence of cancer (see also Chapter 2). But the occurrence of cancer in captive, undomesticated animals could be influenced by other factors, such as greater exposures of captive populations to oncogenic pathogens or chemicals. Moreover, the common occurrence of cancer in captive animals relative to wild animals does not mean that cancer is actually higher in captive animals, because detection of cancer in captive animals may be more comprehensive. Most of the reports of cancer in captive undomesticated animals come from zoos, where causes of death can be quantified for the entire population.

FINDINGS SUGGESTIVE OF PATHOGEN-INDUCED CANCER IN NATURE

Oncogenicity in Closely Related Host/Pathogen Combinations

The sophisticated mechanisms by which viruses abrogate barriers to cancer must have taken substantial time to evolve. Moreover, the targets of the oncogenic proteins of pathogens tend to be conserved among closely related species, probably because these proteins have precise functions that are not species-specific. The p53 protein, for example, participates in cell cycle arrest and cell suicide (apoptosis) in response to DNA damage or infection within a cell (Williams and Schumacher, 2016). The p53 protein tends to be conserved evolutionarily (May and May, 1999), and very similar in closely related species, such as humans and chimpanzees (Puente et al., 2006). It is targeted by oncogenic proteins of human tumor viruses, which can thus block an infected cell's ability to arrest its cell cycle or undergo apoptosis (Ewald and Swain Ewald, 2012). If a virus that is capable of infecting a human has oncogenic effects by targeting p53 and a closely related virus can infect a chimpanzee, then it would be reasonable to expect that the chimpanzee virus could also have a prooncogenic effect in the chimpanzee by targeting p53. More generally, if one knows that a virus in a human or domestic animal uses such mechanism, then an association of a very similar virus with a similar host in nature suggests that an oncogenic effect is likely.

T-lymphotropic viruses in humans and simians provide an illustration. They are referred to as human T-lymphotropic viruses (HTLVs) when isolated from humans and simian T-lymphotropic viruses (STLVs) when isolated from monkeys or apes (i.e., simians). STLV type 1 (STLV-1) has been transmitted between simians and humans several times in recent millennia, and are therefore closely related phylogenetically (Niphuis et al., 2003). STLV-1 can be common in natural populations, with prevalences of over 50% being reported in chimpanzees and monkeys (Peeters et al., 2014). They have been associated with anecdotal reports of leukemia or lymphoma in African green monkeys, baboons, macaques, and gorillas (Blakeslee et al., 1987; Homma et al., 1984; Lee et al., 1985; Mone et al., 1992). These reports of illness are based mostly on captive animals, suggesting that these cancers are most apparent when animals can be closely monitored. HTLV-1 proteins compromise all of the main barriers to oncogenesis (Ewald and Swain Ewald, 2012). In humans, adult T-cell leukemias or lymphomas occur in about 5% of HTLV-1 infections. These cancers arise after many decades of asymptomatic infection with death generally occurring within several months of the onset of cancer (Nicot, 2005). The presence of HTLV-induced cancer is therefore short relative to the presence of HTLV infection. If STLV-induced cancers are similarly rare and transient in simian hosts relative to STLV infection, the cancers would be difficult to detect except in captivity. Taken together these considerations suggest that STLVs probably play a role in the simian cancers with which they are associated even though better evidence is still needed.

In domestic cats, feline papillomaviruses cause tumors and are associated with invasive squamous cell carcinomas (Munday et al., 2008; Nespeca et al., 2006). Benign tumors and squamous cell carcinomas positive for feline papillomaviruses have been found in snow leopards (Sundberg et al., 2000). A case of squamous cell carcinoma in a wild lion has been reported (Mwase et al., 2013), although it was not tested for feline papillomavirus. Improvements in identifying papaillomavirus-infected tissue should help clarify the role and scope of feline papilloma viruses in squamous cell carcinomas of wild felids (Mitsouras et al., 2011; Munday and Aberdein, 2012).

High Prevalence of Oncogenic Pathogens

In the absence of widespread monitoring of cancer in wild populations, the potential importance of infection-induced cancers in nature can be assessed by quantifying prevalences of pathogens that are known to contribute to oncogenesis. For example, infections with feline immunodeficiency viruses (FIVs) and the HIVs are similar with respect to immune manifestations, course, and cell tropism. Like HIV in humans, FIV is associated with increased risk of cancer in domestic cats (Bienzle, 2014; Hartmann, 2012; Magden et al., 2011), apparently through decreased immune surveillance rather than direct cellular transformation (Beatty, 2014). In captive and wild lions, FIV has been associated with depletion of helper T cells (Pecon-Slattery et al., 2008; Roelke et al., 2009). More than 95% of noncaptive lions were positive for FIV in three of the four sampled populations (Roelke et al., 2009). These high prevalences, together with the associations of FIV with lymphoma in domestic cats (Hartmann, 2012), HIV-like immune suppression in noncaptive African lions, and its occurrence in a lymphoma in a captive lion (Poli et al., 1995), suggest that FIV may contribute to cancer in wild populations of lions.

A retrospective study in US zoos of captive-born African lions with lymphoma found no FIV in tumor samples, nor any serological positivity (Harrison et al., 2010). This comparison emphasizes the probability that lymphomas in lions will have multiple causes, and, more generally, that infectious causes of cancer might be variable across populations; they sometimes may be more prevalent in nature because pathogens may not persist in small, captive populations (e.g., in zoos) where opportunities for transmission are limited. In the case of FIV, transmission appears to be mainly through biting, a route that is undoubtedly suppressed in captive situations relative to in nature.

Assessments of the potential for infection-induced cancer in wild populations need to account for introductions into the population through contact with humans or domesticated animals. The lentivirus feline leukemia virus (FeLV) provides an example of an outbreak that can be traced to recent transfer from a domestic animal. Domestic cats (*Felis catus*) infected with FeLV were 60 times more likely than uninfected cats to develop lymphoma or leukemia (Beatty, 2014; Hartmann, 2012). Transmission from domestic cats caused an outbreak of severe disease in the Florida panther (*Puma concolor*) population. The manifestations of *P. concolor* were consistent with those seen in the *F. catus*, including lymphoid hyperplasia and lymphadenopathy (Brown et al., 2008). Studies of *P. concolor* populations generally show no FeLV infection, although an anecdotal case of a FeLV-infected *P. concolor* with lymphoproliferative disease was reported from California (Brown et al., 2008; Jessup, 1993). Recognition that this outbreak was caused by a transfer of FeLV from domestic cats helps place the severity of this outbreak in perspective. Cancer caused by FeLV may occur in *P. concolor* but appears to be rare because of the transitory maintenance of FeLV in this species. However, in contrast with this transitory presence of FeLV, phylogenetic associations indicate that FIV has been transmitted from *F. catus* to *P. concolor* distantly in the past and is maintained in *P. concolor* throughout the Americas (Pecon-Slattery et al., 2008).

In another variation on this theme, interspecific transmission from domestic cats is apparently responsible for moderate positivity of both FeLV (20%) and FIV (13%) in noncaptive guignas (*Leopardus guigna*), a felid found in Chile (Mora et al., 2015). This transmission was attributed to human activities that disturbed habitats and allowed increased contacts between *F. catus* and *L. guignas*.

These examples of lentiviral infections in domestic and wild felids show how prevalence can be used as a clue to the presence of infection-associated cancers in wild populations, but also illustrate the need for population-specific assessments. This point has been directly demonstrated by the documentation of interpopulation variation of FIV prevalence among felids in the United States (Carver et al., 2016). A comparison of domestic cats in different settings (Little, 2005) illustrates how the characteristics of the local population can be associated with variation in FIV prevalence. Positivity was 23% of stray urban cats, 5% in feral cats, and 6% in client-owned cats, variation that may reflect more biting among stray urban cats than among cats in the other two categories.

Phylogenetic comparisons of pathogens in humans, domestic animals, and wild animals provide insight into the direction of the interspecies transmission and duration of an association between a virus and a host, under the assumption that viruses from the source host species will be more genetically diverse and that viruses in the recipient hosts will be embedded within clades from the source hosts. The phylogenetic pattern of FIV indicates that it has had long histories with domestic cats and its most well-studied wild hosts (Pecon-Slattery et al., 2008). The FIV from lions (*Panthera leo*) is distinct from the main clade of domestic cats indicating a long evolutionary association with lions, but viral recombination in the subtype E clade from lions indicates transmission from domestic cats to lions after the more ancient divergence of these clades in these two species (Pecon-Slattery et al., 2008).

The greater diversity of simian immunodeficiency viruses (SIVs) in chimpanzees relative to type 1 HIVs in humans indicates that the HIV-1 pandemic arose through the transfer of SIVs from chimpanzees about a century ago. The diversity of simian T-lymphotropic viruses in apes indicates that HTLV-1 arose in humans by similar transfers from apes over a much longer period of time. The longer amount of time for HTLV-1 evolution in humans raises the possibility that the oncogenic characteristics of HTLV-1 might have arisen recently as a result of adaptation to humans and therefore draws attention to the value of assessing whether the STLV-1 proteins have oncogenic interactions with tumor suppressor proteins, such as p53 that are similar to those documented for HTLV-1.

ASSOCIATIONS BETWEEN NONMALIGNANT NEOPLASIA AND CANCER

In wild and domesticated vertebrates, nonmalignant neoplasms are common and often associated with infection (Granoff, 1973; Theilen and Madewell, 1987). Studying such nonmalignant neoplasias may help resolve whether cancer is pervasive in nature. Oncogenesis generally begins with nonmalignant growth. In some infection-induced cancers the developmental connection between nonmalignant and malignant neoplasias is so well recognized that neoplasias are referred to as precancerous rather than benign. Cervical intraepithelial neoplasias, for example, are often referred to as a precancerous state for cervical cancer, which is caused primarily by human papillomaviruses. For any given cancer, the nonmalignant precursors should be more apparent in a population because nonmalignant neoplasias are present in an individual for a longer time and are more numerous in the population (because the proportion of nonmalignant neoplasias progressing to malignancy may be small).

For both humans and domestic animals, recognition of infectious causation and links between benign and malignant infectious tumors has been protracted and is still expanding (e.g., Alberti et al., 2010; Munday, 2014a,b). In domestic animals this process has linked actinic keratosis and papillomas with squamous cell carcinomas, hematopoietic dysplasia with acute myelogenous leukemias, endometrial hyperplasia with adenocarcinoma, and "benign" mammary tumors with malignant carcinomas

(Madewell, 1987; Sorenmo et al., 2009). These findings parallel analogous associations in human cancer pertaining to, for example, cervical cancer.

The linking of benign infectious neoplasias with squamous cell carcinomas of cows provides an illustration. Over a half century, bovine papillomaviruses were linked to papillomas and squamous cell carcinomas of the alimentary tract and bladder (Campo, 1997; Jarrett et al., 1978; Madewell and Theilen, 1987). This connection between papillomas and cancer of the alimentary tract was noticed as a result of particular permissive environment (Campo, 1997; Jarrett et al., 1978). Cattle grazing on bracken fern in the Scottish highlands showed high progression of papilloma to cancer, which was attributed to the bracken fern causing immune suppression and mutations in cows (Campo, 1997; Nasir and Campo, 2008). A similar association of papillomavirus infection with immune suppression and mutation was later expanded to explain bovine urinary bladder cancer (Nasir and Campo, 2008).

Historically, squamous cell carcinoma of horses had been attributed to oncogenic effects of chronic inflammation, exposure to ultraviolet light, and smegma (a sebaceous genital secretion) accumulation (Knight et al., 2013). Recently, *Equus caballus* papillomavirus-2 (Scase et al., 2010) was associated with both equine penile squamous cell penile carcinoma and benign penile papillomas (Knight et al., 2013; Scase et al., 2010). The histopathological linkage of the two neoplasias indicated that this papillomavirus was the cause of the cancer (Lange et al., 2013; Torres and Koch, 2013; Van Den Top et al., 2010).

Since the early 1990s, feline papillomaviruses have been isolated from benign neoplasias in domestic cats (Sundberg et al., 2000). Eventually these papilloma-infected neoplasias were shown to transform into in situ carcinoma (Ravens et al., 2013; Wilhelm et al., 2006) and a range of oral and cutaneous squamous cell carcinomas in domestic cats (Munday, 2014a,b).

One problem with understanding the spectrum of cancer in nature is definitional. Although clearly metastatic or invasive tumors can be distinguished from tumors that have no capacity to spread from the original site, the dividing line between benign and malignant neoplasias can be indistinct, especially in natural settings where the histological basis is generally less complete. Some uncertainty exists in the assignment of the accepted transition points from precancerous to cancerous neoplasia (Sleeckx et al., 2011; Sorenmo et al., 2009). The ambiguity is more problematic for neoplasias that are not yet linked to a malignancy, as illustrated by cutaneous fibromas of ungulates. These tumors, caused by papillomaviruses, commonly occur in wild populations of deer and elk (O'Banion and Sundberg, 1987; Stenlund et al., 1983), being reported from about 1% to 10% of white-tailed deer killed by hunters in the northeastern United States (Sundberg and Nielsen, 1982). These neoplasias are often referred to as benign (Moreno-Lopez et al., 1986) but can be lethal when present at many cutaneous sites in an animal as well as in the lungs (Hubbard et al., 1983a,b; Moreno-Lopez et al., 1986), where they have been interpreted as metastatic by some investigators (Hubbard et al., 1983a,b; Koller and Olson, 1971).

Current understanding of infection-associated cancers in populations of nonmammalian vertebrates is minimal. Neoplasias have been reported in 1–6% of captive lizards, often occurring in nongeriatric animals (Hernandez-Divers and Garner, 2003). They were implicated in 23% of the deaths in autopsied zoo specimens (Ramsay et al., 1996). About three-quarters of the neoplasias were considered malignant and incidence varied greatly over time, suggesting an environmental influence (Ramsay et al., 1996).

Etiologies of neoplasias in reptiles are largely unknown, but associations between viruses and neoplasias have been reported in at least two species each of snakes, lizards, and turtles (reviewed by Hernandez-Divers and Garner, 2003) (see also Chapter 2). Fibropapillomatoses in sea turtles, caused by a herpes virus, are generally considered to be nonmalignant, but tumors can be present in many different areas of the body and potentially fatal (Aguirre and Lutz, 2004). These studies draw attention to the possibility that infectious cancers could be prevalent but largely unreported in reptiles in nature.

Nonmalignant neoplasias are common in fish species. They have been associated with viral infections (Anders and Yoshimizu, 1994; Granoff, 1973; Robert, 2010), but the extent to which these apparently benign tumors progress to malignancy is poorly understood.

Invertebrate neoplasms tend to be difficult to categorize because criteria for recognizing malignancy in vertebrates may not be directly transferrable from vertebrates, and the dividing line between cancer and nonmalignant neoplasias may therefore be ambiguous (Robert, 2010). Cancer or cancer-like tumors have been reported in cnidarians, flatworms, roundworms, crustaceans, and insects (Domazet-Loso et al., 2014; Hall et al., 1986; Lightner and Hedrick, 1987; Robert, 2010; Tar and Torok, 1964). In shrimp, a metastatic, lymophma-like neoplasm has been associated with viral infection (Lightner and Brock, 1987; Lightner and Hedrick, 1987; Owens et al., 1991; Robert, 2010). Another cancer-like neoplasm of shrimp occurs in embryos, and is characterized by intranuclear inclusion bodies typical of viral infections (Lightner and Hedrick, 1987). The small number of cells and short timespan needed to generate this cancer accords with the idea that infection-induced cancers could be pervasive in nature, even in young individuals.

Pathogen-associated tumors have long been recognized in plants and algae (Braun and Stonier, 1958; Tsekos, 1982), but cancers have not been reported. The absence in this case is likely real rather than just a reflection of a reporting bias, being attributable to the absence of underlying genetic instructions for cell mobility and invasiveness in the normal functioning of plants (Ewald and Swain Ewald, 2014). In plants, pathogens therefore cannot simply unlock instructions for mobility and invasiveness as they can in animals.

JOINT CONTRIBUTIONS TO ONCOGENESIS

The identification of one cause of a cancer does not mean that other causes are not acting in concert. The joint causes could be other parasites (Ewald and Swain Ewald, 2014) or noninfectious environmental variables. Some cancers result from transmission of the cancer cell to new hosts (Murchison, 2008; Ujvari et al., 2016; Chapter 12 of this volume). Because oncogenic pathogens may facilitate the transition from a normal cell to a cancer cell, they may play a role in the evolution from normal cells to transmissible cancer cells. This possibility tends to be overlooked in studies of transmissible cell tumors, probably because pathogens are not necessary to explain the spread of a cancer in the host population. HeLa cells provide an analogy. Although they evolved from a cancer that was instigated by a papillomavirus infection, they evolved into what is essentially a transmissible cancer cell that uses lab media as its host. It still maintains much of the viral genome that made it a cancer before it was a cell-transmissible cancer. Transmissible cancer cells may similarly maintain viral genes that contribute to their oncogenic characteristics. One caveat is that immune surveillance would favor a greater reduction in expression of viral proteins in transmissible cancer cells than in HeLa cells, making any presence of infection more cryptic than in HeLa cells.

Another aspect of joint causation involves interactions between parasites and noninfectious environmental agents, such as carcinogenic chemicals and radiation. Noninfectious agents have been linked to cancer in nature (Falkmer et al., 1978; Martineau et al., 2002). They have generally been studied independently of infectious causes of cancer, but they could contribute to oncogenesis jointly with infectious agents by generating mutations or through immune suppression.

Although evidence on this matter is limited, some research suggests that noninfectious agents do interact with pathogens to increase neoplasias in wildlife. Sea turtle fibropapillomatosis, caused by a herpes virus, is more prevalent in polluted areas (Foley et al., 2005; Herbst and Klein, 1995). Polychlorinated biphenyls have immunosuppressive effects (Serdar et al., 2014) and are elevated in the blubber of sea lions with genital carcinomas, which are also caused by a herpes virus (McAloose and Newton, 2009; Ylitalo et al., 2005).

By focusing too narrowly on associations between human activities and cancer in wild animals, observers could mistakenly conclude that human influences on cancer in wild animals (e.g., through chemical pollution) are the predominant causes of cancer in nature. The recognition that both infection and environmental degradation can contribute to particular cancers leads to the conclusion that cancer may be elevated in areas of environmental degradation but may still occur in areas that are not degraded.

A BROADENING SCOPE FOR INTERVENTIONS

The recognition of infectious contributions to oncogenesis in nature widens the scope for safe and effective interventions, particularly through vaccination and interference with transmission. The value of such interventions is especially apparent for endangered species or populations of species. Both approaches have proved useful in controlling human cancers. Prophylactic vaccines, for example, are now in use to prevent infectious causes of cervical and liver cancer (Kim et al., 2015; McKee et al., 2015) and interruption of transmission has been useful in control of blood-borne and sexually transmitted oncogenic pathogens that contribute directly or indirectly to cancer, such as hepatitis viruses, human papillomaviruses, and HIV.

Vaccines have been used in domestic animals to protect against oncogenic infections (e.g., with FeLV; Louwerens et al., 2005) and could be used to protect vulnerable populations, especially when the populations are sufficiently small and/or vaccine administration sufficiently efficient. For example, vaccination of Florida panther populations against FeLV to protect against future cross-species transmissions from cats could be feasible. Interruption of transmission seems generally less feasible for animals than for humans, but could be effective. The papillomavirus that causes sarcoids in horses and cattle appears to be transmitted largely by biting flies (Finlay et al., 2009). If so, fly-control measures may therefore reduce the incidence of sarcoid in these domestic animals. If vector transmission of oncogenic pathogens occurs in wild populations, vector-control strategies may also be feasible.

More generally, the emerging understanding of the evolution of pathogen-induced oncogenesis as a byproduct of selection for pathogen persistence within hosts suggests that much of the pathogen-induced oncogenesis that will be found in natural ecosystems will involve direct effects of a relatively small subset of infectious agents rather than indirect effects from general responses to infection, such as inflammation-induced mutagenesis. If so, efforts to understand the presence and causes of cancer across species offer promise for feasible interventions to control cancer across species.

Acknowledgments

This study was supported by a fellowship from the Wissenschaftskolleg zu Berlin (Wiko) and a grant from the Rena Shulsky Foundation awarded to P.W.E.

References

Aguirre, A.A., Lutz, P.L., 2004. Marine turtles as sentinels of ecosystem health: is fibropapillomatosis an indicator? Ecohealth 1, 275–283.
Alberti, A., Pirino, S., Pintore, F., Addis, M.F., Chessa, B., Cacciotto, C., Cubeddu, T., Anfossi, A., Benenati, G., Coraddzza, E., Lecis, R., Antuofermo, E., Carcangiu, L., Pittau, M., 2010. Ovis aries papillomavirus 3: a prototype of a novel genus in the family Papillomaviridae associated with ovine squamous cell carcinoma. Virology 407, 352–359.
Anders, K., Yoshimizu, M., 1994. Role of viruses in the induction of skin tumours and tumourlike proliferations of fish. Dis. Aquat. Organ. 19, 215–232.
Barbosa, T., Zavala, G., Cheng, S., Villegas, P., 2007. Pathogenicity and transmission of reticuloendotheliosis virus isolated from endangered prairie chickens. Avian Dis. 51, 33–39.
Beatty, J., 2014. Viral causes of feline lymphoma: retroviruses and beyond. Vet. J. 201, 174–180.
Bienzle, D., 2014. FIV in cats—a useful model of HIV in people? Vet. Immunol. Immunopathol. 159, 171–179.
Bittner, J.J., 1942. The milk-influence of breast tumors in mice. Science 95, 462–463.
Blakeslee, Jr., J.R., Mcclure, H.M., Anderson, D.C., Bauer, R.M., Huff, L.Y., Olsen, R.G., 1987. Chronic fatal disease in gorillas seropositive for simian T-lymphotropic virus I antibodies. Cancer Lett. 37, 1–6.
Braun, A.C., Stonier, T., 1958. Morphology and physiology of plant tumors. Protoplasmatologia 10, 1–93.
Brown, M.A., Cunningham, M.W., Roca, A.L., Troyer, J.L., Johnson, W.E., O'Brien, S.J., 2008. Genetic characterization of feline leukemia virus from Florida panthers. Emerg. Infect. Dis. 14, 252–259.
Campo, M.S., 1997. Bovine papillomavirus and cancer. Vet. J. 154, 175–188.
Carver, S., Bevins, S.N., Lappin, M.R., Boydston, E.E., Lyren, L.M., Alldredge, M., Logan, K.A., Sweanor, L.L., Riley, S.P., Serieys, L.E., Fisher, R.N., Vickers, T.W., Boyce, W., Mcbride, R., Cunningham, M.C., Jennings, M., Lewis, J., Lunn, T., Crooks, K.R., Vandewoude, S., 2016. Pathogen exposure varies widely among sympatric populations of wild and domestic felids across the United States. Ecol. Appl. 26, 367–381.
Caulin, A.F., Maley, C.C., 2011. Peto's paradox: evolution's prescription for cancer prevention. Trends Ecol. Evol. 26, 175–182.
Domazet-Loso, T., Klimovich, A., Anokhin, B., Anton-Erxleben, F., Hamm, M.J., Lange, C., Bosch, T.C., 2014. Naturally occurring tumours in the basal metazoan Hydra. Nat. Commun. 5, 4222.
Dorn, C.R., Priester, W.A., 1987. Epidemiology. In: Theilen, G.H., Madewell, B.R. (Eds.), Veterinary Cancer Medicine. second ed. Williams & Wilkins, New York, pp. 27–52.
Effron, M., Griner, L., Benirschke, K., 1977. Nature and rate of neoplasia found in captive wild mammals, birds, and reptiles at necropsy. J. Natl. Cancer Inst. 59, 185–198.
Ewald, P.W., Swain Ewald, H.A., 2012. Infection, mutation, and cancer evolution. J. Mol. Med. (Berl.) 90, 535–541.
Ewald, P.W., Swain Ewald, H.A., 2013. Toward a general evolutionary theory of oncogenesis. Evol. Appl. 6, 70–81.
Ewald, P.W., Swain Ewald, H.A., 2014. Joint infectious causation of human cancers. Adv. Parasitol. 84, 1–26.
Ewald, P.W., Swain Ewald, H.A., 2015. Infection and cancer in multicellular organisms. Philos. Trans. R. Soc. B 370, 1–11.
Ewald, P.W., Swain Ewald, H.A., 2016. Evolution, infection and cancer. In: Alvergne, A., Jenkinson, C., Faurie, C. (Eds.), Evolutionary Thinking in Medicine—From Research to Policy and Practice. Springer International, Switzerland, pp. 191–207.

Falkmer, S., Marklund, S., Mattsson, P.E., Rappe, C., 1978. Hepatomas and other neoplasms in the Atlantic hagfish (*Myxine glutinosa*): a histopathologic and chemical study. Ann. N. Y. Acad. Sci. 298, 342–355.

Ferlay, J., Shin, H.R., Bray, F., Forman, D., Mathers, C., Parkin, D.M., 2010. Estimates of worldwide burden of cancer in 2008: GLOBOCAN 2008. Int. J. Cancer 127, 2893–2917.

Finlay, M., Yuan, Z., Burden, F., Trawford, A., Morgan, I.M., Campo, M.S., Nasir, L., 2009. The detection of bovine papillomavirus type 1 DNA in flies. Virus Res. 144, 315–317.

Foley, A.M., Schroeder, B.A., Redlow, A.E., Fick-Child, K.J., Teas, W.G., 2005. Fibropapillomatosis in stranded green turtles (*Chelonia mydas*) from the eastern United States (1980–98): trends and associations with environmental factors. J. Wildl. Dis. 41, 29–41.

Granoff, A., 1973. Herpesvirus and the Lucke tumor. Cancer Res. 33, 1431–1433.

Gross, L., 1951. "Spontaneous" leukemia developing in C3H mice following inoculation in infancy, with AK-leukemic extracts, or AK-embryos. Proc. Soc. Exp. Biol. Med. 76, 27–32.

Grulich, A.E., Vajdic, C.M., 2015. The epidemiology of cancers in human immunodeficiency virus infection and after organ transplantation. Semin. Oncol. 42, 247–257.

Hall, F., Morita, M., Best, J.B., 1986. Neoplastic transformation in the planarian: I. Cocarcinogenesis and histopathology. J. Exp. Zool. 240, 211–227.

Harrison, T.M., Mcknight, C.A., Sikarskie, J.G., Kitchell, B.E., Garner, M.M., Raymond, J.T., Fitzgerald, S.D., Valli, V.E., Agnew, D., Kiupel, M., 2010. Malignant lymphoma in african lions (panthera leo). Vet. Pathol. 47, 952–957.

Hartmann, K., 2012. Clinical aspects of feline retroviruses: a review. Viruses 4, 2684–2710.

Herbst, L.H., Klein, P.A., 1995. Green turtle fibropapillomatosis: challenges to assessing the role of environmental cofactors. Environ. Health Perspect. 103 (Suppl. 4), 27–30.

Hernandez-Divers, S.M., Garner, M.M., 2003. Neoplasia of reptiles with an emphasis on lizards. Vet. Clin. North Am. Exot. Anim. Pract. 6, 251–273.

Homma, T., Kanki, P.J., King, Jr., N.W., Hunt, R.D., O'Connell, M.J., Letvin, N.L., Daniel, M.D., Desrosiers, R.C., Yang, C.S., Essex, M., 1984. Lymphoma in macaques: association with virus of human T lymphotrophic family. Science 225, 716–718.

Hubbard, G.B., Fletcher, K.C., Schmidt, R.E., 1983a. Fibrosarcoma in a Pere David's deer. Vet. Pathol. 20, 779–781.

Hubbard, G.B., Schmidt, R.E., Fletcher, K.C., 1983b. Neoplasia in zoo animals. J. Zoo Anim. Med. 14, 33–40.

Jarrett, W.F., Mcneil, P.E., Grimshaw, W.T., Selman, I.E., Mcintyre, W.I., 1978. High incidence area of cattle cancer with a possible interaction between an environmental carcinogen and a papilloma virus. Nature 274, 215–217.

Jessup, D., 1993. Feline leukemia virus infection and renal spirochetosis in a free-ranging cougar (*Felis concolor*). J. Zoo Wildl. Med. 24, 73–79.

Kim, M.N., Han, K.H., Ahn, S.H., 2015. Prevention of hepatocellular carcinoma: beyond hepatitis B vaccination. Semin. Oncol. 42, 316–328.

King, D.P., Hure, M.C., Goldstein, T., Aldridge, B.M., Gulland, F.M., Saliki, J.T., Buckles, E.L., Lowenstine, L.J., Stott, J.L., 2002. Otarine herpesvirus-1: a novel gammaherpesvirus associated with urogenital carcinoma in California sea lions (*Zalophus californianus*). Vet. Microbiol. 86, 131–137.

Knight, C.G., Dunowska, M., Munday, J.S., Peters-Kennedy, J., Rosa, B.V., 2013. Comparison of the levels of *Equus caballus* papillomavirus type 2 (EcPV-2) DNA in equine squamous cell carcinomas and non-cancerous tissues using quantitative PCR. Vet. Microbiol. 166, 257–262.

Koller, L.D., Olson, C., 1971. Pulmonary fibroblastomas in a deer with cutaneous fibromatosis. Cancer Res. 31, 1373–1375.

Lange, C.E., Tobler, K., Lehner, A., Grest, P., Welle, M.M., Schwarzwald, C.C., Favrot, C., 2013. EcPV2 DNA in equine papillomas and in situ and invasive squamous cell carcinomas supports papillomavirus etiology. Vet. Pathol. 50, 686–692.

Lee, R.V., Prowten, A.W., Satchidanand, S.K., Srivastava, B.I., 1985. Non-Hodgkin's lymphoma and HTLV-1 antibodies in a gorilla. N. Engl. J. Med. 312, 118–119.

Leroux, C., Girard, N., Cottin, V., Greenland, T., Mornex, J.F., Archer, F., 2007. Jaagsiekte sheep retrovirus (JSRV): from virus to lung cancer in sheep. Vet. Res. 38, 211–228.

Lightner, D.V., Brock, J.A., 1987. A lymphoma-like neoplasm arising from hematopoietic tissue in the white shrimp, *Penaeus vannamei* Boone (Crustacea: Decapoda). J. Invertebr. Pathol. 49, 188–193.

Lightner, D.V., Hedrick, R.P., 1987. Embryonal carcinoma of developing embryos of grass shrimp *Palaemon orientis* (Crustacea: Decapoda). Dis. Aquat. Organ. 3, 101–106.

Little, S.E., 2005. Feline immunodeficiency virus testing in stray, feral, and client-owned cats of Ottawa. Can. Vet. J. 46, 898–901.

Louwerens, M., London, C.A., Pedersen, N.C., Lyons, L.A., 2005. Feline lymphoma in the post-feline leukemia virus era. J. Vet. Intern. Med. 19, 329–335.

Madewell, B.R., 1987. Cancer diagnosis. In: Theilen, G.H., Madewell, B.R. (Eds.), Veterinary Cancer Medicine. second ed. Williams & Wilkins, New York, pp. 3–12.

Madewell, B.R., Theilen, G.H., 1987. Etiology of cancer in animals. In: Theilen, G.H., Madewell, B.R. (Eds.), Veterinary Cancer Medicine. second ed. Williams & Wilkins, New York, pp. 13–25.

Magden, E., Quackenbush, S.L., Vandewoude, S., 2011. FIV associated neoplasms—a mini-review. Vet. Immunol. Immunopathol. 143, 227–234.

Mantovani, A., Allavena, P., Sica, A., Balkwill, F., 2008. Cancer-related inflammation. Nature 454, 436–444.

Martineau, D., Bowser, P.R., Renshaw, R.R., Casey, J.W., 1992. Molecular characterization of a unique retrovirus associated with a fish tumor. J. Virol. 66, 596–599.

Martineau, D., Lemberger, K., Dallaire, A., Labelle, P., Lipscomb, T.P., Michel, P., Mikaelian, I., 2002. Cancer in wildlife, a case study: beluga from the St. Lawrence estuary, Quebec, Canada. Environ. Health Perspect. 110, 285–292.

May, P., May, E., 1999. Twenty years of p53 research: structural and functional aspects of the p53 protein. Oncogene 18, 7621–7636.

McAloose, D., Newton, A.L., 2009. Wildlife cancer: a conservation perspective. Nat. Rev. Cancer 9, 517–526.

McCallum, H., Jones, M., 2012. Infectious cancers in wildlife. In: Aguire, A.A., Ostfeld, R.S., Dasak, P. (Eds.), New Directions in Conservation Medicine. Applied Cases of Ecological Health. Oxford, New York, pp. 270–283.

McKee, S.J., Bergot, A.S., Leggatt, G.R., 2015. Recent progress in vaccination against human papillomavirus-mediated cervical cancer. Rev. Med. Virol. 25 (Suppl. 1), 54–71.

Mitsouras, K., Faulhaber, E.A., Hui, G., Joslin, J.O., Eng, C., Barr, M.C., Irizarry, K.J., 2011. Development of a PCR assay to detect papillomavirus infection in the snow leopard. BMC Vet. Res. 7, 1–11.

Mone, J., Whitehead, E., Leland, M., Hubbard, G., Allan, J.S., 1992. Simian T-cell leukemia virus type I infection in captive baboons. AIDS Res. Hum. Retroviruses 8, 1653–1661.

Mora, M., Napolitano, C., Ortega, R., Poulin, E., Pizarro-Lucero, J., 2015. Feline immunodeficiency virus and feline leukemia virus infection in free-ranging guignas (*Leopardus guigna*) and sympatric domestic cats in human perturbed landscapes on Chiloe Island, Chile. J. Wildl. Dis. 51, 199–208.

Moreno-Lopez, J., Morner, T., Pettersson, U., 1986. Papillomavirus DNA associated with pulmonary fibromatosis in European elks. J. Virol. 57, 1173–1176.

Moss, S.F., Blaser, M.J., 2005. Mechanisms of disease: inflammation and the origins of cancer. Nat. Clin. Pract. Oncol. 2, 90–97.

Munday, J.S., 2014a. Bovine and human papillomaviruses: a comparative review. Vet. Pathol. 51, 1063–1075.

Munday, J.S., 2014b. Papillomaviruses in felids. Vet. J. 199, 340–347.

Munday, J.S., Aberdein, D., 2012. Loss of retinoblastoma protein, but not p53, is associated with the presence of papillomaviral DNA in feline viral plaques, Bowenoid in situ carcinomas, and squamous cell carcinomas. Vet. Pathol. 49, 538–545.

Munday, J.S., Kiupel, M., French, A.F., Howe, L., 2008. Amplification of papillomaviral DNA sequences from a high proportion of feline cutaneous in situ and invasive squamous cell carcinomas using a nested polymerase chain reaction. Vet. Dermatol. 19, 259–263.

Murchison, E.P., 2008. Clonally transmissible cancers in dogs and Tasmanian devils. Oncogene 27 (Suppl. 2), S19–S30.

Mwase, M., Mumba, C., Square, D., Kawarai, S., Madarame, H., 2013. Cutaneous squamous cell carcinoma presenting as a wound with discharging sinus tracts in a wild African lion (Panthera leo). J. Comp. Pathol. 149, 520–523.

Nasir, L., Campo, M.S., 2008. Bovine papillomaviruses: their role in the aetiology of cutaneous tumours of bovids and equids. Vet. Dermatol. 19, 243–254.

Nath, G., Gulati, A.K., Shukla, V.K., 2010. Role of bacteria in carcinogenesis, with special reference to carcinoma of the gallbladder. World J. Gastroenterol. 16, 5395–5404.

Nespeca, G., Grest, P., Rosenkrantz, W.S., Ackermann, M., Favrot, C., 2006. Detection of novel papillomaviruslike sequences in paraffin-embedded specimens of invasive and in situ squamous cell carcinomas from cats. Am. J. Vet. Res. 67, 2036–2041.

Niccoli, T., Partridge, L., 2012. Ageing as a risk factor for disease. Curr. Biol. 22, R741–R752.

Nicot, C., 2005. Current views in HTLV-I-associated adult T-cell leukemia/lymphoma. Am. J. Hematol. 78, 232–239.

Niphuis, H., Verschoor, E.J., Bontjer, I., Peeters, M., Heeney, J.L., 2003. Reduced transmission and prevalence of simian T-cell lymphotropic virus in a closed breeding colony of chimpanzees (Pan troglodytes verus). J. Gen. Virol. 84, 615–620.

O'Banion, M.K., Sundberg, J.P., 1987. Papillomavirus genomes in experimentally induced fibromas in white-tailed deer. Am. J. Vet. Res. 48, 1453–1455.

Owens, L., De Beer, S., Smith, J., 1991. Lymphoidal parvovirus-like particles in Australian penaeid prawns. Dis. Aquat. Organ. 11, 129–134.

Paul, T.A., Quackenbush, S.L., Sutton, C., Casey, R.N., Bowser, P.R., Casey, J.W., 2006. Identification and characterization of an exogenous retrovirus from Atlantic salmon swim bladder sarcomas. J. Virol. 80, 2941–2948.

Pecon-Slattery, J., Troyer, J.L., Johnson, W.E., O'Brien, S.J., 2008. Evolution of feline immunodeficiency virus in Felidae: implications for human health and wildlife ecology. Vet. Immunol. Immunopathol. 123, 32–44.

Peeters, M., D'Arc, M., Delaporte, E., 2014. Origin and diversity of human retroviruses. AIDS Rev. 16, 23–34.

Poli, A., Abramo, F., Cavicchio, P., Bandecchi, P., Ghelardi, E., Pistello, M., 1995. Lentivirus infection in an African lion: a clinical, pathologic and virologic study. J. Wildl. Dis. 31, 70–74.

Puente, X.S., Velasco, G., Gutierrez-Fernandez, A., Bertranpetit, J., King, M.C., Lopez-Otin, C., 2006. Comparative analysis of cancer genes in the human and chimpanzee genomes. BMC Genomics 7, 15.

Rabson, A.S., Kirschstein, R.L., 1962. Induction of malignancy in vitro in newborn hamster kidney tissue infected with simian vacuolating virus (SV40). Proc. Soc. Exp. Biol. Med. 111, 323–328.

Ramsay, E.C., Munson, L., Lowenstein, L., Fowler, M.E., 1996. A retrospective study of neoplasia in a collection of captive snakes. J. Zoo Wildl. Med. 27, 28–34.

Ravens, P.A., Vogelnest, L.J., Tong, L.J., Demos, L.E., Bennett, M.D., 2013. Papillomavirus-associated multicentric squamous cell carcinoma in situ in a cat: an unusually extensive and progressive case with subsequent metastasis. Vet. Dermatol. 24, 642–645, e161–e162.

Robert, J., 2010. Comparative study of tumorigenesis and tumor immunity in invertebrates and nonmammalian vertebrates. Dev. Comp. Immunol. 34, 915–925.

Roche, B., Hochberg, M.E., Caulin, A.F., Maley, C.C., Gatenby, R.A., Misse, D., Thomas, F., 2012. Natural resistance to cancers: a Darwinian hypothesis to explain Peto's paradox. BMC Cancer 12, 387.

Roelke, M.E., Brown, M.A., Troyer, J.L., Winterbach, H., Winterbach, C., Hemson, G., Smith, D., Johnson, R.C., Pecon-Slattery, J., Roca, A.L., Alexander, K.A., Klein, L., Martelli, P., Krishnasamy, K., O'brien, S.J., 2009. Pathological manifestations of feline immunodeficiency virus (FIV) infection in wild African lions. Virology 390, 1–12.

Rous, P., 1910. A transmissible avian neoplasm (sarcoma of the common fowl). J. Exp. Med. 12, 696–705.

Rous, P., 1911. A sarcoma of the fowl transmissible by an agent separable from the tumor cells. J. Exp. Med. 13, 397–411.

Scase, T., Brandt, S., Kainzbauer, C., Sykora, S., Bijmholt, S., Hughes, K., Sharpe, S., Foote, A., 2010. Equus caballus papillomavirus-2 (EcPV-2): an infectious cause for equine genital cancer? Equine Vet. J. 42, 738–745.

Serdar, B., Leblanc, W.G., Norris, J.M., Dickinson, L.M., 2014. Potential effects of polychlorinated biphenyls (PCBs) and selected organochlorine pesticides (OCPs) on immune cells and blood biochemistry measures: a cross-sectional assessment of the NHANES 2003–2004 data. Environ. Health 13, 114.

Shope, R.E., Hurst, E.W., 1933. Infectious papillomatosis of rabbits: with a note on the histopathology. J. Exp. Med. 58, 607–624.

Sleeckx, N., De Rooster, H., Veldhuis Kroeze, E.J., Van Ginneken, C., Van Brantegem, L., 2011. Canine mammary tumours, an overview. Reprod. Domest. Anim. 46, 1112–1131.

Sorenmo, K.U., Kristiansen, V.M., Cofone, M.A., Shofer, F.S., Breen, A.M., Langeland, M., Mongil, C.M., Grondahl, A.M., Teige, J., Goldschmidt, M.H., 2009. Canine mammary gland tumours; a histological continuum from benign to malignant; clinical and histopathological evidence. Vet. Comp. Oncol. 7, 162–172.

Stenlund, A., Moreno-Lopez, J., Ahola, H., Pettersson, U., 1983. European elk papillomavirus: characterization of the genome, induction of tumors in animals, and transformation in vitro. J. Virol. 48, 370–376.

Stroud, R.K., Amundson, T.E., 1983. Squamous cell carcinoma in a free-ranging white-tailed deer (Odocoileus virginianus). J. Wildl. Dis. 19, 162–164.

Summers, J., Smolec, J.M., Snyder, R., 1978. A virus similar to human hepatitis B virus associated with hepatitis and hepatoma in woodchucks. Proc. Natl. Acad. Sci. USA 75, 4533–4537.

Sundberg, J.P., Nielsen, S.W., 1982. Prevalence of cutaneous fibromas in white-tailed deer (Odocoileus virginianus) in New York and Vermont. J. Wildl. Dis. 18, 359–360.

Sundberg, J.P., Van Ranst, M., Montali, R., Homer, B.L., Miller, W.H., Rowland, P.H., Scott, D.W., England, J.J., Dunstan, R.W., Mikaelian, I., Jenson, A.B., 2000. Feline papillomas and papillomaviruses. Vet. Pathol. 37, 1–10.

Tar, E., Torok, L.J., 1964. Investigations on somatic twin formation benign + malignant tumours in species Dugesia tigrina (Planariidae, Turbellaria). Acta Biol. Hung. 15, 34.

Theilen, G.H., Madewell, B.R., 1987. Veterinary Cancer Medicine. Williams & Wilkins, New York.

Torres, S.M., Koch, S.N., 2013. Papillomavirus-associated diseases. Vet. Clin. North Am. Equine Pract. 29, 643–655.

Trinchieri, G., 2012. Cancer and inflammation: an old intuition with rapidly evolving new concepts. Annu. Rev. Immunol. 30, 677–706.

Tsekos, I., 1982. Tumor-like growths induced by bacteria in the thallus of a red alga, Gigartina teedii (Roth) Lamour. Ann. Bot. (Lond.) 49, 123–126.

Ujvari, B., Gatenby, R.A., Thomas, F., 2016. The evolutionary ecology of transmissible cancers. Infect. Genet. Evol. 39, 293–303.

Van Den Top, J.G., Ensink, J.M., Grone, A., Klein, W.R., Barneveld, A., Van Weeren, P.R., 2010. Penile and preputial tumours in the horse: literature review and proposal of a standardised approach. Equine Vet. J. 42, 746–757.

Vittecoq, M., Roche, B., Daoust, S.P., Ducasse, H., Misse, D., Abadie, J., Labrut, S., Renaud, F., Gauthier-Clerc, M., Thomas, F., 2013. Cancer: a missing link in ecosystem functioning? Trends Ecol. Evol. 28, 628–635.

Wilhelm, S., Degorce-Rubiales, F., Godson, D., Favrot, C., 2006. Clinical, histological and immunohistochemical study of feline viral plaques and bowenoid in situ carcinomas. Vet. Dermatol. 17, 424–431.

Williams, A.B., Schumacher, B., 2016. p53 in the DNA-damage-repair process. Cold Spring Harb. Perspect. Med. 6.

Woolford, L., O'Hara, A.J., Bennett, M.D., Slaven, M., Swan, R., Friend, J.A., Ducki, A., Sims, C., Hill, S., Nicholls, P.K., Warren, K.S., 2008. Cutaneous papillomatosis and carcinomatosis in the Western barred bandicoot (*Perameles bougainville*). Vet. Pathol. 45, 95–103.

Ylitalo, G.M., Stein, J.E., Hom, T., Johnson, L.L., Tilbury, K.L., Hall, A.J., Rowles, T., Greig, D., Lowenstine, L.J., Gulland, F.M., 2005. The role of organochlorines in cancer-associated mortality in California sea lions (*Zalophus californianus*). Mar. Pollut. Bull. 50, 30–39.

4

Pseudohypoxia: Life at the Edge

Shonagh Russell*,**, Robert A. Gatenby[†,‡], Robert J. Gillies**,[†],
Arig Ibrahim-Hashim**

*Department of Cancer Biology Ph.D. Program, University of South Florida, Tampa, FL, United States
**Department of Cancer Imaging and Metabolism, H. Lee Moffitt Cancer Center and Research
Institute, Tampa, FL, United States
[†]Department of Radiology, H. Lee Moffitt Cancer Center and Research Institute,
Tampa, FL, United States
[‡]Department of Integrated Mathematical Oncology, H. Lee Moffitt Cancer Center and Research
Institute, Tampa, FL, United States

INTRODUCTION

Pseudohypoxia is a relatively uninvestigated phenomenon wherein cells and tissues exhibit a hypoxic phenotype even in the presence of oxygen. We propose that pseudohypoxia provides an evolutionary advantage that allows tumor cells to achieve fitness maxima in stressful, often transient, microenvironments (see also the Introductory Chapter). Pseudohypoxia is often observed at the interface between invasive tumors and the surrounding stroma. Hence, characterization of the tumor edge, for example, the outline of the tumor on a magnetic resonance imaging (MRI) image (Sorensen et al., 2001), may have diagnostic and prognostic value. This region at the edge of the tumor is often highly variable both phenotypically and genotypically from the core. This is likely due to different microenvironments at the edge and core, leading to different evolutionary selection forces. Pseudohypoxia is energetically costly because it is associated with a reliance on fermentative glucose metabolism, yet cells appear to be willing to expend this cost, which presumably increases their fitness maximum and survival in hostile environments. Analyzing imaging features, for example, in computed tomography (CT) scans, has shown that not only the edge of a tumor is variable but also features can be extracted to predict prognosis, with higher variability (e.g., entropy) at the edge showing poorer prognosis (Grove et al., 2015). Radiographic imaging is a convenient modality to investigate pseudohypoxia as images are invariably obtained from every cancer patient as an aid in cancer diagnosis. Differential expression patterns at the tumor edge are additionally observed through mathematical models based on in vivo and clinical studies (Eissa et al., 2005; Estrella et al., 2013; Koelzer and Lugli, 2014; Lloyd et al., 2016). Although there are differences between primary tumors, metastases from the same primary tumor, and inter- and intratumor heterogeneity, the edge is often a unique feature in all of these presentations. The edge displays phenotypically different attributes compared to the core and these features impact on the surrounding stromal and immune components of the microenvironment. The "acid-mediated invasion hypothesis" states that expression of glycolytic proteins at the edge creates an acidic microenvironment, which untransformed stromal cells are not equipped to withstand, inducing apoptosis and resulting in space being created into which tumor cells can proliferate (Martin et al., 2010). The immune system is also affected, for example, T cells cannot switch to a glycolytic phenotype in acidic conditions and this switch is necessary for their activity (Pilon-Thomas et al., 2016). One way in which tumor cells create this acidic environment is through the Warburg effect and this acidity is enhanced by increased carbonic anhydrase nine (CA-IX) expression (Swietach et al., 2010; Warburg, 1956). The edge will employ protective strategies by upregulating sodium–hydrogen exchangers to maintain a neutral

Ecology and Evolution of Cancer. http://dx.doi.org/10.1016/B978-0-12-804310-3.00004-1

internal pH as even cancer cells cannot have an internal pH, which strays far from neutral (Ammith et al., 2015; Hulikova et al., 2013).

The involvement of acidosis in invasion appears to be confirmed by studies using buffer therapy that directly neutralizes the acidic environment and reduces the relative fitness of tumor cells, resulting in decreased metastasis and tumor burden (Ibrahim-Hashim et al., 2012). We propose that understanding of the effect of this pseudohypoxic phenotype can be used for prognosis, that it may be a universal prognostic biomarker, and furthermore that it may identify novel drug targets. The edge is playing an expensive game of evolutionary cat and mouse and if we can stunt this, its fitness will be decreased due to finding itself in an environment not suitable for proliferation and migration. Pseudohypoxia is widespread; we just need to look for it.

DISTINGUISHING PSEUDOHYPOXIA FROM HYPOXIA

As tumors invade normal stroma, they have access to the normal stromal vasculature and, hence, are generally well perfused. Despite this, cells at the invading edge display characteristics of cells that are associated with a nutrient- and oxygen-depleted state. This results in three major features being displayed by life at the edge of the tumor: (1) a phenotypic switch to an acid-producing, more proliferative cell type; (2) increased morphological variability; and (3) differential expression of proteins from those expressed by cells in the core.

Although atmospheric oxygen concentration has a partial pressure of 21% (~150 mmHg) (Poulsen et al., 2016), oxygen levels in tissues are considerably less, ca. 50 mmHg. In hypoxic tissues, oxygen levels can be immeasurably low, that is, less than 2 mmHg (0.3%). Tumors are hypoxic due to poor perfusion caused by a limited number of blood vessels or poorly functioning, leaky blood vessels (Dudley, 2012; Gillies et al., 1999). Cancer cells have adapted to survive in these hostile environments, whereas normal tissues generally have not; they have limited flexibility, whereas tumor cells have more plasticity to ensure survival in multiple different fitness landscapes. Hypoxia induces stress responses in tumor cells mediated by stabilization and accumulations of the hypoxia-inducible transcription factor 1 (HIF-1), leading to activation of its subsequent transcription targets and a switch in metabolism from oxidative to fermentative (Bensaad et al., 2014; Rodriguez-Enriquez et al., 2010; Semenza, 2000). Although tumor hypoxia is more prolonged, this switch in metabolism bears some similarities to the acute (Pasteur) effect, wherein cells switch to fermentative metabolism in the absence of oxygen. For example, such a metabolic switch is observed during

exercise in skeletal muscle when oxygen consumption exceeds supply, forcing muscle cells to switch to glycolysis in order to produce energy, resulting in a buildup of lactic acid (Adams et al., 1990; Okail, 2010). Although this acute adaptation is common throughout physiology, cancer cells express a glycolytic phenotype even in the presence of oxygen, known as the Warburg effect. When first described in 1926, Warburg hypothesized that this altered metabolic state was due to mutations within the mitochondria of cancer cells, resulting in impaired oxidative phosphorylation and forcing the switch to glycolysis for survival (Warburg et al., 1926). While this is sometimes the case, it is rare, and now it is appreciated that although glycolysis appears energy deficient on paper, it is a much faster process and produces macromolecules needed for amino acid, nucleotide, and lipid synthesis; all of these macromolecules are needed for cancer cell proliferation (Chatterjee et al., 2006; DeBerardinis et al., 2008). Cancer cells often will be glycolytic in oxygen-rich environments and therefore glycolysis can be termed one feature of a pseudohypoxic phenotype.

Pseudohypoxia is a term first coined to describe effects of hyperglycemia in diabetes, used to define the complications occurring in diabetes due to metabolites acting on the blood vessels, causing them to behave as if they were being starved of oxygen. It is also occasionally used to refer to a chemical substrate, for example, cobalt chloride (Okail, 2010) that can increase HIF-1α as if the cells were in chemical hypoxia. In the current work, we use pseudohypoxia to describe a cellular phenotype that results from evolutionary selection and, hence, must confer a selective advantage. In this review *pseudohypoxia refers to a state in which cells or tissues express hypoxia-related genes and proteins, even when there is ample oxygen present.* This pseudohypoxic phenotype is commonly seen at the edge of a tumor (the clinically defined and histological border of a tumor). The pseudohypoxic phenotype is not a single feature; it comprises multiple features including the phenotypic environment generated, morphology of the edge compared to the core, and the microscopic composition of the edge; each feature set will be discussed in turn in this review.

The clinical relevance and physiological impact that a pseudohypoxic edge has on tumor behavior can be profound. A pseudohypoxic phenotype is commonly observed at the leading edges of invasive tumors and, hence, must provide a selective advantage to these cells and, by extrapolation, to the tumor as a whole. One possible advantage could be that, during invasion, periods of hypoxia may be encountered during tissue remodeling that is inherent in invasion. It may be possible that pseudohypoxic cells are more likely to survive as they are already functioning as if they were being exposed to a lower-than-normal oxygen concentration. Pseudohypoxic phenotypes are unlikely to change during hypoxic conditions, meaning cellular resources would not need to be expended to alter responses under times of fluctuating oxygen concentration in the microenvironment.

The initiation or mechanism behind the induction of a pseudohypoxic phenotype does not appear to be due to one specific event in every tumor. We have previously observed that intermittent (but not chronic) hypoxia is selected for a pseudohypoxic phenotype where cells are more proliferative, more invasive, and more resistant to chemotherapies. This phenotype was fixed even after cells were cultured for multiple passages in normoxic conditions. Intermittent hypoxia is likely to naturally occur in tumors while they are growing and blood vessels are forming and collapsing (Robertson-Tessi et al., 2015). It is not known whether pseudohypoxic cells may already be present in a mixed tumor population and grow out via clonal selection, or whether random mutations induced by fluctuating hypoxia generate this phenotype and subsequently select for it. Pseudohypoxic phenotypes can occur through specific mutations that upregulate HIF-1α expression or prevent its degradation. In glioma, isocitrate dehydrogenase (IDH) mutants result in the reduction of alpha-ketoglutarate that is involved in degradation of HIF-1α and as a result a more pseudohypoxic, invasive phenotype is formed (Fu et al., 2012). Other mutations linked to upregulation of HIF-1α in normoxia include, inter alia, phosphatase and tensin homolog (PTEN), von Hippel–Lindau protein (pVHL), p53, epidermal growth factor (EGF), and mutant Ras and src (Pugh and Ratcliffe, 2003). It has been suggested that metabolite balance within the cell can induce a pseudohypoxic phenotype. For example, increased succinate and fumarate in succinate dehydrogenase–deficient and fumarate dehydrogenase–deficient cells has been shown to activate HIF-1α (Fu et al., 2012; Ward et al., 2012). Imbalances in nicotinamide adenine dinucleotide (NAD$^+$) levels trigger HIF-1α and downstream metabolic reprogramming and this occurs during aging but this is also possible in tumors (Menendez et al., 2014). All of these different mechanisms result in HIF-1α upregulation or stabilization. However, it must be remembered that although stabilization of HIF can be a proximal cause of pseudohypoxia, other factors come into play between this activation and the resulting phenotype of increased aerobic glycolysis.

Among other phenotypic characteristics of pseudohypoxia, a predominant pathway is that of the Warburg effect, wherein cells preferentially use glucose as a substrate to feed into glycolysis, producing lactic acid even in the presence of oxygen. Studies have shown that cells at the tumor edge that produce acid are much more likely to invade compared to non–acid-producing cells at the core (Lloyd et al., 2016). These core and edge cell types have been broadly termed "pioneers" and "engineers." Pioneers are more invasive, aggressive, proliferative,

and at a higher risk of cell death, whereas engineers are more quiescent, induce angiogenesis, and have limited resources. The pioneers are more glycolytic and this produces a more acidic microenvironment at the tumor edge (Lloyd et al., 2016).

There are repercussions, benefits, and trade-offs associated with having an acidic microenvironment surrounding the tumor. An acidic microenvironment affects the tumor cells themselves, the immune system, and the stroma, including fibroblasts. Estrella et al. (2013) studied the ability of an acidic environment to increase invasion of tumor using dorsal window chambers. Increased regional acidity was correlated with an increased regional invasive potential of the tumor cells. Importantly, the invasive behavior was abrogated when pH was neutralized with buffers. This increased invasion of tumor cells seen in low-pH conditions is thought to occur because acid will induce apoptosis in the surrounding stromal cells, inhibit immune attack, and provide favorable conditions for tumor cells to proliferate.

Acid has been shown to induce release of lysosomal proteases (cathepsins) and induce matrix metalloproteinases (MMPs), resulting in degradation of the matrix and space into which tumor cells can proliferate (Kato et al., 2005; Rothberg et al., 2013). Extracellular matrix remodeling through cell apoptosis and degradation results in increased space, oxygen, and nutrients for the invading cancer cells. However, in order to successfully invade, the invading edge must also inhibit attack from the immune system.

Multiple cell types in the immune system are affected by pH. For example, T cells require glycolysis for their activity and it has been shown that acidic pH will prevent a T cell from metabolically switching from oxidative phosphorylation to glycolysis (Pilon-Thomas et al., 2016). Notably, the acidic microenvironment does not affect the intracellular pH and, as yet, the mechanism by which cells sense an acidic pH is unknown. Lactic acid has been shown to polarize macrophages to a tumor-promoting (M2) phenotype, which assists in tumor invasion (Colegio et al., 2014). The immune component of the tumor is frequently observed to be hampered by the tumor microenvironment itself and further studies are required on all individual components of the immune system, especially when immunotherapy is emerging as an important cancer treatment. Therefore, this state of pseudohypoxia and its identification within tumors could be extremely clinically relevant, not only to target but also for understanding where our limitations with current immunotherapy treatments lie. How can we identify pseudohypoxia and differences at the tumor edge and core without patient histological samples or in vivo studies? Two tools at our disposal are mathematical models and imaging modalities to observe the physical difference between the tumor edge and the core.

PSEUDOHYPOXIA: TUMOR EDGE VERSUS CORE

Although tumors may be initially spherical, over time and with clonal expansion driven by selective pressures, tumors often become irregular in shape and contain multiple clones (Marusyk et al., 2012). Heterogeneity within tumors is relevant to outcome, and may be a proximal cause of therapy resistance. Furthermore, tumor heterogeneity explicitly dictates that single biopsies may not accurately reflect the complexity of a tumor as a whole (Janku, 2014). Nevertheless, although this intratumor heterogeneity is vitally important, the difference between the edge of the tumor and the core is an undervalued piece of data currently not being utilized to its full potential. In early stage cancers, for example, ductal carcinoma in situ (DCIS), mathematical models and image analytics may have the potential to distinguish those that progress from those that do not. In the clinic there are limitations with regards to DCIS prognosis (Nofech-Mozes et al., 2005) as there are no reliable biomarkers that can be used to distinguish progression. We propose that expression of pseudohypoxic markers in subregions of DCIS can predict the existence of a subsequently invasive phenotype. Identifying DCIS progressors is only one potential outcome; there are numerous different tumor types that could benefit from this analysis. Often tumors have variable tumor edges, some being smoother and rounder, others being more irregular, and within that tumor the cell density varies as a result of necrosis, proliferation, and nutrient and gaseous gradients throughout the tumor. We know that larger tumors can become necrotic in the center due to inadequate nutrient supply, a hypoxic environment, and an inability to remove waste and these larger tumors are more metabolically heterogeneous.

It is important to realize that there is still heterogeneity at the tumor edge, not every region of the tumor is exposed to the same microenvironmental stress, and each region will evolve according to its fitness and surrounding environmental pressure. This allows for different regions of the tumor to expand until conditions are favorable for other sections of the same tumor. These irregularities result in different adaptations resulting in an irregular edge that can be observed in magnetic resonance (MR) and CT images. Using CT as a way to identify whether this is clinically feasible, a few studies have shown that using defined features extracted from CT images, lung tumors could be quantitatively described using two features: tumor shape complexity and intratumor density variation (Grove et al., 2015; Suo et al., 2016). These independent studies observed that non-small cell lung cancer (NSCLC) tumors with a more irregular shape and a large tumor shape complexity feature score had a poorer prognosis.

These studies simply compared the boundary (edge) of the tumor to the core. CT imaging is a modality that can be used to describe habitats and tumor heterogeneity through specific features. Asking simple questions about shape (e.g., does a more irregular shape result in poorer prognostic outcome?) is something that could be initiated as a feature of all CT scans. They identified the following through this study: features at different lung cancer stages were significantly different from each other, tumor cores with high entropy were likely due to necrosis and heterogeneity, and tumor edges with relatively higher heterogeneity (entropy) resulted in decreased patient survival. These simple quantitative imaging biomarkers comparing the edge and the edge versus core of the tumor could be used in the clinic as additional diagnostic and prognostic tools.

Although CT scans can show an irregular tumor edge, histology is needed to definitively define stage and subtype of tumor, in addition to genetic screening for targeted therapies. This ability to dissect images for prognosis is not meant to be a tool to replace but a tool to aid with patient prognosis and diagnosis. It has the potential to be expanded to those that treatment does not work for; it is possible that nonresponders have varying physiological features that play a role in drugs being ineffective or in developing drug resistance. This is an intriguing prospect that needs to be followed up on.

Other studies have investigated solitary pulmonary nodules and could distinguish, using CT textural features, which nodules would most likely be malignant based on the edge-to-core heterogeneity (Suo et al., 2016). A study in 1991 showed T2-weighted MR images of uterine leiomyomas had a greater signal intensity at the border and this was related to a more blood vessel, lymphatic system–rich region around the tumor edge (Mittl et al., 1991). This could suggest that those with a high-intensity tumor edge are more likely to be aggressive and progress. Ultrasound has even been used to distinguish breast tumors with rough edges to suggest a malignant phenotype over a smooth edge, suggesting a benign tumor (Su et al., 2011).

Tumors are not altering the edge without benefit. Therefore there is an evolutionary benefit to these cells to have a pseudohypoxic edge that is different from the core. It allows the tumor to grow and survive in different conditions. If we take a look at examples in nature, we can see that they parallel extremely well with tumors. The invasive delicate skink was introduced to Lord Howe Island in the 1980s and subsequently over time, seven distinct haplotypes were observed. These haplotypes were usually found in specific regions of the island, likely due to the genetic fitness advantage gained in the surrounding environment (Moule et al., 2015). This is similarly seen in tumors with evolutionary pressure resulting in expression of genotypes and phenotypes suited to the surrounding microenvironment because if the cells do not evolve, they will die. The edge of the tumor is derived under different selective pressures compared to the core. This pressure drives selection for a phenotype that favors migration and invasion, allowing the edge to grow faster and invade into the stroma and surrounding tissue. Cane toads (native to Central and South America) were introduced to Australia over 70 years ago (Shine, 2012). They subsequently thrived due to a lack of natural predators or disease susceptibilities. The cane toad is toxic and, in a similar way, tumors keep their predators, that is, the immune system, at bay. One of the "toxins" secreted by tumors is an acidic barrier that differentially harms the local environment in favor of tumor growth.

The core, unlike the edge, is a more stable region of a tumor; it does not appear to evolve at a fast enough rate to withstand pressures, such as hypoxia and often this will induce cell death. The core is more oxidative so protection from acidosis is not as vital. Thus, we speculate that energy is not wasted in producing proteins to maintain a neutral intracellular pH due to a low external pH. Often invasive species will induce genetic and phenotypic changes in the surrounding species, for example, on small islands in the Bahamas, introduction of an invasive rat species caused loss of genetic diversity within the endemic brown anole lizard population, over a single generation (Gasc et al., 2010). This is mimicked in cancer with the acidic environment stimulating rapid adaptation in fibroblasts that can result in metabolic cooperation between cancer cells and cancer-associated fibroblasts (Rattigan et al., 2012).

The edge consumes large amounts of glucose and produces acid, outcompeting other cells in the vicinity. Studies have shown that vasculature is increased at the tumor edge (Naito et al., 2012), which would not only provide nutrients but also remove some of the lactic acid waste produced by glycolysis. Interpreting these observations in light of evolutionary dynamics, such a microenvironment might represent stromal reactivity in an attempt to reduce the invasive potential of tumors.

Even tumors have a limit to how much acid they can stand before the cost is too great. Environmental pressures are known to induce species to adapt and evolve in much a similar way that microenvironmental pressures cause cancer to evolve. The peppered moth is one such example; in England during the 1950s, coal factories produced a large amount of black smoke that caused the habitats of a predominantly white species of moth to become black. This change in habitat resulted in predators having an increased chance of being successful in hunting the moths. However, a black melanic form of the same moth was then selected for as it would blend into the background of the blackened trees. The pressures from the environment selected for this color

as it provided a fitness advantage (Cook and Saccheri, 2013). This is observed at the edge of tumors as a glycolytic, acid-producing phenotype is selected for over an oxidative one, with multiple benefits including, but not limited to, suppression of immune attack. Although nature provides great examples that can be employed when looking at tumor evolution, in order to study and understand this tumor edge variation and the production of a pseudohypoxic phenotype, mathematical modeling must be employed to gain clinical efficacy.

MODELING OF PSEUDOHYPOXIA

Mathematical modeling—whether it is ordinary differential equations, partial differential equations, agent-based models, or another model—can be heralded for aiding in understanding multiple parameters of tumor growth, metastasis, and drug resistance (Beerenwinkel et al., 2015). Although in vivo and clinical studies are the gold standard in cancer research, these are costly and complex; models can provide a range of scenarios to understand multiple parameters within the tumor and its microenvironment in a high-throughput manner that can interrogate a wide range of starting conditions and responses. Moreover, modeling can not only inform studies but also identify nuances that may not be readily apparent. Evolutionary game theory predicts that there must be a benefit to cost trade-off that results in cellular and tissue fitness. Thus, a tumor evolving a pseudohypoxic phenotype must have a benefit, such as that seen in glioma (Basanta et al., 2008).

In order for a tumor to evolve to a fitness maximum, clonal evolution occurs through *heritable* phenotypic diversity in combination with microenvironmental selection. Phenotypic diversity can result from genetic or epigenetic alterations and microenvironmental selection pressures change during tumor growth and with the addition of therapy. The tumor–stromal interface provides a unique microenvironment that selects for cells at the tumor edge to be more proliferative and loosely packed with an invasive potential. In contrast, the environment at the core selects for cells that grow to a greater cell density and are less proliferative. These phenotypes result in a Darwinian trade-off for cells in tumors depending on their location: fecundity (edge) versus survivorship (core), and this variation produces a variable tumor–host interface, one being a pseudohypoxic phenotype tumor edge (Lloyd et al., 2016). Other game theory models have shown that development of a glycolytic phenotype can result in invasive subclones, through frequency-dependent clonal selection, and these mutants (likely pseudohypoxic) have increased motility promoting an invasive phenotype (Archetti, 2015). This model lends itself well to the argument of pseudohypoxia being a combination of features that drive the invasive edge to be more variable compared to the core.

Other cell-based mathematical models, including a study using a hybrid cellular automata, have strengthened the argument that selective pressure is generated over time due to persistent environmental pressure, resulting in an aggressively invasive, glycolytic phenotype. This cellular phenotype will migrate to the tumor edge and aid invasion into the tissue (Robertson-Tessi et al., 2015). Glycolysis results in an acidic microenvironment and decreasing acidity in tumor models results in decreased overall tumor fitness, including reduced cellular proliferation and invasion. This decreased fitness is likely due to the loss of protection at the edge and the resulting internal core being exposed to immune infiltration. These models have been shown to gain greater understanding of the effects of oxygen concentration on tumors, avascular tumors, angiogenesis, and the role of the immune system (Bellomo et al., 2004; Delgado-SanMartin et al., 2015; Jiang et al., 2005; Kolobov and Kuznetsov, 2013; Martins et al., 2007). If we can further understand the pseudohypoxic phenotype, parsimonious models could be used to identify a minimum set of features necessary for the pseudohypoxic phenotype, which features are the most beneficial to target, and which cancers would be most susceptible to tumor death as a result of loss of tumor edge integrity. To inform such models and gain greater understanding, we need to define the microscopic features of the edge compared to those of the core.

PSEUDOHYPOXIA AT THE MICROSCOPIC LEVEL

Mathematical models need to be informed and validated, pharmaceutical companies need targets, and clinicians need prognostic and diagnostic features. Microscopically, the edge is a unique landscape that can address all of these needs. Through mathematical modeling, game theory, experimental studies, and histological analysis of patient samples, it has been shown that there are certain features and proteins that are more often observed at the edge and conversely the core in multiple tumor types. We expect that these differences provide an evolutionary advantage to the tumor, as they are commonly observed, enabling survival in a hostile environment with protection for the core.

Proteins of interest commonly observed at the edge of invading tumors include glucose transporter 1 (GLUT-1), CA-IX, and Ki-67. Notably, GLUT-1 and CA-IX are HIF client proteins, although the tumor edge is well vascularized and, hence, not hypoxic (Figs. 4.1 and 4.2). The core, on

Protein	Location	Resulting tumor phenotype
HIF-1α	Edge and core	• Aggressive • Metastatic • Glycolytic
CA-IX	Edge	• Acid producing • Intracellular buffering • Invasive
GLUT-1	Edge	• Increased glucose uptake • Glycolytic • Acid producing
CA-XII	Core	• Acid production • Decreased invasiveness
Cleaved caspase 3	Core	• Apoptotic • Necrotic tissue

FIGURE 4.1 Proteins observed in a pseudohypoxic phenotype, the location found, and the phenotype exhibited by the tumor expressing the protein. *CA-IX*, Carbonic anhydrase nine; *CA-XII*, carbonic anhydrase XII; *GLUT-1*, glucose transporter 1; *HIF-1α*, hypoxia-inducible transcription factor 1α.

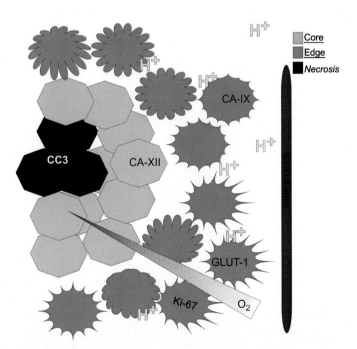

FIGURE 4.2 Schematic representing a pseudohypoxic phenotype tumor with a necrotic core, high acidity, and invasive edge. *CA-IX*, Carbonic anhydrase nine; *CA-XII*, carbonic anhydrase XII; *CC3*, cleaved caspase-3; *GLUT-1*, glucose transporter 1.

the other hand, is enriched for proteins, such as carbonic anhydrase XII (CA-XII) and cleaved caspase-3 (CC3).

TUMOR EDGE

Carbonic Anhydrase IX

CA-IX is a zinc metalloenzyme that catalyzes the reversible hydration of CO_2 into H^+ and HCO_3^- (Tafreshi et al., 2012). It is often upregulated and a negative prognostic factor in a number of cancers, including ovarian, pancreatic, medullary thyroid, breast, lung, glioblastoma, and multiple other cancer types (Chafer and Dedhar, 2015; Choschzick et al., 2011; Grandane et al., 2015; Ihnatko et al., 2006; Li et al., 2016; Lou et al., 2011; Pore et al., 2015; Tafreshi et al., 2012; Takacova et al., 2014). In pancreatic cancer, short hairpin RNA (shRNA) targeted toward CA-IX blocks tumor proliferation and initiation, making CA-IX a very attractive clinical target, not only in pancreatic but in the aforementioned cancer types as well (Pore et al., 2015). CA-IX is not typically expressed in normal tissues, except for the gastrointestinal tract, and it is an acidifying enzyme that has the ability to function at a lower pH than other CA isoforms, which is probably why it is useful in the gastrointestinal tract. CA-IX's protein structure is similar to the rest of the carbonic anhydrase family except for an exofacial proteoglycan (PG)-like domain, which presumably acts as a buffer for its enzymatic active site, lowering the pK_a and enabling it to function at an acidified pH (Alterio et al., 2009). Recently CA-IX, through modeling and histological analysis, has been observed at the edge of tumors, adding to the pseudohypoxic phenotype. This is thought to collaborate in generating and maintaining an acidified tumor microenvironment in concert with upregulated glycolysis through GLUT-1 overexpression (Lloyd et al., 2016). Not only does CA-IX stimulate extracellular acid production, but it also has been suggested that the bicarbonate ions produced can be shuttled back into the cell to neutralize the intracellular pH (Svastova et al., 2012).

CA-IX is often used as a hypoxia marker, as it is a HIF client protein. However, both published and unpublished data have shown that CA-IX is not invariably associated with hypoxia and, hence, can be differentially classified as a marker of hypoxia or pseudohypoxia, depending on its location (Mayer et al., 2005). CA-IX can be upregulated by multiple mechanisms; most prominent is upregulation by HIF-1α; however, HIF-1α can also be stabilized in the absence of hypoxia, such as in conditions of αKG depletion described earlier. CA-IX is observed in both the hypoxic periluminal region and the nonhypoxic basement membrane-adjacent regions of DCIS, which are thus pseudohypoxic (Verduzco et al., 2015). CA-IX can be upregulated by phosphoinositide 3-kinase (PI3K) and mitogen-activated protein kinase (MAPK) pathways, irrespective of hypoxia status (Choschzick et al., 2011). Studies in ovarian cancer, where PI3K is often amplified, showed CA-IX overexpression in endometrioid and mucinous ovarian cancer subtypes. This increased CA-IX expression was associated with a decreased overall survival (Choschzick et al., 2011). Extracellular signal-regulated kinase (ERK) specifically has been shown to be involved in CA-IX promoter activation and induction of transcription at the hypoxia response element (HRE) and pathogenesis-related (PR1) gene promoter regions (Kaluz et al., 2006). Further analysis of these data and other histological studies is needed to provide insight into how widespread these pseudohypoxic and CA-IX–expressing regions are. Even simple cell:cell contact through increased density has been shown to induce CA-IX expression; if the tumor core is very densely packed, this could result in induction of CA-IX expression and then migration to the tumor periphery forming the CA-IX pseudohypoxic edge (Kaluz et al., 2006). CA-IX has additionally been shown to be regulated by microrchidia-2 (MORC-2) by decreasing acetylation of histone H3 at the CA-IX promoter, overall decreasing mRNA and protein expression. Therefore, any mutations in MORC-2 have the potential to upregulate CA-IX expression (Shao et al., 2010). Within medullary thyroid cancer, Ret protooncogene (RET) is frequently upregulated or amplified and RET is linked to HIF-1α activation and downstream increased CA-IX expression. Moreover, c-Src proto-oncogene (Src) has been linked to the regulation of CA-IX (Takacova et al., 2010); both Src and CA-IX have been associated with more metastatic phenotypes (Irby and Yeatman, 2000). Finally CA-IX is known to be upregulated when cells are exposed to acidic conditions in glioblastoma (Ihnatko et al., 2006). Thus, at the tumor edge where cells are glycolytic and producing acid, the decreased pH could result in a phenotypic switch inducing CA-IX expression.

Although CA-IX can be upregulated and maintained by multiple mechanisms, why is it commonly observed at the leading edge? Importantly, CA-IX is associated with a more aggressive and invasive phenotype. Studies have linked CA-IX to be involved with migration at the leading edge of invading tumors and a decreased extracellular pH can influence the focal contacts of the tumor, increasing invasiveness (Csaderova et al., 2013). When CA-IX is inhibited, it results in reduced cellular proliferation leading to decreased tumor growth (Chiche et al., 2009). CA-IX is indicative of a pioneer phenotype as is GLUT-1. One of the reasons our group believes that CA-IX is a relevant target is due to studies using buffer therapy to control and inhibit tumor growth. For example, a study in transgenic prostate cancer (TRAMP) mice showed that bicarbonate-containing water, systemic "buffer therapy," could neutralize tumor acidity and inhibit both tumor growth and metastasis, with the forethought being that specifically targeting CA-IX with a drug could potentially have the same effect as buffer therapy (Ibrahim-Hashim et al., 2012). Clinical trials with sodium bicarbonate buffer therapy have been unsuccessful due to poor patient compliance; a targeted therapy is likely to have more success. Multiple compounds have been produced targeting CA-IX with promising results (Dubois et al., 2013; Lou et al., 2011; Mahon et al., 2015; Moeker et al., 2014; Pala et al., 2014; Rami et al., 2013; Wichert and Krall, 2015; Wichert et al., 2015; Yang et al., 2015); however, clinical trials for these have not yet been completed.

A recent metaanalysis has shown that high CA-IX expression over multiple cancer types resulted in worse treatment outcome, worse overall survival, decreased disease-free survival, decreased metastasis and progression-free survival, and an increase in the likelihood of a locoregional occurrence (van Kuijk et al., 2016). An interesting study by Ledaki et al. (2015) lends itself toward the pseudohypoxic phenotype, as they found that hypoxia promotes epigenetic regulation of CA-IX, resulting in formation of both inducible and noninducible CA-IX–expressing populations. This noninducible, constitutive form is linked to the pseudohypoxic phenotype, likely brought on by intermittent hypoxia exposure, as we have observed previously (Verduzco et al., 2015). Cells with constitutive CA-IX exhibited stem-cell–like markers with an increased self-renewal capacity and may be related to increased survival at the tumor edge more readily than the inducible CA-IX–expressing populations (Ledaki et al., 2015).

Interestingly, we argue that CA-IX at the tumor edge promotes a more aggressive phenotype and that the resulting acidosis can affect neighboring stromal populations and repel an immune attack. Expression of CA-IX is also associated with granulocyte-colony stimulating factor (G-CSF) mobilization through upregulation of the NF-kappa B pathway and results in mobilization of myeloid-derived suppressor cells to lung metastatic niches, priming them for tumor formation (Chafer and Dedhar, 2015). Interestingly, CA-IX expression is higher in metastatic tumors compared to that in primary tumors and this may be associated with CA-IX priming metastatic sites and exhibiting a more invasive, aggressive phenotype compared to the core of the tumor (Pertega-Gomes et al., 2015).

Glucose Transporter 1

The acidic environment is in large part the result of lactic acid excretion from the tumor generated by glycolysis. To upregulate glycolysis, cells need to increase their glucose uptake by upregulating the expression of GLUT-1. Tumor growth at the leading edge of tumors requires not only sufficient energy but also the macromolecules to build and move forward. Glycolysis provides both of these at a high rate compared to oxidative phosphorylation (Zheng, 2012). Indeed, a study by Young et al. (2011) showed that decreased GLUT-1 transporter expression did not affect cellular ATP levels, but significantly affected lipid synthesis and growth on plastic, suggesting a role of glycolysis in providing building blocks for macromolecular synthesis.

GLUT-1 is a HIF client gene product that is observed in both hypoxic and oxygenated conditions, making it a relevant marker for hypoxia/pseudohypoxia, depending on its location. As mentioned earlier, cancer cells often metabolize glucose at high rates, even in the presence of sufficient oxygen, known as aerobic glycolysis or the Warburg effect. A key regulatory component of glycolytic flux is upregulation of glucose transporters so that membrane transport will not become rate limiting in times of high demand. GLUT-1 is the most commonly observed glucose transporter isoform observed to be highly expressed at the edge of tumors in both models and by immunohistochemistry (IHC) (Carvalho et al., 2011; Lloyd et al., 2016). GLUT-1 expression is higher at the leading edge of the tumor and localizes to invasive regions of the tumor periphery (Estrella et al., 2013; Lloyd et al., 2016). Furthermore, GLUT-1 expression is often found adjacent to stromal regions with high CD31 staining, a blood vessel marker, suggesting that these regions are not hypoxic and yet the pseudohypoxic phenotype is observed.

Elevated expression can be attributed to at least five different intrinsic pathways, including, inter alia, HIF, myc, src, p53/TIGAR, and mitochondrial defects (Gillies et al., 2008). Activation of PI3K and MAPK pathways also can lead to HIF-1α stabilization, resulting in elevated GLUT-1 expression (Fraga et al., 2015). GLUT-1 is often observed in more aggressive tumors (Kunkel et al., 2003), where it is associated with poor prognosis. However, other glucose transporter family members, notably GLUT-3, can facilitate glucose uptake and promote a pseudohypoxic phenotype, as GLUT-1 is not seen in all types of cancer (Carvalho et al., 2011). GLUT-1 can be extrinsically regulated by a number of hormones, including EGF, estrogen (Macheda et al., 2005), insulin (Zorzano et al., 1989), insulin-like growth factor, and follicle-stimulating hormone (FSH), which could be stimulated or coopted by the tumor to stimulate GLUT-1 upregulation (Kodaman and Behrman, 1999; Macheda et al., 2005; Zorzano et al., 1989).

Clinically, GLUT-1 expression correlates highly with increased 18-fluorodeoxyglucose (FDG) uptake in tumors by positron emission tomography (PET) imaging (Li et al., 2015a). Elevated uptake of FDG and upregulation of GLUT-1 are observed in a number of cancers including bladder, prostate, breast, and pancreatic and are associated with poorer prognosis (Chan et al., 2016; Chandler et al., 2003; Hussein et al., 2011; Whyard et al., 2016). Therefore, through multiple mechanisms GLUT-1 is already seen to be a clinically relevant target and could be one of the most important features in a pseudohypoxic phenotype by generating an acidic microenvironment.

Ki-67

Ki-67 is a commonly used stain for cellular proliferation and is often seen elevated at the edge of invasive tumors by IHC (Dowsett et al., 2011; Li et al., 2015b; Lloyd et al., 2016; Rubio, 2006). Increased Ki-67 expression has been linked to worse prognosis and increased likelihood of metastasis in prostate cancer (Green et al., 2016). In breast cancer, expression of Ki-67 in the primary tumor has been suggested as an independent significant factor for reduced overall survival (Nishimura et al., 2014). Hence, a tumor edge with high Ki-67 staining may be more likely to be invasive, metastatic, and chemoresistant. Thus, there are many, likely interrelated components expressed at the edge of invasive tumors, making it an attractive target to overcome resistance and sensitize to immune responses. Targeting one or multiple components of this unique microenvironment could reduce fitness of cells at the leading edge, driving the tumor into an evolutionary corner.

TUMOR CORE

In parallel to the tumor edge the core exhibits a very different phenotype, one that does not appear to have the ability to be fit in more adverse environments. Indeed, the core can be characterized by regions that are well perfused, adjacent to regions with poor perfusion and high necrosis. The interface between these two populations may impact high evolutionary selection for cells with edge-like characteristics.

Carbonic Anhydrase XII

CA-XII is in the same family as CA-IX and catalyzes the same reversible reaction, reversible hydration of CO_2 into H^+ and HCO_3^-. It is more active at neutral pH as it does not have a PG-like domain to shield its active site. It is known to form homodimers in cellular membranes and appears to be ubiquitously expressed throughout tumors, including the core (Lloyd et al., 2016; Whittington et al., 2001). Unlike CA-IX, CA-XII has been seen to be a positive prognostic factor in certain types of cancer, including breast and lung (Ilie et al., 2011; Watson et al., 2003). However, this is not universal, as CA-XII can also be a negative prognostic factor in, for example, gliomas and oral squamous cell carcinomas (Chien et al., 2012; Haapasalo et al., 2008; Ilie et al., 2011; Watson et al., 2003). It may be that the prognostic value for CA-XII depends on its intratumoral location, which has not been adequately explored, that is, its spatial relationship in conjunction with other pseudohypoxic phenotypes, such as increased glycolysis, acidification, and upregulation of Ki-67 at the tumor edge.

Caspase-3

Proteolytic cleavage, and hence activation of the protease, CC3, occurs during programmed cell death (apoptosis) and is often seen in tumors under microenvironmental or genotoxic stress, often in areas adjacent to "necrotic" cores. Necrosis and CC3 have been linked to increased aggressiveness and poor prognosis in, for example, endometrial cancer (Bredholt et al., 2015).

CONCLUSIONS

This review has focused on the concept of pseudohypoxia, which has not commonly been applied to tumor growth. Evidence presented illustrates that pseudohypoxia can be relevant to understanding cancer progression, prognosis, diagnosis, and treatment. Importantly, the relevance of pseudohypoxia can be appreciated only when considering its intratumoral location. Thus, the mere presence of a gene product in a tumor can have variable prognostic and predictive power depending on the microenvironment in which it is expressed. Investigating this will require a significant amount of further work, combining mesoscopic radiographic imaging, microscopic IHC, and mathematical models. Such studies are needed to identify how widespread this pseudohypoxic phenotype is and how to exploit this information for improved prognosis and to identify novel therapeutic strategies. It is also not known whether pseudohypoxic states are seen in liquid cancer tumors. On one hand, it is unlikely as they are often less heterogeneous and do not require an invasive leading edge for success but, on the other hand, their "home" is in the bone marrow, which in cancers is highly heterogeneous in oxygenation and perfusion. Greater understanding of the relevance of this variable edge among different tumor types could identify which tumors would benefit from a loss of tumor edge integrity or which would survive.

References

Adams, G.R., Foley, J.M., Meyer, R.A., 1990. Muscle buffer capacity estimated from pH changes during rest-to-work transitions. J. Appl. Physiol. 69, 968–972.

Alterio, V., et al., 2009. Crystal structure of the catalytic domain of the tumor-associated human carbonic anhydrase IX. Proc. Natl. Acad. Sci. USA 106 (38), 16233–16238.

Ammith, S.R., et al., 2015. Na+/H+ exchange in the tumour microenvironment: does NHE1 drive breast cancer carcinogenesis? Int. J. Dev. Biol. 59, 367–377.

Archetti, M., 2015. Heterogeneity and proliferation of invasive cancer subclones in game theory models of the Warburg effect. Cell Prolif. 48 (2), 259–269.

Basanta, D., et al., 2008. Evolutionary game theory elucidates the role of glycolysis in glioma progression and invasion. Cell Prolif. 41 (6), 980–987.

Beerenwinkel, N., et al., 2015. Cancer evolution: mathematical models and computational inference. Syst. Biol. 64 (1), e1–e25.

Bellomo, N., Bellouquid, A., Delitala, M., 2004. Mathematical topics on the modelling complex multicellular systems and tumor immune cells competition. Math. Models Methods Appl. Sci. 14, 1683.

Bensaad, K., et al., 2014. Fatty acid uptake and lipid storage induced by HIF-1alpha contribute to cell growth and survival after hypoxia–reoxygenation. Cell Rep. 9 (1), 349–365.

Bredholt, G., Mannelqvist, M., Stefansson, I.M., Birkeland, E., Bø, T.H., Øyan, A.M., Trovik, J., Kalland, K.H., Jonassen, I., Salvesen, H.B., Wik, E., Akslen, L.A., 2015. Tumor necrosis is an important hallmark of aggressive endometrial cancer and associates with hypoxia, angiogenesis and inflammation responses. Oncotarget 6, 39676–39691.

Carvalho, K.C., et al., 2011. GLUT1 expression in malignant tumors and its use as an immunodiagnostic marker. Clinics 66 (6), 965–972.

Chafer, S., Dedhar, S., 2015. Carving out its niche: a role for carbonic anhydrase IX in pre-metastatic niche development. Oncoimmunology 4 (12), e1048955.

Chan, A.K., Bruce, J., Siriwardena, A.K., 2016. Glucose metabolic phenotype of pancreatic cancer. World J. Gastroenterol. 22 (12), 3471–3485.

Chandler, J.D., et al., 2003. Expression and localization of GLUT1 and GLUT12 in prostate carcinoma. Cancer 97 (8), 2035–2042.

Chatterjee, A., Mambo, E., Sidransky, D., 2006. Mitochondrial DNA mutations in human cancer. Oncogene 25 (34), 4663–4674.

Chiche, J., et al., 2009. Hypoxia-inducible carbonic anhydrase IX and XII promote tumor cell growth by counteracting acidosis through the regulation of the intracellular pH. Cancer Res. 69 (1), 358–368.

Chien, M.H., et al., 2012. Tumor-associated carbonic anhydrase XII is linked to the growth of primary oral squamous cell carcinoma and its poor prognosis. Oral Oncol. 48 (5), 417–423.

Choschzick, M., et al., 2011. Overexpression of carbonic anhydrase IX (CA-IX) is an independent unfavorable prognostic marker in endometrioid ovarian cancer. Virchows Arch. 459 (2), 193–200.

Colegio, O.R., et al., 2014. Functional polarization of tumour-associated macrophages by tumour-derived lactic acid. Nature 513 (7519), 559–563.

Cook, L.M., Saccheri, I.J., 2013. The peppered moth and industrial melanism: evolution of a natural selection case study. Heredity (Edinb.) 110 (3), 207–212.

Csaderova, L., et al., 2013. The effect of carbonic anhydrase IX on focal contacts during cell spreading and migration. Front. Physiol. 4, 271.

DeBerardinis, R.J., et al., 2008. The biology of cancer: metabolic reprogramming fuels cell growth and proliferation. Cell Metab. 7 (1), 11–20.

Delgado-SanMartin, J.A., et al., 2015. Oxygen-driven tumour growth model: a pathology-relevant mathematical approach. PLoS Comput. Biol. 11 (10), e1004550.

Dowsett, M., et al., 2011. Assessment of Ki67 in breast cancer: recommendations from the International Ki67 in Breast Cancer working group. J. Natl. Cancer Inst. 103 (22), 1656–1664.

Dubois, L., et al., 2013. Targeting carbonic anhydrase IX by nitroimidazole based sulfamides enhances the therapeutic effect of tumor irradiation: a new concept of dual targeting drugs. Radiother. Oncol. 108 (3), 523–528.

Dudley, A.C., 2012. Tumor endothelial cells. Cold Spring Harb. Perspect. Med. 2 (3), a006536.

Eissa, S., et al., 2005. Real-time PCR hTERT mRNA pattern in tumor core, edge, resection margin, and lymph nodes in laryngeal tumors: relation to proliferative index and impact on prognosis. Clin. Biochem. 38 (10), 873–878.

Estrella, V., et al., 2013. Acidity generated by the tumor microenvironment drives local invasion. Cancer Res. 73 (5), 1524–1535.

Fraga, A., et al., 2015. Hypoxia and prostate cancer aggressiveness: a tale with many endings. Clin. Genitourin. Cancer 13 (4), 295–301.

Fu, Y., et al., 2012. Glioma derived isocitrate dehydrogenase-2 mutations induced up-regulation of HIF-1alpha and beta-catenin signaling: possible impact on glioma cell metastasis and chemo-resistance. Int. J. Biochem. Cell Biol. 44 (5), 770–775.

Gasc, A., et al., 2010. Invasive predators deplete genetic diversity of island lizards. PLoS One 5 (8), e12061.

Gillies, R.J., et al., 1999. Causes and effect of heterogeneous perfusion in tumors. Neoplasia 1 (3), 197–207.

Gillies, R.J., Robey, I., Gatenby, R.A., 2008. Causes and consequences of increased glucose metabolism of cancers. J. Nucl. Med. 49 (Suppl. 2), 24S–42S.

Grandane, A., et al., 2015. 6-substituted sulfocoumarins are selective carbonic anhydrase IX and XII inhibitors with significant cytotoxicity against colorectal cancer cells. J. Med. Chem. 58, 3975–3983.

Green, W.J., et al., 2016. KI67 and DLX2 predict increased risk of metastasis formation in prostate cancer—a targeted molecular approach. Br. J. Cancer 115, 236–242.

Grove, O., et al., 2015. Quantitative computed tomographic descriptors associate tumor shape complexity and intratumor heterogeneity with prognosis in lung adenocarcinoma. PLoS One 10 (3), e0118261.

Haapasalo, J., et al., 2008. Identification of an alternatively spliced isoform of carbonic anhydrase XII in diffusely infiltrating astrocytic gliomas. Neuro Oncol. 10 (2), 131–138.

Hulikova, A., et al., 2013. Regulation of intracellular pH in cancer cell lines under normoxia and hypoxia. J. Cell. Physiol. 228 (4), 743–752.

Hussein, Y.R., et al., 2011. Glut-1 expression correlates with basal-like breast cancer. Transl. Oncol. 4 (6), 321–327.

Ibrahim-Hashim, A., et al., 2012. Systemic buffers inhibit carcinogenesis in TRAMP mice. J. Urol. 188 (2), 624–631.

Ihnatko, R., et al., 2006. Extracellular acidosis elevates carbonic anhydrase IX in human glioblastoma cells via transcriptional modulation that does not depend on hypoxia. Int. J. Oncol. 29, 1025–1033.

Ilie, M.I., et al., 2011. Overexpression of carbonic anhydrase XII in tissues from resectable non-small cell lung cancers is a biomarker of good prognosis. Int. J. Cancer 128 (7), 1614–1623.

Irby, R.B., Yeatman, T.J., 2000. Role of src expression and activation in human cancer. Oncogene 19, 5636–5642.

Janku, F., 2014. Tumor heterogeneity in the clinic: is it a real problem? Ther. Adv. Med. Oncol. 6 (2), 43–51.

Jiang, Y., et al., 2005. A multiscale model for avascular tumor growth. Biophys. J. 89 (6), 3884–3894.

Kaluz, S., Kaluzova, M., Stanbridge, E.J., 2006. The role of extracellular signal-regulated protein kinase in transcriptional regulation of the hypoxia marker carbonic anhydrase IX. J. Cell. Biochem. 97 (1), 207–216.

Kato, Y., et al., 2005. Acidic extracellular pH induces matrix metalloproteinase-9 expression in mouse metastatic melanoma cells through the phospholipase D-mitogen-activated protein kinase signaling. J. Biol. Chem. 280 (12), 10938–10944.

Kodaman, P.H., Behrman, H.R., 1999. Hormone-regulated and glucose-sensitive transport of dehydroascorbic acid in immature rat granulosa cells. Endocrinology 140, 3659–3665.

Koelzer, V.H., Lugli, A., 2014. The tumor border configuration of colorectal cancer as a histomorphological prognostic indicator. Front. Oncol. 4, 29.

Kolobov, A.V., Kuznetsov, M.B., 2013. The study of angiogenesis effect on the growth rate of an invasive tumor using a mathematical model. Russ. J. Numerical Anal. Math. Model. 28, 471–483.

Kunkel, M., et al., 2003. Overexpression of Glut-1 and increased glucose metabolism in tumors are associated with a poor prognosis in patients with oral squamous cell carcinoma. Cancer 97 (4), 1015–1024.

Ledaki, I., et al., 2015. Carbonic anhydrase IX induction defines a heterogeneous cancer cell response to hypoxia and mediates stem cell-like properties and sensitivity to HDAC inhibition. Oncotarget 6, 19413–19427.

Li, J, et al., 2015a. Is carbonic anhydrase IX a validated target for molecular imaging of cancer and hypoxia? Future Oncol. 11, 1531–1541.

Li, L.T., Chen, Q., Zheng, J.N., 2015b. Ki67 is a promising molecular target in the diagnosis of cancer (review). Mol. Med. Rep. 11, 1566–1572.

Li, Y., et al., 2016. Roles of carbonic anhydrase IX in development of pancreatic cancer. Pathol. Oncol. Res. 22, 277–286.

Lloyd, M.C., et al., 2016. Darwinian dynamics of intratumoral heterogeneity: not solely random mutations but also variable environmental selection forces. Cancer Res. 76 (11), 3136–3144.

Lou, Y., et al., 2011. Targeting tumor hypoxia: suppression of breast tumor growth and metastasis by novel carbonic anhydrase IX inhibitors. Cancer Res. 71 (9), 3364–3376.

Macheda, M.L., Rogers, S., Best, J.D., 2005. Molecular and cellular regulation of glucose transporter (GLUT) proteins in cancer. J. Cell. Physiol. 202 (3), 654–662.

Mahon, B.P., Pinard, M.A., McKenna, R., 2015. Targeting carbonic anhydrase IX activity and expression. Molecules 20 (2), 2323–2348.

Martin, N.K., et al., 2010. Tumour-stromal interactions in acid-mediated invasion: a mathematical model. J. Theor. Biol. 267 (3), 461–470.

Martins, M.L., Ferreira, S.C., Vilela, M.J., 2007. Multiscale models for the growth of avascular tumors. Phys. Life Rev. 4 (2), 128–156.

Marusyk, A., Almendro, V., Polyak, K., 2012. Intra-tumour heterogeneity: a looking glass for cancer? Nat. Rev. Cancer 12 (5), 323–334.

Mayer, A., Hockel, M., Vaupel, P., 2005. Carbonic anhydrase IX expression and tumor oxygenation status do not correlate at the microregional level in locally advanced cancers of the uterine cervix. Clin. Cancer Res. 11 (20), 7220–7225.

Menendez, J.A., Alarcon, T., Joven, J., 2014. Gerometabolites: the pseudohypoxic aging side of cancer oncometabolites. Cell Cycle 13 (5), 699–709.

Mittl, Jr., R.L., Yeh, I.T., Kressel, H.Y., 1991. High signal intensity rim surrounding uterine leiomyomas on MR images: pathologic correlation. Radiology 180, 81–83.

Moeker, J., et al., 2014. Structural insights into carbonic anhydrase IX isoform specificity of carbohydrate-based sulfamates. J. Med. Chem. 57 (20), 8635–8645.

Moule, H., et al., 2015. A matter of time: temporal variation in the introduction history and population genetic structuring of an invasive lizard. Curr. Zool. 61, 456–464.

Naito, H., et al., 2012. Changes in blood vessel maturation in the fibrous cap of the tumor rim. Cancer Sci. 103 (3), 433–438.

Nishimura, R., et al., 2014. Prognostic significance of Ki-67 index value at the primary breast tumor in recurrent breast cancer. Mol. Clin. Oncol. 2 (6), 1062–1068.

Nofech-Mozes, S., et al., 2005. Prognostic and predictive molecular markers in DCIS. Adv. Anat. Pathol. 12, 256–264.

Okail, M.S.A., 2010. Cobalt chloride, a chemical inducer of hypoxia-inducible factor-1α in U251 human glioblastoma cell line. J. Saudi Chem. Soc. 14 (2), 197–201.

Pala, N., et al., 2014. Carbonic anhydrase inhibition with benzenesulfonamides and tetrafluorobenzenesulfonamides obtained via click chemistry. ACS Med. Chem. Lett. 5 (8), 927–930.

Pertega-Gomes, N., et al., 2015. A glycolytic phenotype is associated with prostate cancer progression and aggressiveness: a role for monocarboxylate transporters as metabolic targets for therapy. J. Pathol. 236, 517–530.

Pilon-Thomas, S., et al., 2016. Neutralization of tumor acidity improves antitumor responses to immunotherapeutic interventions. Cancer Res. 76, 1381–1390.

Pore, N., et al., 2015. In vivo loss of function screening reveals carbonic anhydrase IX as a key modulator of tumor initiating potential in primary pancreatic tumors. Neoplasia 17 (6), 473–480.

Poulsen, C.J., Tabor, C., White, J.D., 2016. Long-term climate forcing by atmospheric oxygen concentrations. Science 353, 132.

Pugh, C.W., Ratcliffe, P.J., 2003. Regulation of angiogenesis by hypoxia: role of the HIF system. Nat. Med. 9, 677–684.

Rami, M., et al., 2013. Hypoxia-targeting carbonic anhydrase IX inhibitors by a new series of nitroimidazole-sulfonamides/sulfamides/sulfamates. J. Med. Chem. 56 (21), 8512–8520.

Rattigan, Y.I., et al., 2012. Lactate is a mediator of metabolic cooperation between stromal carcinoma associated fibroblasts and glycolytic tumor cells in the tumor microenvironment. Exp. Cell Res. 318 (4), 326–335.

Robertson-Tessi, M., et al., 2015. Impact of metabolic heterogeneity on tumor growth, invasion, and treatment outcomes. Cancer Res. 75 (8), 1567–1579.

Rodriguez-Enriquez, S., et al., 2010. Oxidative phosphorylation is impaired by prolonged hypoxia in breast and possibly in cervix carcinoma. Int. J. Biochem. Cell Biol. 42 (10), 1744–1751.

Rothberg, J.M., et al., 2013. Acid-mediated tumor proteolysis: contribution of cysteine cathepsins. Neoplasia 15 (10), 1125–1137.

Rubio, C.A., 2006. Cell proliferation at the leading invasive front of colonic carcinomas. Preliminary observations. Anticancer Res. 26, 2275–2278.

Semenza, G.L., 2000. HIF-1: mediator of physiological and pathophysiological responses to hypoxia. J. Appl. Physiol. 88, 1474–1480.

Shao, Y., et al., 2010. Involvement of histone deacetylation in MORC2-mediated down-regulation of carbonic anhydrase IX. Nucleic Acids Res. 38 (9), 2813–2824.

Shine, R., 2012. Invasive species as drivers of evolutionary change: cane toads in tropical Australia. Evol. Appl. 5 (2), 107–116.

Sorensen, A.G., et al., 2001. Comparison of diameter and perimeter methods for tumor volume calculation. J. Clin. Oncol. 19, 551–557.

Su, Y., et al., 2011. Automatic detection and classification of breast tumors in ultrasonic images using texture and morphological features. Open Med. Inform. J. 5, 26–37.

Suo, S., et al., 2016. Assessment of heterogeneity difference between edge and core by using texture analysis: differentiation of malignant from inflammatory pulmonary nodules and masses. Acad. Radiol. 23, 1115–1122.

Svastova, E., et al., 2012. Carbonic anhydrase IX interacts with bicarbonate transporters in lamellipodia and increases cell migration via its catalytic domain. J. Biol. Chem. 287 (5), 3392–3402.

Swietach, P., et al., 2010. New insights into the physiological role of carbonic anhydrase IX in tumour pH regulation. Oncogene 29 (50), 6509–6521.

Tafreshi, N.K., et al., 2012. Noninvasive detection of breast cancer lymph node metastasis using carbonic anhydrases IX and XII targeted imaging probes. Clin. Cancer Res. 18 (1), 207–219.

Takacova, M., et al., 2010. Src induces expression of carbonic anhydrase IX via hypoxia-inducible factor 1. Oncol. Rep. 23, 869–874.

Takacova, M., et al., 2014. Expression pattern of carbonic anhydrase IX in medullary thyroid carcinoma supports a role for RET-mediated activation of the HIF pathway. Am. J. Pathol. 184 (4), 953–965.

van Kuijk, S.J., et al., 2016. Prognostic significance of carbonic anhydrase IX expression in cancer patients: a meta-analysis. Front. Oncol. 6, 69.

Verduzco, D., et al., 2015. Intermittent hypoxia selects for genotypes and phenotypes that increase survival, invasion, and therapy resistance. PLoS One 10 (3), e0120958.

Warburg, O., 1956. On the origin of cancer cells. Science 123 (3191), 309–314.

Warburg, O., Wind, F., Negelein, E., 1926. Liber den Stoffwechsel von Tumoren im Korper. Klin. Wochenschr. 5, 829–832.

Ward, P.S., et al., 2012. Identification of additional IDH mutations associated with oncometabolite $R(-)$-2-hydroxyglutarate production. Oncogene 31 (19), 2491–2498.

Watson, P.H., et al., 2003. Carbonic anhydrase XII is a marker of good prognosis in invasive breast carcinoma. Br. J. Cancer 88 (7), 1065–1070.

Whittington, D.A., et al., 2001. Crystal structure of the dimeric extracellular domain of human carbonic anhydrase XII, a bitopic membrane protein overexpressed in certain cancer tumor cells. Proc. Natl. Acad. Sci. USA 98 (17), 9545–9550.

Whyard, T., et al., 2016. Metabolic alterations in bladder cancer: applications for cancer imaging. Exp. Cell Res. 341 (1), 77–83.

Wichert, M., Krall, N., 2015. Targeting carbonic anhydrase IX with small organic ligands. Curr. Opin. Chem. Biol. 26, 48–54.

Wichert, M., et al., 2015. Dual-display of small molecules enables the discovery of ligand pairs and facilitates affinity maturation. Nat. Chem. 7 (3), 241–249.

Yang, X., et al., 2015. Imaging of carbonic anhydrase IX with a 111In-labelled dual motif inhibitor. Oncotarget 6, 33733–33742.

Young, C.D., et al., 2011. Modulation of glucose transporter 1 (GLUT1) expression levels alters mouse mammary tumor cell growth in vitro and in vivo. PLoS One 6 (8), e23205.

Zheng, J., 2012. Energy metabolism of cancer: glycolysis versus oxidative phosphorylation (review). Oncol. Lett. 4 (6), 1151–1157.

Zorzano, A., et al., 1989. Insulin-regulated glucose uptake in rat adipocytes is mediated by two transporter isoforms present in at least two vesicle populations. J. Biol. Chem. 264, 12358–12363.

5

The Genomic Landscape of Cancers

Henry H. Heng

Center for Molecular Medicine and Genetics, Wayne State University School of Medicine,
Detroit, MI, United States; Department of Pathology, Wayne State University School of Medicine,
Detroit, MI, United States; Karmanos Cancer Institute, Detroit, MI, United States

INTRODUCTION

With the rapid progress of the cancer genome sequencing project, the "cancer genome landscape" has been mainly described using DNA-based molecular characterization (Wood et al., 2007). Typically, the cancer genome landscape refers to the distributional pattern of gene mutations across the genome in a given cancer type. Even though such a landscape is often illustrated based on data collected from patient populations, which reflect the intertumor pattern, it is increasingly being used to profile the intratumor pattern. Furthermore, metaanalyses and/or pan-cancer analyses are now available with which to compare mutation landscapes across all cancer types (Williams et al., 2016), and to integrate various-omics data (The Cancer Genome Atlas Research Network, 2013).

For the majority of cancer types, the overall landscape consists of a small number of "mountains" (commonly altered genes detected in a high percentage of tumors) and a much larger number of "hills" (genes altered in limited cases) (Vogelstein et al., 2013), plus a massive amount of rare mutations (nonclonal type), many of which are often under the radar due to their low frequencies.

Due to the fact that (1) different individual tumors display different landscapes, (2) different parts of the same tumor vary, (3) different stages of the same part vary, and (4) the overall genomic landscape is often drastically changed following drug treatment, there seems to be no fixed cancer genome. The dynamics of cancer genomic landscapes suggest that cancers are complex, adaptive systems, which ultimately challenge efforts to identify the targetable patterns of the cancer genome landscape (Heng, 2007a, 2016a). Clearly, the immediate task is to establish a full picture of the multiple levels of genomic landscapes of cancer, using the framework of a complex adaptive system and genome-based cancer ecology and evolution. To achieve this goal, some basic concepts regarding the nature of inheritance, the pattern of somatic evolution, and the mechanism of genetic heterogeneity need to be integrated.

KEY FEATURES AND CHALLENGES OF THE CANCER GENOMIC LANDSCAPE

The Gene Mutation Landscape

To highlight some of the most interesting and surprising findings of the gene mutation landscape revealed by current genome sequencing efforts, the key aspects given in the next subsections are briefly summarized and discussed. Further analyses can be found in Heng (2016a).

Too Many Gene Mutations but Too Few Common Driver Mutations

Across all cancer types, there are many gene mutations. However, the majority of them are diverse "passenger mutations." Only a very few common gene mutations ("mountains") are shared by various cancer types. Similarly, a "long-tailed pattern" of "rare" mutation is observed across the majority of individual cancer types. Such highly heterogeneous findings of gene mutation patterns are one of the biggest disappointments for the Cancer Genome Atlas project, as its initial goal was to identify common key mutations (The Cancer Genome Atlas Network, 2015).

To deal with this puzzling abundance of mutations, detected mutations are divided as "drivers" and "passengers." Driver aberrations confer a selective advantage. In contrast, "incidental" passenger aberrations do not have growth advantage or have only a negligible effect (Gerlinger et al., 2014; Stratton et al., 2009). Such

Ecology and Evolution of Cancer. http://dx.doi.org/10.1016/B978-0-12-804310-3.00005-3

classification, however, is based on the stepwise evolutionary pattern of cancer, and needs to be reexamined if the cancer evolutionary pattern is not stepwise (Heng, 2007a, 2009, 2016a).

In addition to the high mutation rate, the overall mutation load is drastically different among different cancer types, and the median frequency of nonsynonymous mutations varied by over 1000-fold among all cancer types (Lawrence et al., 2013). Adult cancers with a clear association with environmental factors (mutagens) display the highest rate of mutations (e.g., lung cancer and melanoma display >130 nonsynonymous mutations per tumor), followed by adult tumors without dominant environmental factors (breast, prostate, and ovarian cancers display 30–80 mutations per tumor), and then by the liquid type, which is similar to pediatric cancers (<10–20 per tumor) (Vogelstein et al., 2013). Despite such a pattern, however, there is a high degree of intertumor and intratumor heterogeneity. Surprisingly, such variations of mutation frequencies across patients within the same cancer type can span three orders of magnitude. For example, for melanoma and lung cancer, which represent high mutation rates, the frequency spans from 0.1 to 100 Mb^{-1}, while for acute myeloid leukemia (AML), which represents a low rate of mutation, the frequencies range from 0.01 to 10 Mb^{-1}. Moreover, the mutation rate can also be drastically variable within the same genome. Based on whole-genome sequencing data obtained from 126 tumor–normal pairs representing 10 tumor types, the difference across the same genome was found to be more than five-fold (Lawrence et al., 2013).

Paradoxically, despite the high mutation rates and diverse mutation loads, for many solid tumors there are fewer driver mutations than expected, and especially a lack of oncogene drivers. This is rather disappointing as a central goal of cancer genome analysis is the identification of cancer driver gene mutations (Stratton et al., 2009). For many individual tumors, there is zero to one driver oncogene. In contrast, for established cancer cell lines, more oncogenes are observed, suggesting a possible selective condition through which cell lines can favor proliferation genes. The lack of key driver mutations also directly contradicts the cancer gene mutation theory, which predicts that five to eight "hits" are required for cancer formation. In fact, the "multiple hits" concept not only fit the age distributions of cancer patients (Armitage and Doll, 1954) but also served as the key conceptual basis for searching for these sequential cancer genes in the first place. More bizarrely, for some normal tissues, there are many typical cancer driver mutations (Martincorena et al., 2015).

Some Interesting Emergent Patterns

The distribution pattern of mutations differs between some well-characterized oncogenes and tumor suppressor genes. Oncogenes display recurrent mutations at identical amino acid positions (influenced by key functional domains). In contrast, tumor suppressor genes are mutated throughout their entire protein, as more positions can be mutated or truncated to cause a loss of function (Vogelstein et al., 2013).

Many reports have highlighted the relative proportions of the six different possible base-pair substitutions (Alexandrov and Stratton, 2014; Garraway and Lander, 2013; Nik-Zainal et al., 2012). The rationale of performing such analysis is based on the idea that the signature of mutation spectrum can offer a mechanistic understanding of how cancer formed, and that approximately 95% of mutations from cancer genomes belongs to this type. Surely, some interesting patterns are observed using certain selection criteria. For example, lung cancers share a mutational spectrum dominated by C>A mutations, which can be explained by their exposure to the polycyclic aromatic hydrocarbons in tobacco smoking (Pleasance et al., 2010b). More impressive is the pattern of mutation observed in melanoma, which uniquely reflects the frequent C>T mutations resulting from the misrepair of ultraviolet-induced covalent bonds between adjacent pyrimidines (Pleasance et al., 2010a). Despite the fact that different mutational spectra can be linked to different cancer types, heterogeneity makes it hard to understand the common mechanism for each cancer type. For lung cancer, there is a disconnection between the overall mutation spectrum and some specific genetic alterations, such as the effects of DNA repair, tandem duplication in a specific gene, and chromosomal translocations. In addition, even though the rate of C>A mutation is high (34%), there are still other types of mutations; together, C>T and T>C mutations account for nearly 40%. Another example is the difficulty of explaining breast cancer using the mutation spectrum. In fact, based on an across-cancer–type comparison, at least 25 types of mutation spectrum (or processes) have been identified. However, half of these patterns cannot be explained (Stratton, 2013). Moreover, even though a single gene mutation spectrum can be established for many individual carcinogens (Nik-Zainal et al., 2015), the real challenge is to establish the informational integration of the whole-genome spectra, where dynamic interactions rather than individual genes are important. The essential but less specific impacts of individual parts on the whole system behavior within macrocellular evolution need to be seriously reconsidered (Heng, 2016a).

The Validation and Clinical Limitation

The cancer genome sequencing project has confirmed most of the previously known cancer genes as driver mutations. The number of true driver gene mutations is less than 200–550, or less than 1–3% of all human genes (based on Bert Vogelstein's and Michael Stratton's estimations, respectively) (Stratton, 2013; Vogelstein, 2011;

Vogelstein et al., 2013). With an ever larger number of additional cases being sequenced, and especially as the concept of a driver gene mutation is modified, we anticipate that the number of contributing gene mutations will continue to increase.

Limited new driver genes have been discovered. However, there are many known cancer genes that have been extended to new cancer types. In addition, many newly discovered driver genes involve metabolism, chromatin remodeling, protein homeostasis, and the immune system. These new classes of genes do not belong to the classic cancer genes, which agrees with the involvement of diverse contributing factors and the importance of system constraint in cancer evolution (Heng, 2007a, 2016a,b). Interestingly, many biological considerations and computational programs have been used to narrow down the true driver gene mutations to solve the problem of large numbers of "false-positive mutations." For example, the pattern of mutation can be more important than mutation frequencies when identifying drivers. According to Vogelstein et al. (2013), a driver oncogene needs to have >20% of the recorded missense mutations in the gene at a recurrent position, while a tumor suppressor driver requires that >20% of the recorded mutations in the gene result in inactivation. A similar effort to reduce the number of drivers can be found in lung cancer analysis by different approaches. A sophisticated method, MutSigCV (which identifies genes that are significantly mutated in cancer genomes, using a model with mutational covariates), was used to select drivers from a large number of gene mutations. By integrating mutation frequency/spectrum, gene-specific background mutation rates, expression profiles, and replication times, this program was able to select 11 genes from a list of 450 (Lawrence et al., 2013). Interestingly, only 1 novel gene was identified, as the remaining 10 genes were previously known. The power of this analysis once again paradoxically reduces the significance of whole-genome sequencing efforts in terms of new cancer gene discovery.

Time is an important dimension in the mutational landscape, despite the fact that it is hard to integrate, since most sequencing data of solid tumors are from end products of cancer evolution. Even though some liquid cancers display more stepwise features, mutation profiles featuring a clear accumulation of gene mutations are much less common than we had believed. For instance, the sequential accumulation of APC, TP53, and KRAS in driving colon cancer accounts for only a small portion of cases, as the frequency of tumors that have all three mutations is lower than frequencies of tumors that have none of these three mutations (Heng, 2016a). When examining the key transitions of cancer initiation and progression, the mechanisms seem very diverse. For example, from preneoplastic phase to tumor, the number of mutations can double, but it is hard to identify the common gene mutations for this transition. For some tissue types, a high degree of gene mutation can be detected from normal tissue. From primary tumor to metastasis, the pattern of gene mutation has yet to be identified despite many intensive efforts, as they often display similar gene mutation profiles. When comparing targeted sequencing on normal, primary, and metastatic tissue from colorectal cancer patients, there is a high degree of concordance between early occurring and recurrent mutations, and these key drivers (KRAS, NRAS, and BRAF mutations) are always identical in the primary and metastatic tumors (Brannon et al., 2014). Not surprisingly, the mutation landscape is often altered following drug treatment, which poses the ultimate challenge for the targeting therapy based on known mutation landscapes (Tan et al., 2015). It should also be pointed out that the time window of cancer formation has been studied using mathematical model based on gene mutation patterns (Tomasetti et al., 2015). Now, with the knowledge of the key involvement of chromosomal changes, further modification is necessary.

To solve the challenge posed by the numerous, diverse gene mutations of cancer, Vogelstein et al. (2013) have linked many driver genes to 12 pathways that provide selective growth advantages for the cell. However, it is one thing to classify different gene mutations into different categories; it is another thing to predict how pathway switching can occur and what the clinical consequences will be based on the highly dynamic landscape. A recent analysis using various genome-wide platforms and 1 proteomic platform on 3527 specimens from 12 cancer types revealed a unified classification into 11 major subtypes (Hoadley et al., 2014). This study suggests that "cell-of-origin" rather than pathway-based features dominate the molecular taxonomy of diverse tumor types. The question is the following: how can this newly identified molecular taxonomy be used for therapy when there are many types of cancer, and many more different responses for even one type? Biologists are not used to appreciating the buildup of complexity based on adding the number of elements involved. When the number of pathways increases, the game of complexity can become too much to handle. Who can ignore the three-body problem (Weinberg, 2014)? The three-body problem represents a well-known challenge in physics: when calculating the specific positions, masses, and velocities of three mutually interacting bodies at a certain point in time, it is nearly impossible to predict the precise motions of these three objects, which are generally nonrepeating. The implication of this concept to cancer is that when many factors are concerned, it is very difficult to predict the dynamic relationships among them. By the way, most of the passenger mutations need to be considered if we truly accept the concept of cancer evolution. As pointed out by Vogelstein's group in the same article, heterogeneity appears largely confined to passenger gene mutations. If we have to deal with the heterogeneity issue, then the key is to understand the behavior of these so-called passengers. It is now known that these so-called passenger mutations can become drivers when the environment changes (Heng, 2016a; Horne et al., 2015b).

There are limited clinical implications of the gene mutation landscape. One key question to ask is how useful these gene mutation patterns can be in the clinic. While there are successful examples of targeting driver mutations in the clinic, so far, many impressive tumor responses have not translated into overall survival. Similarly, the commonly mutated gene profile in pancreatic cancer is unlikely to be the primary determinant of very long-term survival following resection (Dal Molin et al., 2015). Furthermore, for some types of cancers, such as ovarian cancer and cutaneous melanoma, there seems to be no significant clinical outcome correlation with genomic landscape or classification (The Cancer Genome Atlas Network, 2015). The limitations from using gene mutation profiles to predict clinical outcomes seem obvious. Mutations in DNA are not the key determining factors of cancer when the macrocellular evolutionary process is needed. Furthermore, the genetic information is fuzzy and there is a loose correlation between the genotype and the phenotype (Heng, 2016a). Interestingly, in cutaneous melanoma studies, samples classified by transcriptomic subclass associated with immune function can be linked to improved patient survival. Based on the recent exciting development of immunotherapy, we need to investigate the good outcomes of immunotherapy based on genomic landscapes.

Epigenomic Landscape

The importance of the epigenetic landscape in cancer is based on its unique features: the connection of tissue types and the "cell of origin," the cellular microenvironment, its sensitivity to various stresses, and, ultimately, the essential linkage to cellular states and genome instability-mediated cancer evolution (Heng et al., 2009; Roadmap Epigenomics Consortium, 2015). Different from the "core" somatic genomes of each individual with relative genetic stability (note that, however, recent studies have uncovered an increased degree of somatic genetic variations; see Heng, 2016a), the epigenomic landscape of each cell can vary substantially (from changes in histone modification and DNA methylation, which alter DNA accessibility, resulting in distinct gene expression programs and cellular states).

It is known that abnormal DNA methylation and histone modifications play an important role in cancer (Jones and Baylin, 2007), and the connection between the epigenomic landscape and states of cancer has been receiving increasing attention (Huang et al., 2009) (see also Chapter 6). By combining the concepts of "epigenetic landscape" and "fitness landscape," Huang (2013) has united development and somatic evolution as the drivers of the relentless increase in malignancy. However, because of the domination of gene mutation theory, it was debatable whether epigenetic alterations were causal drivers or simply a noncausal correlation of the cancerous state. With the identification or confirmation of a few dozen genes of epigenomic regulators that display highly recurrent somatic variations across multiple cancer types, this issue seems settled (Garraway and Lander, 2013). According to the criterion of defining a driver gene mutation, potential epigenetic driver genes have been characterized in tumor types and subtypes. For example, the epigenetic silencing of key tumor suppressor, regulatory, and repair genes has been identified in various cancers, which are also associated with the interruption of some important cellular signaling pathways (DNA repair, RB1/CDK4 cell cycle regulation, Wnt/β-catenin, TGF-β, and cellular differentiation pathways), control of replication timing and nuclear architecture, and epigenetic regulation of repeat elements (Podlaha et al., 2012; Weisenberger and Liang, 2015).

There are large numbers of publications focusing on the epigenetic landscape of cancer. In addition to histone modification, chromatin remodeling/compaction, RNA splicing, noncoding RNA characterization, and the activity of endogenous retroviruses have generated high interest (for further details see Chapter 6). There are some issues that need to be addressed to further advance the field, however:

1. Despite the fact that epigenetic regulation plays an important role in global gene regulation, the majority of analyses are still linked in their scope to individual genes or specific pathways (by characterizing epigenetic effects on enhancers, promotors, gene bodies, and noncoding regions). While the specific gene-related epigenetic signature is useful to study some pathways in relative isolation, making sense of the global epigenetic pattern is more important and challenging, as evolutionary selection is based on the "package of an entire system."

2. Epigenetic changes are much more sensitive than gene mutations to environmental influence, and the connection between epigenetic changes and chromosomal instability (CIN) is obvious (Heng et al., 2009). Both gene mutations and epigene aberrations can contribute to cancer evolution by providing the necessary genetic and nongenetic variations. In fact, years before a tumor enters into the invasive stage, the epigenome is drastically altered already. However, to generate a new system,

genome-level change must be achieved, as simply modifying the system by gene mutation and/or epigene regulation is not enough. To date, limited study has been done to illustrate the mechanisms of how epigenomic landscape contributes to karyotype evolution. For example, there is a misconception based on the explanation of Mintz's classic observations that the cancer genome can be reversed or corrected simply by changing the microenvironment (Heng, 2016a; Illmensee and Mintz, 1976). Yes, the reversal to normal phenotype from single teratocarcinoma cells can be achieved using nuclear transfer technology, and teratoma-free animals can be produced with many tumor-derived normal tissues. This amazing observation strongly supports the notion that epigenetic reprograming is the key for understanding cancer. However, as pointed out by the authors, a near-normal chromosome complement is sufficient (important) for total restoration of orderly gene expression in a normal embryonic condition (Illmensee and Mintz, 1976). This later conclusion is highly significant, as it points out the genome-level constraint for epigenetic plasticity (Heng, 2016a). Sure enough, nearly 30 years later, with reanalysis of a similar reversal phenomenon with some of these original cell lines, genetic lesions rather than epigenetic restrictions are highlighted (Blelloch et al., 2004). Clearly, when discussing the epigenetic impact on cancer, there is an elephant in the room—the karyotype, which defines the global landscape (Heng et al., 2011b, 2013a; Huang, 2013). More meaningful integration is thus necessary.

3. Increased attention is being paid to study the high-order structure of the chromosomes and its linkage to epigenetic function. Various high-resolution maps of three-dimensional chromatin interaction have been generated to illustrate the chromatin contribution to epigenetics (Jin et al., 2013). Despite these technical advances [including chromatin immunoprecipitation (ChIP), methylation-sensitive restriction enzyme digestion (MRE), methylated DNA immunoprecipitation (MeDIP), and various RNA profiling methods], the challenge is to gain a systematic and a holistic understanding of how the epigenomic landscape contributes to lineage specification and heterogeneity-mediated plasticity; it is also necessary to better understand cellular circuitry and its switching, and how epigenetic variation benefits cellular adaptation but at the same time leads to cancer as a trade-off (Horne et al., 2014). Obviously, focusing on the entire genome is the key. To achieve this goal, a new evolutionary conceptual framework is urgently needed. There is also a call to target epigenetic programming in cancer therapy

rather than focusing on specific genes/pathways. Again, a multiplatform analysis of 12 cancer types suggests that "cell-of-origin" rather than pathway-based features dominate the molecular taxonomy of diverse tumor types (Hoadley et al., 2014).

Recently, the first integrative analysis of all reference epigenomes was published (Roadmap Epigenomics Consortium, 2015), which represents the most comprehensive map of the human epigenomic landscape to date. Since this collection is based on data of primary cells and tissues, it is important to systematically compare them with epigenetic profiles of various types of cancer, and to study the relationship between epigenetics and the gene mutation landscape.

Genome/Karyotype Landscape

The terminology of the genome has been misused by many. The sequencing phase of genome research has been called the genome project, as if characterizing the genetic parts list is equal to understanding of the system's function. While the current sequencing project has confirmed that most solid tumors have dozens of translocations and aneuploidy, the significance of these alterations has been downplayed.

The attitude of ignoring the chromosomes in cancer is based on the following viewpoints: (1) the majority of translocations are considered to be passengers rather than drivers; (2) there are roughly 10 times fewer genes affected by chromosomal changes than by point mutations (Vogelstein et al., 2013) (possibly judged by the number of directly impacted fusion genes); and (3) chromosomal variations are incidental, as chromosomes serve only as the vehicle of genes, and most chromosomal aberrations reflect cancer formation but do not cause it. When the topic of "the genomic architecture of cancer" is discussed, the majority of information is about the gene mutation landscape, and the landscape of chromosomal aberrations, the most important part of the genomic architecture, is often briefly mentioned in the Q/A session, without being discussed in depth. Even in many cancer genome sequencing papers that recorded a high level of chromosomal aberrations, this fact is only mentioned, without detailed analyses.

Nevertheless, cancer genome sequencing begins to slowly challenge these assumptions. First, the sequencing data confirm that chromosomal changes are overwhelming in the majority of cancer types without cell culture and cytogenetic preparation (Baca et al., 2013). The high degree of large-scale genomic rearrangements is the general rule, rather than the exception, for many cancer types (Stephens et al., 2009; Heng, 2007c; Heng et al., 2016a). Of equal importance, the failure to identify the long-expected key common cancer gene mutations opens the door to considering the potential role of chromosomal changes in cancer.

Second, chromosomal changes seem to be the key incidents for many transitions of cancer evolution, from metastasis to drug resistance, as the mutation and epigenetic landscapes are not able to explain these key transitions, especially due to the lack of a pattern of gene mutations.

Third, the "rediscovery" of genome chaos by sequencing has generated a wave of excitement (Baca et al., 2013; Horne and Heng, 2014; Stephens et al., 2011). For example, the phenomena of chromothripsis and chromoplexy are detected from the majority of cancer types, and for some cancers, massive genome changes are the key characteristics. These observations have forcefully suggested that cancer evolution can be achieved by "sudden" genome-level changes rather than the accumulation of a series of small genetic alterations. However, even though chaotic genomes detected from patients have generated many high-profile publications (Baca et al., 2013; Stephens et al., 2011), there are some important misunderstandings: (1) it was suggested that these chaotic genomes are generated by one catastrophic event. By tracing macrocellular evolution using an in vitro model, it has become clear that chaotic genomes can occur much more frequently when unstable genomes are challenged by a high level of stress. They can also be induced multiple times, depending on the treatment (Liu et al., 2014). Furthermore, these transitional structures are highly dynamic and continue to change until survivable karyotypes emerge and become dominant. These survivors often display much simpler karyotypes. (2) Due to the instability of these chaotic genomes, many highly reorganized karyotypes will be eliminated prior to the later stage of this process, even though they are essential for generating more stable karyotypes. In addition, many chaotic genomes exist in the nonclonal chromosome aberration (NCCA) form (Heng et al., 2016a,b; Stepanenko et al., 2015), as it is difficult for chaotic genomes to be clonal, except in some cases of chromothripsis. Thus, the lower frequencies of such structures detected from some cancer types do not diminish the importance of chaotic genomes. It is likely that most cancers involve the chaotic genome, even though the observed frequencies are much lower in formed tumors (Heng, 2016a; Heng et al., 2006). (3) Chromothripsis represents only one subtype of chaotic genome, and chromothripsis might provide better survival opportunities than other types with more complicated and massive genome-level changes might. The latter type might destroy some important modules, leading to the cell death. (4) To illustrate the mechanism of the chaotic genome, some key cancer genes or specific processes are used. For example, some have focused on DNA replication (Zhang et al., 2015). However, according to the genome theory of cancer, the key is to change the

karyotype-defined system inheritance. (5) Current diverse research efforts are trying to link genome chaos to specific molecular causative factors, such as p53 mutations, DNA replication errors, and other factors. However, the formation of the chaotic genome can be contributed by diverse molecular mechanisms, which can be unified as high levels of stress-induced cellular survival strategies. To effectively survive, rapid and massive genome reorganization becomes the most effective way. (6) The chaotic genome was not originally discovered by sequencing, but by molecular cytogenetic analyses during the study of karyotype dynamics in punctuated cancer evolution (Heng et al., 2004b, 2006). These structures were named as "karyotype chaos" or "chromosomal/genome chaos" (Duesberg, 2007). Subsequent investigations have linked genome chaos to elevated CIN, the survival strategy of cells, and evolutionary potential reflected by transcriptome dynamics (Heng et al., 2011b; Liu et al., 2014; Stevens et al., 2013, 2014). Now, with the confirmation of this information by sequencing of patient samples, these structures should no longer be viewed as in vitro artifacts, as they are important players in cancer evolution. It is important to point out that genome chaos has been linked to generations of cells with reduced genome sizes and stem cell features, and therefore plays an important role in rapid cancer evolution (genome-mediated macrocellular evolution) (Heng et al., 2016a; Ye et al., submitted for publication).

Somatic Copy Number Alteration Landscape

Copy number variation (CNV) represents one important type of somatic aberration (Feuk et al., 2006). It was originally defined as the amplification or deletion of DNA in the size of >1 kb, and it was later widened to include much smaller sizes (>50 bp) due to methodology development (Girirajan et al., 2011). Recently, the cytogenetically visible copy number variations (CG-CNVs) were discussed (Liehr, 2016). Since the discovery of CNV, there were high hopes that a large number of CNVs could be linked to various common and complex diseases, as it has been very challenging to identify common gene mutations as the key causation factors. Similarly, illustrating the CNV landscape (a type of genomic signature of a driver gene) has become an essential part of cancer genome sequencing (Akavia et al., 2010), as somatic copy number alterations (SCNAs, which are different from germline CNVs) are extremely common in cancer; the expectation is that the amplification of oncogenes or deletion of tumor suppressor genes can be linked to tumorigenesis, and common cancer SCNAs or cancer type–specific or pathway-specific SCNAs can be identified. Unfortunately, the interpretation of the CNV landscape is not that straightforward, especially with regards to distinguishing driver SCNAs from numerous SCNAs that randomly accumulate during tumorigenesis. Following high-resolution analyses of SCNAs from 3131 cancer samples representing 26 histological types, 158 regions of focal SCNA were identified with significant frequency across multiple cancer types, and some linkages between SCNAs and known cancer genes, apoptotic pathways, and the NF-κB pathway have been established. However, among 158 SCNAs, over 70% of them (122) cannot be explained by the presence of a known cancer target gene located within these regions (Beroukhim et al., 2010). Nearly 5000 samples from the Cancer Genome Atlas Pan-Cancer data set were later analyzed. Some interesting patterns of SCNAs are observed. For example, the sizes of internal chromosomal SCNAs tended to be shorter than telomere-bounded SCNAs; 37% of cancers with whole-genome doubling were linked with elevated rates of every other type of SCNA, as well as some mutations, amplifications, and alterations (in *TP53*, *CCNE1*, and PPP2R); amplified regions without known oncogenes were enriched in epigenetic regulating genes. Interestingly, among 140 significantly recurrent focal SCNAs, the 102 of these without known cancer genes still accounted for over 70% (Zack et al., 2013), illustrating the challenge of making sense of the large number of SCNAs that are not directly related to the cancer gene mutations.

It is interesting to compare the signatures of CNVs and SCNAs. Due to the high level of somatic genetic dynamics, the degree of SCNAs is likely much higher than that of germline CNVs, which can further complicate the relationship between copy number and expression profile, as well as other features. For example, the large size of CNVs could be eliminated by sexual reproduction (Heng, 2007b, 2016a), while at somatic cell level, without this "filter" to purify the altered genome, the size of SCNAs could be much larger, and so could be the impact on overall gene expression. As for observations that genes in CNV regions are expressed at lower and more variable levels than genes mapping elsewhere (Henrichsen et al., 2009), and that CNV genes are significantly enriched within transcripts that show variable time courses of expression (Zhou et al., 2011), we anticipate a more dominant degree of variation to be observed in cancer cell populations, especially when the genome is highly unstable.

Ultimate Challenges Beyond the Genomic Landscape of Cancer

In addition to the many specific difficulties that each type of landscape faces, there are some common challenges: first, for studying the function of a specific cancer gene mutation landscape, what will be the influence from other levels of genetic/epigenetic organization, especially when different landscapes function in conflicting styles? To use karyotype alteration as an example,

does the same gene mutation display the same function or anticipate the same pathway when it exists in different karyotypes? Recently, based on published data sets, the gene expression signature of primary stress sensors and major mediators of unfolded protein response (UPR) pathways in various human tissues/organs was compared against that in mice. The results revealed that the expression pattern of UPR significantly differs between human and mouse, between normal and disease tissue, and among different tissue types. It is thus necessary to connect specific pathways to the involvement of different landscapes.

Second, how does cancer evolution work? Is it based on the gene mutation landscape, epigenetic landscape, or karyotype landscape? What do the multiple levels of landscape look like? Does the pattern of cancer evolution matter? Which types of landscapes are more important in a defined context? Is gene mutation accumulation more important when the new cancer genome has already formed (as genes with growth advantages could promote population growth)? To address these questions, a more holistic evolutionary platform is needed.

Third, what is the difference between germline genomics and somatic cell genomics? What defines the system inheritance? Is genetic information passed among cells precise or fuzzy given the observations that there are high degrees of genetic/nongenetic variations even in normal tissues? How can fuzzy inheritance serve as a mechanism of genetic/nongenetic heterogeneity (this based on the observation that a single cell can potentially pass down a range of genetic alterations—the concept of fuzzy inheritance was introduced to describe the internal fuzziness of genetic information; see Heng, 2016a,b)? Answering these questions will connect stress-induced adaptation and its trade-off within the context of genomics and evolution.

GENOME THEORY OF SOMATIC CELL EVOLUTION

Fortunately, the search for a new genome-based conceptual framework of cancer evolution has made significant progress. Considering cancer as a process of new system emergence through genome reorganization during the two phases of cancer evolution, genome theory aims to depart from the gene-centric concept by distinguishing the function of gene/epigene and genome for both genetics and evolution (Heng, 2009, 2016a).

Redefine the Cancer Genome

Despite the efforts to describe features of cancer genome, there is no solid concept of the cancer genome in the first place. It is known that most cancer cases display different karyotypes, such that there is no standardized cancer karyotype or genome for the majority of cancers; both the types and numbers of genes involved are also very different (the composition of genes differs due to the altered karyotypes). More challengingly, the overwhelming genetic/nongenetic heterogeneity observed in cancer leads to the fact that no real average population exists in most cases, as the clonal concept in cancer is not a general characteristic of most cases (which, again, differs from the normal somatic genome, where genome variability is minimal) (Abdallah et al., 2013; Horne et al., 2015a,b,c). Unfortunately, the true cancer landscape of heterogeneity has been largely ignored based on the rationale of profiling the average cancer genome, despite the fact that these challenges have been long realized (Heng, 2007a; Heppner, 1984). Following over a decade of sequencing, and the discovery/rediscovery of the high degree of genetic heterogeneity in cancer, it is time to restate and to synthesize the concept of the cancer genomic landscape based on the correct understanding of the genome.

What Is the Genome?

One of the biggest confusions in current genomic research is regarding the concept of the genome. For many molecular biologists, the term "genome" simply represents a name for the DNA sequences, including genes, of a given species. Based on a quick search of the literature, typical definitions include the following: "A genome is the complete set of genes or genetic material of an organism"; "A genome is an organism's complete set of genetic instructions"; and "a genome is the full set of instructions needed to make every cell, tissue, and organ in your body. Almost every one of your cells contains a complete copy of these instructions, written in the four-letter language of DNA (A, C, T, and G) (http://www.broadinstitute.org/education/glossary/genome)." By comparing all of these different definitions, most of them simply refer to all genes or all genetic materials, and some of them vaguely mention genetic instructions, while very few of them involve the chromosome.

The term genome was introduced in 1920, a blend of the two words "gene" and "chromosome" (Winkler, 1920). It seems, then, that the chromosome portion of the genome has been lost. According to the gene-centric view, the chromosome is just the vehicle of the gene; if that is that case, what further significance could it hold? On the other hand, with the increased difficulty of relying on the gene to explain inheritance, is it possible that the missing link is the chromosome itself? According to genome theory (Heng, 2016a), the genome is not simply the collection of all genes. The genome context is composed of gene content/DNA sequences, plus the important "genomic topology," the missing link. Similar genome content can form different species with different

genomic topologies, and without genomic topology, the genes are just some materials (Heng, 2009; Heng et al., 2011a). The system viewpoint of the definition of genome is as follows:

> A genome is the complete set of genetic material (including gene content) of an organism, which is organized by the unique composition of chromosomes or karyotypes. While genes represent parts inheritance (how individual genes code/regulate individual proteins), the karyotype determines the topological order of genes along and among chromosomes, which ultimately defines interactive relationship among genes, representing the system inheritance (how the blueprint works to instruct the protein network).

Why Does the Whole Set of Chromosomes (Karyotype) of a Given Species Represent a New Coding System?

To understand the importance of chromosomes in defying genomic information, one needs to accept the concept that chromosomes code the system inheritance (Heng, 2009; Heng et al., 2011a, 2013b). Again, the metaphor of building materials and architecture is useful. To make bricks, tiles, window glass, etc., individual instructions are clearly needed (similar to a manual for making specific proteins from genes). However, the architecture is designed based on a blueprint, a topological set of instructions on how to put the parts together (similar to a plan which allows proteins to interact in a specific fashion).

What, then, is special about the chromosomes for different systems (or species)? Investigating the potential function of karyotypes was a logical step. To connect the dots, the following key facts have been considered: (1) karyotype is an important feature of organismal evolution. Using mammals as an example, different species have similar gene content, but different karyotypes (Ye et al., 2007). In fact, across all species, genome reorganization is a general rule for evolution. (2) The physical relationship among many genes is conserved within well-defined syntenies of chromosomal regions among many species, suggesting that the physical relationship of groups of genes along a chromosome might represent an important type of genetic information. (3) In the somatic evolution of cancer, karyotype alteration is also a key feature, especially within the macrocellular evolutionary phase, while gene mutation is dominant in the microcellular evolution phase (Heng et al., 2006, 2011b). (4) It is known that there is a topological order of different chromosomes within the nuclei, and chromatin domain configuration is important for regulating gene expression or genetic recombination (Heng et al., 1996, 2001, 2004a). (5) Rapid and massive genome reorganization, also known as genome chaos, can be induced by high levels of stress, in both pathological and experimental conditions, and the chaotic genome displays elevated transcriptome dynamics.

On considering these points, the karyotype becomes a strong candidate for carrying system inheritance. On the surface, a karyotype is the number, size, and shape of chromosomes in an organism. Deep down, however, each chromosome maintains the physical order of genes/DNA sequences, and all chromosomes are likely to provide a physical platform for all genes to interact within nuclei (possibly through some genomic topological elements). In contrast to gene-defined "parts inheritance," the genome-defined "system inheritance" is the blueprint (Fig. 5.1) (Heng, 2009, 2010; Heng et al., 2011a).

Increased experimental evidence and syntheses support this insight. Examples include the following: (1) the change of location of a specific gene within the genome can lead to the emergence of key features (Blount et al., 2012); (2) chromosomal number change can result in the cell regaining key features previously contributed by a specific key gene (Pavelka et al., 2010); (3) karyotype alterations lead to overall changes of the transcriptome (Stevens et al., 2013, 2014); (4) chromosomal rearrangement, rather than genome duplication, is responsible for the macroevolution of angiosperms (Dodsworth et al., 2016); and (5) most excitingly, the recently developed chromosome conformation capture technology (high-C) has revealed that topological associated domains (TADs) are essential for gene regulation (Dixon et al., 2012; Sexton et al., 2012). A highly similar pattern of TADs is observed among different types of tissues of the same species, suggesting that the pattern of TADs could serve as a structural matrix for all cell types to regulate genes. We have suggested that a specific pattern of chromatin interaction in interphase chromatin can be determined simply by the order of the gene/genetic topological elements along an individual chromosome, and among all chromosomes (Heng, 2009, 2016a; Heng et al., 2011a,b). TADs clearly represent one such example.

How Is the System Inheritance Passed, and Why Is it Challenging to Define the Cancer Genome?

Following a comparison of evolutionary patterns of sexual and asexual reproduction, it was surprisingly realized that the main function of sexual reproduction is to maintain the same karyotypes for a given species, rather than just producing diversity at gene level, as the meiotic process and other related stages can eliminate drastically altered genomes (Gorelick and Heng, 2011; Heng, 2007b, 2016a,b; Heng et al., 1996; Wilkins and Holliday, 2009). This important realization also explains the high degree of genomic diversity observed in asexual species (Ellstrand and Roose, 1987; Konstantinidis and Tiedje, 2005). In the case of cancer, due to the occurrence of the somatic process in the absence of the sexual

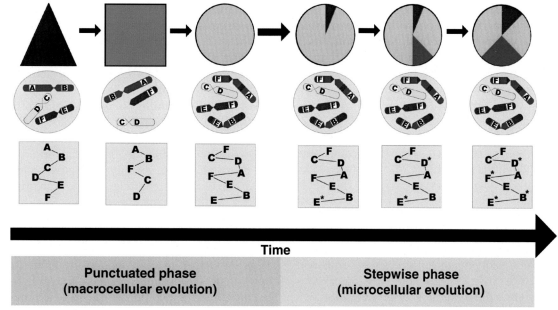

Time

Punctuated phase (macrocellular evolution) **Stepwise phase (microcellular evolution)**

FIGURE 5.1 The relationship between karyotypes, system inheritance, and the pattern of cancer evolution. Cancer evolution can be divided into two phases, the punctuated phase (or macroevolutionary phase) and the stepwise gradual phase (or microevolutionary phase). Punctuated phases are characterized by rapid genome changes, with each shape representing a unique genome system (top panel). Different chromosomes are drawn within the nucleus below the corresponding system. Genes are designated A, B, C, D, E, and F within the chromosomes, and corresponding protein networks are illustrated below by the relationships between proteins A, B, C, D, E, and F. Different karyotypes determine the network structure. The stepwise phase is characterized by gene mutations/epigenetic alterations that aid in adaptation. Genetic/epigenetic alteration is indicated by asterisks in the protein network. *Source: Reproduced from Horne, S.D., Pollick, S.A., Heng, H.H., 2015a. Evolutionary mechanism unifies the hallmarks of cancer. Int. J. Cancer 136, 2012–2021; Horne, S.D., Ye, C.J., Abdallah, B.Y., Liu, G., Heng, H.H., 2015b. Cancer genome evolution. Transl. Cancer Res. 4, 303–313; Horne, S.D., Ye, C.J., Heng, H.H., 2015c. Chromosomal instability (CIN) in cancer. eLS 1–9.*

filter, along with the fact that genome-level complexity is much higher than in bacteria due to the involvement of multiple chromosomes, the karyotypic changes are often overwhelming; this is especially true during the macrocellular evolutionary phase, in which genome instability dominates (Horne et al., 2015a,c). That is the reason why it is hard to get a fixed cancer genome. Obviously, the concept of the germline genome cannot be applied to the highly dynamic cancer genomes, especially during the macrocellular evolutionary phase, in which the chaotic genome dominates (Liu et al., 2014). This synthesis further redefined the nature of inheritance. Different from the classical view of precise genetic information, overall genetic information is fuzzy and forms the basis for genetic heterogeneity, the essential feature of biological systems that are adapting in highly dynamic environments (Heng, 2016a; Heng et al., 2016a,b; Horne et al., 2015c).

Redefine the Pattern of Cancer Evolution

With the recent molecular confirmation that cancer genomes are highly dynamic/heterogeneous, a timely question is how this reality will impact the general understanding of the cancer, including the evolutionary and ecological frameworks used to explain how cancer works. Immediate efforts should be made to identify the true pattern of cancer evolution, as the traditional concept of cancer evolution is simply borrowed from neo-Darwinian evolutionary theory, which is gene-centric (ignoring the karyotypes alterations). Focus should also be placed upon tracing the common patterns that accumulate small genetic changes over time.

The generally accepted pattern of cancer evolution has been "stepwise," which was anticipated by waves of cellular clonal events (Nowell, 1976). Despite the fact that the anticipated common pattern has been demonstrated in a few exceptions (such as the Ph chromosome in chronic myelogenous leukemia (CML), the "two hits" of *RB* gene mutations in retinoblastoma, and the APC–TP53–Ras mutation accumulation model in colon cancer), it remains challenging to explain most cancer types (Heng, 2007a; Horne et al., 2013). Moreover, even in these exceptional cases, the reality is much more complicated, as cancer genome sequencing data recently confirmed (Horne et al., 2013, 2015a). Clearly, the only logical conclusion is to redefine the pattern of cancer evolution.

Two Phases (Macro- and Microcellular) Cancer Evolution

To examine the pattern of cancer evolution, one effective method is to dissect the cellular process itself. Based on the clinical observation that most cancers display altered karyotypes, and the fact that many evolutionary studies of cancer have largely focused on gene mutations

and ignored karyotypes, we have refocused on the pattern of karyotype evolution (Heng et al., 2006). Using an in vitro model of immortalization, we have watched evolution in action throughout the entire process, as it began from a cell population with normal karyotype and resulted in an immortalized cell population displaying altered genomes. Prior to immortalization, the karyotypes drastically changed from one passage to the next; coupled with a high degree of stochastic NCCAs, this suggested that there is no continuous and traceable karyotypic evolutionary pattern among these passages. In contrast, however, starting from the later passage when immortalization was established, the same karyotype was shared by the population and was passed among generations, obviously suggesting stepwise karyotype evolution (Heng et al., 2006). Simply put, the punctuated phase of evolution describes the period of new genome or new cellular system creation without passing a defined system inheritance. In contrast, the stepwise phase describes accumulating modifications at the gene level (in essence, parts inheritance) while maintaining the same genome or cellular system. Clearly, the two phases of karyotype evolution mimic punctuated and Darwinian stepwise evolution, respectively (Eldredge and Gould, 1972; Heng, 2007a,b,c, 2016a,b), even though the punctuated and stepwise cycles were initially introduced to explain historical patterns in fossils. Further synthesis has linked the punctuated phase to the macrocellular evolutionary phase, while the stepwise phase is linked to the microcellular evolutionary phase. Transcriptome analysis also linked elevated dynamics to the punctuated phase, indicating that the discontinuous phase has the highest evolutionary potential (Stevens et al., 2013, 2014).

Not only have multiple parallel runs of immortalization displayed the same two phases of evolution, but also all of the most important transitions of cancer evolution, including transformation, metastasis, and drug resistance, display the same patterns (Heng et al., 2011b, 2013a) (see Fig. 5.1). Unfortunately, however, for traditional cancer evolutionary studies, punctuated macrocellular evolution has been ignored, and most efforts have been focused on searching for the pattern of clonal evolution.

Recent cancer genome sequencing efforts, and especially single cell sequencing, have forcefully supported the concept of the two phases of cancer evolution. It was demonstrated that even at the DNA level, there is a punctuated phase of cancer evolution (Navin et al., 2011). Such a phase has also been observed from different cancer types, and has been referred to as "big bang" (Sottoriva et al., 2015). Moreover, when different parts of the same tumor are sequenced, the clonality is generally limited for most solid tumors. As a result, the concept of "macroevolution in cancer" is gaining acceptance (Gerlinger et al., 2014; Klein, 2013), despite the initial resistance that arose when we illustrated it (Heng et al., 2006).

CIN Is the Driving Force for New System Emergence

What factor causes the transition of the two phases? Based on the observation of high frequencies of NCCAs within the punctuated phase, and the direct linkage between NCCAs and genome instability, genome instability has been identified as the trigger factor (Heng et al., 2006, 2011b). Furthermore, system instability can be generated from both internal (genetic) and environmental factors.

The discontinuous nonclonal evolution and stepwise clonal evolution observed in our in vitro model were referred to as punctuated and stepwise evolution, respectively (Heng, 2007a,b,c; Heng et al., 2006), despite the fact that the original concept of punctuated equilibrium was used to describe the pattern of phenotype (Eldredge and Gould, 1972). The punctuated equilibrium of the pattern of genotype has not been analyzed due to a lack of good models. The "watching evolution in action" experiments have not only linked the "genome type" and phenotype (i.e., the altered karyotype is linked to immortalization, invasiveness, and drug resistance) but also pinpointed stress-induced system instability as the trigger of phase switching. The illustrated relationship between stress, genome instability, gene/epigene mutation-mediated microcellular evolution, and karyotype alteration-mediated macrocellular evolution can also be used to explain the history of evolution.

Since the genome context (karyotype) defines the biosystem, while genes define some features of a given system, the result of macrocellular evolution is the emergence of new systems. Even though multiple levels of biosystems are involved (i.e., tissue level and higher levels of organization), all somatic evolution is dependent on inheritance. Despite the fact that many genetic/nongenetic factors (including those at the epigenetic level), as well as environmental factors, can contribute to and even initiate cancer evolution, the key is new system formation and domination. In addition, stable genomes function as a constraint for lower levels of genetic variations. For example, many unstable genomes can display a higher rate of gene mutations. The genome theory of cancer evolution predicts that, following the formation of a new system with a new genome, specific gene mutations can be accumulated during the domination process of cancer cells. In this case, oncogenes might be more important at later stages than at earlier stages of cancer evolution. Note that a typical stepwise evolutionary pattern is not common for most cancer types.

Cancer cells are constantly evolving, and different stages of cancer cells represent virtually different systems (as evidenced by different karyotypes) (Heng

et al., 2011b). This concept dismisses the common expression that cancer cells always learn new tricks. In fact, different stages of the cancer cells are different in that they represent new systems. Understanding this point is important for appreciating the dynamics of cancer evolution, and why the targets are constantly moving. It also supports the model that the success of macroevolution (forming new genome systems) is often prior to microevolution (aggressive accumulation of some cancer genes), which enlarges the population.

Why we Need the Concept of Multiple Levels of the Cancer Genomic Landscape

Knowing the different genetic and evolutionary patterns involved in cancer initiation and progression, it is essential to modify the viewpoint of the cancer genomic landscape. Currently, the genomic landscape model is mainly based on gene mutations and epigenetic regulation (when related to cancer gene mutations), and genome-level changes have so far been left out. Such a practice is caused by the failed realization that cancer is fundamentally a disease of genome alterations, and the reality that it is technically challenging to quantitatively integrate karyotype information. Moreover, many evolutionary landscape models are based on the continuous stepwise concept, reflecting the belief that the accumulation of small genetic changes leads to big changes over time; these models do not distinguish between macro- and microevolution, and consider the genotype as simply equal to individual genes or the collection of genes. The challenge, therefore, is to establish two different types of landscapes that involve multiple levels of genomic profiles.

The first task is to quantitatively compare and integrate different levels of genomic alterations in cancer evolution, both in general and in some specific cases. Such analyses need to focus on the potential different meanings of the same gene mutation within different karyotype backgrounds. The potential conflict among different genomic level is of particular interest. This effort might lead to the appreciation of package-based cancer evolutionary selection. Furthermore, such a real landscape should have more clinical implications, as so far, the much better prediction profile for a number of cancer types is the cytogenetic profile rather than the sequencing profile.

Second, when conceptualizing the genomic landscape and evolutionary fitness landscape, it is necessary to separate genome-level aberrations and gene mutations, as the relationship between them is not based on the gradual accumulation of the latter. The concept of micro- and macroevolution must be integrated into the landscape. To achieve this, we have introduced a multiple-level landscape model (Heng et al., 2011b, 2013a) (Fig. 5.2).

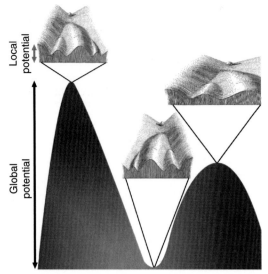

FIGURE 5.2   The multiple-level landscape model. The integrated relationship between gene, epigene, and genome landscapes can be illustrated by this model. Genes and epigenes are illustrated by local landscapes (tiny magnified dips and crevasses on mountaintops or in valleys) and the genomes are illustrated by global landscapes (two large mountains). In this model, evolutionary potential can be classified into different levels. Local potential refers to adaptive potential provided primarily by gene-level or epigenetic changes displaying gene mutations and multiple stable and unstable cellular states (Brock et al., 2009; Huang 2013), while the global potential refers to the overall survival potential achieved and exhibited by genome-level dynamics displaying different karyotype-defined systems (Heng, 2009; Heng et al., 2013a). Despite the importance of microcellular evolution, local adaptive landscapes do not typically drive the macrocellular evolutionary process of cancer. It should be noted that local landscapes are found throughout the global landscape, but individual local landscapes do not directly contribute to changing different global landscapes. The images of local landscape were adapted from Waddington. *Source: Reproduced from Heng, H.H., Stevens, J.B., Bremer, S.W., Liu, G., Abdallah, B.Y., Ye, C.J., 2011b. Evolutionary mechanisms and diversity in cancer. Adv. Cancer Res. 112, 217–253.*

In this model, the local landscape refers to the gene/epigene landscape, while the global landscape refers to the genome or karyotype landscape. The important message is that the accumulation of local fitness is not going to lead to a change of the global landscape. This idea has gained support from a system biologist, who has nicely further articulated the relationship between genome/gene and cancer states (Huang, 2013).

Such a model suggests the context of using ecological concepts to explain cancer evolution. In most ecological conditions, there are interactions within species and among different species. Due to the drastic karyotype variations within cancer cell populations, the genetic differences among these cells are often greater than the differences between species. This fact thus suggests that different cancer cells should be considered as different species. Nevertheless, when applying ecology concepts to cancer, karyotype variation must be included as the most dominant factor.

Moreover, the multiple-level landscape models explain the importance of fuzzy inheritance-mediated evolution. As there are many paths that can change the landscapes, and reduced predictability is directly caused by fuzziness of inheritance and the process of somatic evolution, a large number of diverse pathways can achieve similar fitness, and there is no need for depending on a fixed genetic profile pattern, as long as the new genome-defined systems emerge and become dominant. In some experimental systems, one can artificially push a linear, dominant pathway (by introducing large numbers of gene copies or strong promoters), and gain increased predictability. However, since it is not representative of the general case in cancer, the implication of this result is rather limited.

Finally, this model echoes well the evolutionary mechanism of cancer (Heng et al., 2010; Ye et al., 2009). The general cancer model can be explained by three key components: stress-induced or cellular adaptation-required system dynamics (i.e., increased stochastic changes), cellular population diversity (i.e., genome heterogeneity), and macrocellular evolutionary selection based on the genome package, which promotes the emergence of a new system, including the breakdown of various constraints. Unfortunately, all of these can be contributed by a large number of individual molecular mechanisms. In other words, so many genes/epigenes/environmental factors can contribute to local landscape changes (microcellular evolution), but as long as the new global landscape is changed (macrocellular evolution), it is a point of no return.

IMPLICATIONS AND FUTURE DIRECTIONS

The cancer genome sequencing project has confirmed many predictions about the highly dynamic processes occurring in cancer, and heterogeneity and the importance of the genome-level alterations (Heng, 2007a). With this new knowledge in hand, two very different approaches are now on the table: one is to continue collecting more sequence data by increasing the sample size and/or focus on epigenetic data as well as noncoding regions. However, the combination of genetic alterations could be nearly infinite in number, according to the evolutionary mechanism of cancer and the reality of the high heterogeneity in clinical samples. Another less obvious but important option is to reexamine the current framework of cancer evolution. Due to the limited space in this chapter, the examples given in the next three subsections will be discussed.

The Genomic Landscape Questions the Theory of the Cancer Gene Mutation

Despite the excitement surrounding their focus, one common message of most cancer genome sequencing papers is often a big surprise, in terms of being against known predictions based on the conceptual framework (Heng, 2016a; Horne et al., 2015b). But perhaps even more surprising is the fact that nearly all of these papers have failed to even question the current cancer theory of the genome mutation, despite the fact that its major predictions are not working. In other scientific fields such as physics, when the facts do not fit the theory, it is the time to search for a new theory. This is not the case at all in cancer biology. Clearly, something fundamental is missing in biological research. Yes, integrating multiple levels of genomic landscapes is difficult, but it is the right thing to do. Yes, it is easier to just sequencing more samples than to seriously examine the theoretical basis of the current conceptual framework, but we have to do it now in order to advance the field.

Distinguishing Adaptive and Survival Landscape

It has been frustrating to not be able to popularize the difference between gene-mediated microcellular evolution and genome-meditated macrocellular evolution in cancer, as population genetics, which have been dominant in cancer evolutionary studies, have more or less ignored this key difference. Such ignorance is due to the confusion of distinguishing between a population of the same species (with an identical karyotype) and the mixed population of various species (with different karyotypes). A cancer cellular population often belongs to the latter category. To study the population of a given species, the fitness of the population can be studied by the frequencies of certain genes that contribute to fitness. However, if the population consists of different species with different karyotypes, then fitness should not simply focus on the gene frequencies in a population, but on the advantages of different types of the karyotypes. A very simple example of why we need to think differently about cancer evolution is genome-mediated drug resistance during high-dosage drug treatment. When harsh treatment is applied, most of the cells do not have the time to gradually evolve, but have to be able to survive, and will otherwise die off in a very short timeframe. For the majority of cells, there is no time to wait for many generations to accumulate certain genes with drug resistance. This situation has been illustrated by drug-induced genome chaos. Within one to a few generations, either a new genome system will emerge or cells will be eliminated. In contrast, however, in physiological conditions, moderate drug treatment could be linked to gene mutation accumulation, displaying the accumulation of drug-resistant mutations.

It is thus necessary to separate the concepts of adaptive landscape and survival landscape. For the survival landscape, just improving the gene frequency in populations is not enough, as the emergence of a new genome is a necessary precondition. According to experimental observations, the emergence process, following the induction of genome chaos, will usually last a few weeks. In such a dynamic process, fuzzy inheritance dominates, and even the basic mechanism of cell division becomes chaotic; 1 chaotic cell can divide into over 10–20 cells at once, and the mixture of division and fusion becomes an effective means by which to create new survivable genome systems. Following the establishment of a survivable genome, the adaptive landscape starts to work, to further improve the fitness of the selected genomes. It is thus possible that the multiple landscape models will distinguish and ultimately unify the survival and adaptive landscapes.

Such a separation is of importance to understand the unique features of cancer ecology. In recent years, increased numbers of researchers have considered cancer an ecological process (Gatenby et al., 2009; Heng, 2016a; Merlo et al., 2006; Vittecoq et al., 2013). Accordingly, it was suggested that ecological concepts/approaches could be used as the basis of cancer therapy (Korolev et al., 2014; Pienta et al., 2008). One of the hidden challenges is integrating the instant survival landscape into the adaptive cancer evolution landscape, because for many cancer cells within many evolutionary transition stages, there is no chance for cellular populations to gradually adapt. They either survive or die in less than a few generations. In contrast, most ecological cancer models do not separate genome-mediated macroevolution from gene-mediated microevolution, as they have paid less attention to the concept that it is the genome rather than gene that defines the species, and that cancer is a genomic rather than genetic disease (Heng, 2007a,b,c, 2009, 2016a).

Is the Cancer Evolution Darwinian?

Cancer has long been considered by many as a Darwinian evolutionary process. Despite the time-scale difference, both organismal evolution and cancer evolution share the identical process of natural selection (Thomas et al., 2013). Three key features are often cited to support this viewpoint: cellular populations display variations, these variations are inheritable, and variations affect survival or reproduction of the cellular population (Crespi and Summers, 2005; Heng, 2007a; Merlo et al., 2006). These features are essential for clonal evolution (survival and expansion), and condone competition among normal cells for space and nutrition (Greaves and Maley, 2012; Heng et al., 2006). Recently, however, in light of genome-mediated system inheritance and the two phases of cancer evolution, questions have been raised regarding the nature of inheritance, the pattern of evolution, and their impact on the cancer evolutionary concepts (Heng, 2016a,b). Many of the previously considered "trivial" differences between organismal and cellular evolution are in fact highly significant. For example, a cancer cell population often displays variable karyotypes; as such, these cells should not be compared to different individuals of the same species defined by the same karyotypes, if karyotypes, rather than individual genes, define species. Although a cancer cellular population could mimic natural ecology, where different species interact with each other, there is a big difference. In cancer, macrocellular evolution is constantly occurring, and is not caused by the accumulation of small changes over generations, but by large and rapid changes at the genome level (e.g., genome chaos). In the current natural ecosystem, for most eukaryotic organisms with sexual reproduction, the speciation event is highly limited and believed to require a large scale of time, which allows for the accumulation of small changes. For microorganismal evolution, the "simpler" genome is more dynamic and is similar to cancer cells, but without the complexity of genome architecture (different chromosomes), the scale of changes, in both the speed and the degree of genome reorganization, is much smaller in microorganismal systems than in human cancer. Therefore, the crucial involvement of genome reorganization-mediated macroevolution is the key. Without it, the population genetics approach is effective for studying microevolution. In contrast, with it, the current approach is no longer effective. Since cancer is an issue of both microevolutionary and (moreover) macroevolutionary processes, neo-Darwinian principles have so far failed to explain the evolutionary principles of cancer, despite considerable hand-waving.

It is thus time to search for the pattern and mechanism of non–neo-Darwinian evolution in cancer. One ideal system is watching cancer evolution in action, to compare the relationship between micro- and macrocellular evolution, and to examine the involvement of genome chaos. For example, by tracing drug-induced genome chaos-mediated rapid macrocellular evolution, the general mechanism of drug resistance has been illustrated (Heng, 2007c, 2016a; Horne et al., 2015a). This study also provided the reasoning to question the practice of using neo-Darwinian principles to guide cancer evolutionary studies. First, there seems to be a limited role of typical natural selection in cancer, given the highly dynamic nature of cancer evolution. The two phases of cancer evolution reduce the significance of the stepwise accumulation of gene mutations (as, e.g., they also limit the scale of clonal expansion). Second, genome-level dynamics reject gradualism. The common existence of genome chaos (i.e., chromothripsis, chromoplexy, structural mutation, chromosome catastrophes, etc.) emphasizes

the importance of punctuated cancer evolution. As it happens in all transitional events in cancer (from immortalization to drug resistance), it represents a general rule. Third, macrocellular evolution is a different process than microcellular evolution, and phase transitions are triggered by stress-induced genome instability, rather than a simple accumulation of small changes over time. And fourth, the concepts of immediate survival and gradual adaptation are different, and require different theories in order to be explained. The fact in cancer is that, under harsh conditions (e.g., high-dosage treatment of anticancer drugs), there is no time for small changes to be accumulated over a long period of time. The transition of key phases in cancer evolution is the game of macrocellular evolution.

The collective conclusion is that further studies are needed to critically compare cancer evolution with typical Darwinian evolution. Many might argue that Darwinian evolution includes the punctuated phase, and that, therefore, there is no real need to emphasize the nonstepwise phase. However, the reality is that slowly accumulating, small changes over generations are the key for the principle of natural selection. In fact, it has been difficult to appreciate the importance of massive and rapid genome-level changes in cancer evolution, as they do not fit the stepwise evolutionary principle. By considering the two phases of cancer evolution, and the limited function of individual genes within the context of the genome, it becomes obvious that cancer evolution is a combination of macro- and microevolution observable in a short timescale. The term "macroevolution" has been increasingly used in cancer evolutionary studies since the discovery of the two phases of cellular evolution (Gerlinger et al., 2014; Heng et al., 2006; Klein, 2013; Navin et al., 2011). However, most of the current tools used to study evolution are effective only when addressing microevolution issues. Developing genome-based tools to study macrocancer evolution is thus ultimately important.

Recently, non-Darwinian and non-Darwinian evolution dynamics was used to explain therapy-induced cancer drug resistance (Pisco et al., 2013; Heng, 2007c, 2016a). More directly, by analyzing hundreds of samples from the same hepatocellular carcinoma tumor, it was concluded that clonal diversity agreed well with the non-Darwinian model with no evidence of positive Darwinian selection. In contrast, genetic diversity under a Darwinian model would generally be orders of magnitude smaller (Ling et al., 2015). Given the fact that some authors are well-known evolutionary biologists, their conclusion is highly significant. Knowing the punctuated phase in cancer, the crucial involvement of fuzzy inheritance in genomics, and the difference between macro- and microevolution (Heng, 2016a), especially with the accumulating sequencing data that support the unique pattern of cancer evolution (despite the fact that different terms have been used, e.g., big bang, punctuated, and major shifts in evolutionary trajectories, the essential message of macroevolution is the same), this conclusion is not surprising at all. Immediate research is urgently needed to examine this issue.

Acknowledgments

This article is part of a series of studies entitled "The mechanisms of somatic cell and organismal evolution." I thank Sarah Regan, Steve Horne, and Julie Heng for their editing assistance.

References

Abdallah, B.Y., Horne, S.D., Steven, J.B., Liu, G., Ying, A.Y., Vanderhyden, B., Krawetz, S.A., Gorelick, R., Heng, H.H., 2013. Single cell heterogeneity: why unstable genomes are incompatible with average profiles. Cell Cycle 12, 3640–3649.

Akavia, U.D., Litvin, O., Kim, J., Sanchez-Garcia, F., Kotliar, D., Causton, H.C., Pochanard, P., Mozes, E., Garraway, L.A., Pe'er, D., 2010. An integrated approach to uncover drivers of cancer. Cell 143, 1005–1017.

Alexandrov, L.B., Stratton, M.R., 2014. Mutational signatures: the patterns of somatic mutations hidden in cancer genomes. Curr. Opin. Genet. Dev. 24, 52–60.

Armitage, P., Doll, R., 1954. The age distribution of cancer and a multistage theory of carcinogenesis. Br. J. Cancer 8, 1–12.

Baca, S.C., Prandi, D., Lawrence, M.S., et al., 2013. Punctuated evolution of prostate cancer genomes. Cell 153, 666–677.

Beroukhim, R., Mermel, C.H., Porter, D., Wei, G., Raychaudhuri, S., Donovan, J., Barretina, J., Boehm, J.S., Dobson, J., Urashima, M., Mc Henry, K.T., Pinchback, R.M., Ligon, A.H., Cho, Y.J., Haery, L., Greulich, H., Reich, M., Winckler, W., Lawrence, M.S., Weir, B.A., Tanaka, K.E., Chiang, D.Y., Bass, A.J., Loo, A., Hoffman, C., Prensner, J., Liefeld, T., Gao, Q., Yecies, D., Signoretti, S., Maher, E., Kaye, F.J., Sasaki, H., Tepper, J.E., Fletcher, J.A., Tabernero, J., Baselga, J., Tsao, M.S., Demichelis, F., Rubin, M.A., Janne, P.A., Daly, M.J., Nucera, C., Levine, R.L., Ebert, B.L., Gabriel, S., Rustgi, A.K., Antonescu, C.R., Ladanyi, M., Letai, A., Garraway, L.A., Loda, M., Beer, D.G., True, L.D., Okamoto, A., Pomeroy, S.L., Singer, S., Golub, T.R., Lander, E.S., Getz, G., Sellers, W.R., Meyerson, M., 2010. The landscape of somatic copy-number alteration across human cancers. Nature 463, 899–905.

Blelloch, R.H., Hochedlinger, K., Yamada, Y., Brennan, C., Kim, M., Mintz, B., Chin, L., Jaenisch, R., 2004. Nuclear cloning of embryonal carcinoma cells. Proc. Natl. Acad. Sci. USA 101, 13985–13990.

Blount, Z.D., Barrick, J.E., Davidson, C.J., Lenski, R.E., 2012. Genomic analysis of a key innovation in an experimental Escherichia coli population. Nature 489, 513–518.

Brannon, A.R., Vakiani, E., Sylvester, B.E., Scott, S.N., McDermott, G., Shah, R.H., Kania, K., Viale, A., Oschwald, D.M., Vacic, V., Emde, A.K., Cercek, A., Yaeger, R., Kemeny, N.E., Saltz, L.B., Shia, J., D'Angelica, M.I., Weiser, M.R., Solit, D.B., Berger, M.F., 2014. Comparative sequencing analysis reveals high genomic concordance between matched primary and metastatic colorectal cancer lesions. Genome Biol. 15, 454.

Brock, A., Chang, H., Huang, S., 2009. Non-genetic heterogeneity—a mutation-independent driving force for the somatic evolution of tumours. Nat. Rev. Genet. 10, 336–342.

Crespi, B., Summers, K., 2005. Evolutionary biology of cancer. Trends Ecol. Evol. 20, 545–552.

Dal Molin, M., Zhang, M., de Wilde, R.F., Ottenhof, N.A., Rezaee, N., Wolfgang, C.L., Blackford, A., Vogelstein, B., Kinzler, K.W., Papadopoulos, N., Hruban, R.H., Maitra, A., Wood, L.D., 2015. Very

long-term survival following resection for pancreatic cancer is not explained by commonly mutated genes: results of whole-exome sequencing analysis. Clin. Cancer Res. 21, 1944–1950.

Dixon, J.R., Selvaraj, S., Yue, F., Kim, A., Li, Y., Shen, Y., Hu, M., Liu, J.S., Ren, B., 2012. Topological domains in mammalian genomes identified by analysis of chromatin interactions. Nature 485, 376–380.

Dodsworth, S., Chase, M.W., Leitch, A.R., 2016. Is post-polyploidization diploidization the key to the evolutionary success of angiosperms? Botanical J. Linnean Soc. 180, 1–5.

Duesberg, P., 2007. Chromosomal chaos and cancer. Sci. Am. 296, 52–59.

Eldredge, N., Gould, S.J., 1972. Punctuated equilibria: an alternative to phyletic gradualism. In: Schopf, T.J.M. (Ed.), Models in Paleobiology. Freeman, Cooper and Company, San Francisco, pp. 82–115.

Ellstrand, N.C., Roose, M.L., 1987. Patterns of genotypic diversity in clonal plant species. Am. J. Bot. 74, 123–131.

Feuk, L., Carson, A.R., Scherer, S.W., 2006. Structural variation in the human genome. Nat. Rev. Genet. 7, 85–97.

Garraway, L.A., Lander, E.S., 2013. Lessons from the cancer genome. Cell 153, 17–37.

Gatenby, R.A., Brown, J., Vincent, T., 2009. Lessons from applied ecology: cancer control using an evolutionary double bind. Cancer Res. 69, 7499–7502.

Gerlinger, M., McGranahan, N., Dewhurst, S.M., Burrell, R.A., Tomlinson, I., Swanton, C., 2014. Cancer: evolution within a lifetime. Annu. Rev. Genet. 48, 215–236.

Girirajan, S., Campbell, C.D., Eichler, E.E., 2011. Human copy number variation and complex genetic disease. Annu. Rev. Genet. 45, 203–226.

Gorelick, R., Heng, H.H., 2011. Sex reduces genetic variation: a multidisciplinary review. Evolution 65, 1088–1098.

Greaves, M., Maley, C.C., 2012. Clonal evolution in cancer. Nature 481, 306–313.

Heng, H.H., 2007a. Cancer genome sequencing: the challenges ahead. Bioessays 29, 783–794.

Heng, H.H., 2007b. Elimination of altered karyotypes by sexual reproduction preserves species identity. Genome 50, 517–524.

Heng, H.H., 2007c. Karyotypic chaos, a form of non-clonal chromosome aberrations, plays a key role for cancer progression and drug resistance. In: FASEB: Nuclear Structure and Cancer. Vermont Academy, Saxtons River, VT.

Heng, H.H., 2009. The genome-centric concept: resynthesis of evolutionary theory. Bioessays 31, 512–525.

Heng, H.H., 2010. Missing heritability and stochastic genome alterations. Nat. Rev. Genet. 11, 813.

Heng, H.H., 2016a. Debating Cancer: The Paradox in Cancer Research. World Scientific Publishing, Singapore.

Heng, H.H., 2016b. Heterogeneity-mediated cellular adaptation and its trade-off: searching for the general principles of diseases. J. Eval. Clin. Pract., (Epub ahead of print).

Heng, H.H., Chamberlain, J.W., Shi, X.M., Spyropoulos, B., Tsui, L.C., Moens, P.B., 1996. Regulation of meiotic chromatin loop size by chromosomal position. Proc. Natl. Acad. Sci. USA 93, 2795–2800.

Heng, H.H., Krawetz, S.A., Lu, W., Bremer, S., Liu, G., Ye, C.J., 2001. Re-defining the chromatin loop domain. Cytogenet. Cell Genet. 93, 155–161.

Heng, H.H., Goetze, S., Ye, C.J., Liu, G., Stevens, J.B., Bremer, S.W., Wykes, S.M., Bode, J., Krawetz, S.A., 2004a. Chromatin loops are selectively anchored using scaffold/matrix-attachment regions. J. Cell Sci. 117, 999–1008.

Heng, H.H., Stevens, J.B., Liu, G., Bremer, S.W., Ye, C.J., 2004b. Imaging genome abnormalities in cancer research. Cell Chromosome 3, e1.

Heng, H.H., Stevens, J.B., Liu, G., Bremer, S.W., Ye, K.J., Reddy, P.V., Wu, G.S., Wang, Y.A., Tainsky, M.A., Ye, C.J., 2006. Stochastic cancer progression driven by non-clonal chromosome aberrations. J. Cell. Physiol. 208, 461–472.

Heng, H.H., Bremer, S.W., Stevens, J.B., Ye, K.J., Liu, G., Ye, C.J., 2009. Genetic and epigenetic heterogeneity in cancer: a genome-centric perspective. J. Cell. Physiol. 220, 538–547.

Heng, H.H., Stevens, J.B., Bremer, S.W., Ye, K.J., Liu, G., Ye, C.J., 2010. The evolutionary mechanism of cancer. J. Cell. Biochem. 220, 538–547.

Heng, H.H., Liu, G., Stevens, J.B., Bremer, S.W., Ye, K.J., Abdallah, B.Y., Horne, S.D., Ye, C.J., 2011a. Decoding the genome beyond sequencing: the new phase of genomic research. Genomics 98, 242–252.

Heng, H.H., Stevens, J.B., Bremer, S.W., Liu, G., Abdallah, B.Y., Ye, C.J., 2011b. Evolutionary mechanisms and diversity in cancer. Adv. Cancer Res. 112, 217–253.

Heng, H.H., Bremer, S.W., Stevens, J.B., Horne, S.D., Liu, G., Abdallah, B.Y., Ye, K.J., Ye, C.J., 2013a. Chromosomal instability (CIN): what it is and why it is crucial to cancer evolution. Cancer Metastasis Rev. 32, 325–340.

Heng, H.H., Liu, G., Stevens, J.B., Abdallah, B.Y., Horne, S.D., Ye, K.J., Bremer, S.W., Chowdhury, S.K., Ye, C.J., 2013b. Karyotype heterogeneity and unclassified chromosomal abnormalities. Cytogenet. Genome Res. 139, 144–157.

Heng, H.H., Horne, S.D., Stevens, J.B., Abdallah, B.Y., Liu, G., Chowdhury, S.K., et al., 2016a. Heterogeneity mediated system complexity: the ultimate challenge for studying common and complex diseases. In: Sturmberg, J.P. (Ed.), The Value of Systems and Complexity Sciences for Healthcare. Springer, New York, pp. 107–120, (Chapter 9).

Heng, H.H., Regan, S.M., Liu, G., Ye, C.J., 2016b. Why it is crucial to analyze non clonal chromosome aberrations or NCCAs? Mol. Cytogenet. 13 (9), 15.

Henrichsen, C.N., Chaignat, E., Reymond, A., 2009. Copy number variants, diseases and gene expression. Hum. Mol. Genet. 18 (R1), R1–R8.

Heppner, H.G., 1984. Tumor heterogeneity. Cancer Res. 44, 2259–2265.

Hoadley, K.A., Yau, C., Wolf, D.M., Cherniack, A.D., Tamborero, D., Ng, S., Leiserson, M.D., Niu, B., McLellan, M.D., Uzunangelov, V., Zhang, J., Kandoth, C., Akbani, R., Shen, H., Omberg, L., Chu, A., Margolin, A.A., Van't Veer, L.J., Lopez-Bigas, N., Laird, P.W., Raphael, B.J., Ding, L., Robertson, A.G., Byers, L.A., Mills, G.B., Weinstein, J.N., Van Waes, C., Chen, Z., Collisson, E.A., Benz, C.C., Perou, C.M., Stuart, J.M., Cancer Genome Atlas Research Network, 2014. Multiplatform analysis of 12 cancer types reveals molecular classification within and across tissues of origin. Cell 158, 929–944.

Horne, S.D., Heng, H.H., 2014. Genome chaos, chromothripsis and cancer evolution. J. Cancer Stud. Ther. 1, 1–6.

Horne, S.D., Chowdhury, S.K., Heng, H.H., 2014. Stress, genomic adaptation, and the evolutionary trade-off. Front Genet. 5 (92), 1–6.

Horne, S.D., Stevens, J.B., Abdallah, B.Y., Liu, G., Bremer, S.W., Ye, C.J., Heng, H.H., 2013. Why imatinib remains an exception of cancer research. J. Cell. Physiol. 228, 665–670.

Horne, S.D., Pollick, S.A., Heng, H.H., 2015a. Evolutionary mechanism unifies the hallmarks of cancer. Int. J. Cancer 136, 2012–2021.

Horne, S.D., Ye, C.J., Abdallah, B.Y., Liu, G., Heng, H.H., 2015b. Cancer genome evolution. Transl. Cancer Res. 4, 303–313.

Horne, S.D., Ye, C.J., Heng, H.H., 2015c. Chromosomal instability (CIN) in cancer. eLS, 1–9.

Huang, S., 2013. Genetic and non-genetic instability in tumor progression: link between the fitness landscape and the epigenetic landscape of cancer cells. Cancer Metastasis Rev. 32, 423–448.

Huang, S., Ernberg, I., Kauffman, S., 2009. Cancer attractors: a systems view of tumors from a gene network dynamics and developmental perspective. Semin. Cell Dev. Biol. 20, 869–876.

Illmensee, K., Mintz, B., 1976. Totipotency and normal differentiation of single teratocarcinoma cells cloned by injection into blastocysts. Proc. Natl. Acad. Sci. USA 73, 549–553.

Jin, F., Li, Y., Dixon, J.R., Selvaraj, S., Ye, Z., Lee, A.Y., Yen, C.A., Schmitt, A.D., Espinoza, C.A., Ren, B., 2013. A high-resolution map of the three-dimensional chromatin interactome in human cells. Nature 503, 290–294.

Jones, P., Baylin, S., 2007. The epigenomics of cancer. Cell 128, 683–692.

Klein, C.A., 2013. Selection and adaptation during metastatic cancer progression. Nature 501, 365–372.

Konstantinidis, K.T., Tiedje, J.M., 2005. Genomic insights that advance the species definition from prokaryotes. Proc. Natl. Acad. Sci. USA 102, 2567–2572.

Korolev, K.S., Xavier, J.B., Gore, J., 2014. Turning ecology and evolution against cancer. Nat. Rev. Cancer 14, 371–380.

Lawrence, M.S., Stojanov, P., Polak, P., Kryukov, G.V., Cibulskis, K., Sivachenko, A., Carter, S.L., Stewart, C., Mermel, C.H., Roberts, S.A., Kiezun, A., Hammerman, P.S., McKenna, A., Drier, Y., Zou, L., Ramos, A.H., Pugh, T.J., Stransky, N., Helman, E., Kim, J., Sougnez, C., Ambrogio, L., Nickerson, E., Shefler, E., Cortés, M.L., Auclair, D., Saksena, G., Voet, D., Noble, M., DiCara, D., Lin, P., Lichtenstein, L., Heiman, D.I., Fennell, T., Imielinski, M., Hernandez, B., Hodis, E., Baca, S., Dulak, A.M., Lohr, J., Landau, D.A., Wu, C.J., Melendez-Zajgla, J., Hidalgo-Miranda, A., Koren, A., McCarroll, S.A., Mora, J., Lee, R.S., Crompton, B., Onofrio, R., Parkin, M., Winckler, W., Ardlie, K., Gabriel, S.B., Roberts, C.W., Biegel, J.A., Stegmaier, K., Bass, A.J., Garraway, L.A., Meyerson, M., Golub, T.R., Gordenin, D.A., Sunyaev, S., Lander, E.S., Getz, G., 2013. Mutational heterogeneity in cancer and the search for new cancer-associated genes. Nature 499, 214–218.

Liehr, T., 2016. Cytogenetically visible copy number variations (CG-CNVs) in banding and molecular cytogenetics of human; about heteromorphisms and euchromatic variants. Mol. Cytogenet. 9, 5.

Ling, S., Hu, Z., Yang, Z., Yang, F., Li, Y., Lin, P., Chen, K., Dong, L., Cao, L., Tao, Y., Hao, L., Chen, Q., Gong, Q., Wu, D., Li, W., Zhao, W., Tian, X., Hao, C., Hungate, E.A., Catenacci, D.V., Hudson, R.R., Li, W.H., Lu, X., Wu, C.I., 2015. Extremely high genetic diversity in a single tumor points to prevalence of non-Darwinian cell evolution. Proc. Natl. Acad. Sci. USA 112, E6496–E6505.

Liu, G., Stevens, J.B., Horne, S.D., Abdallah, B.Y., Ye, K.J., Bremer, S.W., Ye, C.J., Chen, D.J., Heng, H.H., 2014. Genome chaos: survival strategy during crisis. Cell Cycle 13, 528–537.

Martincorena, I., Roshan, A., Gerstung, M., Ellis, P., Van Loo, P., McLaren, S., et al., 2015. Tumor evolution. High burden and pervasive positive selection of somatic mutations in normal human skin. Science 348, 880–886.

Merlo, L.M., Pepper, J.W., Reid, B.J., Maley, C.C., 2006. Cancer as an evolutionary and ecological process. Nat. Rev. Cancer 6, 924–935.

Navin, N., Kendall, J., Troge, J., et al., 2011. Tumour evolution inferred by single-cell sequencing. Nature 472, 90–94.

Nik-Zainal, S., Alexandrov, L.B., Wedge, D.C., Van Loo, P., Greenman, C.D., Raine, K., Jones, D., Hinton, J., Marshall, J., Stebbings, L.A., Menzies, A., Martin, S., Leung, K., Chen, L., Leroy, C., Ramakrishna, M., Rance, R., Lau, K.W., Mudie, L.J., Varela, I., McBride, D.J., Bignell, G.R., Cooke, S.L., Shlien, A., Gamble, J., Whitmore, I., Maddison, M., Tarpey, P.S., Davies, H.R., Papaemmanuil, E., Stephens, P.J., McLaren, S., Butler, A.P., Teague, J.W., Jönsson, G., Garber, J.E., Silver, D., Miron, P., Fatima, A., Boyault, S., Langerød, A., Tutt, A., Martens, J.W., Aparicio, S.A., Borg, Å., Salomon, A.V., Thomas, G., Børresen-Dale, A.L., Richardson, A.L., Neuberger, M.S., Futreal, P.A., Campbell, P.J., Stratton, M.R., Breast Cancer Working Group of the International Cancer Genome Consortium, 2012. Mutational processes molding the genomes of 21 breast cancers. Cell. 149, 979–993.

Nik-Zainal, S., Kucab, J.E., Morganella, S., Glodzik, D., Alexandrov, L.B., Arlt, V.M., Weninger, A., Hollstein, M., Stratton, M.R., Phillips, D.H., 2015. The genome as a record of environmental exposure. Mutagenesis 30, 763–770.

Nowell, P.C., 1976. The clonal evolution of tumor cell populations. Science 194, 23–28.

Pavelka, N., Rancati, G., Zhu, J., Bradford, W.D., Saraf, A., Florens, L., Sanderson, B.W., Hattem, G.L., Li, R., 2010. Aneuploidy confers quantitative proteome changes and phenotypic variation in budding yeast. Nature 468, 321–325.

Pienta, K.J., McGregor, N., Axelrod, R., Axelrod, D.E., 2008. Ecological therapy for cancer: defining tumors using an ecosystem paradigm suggests new opportunities for novel cancer treatments. Transl. Oncol. 1, 158–164.

Pisco, A.O., Brock, A., Zhou, J., Moor, A., Mojtahedi, M., Jackson, D., Huang, S., 2013. Non-Darwinian dynamics in therapy-induced cancer drug resistance. Nat. Commun. 4, 2467.

Pleasance, E.D., Cheetham, R.K., Stephens, P.J., McBride, D.J., Humphray, S.J., Greenman, C.D., Varela, I., Lin, M.L., Ordóñez, G.R., Bignell, G.R., Ye, K., Alipaz, J., Bauer, M.J., Beare, D., Butler, A., Carter, R.J., Chen, L., Cox, A.J., Edkins, S., Kokko-Gonzales, P.I., Gormley, N.A., Grocock, R.J., Haudenschild, C.D., Hims, M.M., James, T., Jia, M., Kingsbury, Z., Leroy, C., Marshall, J., Menzies, A., Mudie, L.J., Ning, Z., Royce, T., Schulz-Trieglaff, O.B., Spiridou, A., Stebbings, L.A., Szajkowski, L., Teague, J., Williamson, D., Chin, L., Ross, M.T., Campbell, P.J., Bentley, D.R., Futreal, P.A., Stratton, M.R., 2010a. A comprehensive catalogue of somatic mutations from a human cancer genome. Nature 463, 191–196.

Pleasance, E.D., Stephens, P.J., O'Meara, S., McBride, D.J., Meynert, A., Jones, D., Lin, M.L., Beare, D., Lau, K.W., Greenman, C., Varela, I., Nik-Zainal, S., Davies, H.R., Ordóñez, G.R., Mudie, L.J., Latimer, C., Edkins, S., Stebbings, L., Chen, L., Jia, M., Leroy, C., Marshall, J., Menzies, A., Butler, A., Teague, J.W., Mangion, J., Sun, Y.A., McLaughlin, S.F., Peckham, H.E., Tsung, E.F., Costa, G.L., Lee, C.C., Minna, J.D., Gazdar, A., Birney, E., Rhodes, M.D., McKernan, K.J., Stratton, M.R., Futreal, P.A., Campbell, P.J., 2010b. A small-cell lung cancer genome with complex signatures of tobacco exposure. Nature 463, 184–190.

Podlaha, O., Riester, M., De, S., Michor, F., 2012. Evolution of the cancer genome. Trends Genet. 28, 155–163.

Kundaje, A., Meuleman, W., Ernst, J., Bilenky, M., Yen, A., Heravi-Moussavi, A., Kheradpour, P., Zhang, Z., Wang, J., Ziller, M.J., Amin, V., Whitaker, J.W., Schultz, M.D., Ward, L.D., Sarkar, A., Quon, G., Sandstrom, R.S., Eaton, M.L., Wu, Y.C., Pfenning, A.R., Wang, X., Claussnitzer, M., Liu, Y., Coarfa, C., Harris, R.A., Shoresh, N., Epstein, C.B., Gjoneska, E., Leung, D., Xie, W., Hawkins, R.D., Lister, R., Hong, C., Gascard, P., Mungall, A.J., Moore, R., Chuah, E., Tam, A., Canfield, T.K., Hansen, R.S., Kaul, R., Sabo, P.J., Bansal, M.S., Carles, A., Dixon, J.R., Farh, K.H., Feizi, S., Karlic, R., Kim, A.R., Kulkarni, A., Li, D., Lowdon, R., Elliott, G., Mercer, T.R., Neph, S.J., Onuchic, V., Polak, P., Rajagopal, N., Ray, P., Sallari, R.C., Siebenthall, K.T., Sinnott-Armstrong, N.A., Stevens, M., Thurman, R.E., Wu, J., Zhang, B., Zhou, X., Beaudet, A.E., Boyer, L.A., De Jager, P.L., Farnham, P.J., Fisher, S.J., Haussler, D., Jones, S.J., Li, W., Marra, M.A., McManus, M.T., Sunyaev, S., Thomson, J.A., Tlsty, T.D., Tsai, L.H., Wang, W., Waterland, R.A., Zhang, M.Q., Chadwick, L.H., Bernstein, B.E., Costello, J.F., Ecker, J.R., Hirst, M., Meissner, A., Milosavljevic, A., Ren, B., Stamatoyannopoulos, J.A., Wang, T., Kellis, M., Roadmap Epigenomics Consortium, 2015. Integrative analysis of 111 reference human epigenomes. Nature 518, 317–330.

Sexton, T., Yaffe, E., Kenigsberg, E., Bantignies, F., Leblanc, B., Hoichman, M., Parrinello, H., Tanay, A., Cavalli, G., 2012. Three-dimensional folding and functional organization principles of the *Drosophila* genome. Cell 148, 458–472.

Sottoriva, A., Kang, H., Ma, Z., Graham, T.A., Salomon, M.P., Zhao, J., Marjoram, P., Siegmund, K., Press, M.F., Shibata, D., Curtis, C., 2015. A big bang model of human colorectal tumor growth. Nat. Genet. 47, 209–216.

Stepanenko, A., Andreieva, S., Korets, K., Mykytenko, D., Huleyuk, N., Vassetzky, Y., Kavsan, V., 2015. Step-wise and punctuated genome

evolution drive phenotype changes of tumor cells. Mutat. Res. 771, 56–69.

Stephens, P.J., McBride, D.J., Lin, M.L., Varela, I., Pleasance, E.D., Simpson, J.T., Stebbings, L.A., Leroy, C., Edkins, S., Mudie, L.J., 2009. Complex landscapes of somatic rearrangement in human breast cancer genomes. Nature 462, 1005–1010.

Stephens, P.J., Greenman, C.D., Fu, B., et al., 2011. Massive genomic rearrangement acquired in a single catastrophic event during cancer development. Cell 144, 27–40.

Stevens, J.B., Horne, S.D., Abdallah, B.Y., Ye, C.J., Heng, H.H., 2013. Chromosomal instability and transcriptome dynamics in cancer. Cancer Metastasis Rev. 32, 391–402.

Stevens, J.B., Liu, G., Abdallah, B.Y., Horne, S.D., Ye, K.J., Bremer, S.W., Ye, C.Y., Krawetz, S.K., Heng, H.H., 2014. Unstable genome elevate transcriptome dynamics. Int. J. Cancer 134, 2074–2087.

Stratton, M.R., 2013. The genome of cancer cells. In: Jean Shanks Lecture.

Stratton, M.R., Campbell, P.J., Futreal, P.A., 2009. The cancer genome. Nature 458, 719–724.

Tan, S.H., Sapari, N.S., Miao, H., Hartman, M., Loh, M., Chng, W.J., Iau, P., Buhari, S.A., Soong, R., Lee, S.C., 2015. High-throughput mutation profiling changes before and 3 weeks after chemotherapy in newly diagnosed breast cancer patients. PLoS One 10, e0142466.

The Cancer Genome Atlas Research Network, 2013. The Cancer Genome Atlas Pan-Cancer analysis project. Nat. Genet. 45, 1113–1120.

The Cancer Genome Atlas Research Network, 2015. Genomic classification of cutaneous melanoma. Cell 161, 1681–1696.

Thomas, F., Fisher, D., Fort, P., Marie, J.P., Daoust, S., Roche, B., Grunau, C., Cosseau, C., Mitta, G., Baghdiguian, S., Rousset, F., Lassus, P., Assenat, E., Grégoire, D., Missé, D., Lorz, A., Billy, F., Vainchenker, W., Delhommeau, F., Koscielny, S., Itzykson, R., Tang, R., Fava, F., Ballesta, A., Lepoutre, T., Krasinska, L., Dulic, V., Raynaud, P., Blache, P., Quittau-Prevostel, C., Vignal, E., Trauchessec, H., Perthame, B., Clairambault, J., Volpert, V., Solary, E., Hibner, U., Hochberg, M.E., 2013. Applying ecological and evolutionary theory to cancer: a long and winding road. Evol. Appl. 6, 1–10.

Tomasetti, C., Marchionni, L., Nowak, M.A., Parmigiani, G., Vogelstein, B., 2015. Only three driver gene mutations are required for the development of lung and colorectal cancers. Proc. Natl. Acad. Sci. USA 112, 118–123.

Vittecoq, M., Roche, B., Daoust, S.P., Ducasse, H., Missé, D., Abadie, J., Labrut, S., Renaud, F., Gauthier-Clerc, M., Thomas, F., 2013. Cancer: a missing link in ecosystem functioning? Trends Ecol. Evol. 28, 628–635.

Vogelstein, B., 2011. Cancer genomes and their implications for curing cancer. Johns Hopkins Advanced Program, June 5.

Vogelstein, B., Papadopoulos, N., Velculescu, V.E., Zhou, S., Diaz, Jr., L.A., Kinzler, K.W., 2013. Cancer genome landscape. Science 339, 1546–1558.

Weinberg, R.A., 2014. Coming full circle-from endless complexity to simplicity and back again. Cell 157, 267–271.

Weisenberger, D.J., Liang, G., 2015. Contributions of DNA methylation aberrancies in shaping the cancer epigenome. Transl. Cancer Res. 4, 219–234.

Wilkins, A.S., Holliday, R., 2009. The evolution of meiosis from mitosis. Genetics 181, 3–12.

Williams, M.J., Werner, B., Barnes, C.P., Graham, T.A., Sottoriva, A., 2016. Identification of neutral tumor evolution across cancer types. Nat. Genet. 48, 238–244.

Winkler, H., 1920. Verbreitung und Ursache der Parthenogenesis im Pflanzen—und Tierreiche. Verlag Fischer, Jena.

Wood, L.D., Parsons, D.W., Jones, S., Lin, J., Sjöblom, T., Leary, R.J., Shen, D., Boca, S.M., Barber, T., Ptak, J., Silliman, N., Szabo, S., Dezso, Z., Ustyanksky, V., Nikolskaya, T., et al., 2007. The genomic landscapes of human breast and colorectal cancers. Science 318, 1108–1113.

Ye, C.J., Liu, G., Bremer, S.W., Heng, H.H., 2007. Cancer genomics: the dynamics of cancer chromosomes and genomes. Cytogenet. Genome Res. 118, 237–246.

Ye, C.J., Stevens, J.B., Liu, G., Bremer, S.W., Jaiswal, A.S., Ye, K.J., Lin, M.F., Lawrenson, L., Lancaster, W.D., Kurkinen, M., Liao, J.D., Gairola, C.G., Shekhar, M.P., Narayan, S., Miller, F.R., Heng, H.H., 2009. Genome based cell population heterogeneity promotes tumorigenicity: the evolutionary mechanism of cancer. J. Cell. Physiol. 219, 288–300.

Zack, T.I., Schumacher, S.E., Carter, S.L., Cherniack, A.D., Saksena, G., Tabak, B., Lawrence, M.S., Zhsng, C.Z., Wala, J., Mermel, C.H., Sougnez, C., Gabriel, S.B., Hernandez, B., Shen, H., Laird, P.W., Getz, G., Meyerson, M., Beroukhim, R., 2013. Pan-cancer patterns of somatic copy number alteration. Nat. Genet. 45, 1134–1140.

Zhang, C.Z., Spektor, A., Cornils, H., Francis, J.M., Jackson, E.K., Liu, S., Meyerson, M., Pellman, D., 2015. Chromothripsis from DNA damage in micronuclei. Nature 522, 179–184.

Zhou, J., Lemos, B., Dopman, E.B., Hartl, D.L., 2011. Copy-number variation: the balance between gene dosage and expression in *Drosophila melanogaster*. Genome Biol. Evol. 3, 1014–1024.

6

The Epigenetic Component in Cancer Evolution

Christoph Grunau

University of Perpignan Via Domitia, IHPE UMR 5244, CNRS, IFREMER,
University of Montpellier, Perpignan, France

THE GENOTYPE × ENVIRONMENT CONCEPT

As outlined in great detail in the previous chapters of this book, cancer can be viewed as an evolutionary process in which cancer cells develop a new phenotype, gain a growth advantage, and, provided that the new phenotypes escape the large number of control processes that multicellular organisms possess, outnumber the normal somatic cells. As in any evolutionary process, selection acts also here only on what is visible to the environment, that is, the phenotype. The origin of phenotypic variants had been a central question in evolutionary biology since Darwinian times. Advances in plant breeding and the generation of pure lines through self-pollination had prompted Johannsen in the beginning of the 20th century to develop the genotype–phenotype concept and to lay the bases for the understanding of heritable and nonheritable phenotypic variation (Johannsen, 1911). Based on the ancient nature versus nurture notion, this concept was later extended (Haldane, 1946), in particular to integrate data from animal breeding programs, which showed that phenotypic traits such as milk, egg, or meat production depend not only on the progenitor line but also on the conditions of upbringing, into what is today known as the genotype × environment (G×E) concept. It means that genotype × epigenotype interactions bring about the phenotype (Bowman, 1972; McBride, 1958). It should be noted that until the 1950s, "gene" was operationally defined as a "unit" or "element" and inheritance was implicit (as "genes" were in the gametes) (Johannsen, 1911). With the raise of molecular biology, this notion changed and today "genes" are considered "… DNA molecules whose specific self-replicating structure can … become translated into … a polypeptide chain"

(Jacob and Monod, 1961). The epigenotype definition followed in some senses the same logic: when the term was coined by Waddington (1942), it was related to epigenesis, that is, how genotypes give rise to phenotypes during development. More than 50 years later, in 1996, Arthur Riggs and colleagues defined epigenetics as "the study of mitotically and/or meiotically heritable changes in gene function that cannot be explained by changes in DNA sequence" (Russo et al., 1996). The interrelation of genotype and epigenotype has been described in very different and sometimes contrasting ways (Bird, 2007; Laland et al., 2014). For the purpose of this chapter and in the context of an ecological view on cancer, I propose to revisit the initial notion of G×E. Since words have a power and cannot easily be replaced even though their meaning has changed from the initial one, I suggest establishing for this chapter an extension to the G×E concept. If we go back to the original, heritability-based genotype definition, then G could be replaced by "inheritance system." This system would then be composed of at least two elements: the currently expected genotype (G) and the epigenotype (I). They interact as a dual inheritance system (G×I) with the environment (E) to bring about the phenotype (P) or (G×I)×E→P. Interaction is here used in the sense that both elements act mutually on each other and that feedback exists. The notion of a dual inheritance system was borrowed from Maynard Smith (1990) who first used it, however in a slightly different way. The elements of this system (G and I) will now be defined using their molecular nature, that is, the DNA for G, and the bearers of epigenetic information for I. The best-studied bearers of epigenetic information in cancer research are certainly DNA methylation and modification of histones, closely followed by noncoding RNA, and eventually the location of genes and chromosomes in the nucleus.

Ecology and Evolution of Cancer. http://dx.doi.org/10.1016/B978-0-12-804310-3.00006-5

DNA METHYLATION

In vertebrates, DNA methylation occurs at the 5-position of the pyrimidine ring of cytosine [5-methyl-cytosine (5mC) and 5-hydroxy-methyl-cytosine (5hmC)] and at adenine [N(6)-methyl-adenine (6mA)]. Since 6mA content is typically below 0.02 mol% in animal DNA, research focused on 5mC (~1 mol%). DNA methylation of cytosines is catalyzed by maintenance DNA methyltransferase DNMT1 and de novo methyltransferases DNMT3A and 3B (Ooi et al., 2009). Failure of methylation after replication leads to passive loss of DNA methylation, but demethylation can also actively occur via a pathway that involves ten–eleven translocation (TET) enzymes (Delatte and Fuks, 2013). In the 1970s it was found that 5mC content of tissues can be different and that immortalized cell lines loose roughly 20% of global methylation (Vanyushin et al., 1973). This rapidly motivated search for methylation changes in tumor tissue (Gama-Sosa et al., 1983) and it was subsequently discovered that cancer is characterized by global DNA hypomethylation (especially in repetitive sequences) (Ehrlich, 2002) and local DNA hypermethylation (Esteller, 2002), notably in the promotor region of tumor suppressor genes. The field of clinical epigenetics was born "… to translate that knowledge [on DNA methylation] into clinical applications" (Laird and Jaenisch, 1996). In 2008 Veridex/LabCorp started commercializing a test for prostate cancer based on methylation of glutathione s-transferase pi 1 (GSTP1) in the blood. In 2014, the US Food and Drug Administration (FDA) approved the first noninvasive test for colon cancer that is in part based on detection of aberrant 5mC patterns (Ther, 2014). In parallel, inhibitors of DNA methylating enzymes were developed to treat cancer and two inhibitors, 5-azacytidine (Vidaza) and 5-aza-2′-deoxycytidine (decitabine, Dacogen), have been approved by the FDA and European Medicines Agency (EMA). While other epigenetic information carriers were also found to be modified in cancer tissue, DNA methylation has been exhaustively studied in cancer cells. It was also recently shown that epigenetic engineering via site-specific DNA methylation can lead to long-term stable repression of the tumor suppressors in breast cancer cells by targeting DNA methyltransferase 3a (DNMT3A) to the promoters of these genes (Rivenbark et al., 2012), which established a clear proof of phenotypic effect of DNA methylation in cancer cells. Thus, while it is clear that DNA methylation can modify the cancer phenotype, and that characteristic patterns of epimutations are associated with cancer type, treatment success, and prognostic, the relative importance of mutations and epimutations in cancerogenesis is still controversial. The suite of events that lead to epimutations is also not clear.

Epimutations and the Cancer Phenotype

Two, mutually nonexclusive, scenarios have been proposed; either: (1) epimutations are a very early step in cancerogenesis and precede mutations or (2) changes in the genome and epigenome occur in parallel, and are interrelated and mutually dependent on each other.

In the first scenario, epimutations would occur with higher frequency than mutations leading to a precancerous phenotype that would eventually be fixed by mutations and evolve into uncontrolled growth. There is theoretical and empirical support for such a suite of events. Klironomos et al. (2013) developed a theoretical model of a two-component inheritance system in which a low-fidelity (epigenetic) system allows a population to explore the fitness landscape and lead to sequential fitness increases. Epigenetically based adaptive phenotypes would occur before adaptive genotypes arise. The model predicts accumulation of neutral mutations during this period and an increase in genetic variance. Eventually, genetic mutations result in high-fitness phenotypes. Once this happens, any further epimutation would dislodge the phenotype form the fitness maximum and lineages that owe their adaptive phenotype to the epigenetic variables would be counterselected. Interestingly, in a theoretical model for cancer progression, Gatenby and Vincent (2003) also predicted a two-stage evolutionary model of cancerogenesis. They did not include epimutations into their model but hypothesized that in the initial premalignant phase the local microenvironment can induce the invasive phenotype in the absence of genetic changes. Given the high sensitivity of the epigenome to environmental changes, it is easy to incorporate epimutations into their model and to keep the first step within the order of magnitude of the 40–60 generations of the Hayflick limit (in the Gatenby and Vincent theoretical model it was around 200 divisions). Cancer cells must acquire the capacity of unlimited growth before reaching this limit (Shay and Wright, 2011). The scenario of "epimutation first" would also fit with observations of epigenetic "field effects." The concept of "field effect" or "field cancerization" was introduced by Slaughter et al. (1953) to describe the occurrence of abnormal, but not yet cancerous, tissue around tumors. It was shown that aberrant methylation occurs in these precancerous cells that are adjacent to tumor cells. Lee et al. (2011) investigated DNA methylation in four genes [homeobox A9 (HOXA9), neurofilament heavy polypeptide (NEFH), ubiquitin c-terminal hydrolase l1 (UCHL1), metallothionein 1m (MT1M)] in 372 patients with squamous cell carcinoma and healthy volunteers. For patients with either head and neck cancer or esophageal cancer, biopsies for normal-appearing mucosae were obtained 3 cm above the squamocolumnar junction in the esophagus, or at least 5 cm away from the neoplastic margin. Methylation levels in the normal esophageal mucosa correlated to cumulative lifetime exposures to alcohol consumption, betel quid use, or cigarette smoking that are known cancer risk factors. Ushijima (2013) used a mouse colitis model and found changes in methylation level in three genes in noncancerous colon mucosa. The author also demonstrated that *Helicobacter pylori* infections lead to increase in methylation in seven CpG islands (e.g., regions with a high frequency of CpG sites) that maintain this state after infection had been eradicated. Both ephemeral and relatively stable (~20 years) changes in DNA methylation in response to smoking were also found in peripheral blood leukocytes (Wan et al., 2012). Interestingly, neither the Ushijima nor the Wan studies could see clear correlation of gene expression with DNA methylation levels. This is somewhat counterintuitive but was also observed in a very large metaanalysis of available next generation sequencing (NGS) data on methylation and transcription (Horvath, 2013, 2015). Other examples of DNA methylation changes in normal-appearing tissue as a result of environmental exposures are asbestos (24 CpG in pleural tissue), alcohol drinking (30 CpG in peripheral blood), tobacco smoking (138 CpG in lung tissue) (Christensen et al., 2009), and virus-associated cancers (reviewed in Minarovits et al., 2016). In summary, cancer field might represent a region with local epigenetic changes that allowed a cancer to emerge and that is then qualified as cancer "surrounding" tissue while being actually the source of the cancer. Cancer field effect does not allow concluding about the cause and effect, but further support comes from studies along the adenoma–carcinoma sequence. The adenoma–carcinoma sequence was proposed in the 1980s and describes the transformation of normal colorectal epithelium to an adenoma and ultimately to an invasive and metastatic tumor (Armaghany et al., 2012). There are known genetic alterations associated with progression into carcinoma, but around 30 hypermethylated genes were also found along the adenoma–carcinoma sequence (Patai et al., 2015). Ten of them are frequently hypermethylated in benign and malign colorectal carcinomas. Interestingly, this study revealed also that the frequency of loci with aberrant methylation was higher in premalignant tissue than in cancer. Here, the authors did not find evidence for an epigenetic field effect (at 1–10 cm away from the tumor margin) enforcing the notion of a temporal rather than spatial relation. In this first scenario, a large number of epimutations would occur in the adenomas, allowing accumulation of mutations that finally lead to emergence of one or few rapidly growing cancer clones with a specific (and less diverse) DNA methylation signature.

The interrelation of epigenotype and genotype is often seen in a way in which the epigenotype provides instructions on how to "use" the genetic information. As we have seen earlier, a general and direct impact of gene methylation on gene transcription seems actually to be an exception. Another form of interrelation in the dual inheritance system would be to directly modify the genotype. In the next subsections I will give three examples for this: long interspersed nuclear elements (LINE) hypomethylation, CpG island methylator phenotype (CIMP), and point mutations in methylated CpG.

LINE Hypomethylation

Less than 5% of the human genome codes for proteins while roughly half of it is composed of repetitive sequences. Repetitive DNA can impact the transcriptional and posttranscriptional regulation of protein expression (Hasler et al., 2007) and if they influence transcription, their mechanism of action is often based on chromatin structure changes (Neguembor and Gabellini, 2010). Repeat DNA hypomethylation is a common feature of all cancers (Ehrlich, 2002; Grunau et al., 2005) and in particular LINE-1 is often concerned (Kim et al., 2011; Pattamadilok et al., 2008). LINE-1s (17% of the human genome) are transposable elements that are in general inactive; however, some retained the ability to retrotranspose and to change the genomic DNA sequence (Lander et al., 2001). In several human cancers, for example, colorectal, lung, prostate, and ovarian cancers, somatic LINE-1 retrotransposition was found (Miousse and Koturbash, 2015). LINE-1 transcription is necessary for reinsertion, and DNA demethylation by DNA-demethylating agents (5-azacytidine) has been associated with its aberrant transcription. DNA methylation is therefore considered as the main mechanism for silencing of LINE-1–based mutagenesis. While it is still not clear whether LINE-1 demethylation and activation is a cause or a result of cancerogenesis, the existence of a LINE-1 demethylation cancer field that was found in sporadic colorectal cancer (Pavicic et al., 2012) could mean that LINE-1 hypomethylation is an early event in the process.

CpG Island Methylator Phenotype

At each mitosis the genome is replicated by a DNA polymerase. This process is constantly controlled by the proofreading function of the DNA polymerase and several other systems such as mismatch repair (MMR). MMR failure leads to microsatellite instability (MSI), that is, length polymorphisms of small stretches of repeated DNA sequences. There is a genetic condition (Lynch syndrome) that favors MSI, but in 15–20% of sporadic colorectal cancers MSI is also responsible for chromosomal instability (Armaghany et al., 2012). In a subset of these cancers an increased methylation in promotor regions of several genes including cyclin dependent kinase inhibitor 2A (CDKN2A) and the MMR gene mutl homolog 1 (MLH1) was found. For this type of cancer the term "CIMP" was coined. About 75% of sporadic MSI-positive colon cancers could be associated with CIMP and MLH1 hypermethylation (Issa, 2004). Three alternative causes for CIMP have been put forward. The first suggests that methylation changes occur initially in specific DNA sequences called methylation centers and then spread in *cis* (Turker, 2002). That would explain selectivity of methylation toward specific genes. An alternative explanation for CIMP is the aforementioned effect of environmental effects that have epimutagenic activity. Finally, CIMP could be the result of methylation changes during aging (Issa, 2004). In any case, CIMP leads via aberrant DNA methylation to genetic modifications on the level of microsatellites.

Point Mutations in Methylated CpG

Methylation of cytosine is known to be a hotspot for mutation. Spontaneous deamination of 5-methylcytosine produces thymidine. Deamination of unmethylated cytosine produces uracil, which can be removed by uracil glycosylase, whereas 5mC deamination generates thymine, which cannot be processed by this enzyme. Despite the existence of a specific repair mechanism that restores G/C mismatch, the mutation rate from 5mC to T is therefore 10-fold to 50-fold higher than that of other transitions, depending on local GC content (Fryxell and Moon, 2005). It was estimated that within 20 years, 0.17% of all 5mC was converted into thymine (Cooper and Krawczak, 1989). Long-term increase in DNA methylation would therefore directly translate into an increased point mutation rate. These mutations would occur only in regions that were initially unmethylated, producing a seemingly targeted effect.

HISTONE MODIFICATIONS

The second best-studied epigenetic information carrier is certainly covalent modifications of histones. Histones are much conserved basic proteins. Two histones H3–H4 form a stable tetramer that is flanked by two histone H2A–H2B dimers. Around this structure 147 bp of DNA is wound forming the fundamental unit of the chromatin, the nucleosome (Kornberg, 1974). Each of the core histones contains 20–30 N-terminal amino acids called the histone tail that protrudes from the nucleosome. Amino acids in these tails can be covalently modified. More than 60 histone modifications were described including acetylation, methylation, phosphorylation, ubiquitinylation, and sumoylation (e.g., Campos and Reinberg, 2009; Jenuwein and Allis, 2001; Strahl and Allis, 2000). Addition and removal of these modifications are catalyzed by histone-modifying enzymes and the

resulting modified proteins are called histone isoforms. These enzymes act in large multiprotein complexes that target and control their enzymatic activity (e.g., Greer and Shi, 2012; Roth et al., 2001; Shiio and Eisenman, 2003; Verdin et al., 2003). In addition to covalent modifications, histone tails can also be cleaved by a proteolytic activity (H3 tail clipping), and this cleavage can have impact on gene expression (Santos-Rosa et al., 2009).

In cancer cells, numerous changes in the distribution of modified histone along the genome have been reported (reviewed in Bojang and Ramos, 2014; Fullgrabe et al., 2011). In addition, the activity of histone modifying enzymes (HME) can profoundly be altered in tumor cells (reviewed in Bojang and Ramos, 2014). It was early noticed that global loss of acetylation of histone H4 at lysine 16 (H4K16Ac), together with loss of trimethylation of histone H4 at lysine 20 (H4K20me3), occurs along with DNA hypomethylation at repetitive DNA sequences (Fraga et al., 2005). In contrast, and reminiscent of DNA methylation, local enrichment of H4K20me3 has also been observed and correlated with the silencing of genes during cancer development (Fullgrabe et al., 2011). Histone modifications are of diagnostic value, and, for instance, prostate cancer can be subtyped using H4K20m1, me2, and me3 ratios (Behbahani et al., 2012). At least 10 further types of aberrant histone modifications were described in cancer, including H4K16ac, H3K4me2, and H3K4me3, the mutually exclusive methylation and acetylation of H3K9, H3K27me3, H3K56ac, H4K12ac, H3K18ac (Fullgrabe et al., 2011), and H3 tail clipping (Howe and Gamble, 2015). It is noteworthy that deregulation of HME activity is common in cancer. For instance, histone deacetylase 2 and 6 (HDAC2 and HDAC 6) deregulation are linked to breast cancer and several sporadic cancers; histone demethylase (HDM) overexpression was associated with prostate cancer progression and drug resistance (reviewed in Bojang and Ramos, 2014; Sawan and Herceg, 2010). As for DNA methylation, until very recently, all studies that linked histone modifications were correlative in nature. And as for DNA methylation, this has changed with the advent of epigenetic engineering, also called epigenome editing. For instance, a novel technique, CRISPR-Cas9-based acetyltransferase, was used to acetylate H3K27 at the enhancer of the human myogenic differentiation 1 (MYOD) locus and led to robust transcriptional activation of the gene (Hilton et al., 2015). To abolish transcription, Rodriguez and coworkers trimethylated H3K9 at the Neu/Her2 gene, and the suppression of the Her2 gene led to a significant reduction in cell proliferation (Hebert and Ward, 1972; Magnani, 2014). In addition to opening the way to new therapeutic approaches, these works demonstrate that epigenetic histone modifications can be the cause for cancer phenotypes. It is therefore obvious that there is close relation of cancer development and aberrant histone modification.

Chromatin Colors

While it is relatively clear what histone modifications are chemically and how they are produced, their epigenetic nature is still quite controversial. Most researchers would agree that, for instance, the transport forms of histones (e.g., H4K5ac, H4K12ac, and H3K56ac; reviewed in Keck and Pemberton, 2013) that cargo them from the cytoplasm to the nucleus would not qualify as epigenetic since these histones are not yet associated with the DNA. To better define which histone isoforms we will consider here, I will again call upon the notion of the inheritance system in which the bearer of epigenetic information interacts with the genetic component to bring about a phenotype under influence of the environment. The epigenetic element must interact with the genetic element (DNA). We will call this system the "chromatin." The meaning of the term has undergone considerable changes since it was introduced by Walther Flemming in 1882 who tentatively (and provisionally) used it as a name for the substances in the nucleus that could be stained with aniline dyes (Flemming, 1882). We also owe to Flemming the terms euchromatin and heterochromatin to describe open or closed chromatin structures, respectively. Today, chromatin is considered as a DNA–protein complex (Annunziato, 2008) whose specific structure controls gene expression (Strahl and Allis, 2000). This definition can lead to confusion since histone modifications that are used in the chromatin marking system are often implicitly considered as epigenetic marks. Using the systems definition then only those histone marks that (1) interact with DNA, (2) contribute to the process that produces a phenotype, and are (3) heritable would be considered as epigenetic. Evidently, here the phenotype in question is cancer and the process cancerogenesis. Since we consider in this book cancerogenesis as an evolutionary process of populations composed of cell individuals, we will need to define the time scale of this process. To do so we can base ourselves on Douglas Futuyma's famous definition of "[b]iological evolution [that] … is change in the properties of populations of organisms that transcend the lifetime of a single individual" (Futuyma, 1998). Consequently, we will consider only those histone modifications that are sufficiently stable over time to contribute to cancer development (and exclude, e.g., some histone modifications such as H3K14ac and H3K9ac that change considerably during the circadian rhythm; Etchegaray et al., 2003). This reduces the complexity of epigenetic histone modifications to roughly a dozen. However, histone modifications are known to interact, for example, H3K4me3 or phosphorylation of H3S10 blocks methylation of histone H3 lysine 9

(Binda et al., 2010; Duan et al., 2008), and clipping of H3 that occurs around position 25–30 naturally abolishes all upstream modifications such as H3K4me3. We will formalize this histone crosstalk in the framework of $(G \times I) \times E \to P$ as a system in the system or $(G \times [\text{histone } 1 \times \text{histone } 2 \times \cdots] \times E) \to P$. Recent work has shown that not all possible histone isoform interactions are equally frequent and important, and the notion of chromatin types or "colors" was developed, nicely coming back to the origins of the chromatin term (Filion et al., 2010). The number of different chromatin colors differs between different species and the way the data have been analyzed. It should be noted that nonhistone chromatin proteins such as polycomb group proteins or HP1 also play an important role in "color" determination. For humans, the Roadmap Epigenomics Consortium defined 15 chromatin colors based on the analysis of 5 core marks (H3K-27me3, H3K9me3, H3K36me3, H3K4me1, H3K4me3) in 111 individual epigenomes: (1) active transcription start sites (TSS), (2) flanking TSS, (3) transcription at genes 5′ and 3′, (4) strong transcription, (5) weak transcription, (6) genic enhancer, (7) enhancer, (8) zinc finger (ZNF) genes and repeats, (9) heterochromatin, (10) bivalent/poised TSS, (11) flanking bivalent TSS/enhancers, (12) bivalent enhancer, (13) repressed polycomb, (14) weak repressed polycomb, and (15) quiescent (absence of marks) (Kundaje et al., 2015). Despite the huge dataset, the "color" classification should be considered as preliminary since the number of marks was relatively low and some important marks were missing.

Given the strong association of histone modifications with cancer phenotypes, it is tempting to speculate that similar to the aforementioned scenario in DNA methylation, cancer cells could emerge from healthy cell populations by small histone epimutations that are favored by selection. This hypothesis is for the moment difficult to test due to important technical limitations. Histone modification profiles along genes and genomes are typically established by a method called chromatin immunoprecipitation (ChIP) (O'Neill and Turner, 2003). The technique requires 10^5–10^7 cells and can be followed by massive sequencing and a bioinformatics treatment (ChIP-seq). Current bioinformatics tools will need a four- to eightfold difference in histone modification to detect a significant difference (typically at a false discovery rate (FDR) of 0.01). This means that already a large number of cells (roughly 25%) must have adopted the new histone modification. In other words, histone modification profiles correspond actually to epiallele frequencies in a population and cannot directly be used to estimate histone modification heterogeneity in cancer tissue, as it was possible with DNA methylation that can be determined on single molecule and single base resolution. There is considerable effort to establish single-cell histone profiles and this technical issue might be solved in the near future.

Interaction Between Chromatin Colors and DNA

Histone-defined chromatin colors interact with the DNA in at least two ways: they can influence expression of the genetic information (transcription and splicing) and they interfere with the perpetuation of the genetic information (DNA replication and repair). As outlined earlier, the transcription start sites and the body of transcribed genes carry a particular chromatin color. Repainting this "color" will abolish or slow down transcription. Evidently, chromatin-mediated gene expression is important for cancer progression but also for drug resistance (Sharma et al., 2010). This is quite intuitive and rejoins the older concept of an "open" and transcriptional active and a "closed" and repressive chromatin structure. However, two examples illustrate the more complex situation with "colors": monoallelic expression (MAE) and poised transcription.

MAE is the expression of a gene that is restricted to one allele in a diploid, heterozygous genome. It occurs systematically in imprinted genes and in X-linked genes in women. Roughly 15% of our genes show random MAE. However, in brain cancer (and potentially other cancer types), MAE is higher than in normal tissue; it increases with specific tumor grade, and oncogenes show higher MAE in high-grade tumors compared with that in low-grade tumors (Walker et al., 2012). MAE is associated with a chromatin color in which H3K36me3 and H3K27me3 occur simultaneously in the gene body (Nag et al., 2013). Another peculiar type of chromatin is found in "bivalent/poised TSS" (see previous discussion). These regions are characterized by the cooccurrence of H3K4me3 and H3K27me (i.e., "bivalent" methylation). This type of chromatin was first described in a subset of genes in embryonic stem cells (Bernstein et al., 2006). The interpretation of this finding was that cells are in a poised state with the transcription machinery ready in place marked by an "active" mark (H3K4me3), but stalled by an "inactive" mark (H3K27me3). This chromatin color was later also found to be associated with drug resistance in ovarian cancer. Investigation of the TSS of 29 bivalent genes showed association with 2 additional histone marks, H3K9me2 and H3K9me3, in colon cancer cell lines (reviewed in Balch et al., 2007). Glioblastoma is a deadly cancer that originates probably from a specific population of brain tumor stem cells (BTSCs). Yoo and Bieda (2014) compared the histone modification profiles of tumor stem cells and normal human fetal neural stem cells and found 137 bivalent promoters that had lost H3K27me3 in BTSCs and 191 promoters that had lost H3K4me3, indicating a shift from "bivalent/poised TSS" toward "active transcription start sites" and "heterochromatin," respectively. This change could explain the different developmental trajectories that BTSC and normal neural stem cells will subsequently follow.

Alternative Splicing

Transcription initiation is certainly the first point that comes to mind when chromatin structure and gene expression are related. However, control of splicing might be at least as important. Splicing is executed by the *trans*-acting splicing factors that remove introns from pre-mRNA and fuse the remaining exons. A genome-wide survey of nucleosome positioning data in human T cells discovered that nucleosome density is higher over exons than over introns (Schones et al., 2008). Presumably, pausing of polymerase II (Pol II) at exons allows more time for the splicing machinery to recognize exons and splice out introns. Later, H3K36me3, H3K27me, H3K79me1/2, H4K20me1, and H2BK5me1 were found to be enriched in exons (reviewed in Zhou et al., 2014). Alternate splicing is one of the "hallmarks" of cancer progression (Oltean and Bates, 2014) and the existence of specific cancer isoforms is of diagnostic and eventually therapeutic interest. Mutations and dysregulation of expression of splice factors in cancer (Sveen et al., 2016) are confounding with a potential impact of *cis*-acting epigenetic modifications. However, both can and probably will contribute to aberrant splicing in tumor cells.

Histone Modification and DNA Replication

In addition to gene expression, the replication of DNA is another process in which histone modifications can interfere. DNA replication is a highly regulated process. For most of the genome, homologous loci replicate at the same time during S phase in a highly coordinated manner in normal cells. One well-documented change in cancer cells is aberrant asynchronous replication. Asynchronous replication can lead to mutations and gene expression changes (reviewed in Donley and Thayer, 2013). The role of chromatin structure in DNA replication was essentially studied in model organisms such as *Saccharomyces cerevisiae* and *Drosophila melanogaster* and it was found that histone acetylation and methylation close to the replication origins can affect replication timing (reviewed in Donley and Thayer, 2013; Sawan and Herceg, 2010). Given the results in these models, it is conceivable that histone modifications (and DNA methylation) can influence replication timing and synchrony in cancer cells. Not only DNA must be replicated, but its integrity is also under constant surveillance by DNA repair mechanisms. Major threats to DNA are double-strand breaks (DBS) and they must be handled by the cell. Two principal pathways are involved in DBS repairs: homologous recombination and nonhomologous end joining (reviewed in Sawan and Herceg, 2010). One of the earliest events in DBS signaling is H2A histone family member x (H2AFX) phosphorylation, which leads to the recruitment of histone acetyltransferases (HAT) to the site of breaks. Defects in H3 acetylation result in increased sensitivity toward DNA-damaging agents. Localized hypermutation due to error-prone DNA repair during replication in late S phase (kataegis) was observed in some cancer cells and provides clear evidence for the link between DNA repair and replication (reviewed in Donley and Thayer, 2013).

Histone Modifications: Cause or Consequence of Cancer?

One might now wonder whether changes in histone modifications are the cause or the consequence of malignant growth. Some could argue that this is without relevance as long as we can use histone modifications to diagnose and treat cancer. Indeed, at least seven HDAC inhibitors are currently in clinical phase I or II or already approved (reviewed in Bojang and Ramos, 2014; Lakshmaiah et al., 2014). Results are variable but encouraging and treatment can even result in complete remission when combined with a DNA methyltransferase inhibitor (Undevia et al., 2004). Inhibitors of histone methyltransferases (HMT) or histone-demethylating enzymes are known but most target cofactors or cofactor-binding sites and are not very specific. Similar to the effect on DNA methylation, environmental clues such as tobacco smoking (Sundar et al., 2014), alcohol consumption (Pan et al., 2014), and oxidative stress (reviewed in Kreuz and Fischle, 2016) can modify the global or local composition of histone isoforms. It is therefore conceivable that, similar to DNA methylation changes, modifications in the environment could lead to histone-based epimutations that trigger cancer development. However, instead of asking whether histone epimutations are the cause or consequence of malignancy, it might again be more useful to see the DNA–histone interaction as a system that can be stabilized or destabilized to direct cells toward a malignant or normal evolutionary process. The potential absence of genetic change collides with what one finds in standard textbooks of biology (e.g., Curtis and Barnes, 1989) in which evolution is defined as "… any change in frequency of alleles within a gene pool …." One should remember that such books were written for undergraduate students in which pedagogical simplifications are necessary. The view of Futuyma (1998) on this topic was that "… changes in populations that are considered evolutionary are those that are inheritable via the genetic material from one generation to the next." Here, "genetic material" is not defined by its molecular nature (DNA, RNA, etc.) but by its very nature of heritability, which includes heritable epimutations.

When clinical tumors become detectable, they measure roughly 1–2 cm in diameter (0.5–4 cm^3) (Del Monte, 2009; Lagios, 1993) containing 10^7–10^9 cells. This allows a rough estimate of the time scale of the process: assuming a single cell origin and no mortality, a detectable tumor would need 25–30 cell divisions or mitotic generations. While this estimation is certainly not entirely exact, it provides us with an order of magnitude in which epialleles must be stably inherited to be of evolutionary importance. Laird et al. (2004) measured the fidelity of cytosine methylation in human lymphocytes and found a maintenance methylation efficiency of approximately 96% per site per cell division (error rate 4/100). For histone modifications we have recently shown that transmission fidelity depends on the histone isoform (Roquis et al., 2016) in organisms that lack strong DNA methylation. Loss of heterochromatic silencing occurs in about 1 in every 1000 cell divisions (Dodson and Rine, 2015). These experiments were done in nonhuman cells, but systematic investigations in humans are missing. Nevertheless, many ChIP-seq experiments are done with human cell lines that find basically identical histone modification profiles so that we can assume roughly identical fidelity for DNA methylation and histone modifications under constant environmental conditions. This temporal stability of epimutations will be important to evaluate the role of the two other bearers of epigenetic information: nuclear topography and noncoding RNAs. While the first stores this information in a 3D coordinate system, the latter uses probably stable feedback loops.

NUCLEAR TOPOGRAPHY AND NONCODING RNA

Nuclear topography and noncoding RNAs have clear epigenetic functions since they interact with DNA and contribute to heritable phenotypes; however, due to different reasons, very little is known about their stability during cancer formation.

The Nuclear Space

A correlation of nuclear localization and chromatin structure was shown in electron micrographs more than 50 years ago (Littau et al., 1964). Further work has confirmed that condensed and transcriptional silent chromatin is preferentially located in the nuclear periphery and transcriptional competent chromatin rather in the center of the interphase nucleus (Sadoni et al., 1999). Chromosome localization can be remarkably stable and it has been proposed that in vertebrates the positions of gene-rich, early replicating and gene-poor, later-replicating chromatin have been evolutionarily conserved for more than 300 million years (Tanabe et al., 2002a,b). Human nuclei are surrounded by a nuclear envelope that consists of an inner and an outer membrane. The nuclear space is organized by a mesh of lamins and lamin-binding proteins. This nuclear lamina plays a role in chromosome positioning, DNA replication, and gene expression (reviewed in Shaklai et al., 2007). Lamins are filament proteins and in mammalian cells there are two types: A and B types. Lamins can interact directly or via adaptor proteins with histone-modifying enzymes, but the exact molecular mechanism by which lamin binding, translocation, and modification of histones are achieved is still elusive (Shaklai et al., 2007). The picture is further complicated by the fact that lamins interact also with a large number of transcription factors and members of regulatory pathways. In cancer cells abnormalities of the nuclear envelop are frequent and its altered morphology is a diagnostic tool for pathologists (reviewed in Bell and Lammerding, 2016). Decreased expression of lamin A is found in many cancers and a clinical sign of worse prognosis. Lamins are also emerging as new targets for cancer therapy (Li et al., 2013). The very nature of the nuclear lamina as structural component with potential epigenetic function could make it an element that destabilizes epigenetic heritability rather than contributing to transmission of epigenetic information through several generations. Invasiveness is one of the abnormal features of cancer cells. Cancer cells move and navigate through spaces in the extracellular matrix. These spaces (~2 µm) are smaller than the nucleus and require substantial reorganization of the nuclear matrix (Bell and Lammerding, 2016). Since the deformability of the nucleus is largely determined by A-type lamins, it makes sense to expect high invasiveness and bad prognosis in cancer types with low lamin-A content. There are several known hereditary diseases linked to mutations in lamins, but they are not associated with elevated cancer risks (Shaklai et al., 2007); it is therefore not likely that dysregulation of the lamins itself is the cause for invasiveness. For the moment it is not known how much of the information stored in the nuclear topography is erased during the migration process and it is therefore for the moment difficult to estimate the contribution of this bearer of epigenetic information to the heritability of the cancer phenotype.

Noncoding RNA

There are more than 50 classes of noncoding RNA in eukaryotic cells (Cech and Steitz, 2014). They all have in common not to code for proteins, but their properties and biological functions are very different. In cancer research, microRNAs (miRNAs) and long noncoding RNA (lncRNA) have received largest attention and both are intimately linked to chromatin structure and therefore potentially implicated in the epigenetic component of malignant evolution. miRNAs are short noncoding RNAs of 20–24 nucleotides. Primary transcripts are generated by polymerase II, cleaved and exported to the cytoplasm to be processed, and incorporated as mature

miRNA into the RNA-induced silencing complex (RISC) where it serves to target mRNA and lead to degradation of the target or repression of translation (reviewed in Kala et al., 2013). In animals, a single miRNA can target hundreds of mRNA and it is estimated that ~60% of all mRNA is under miRNA control (Bartel, 2009). miRNAs are generally downregulated in cancer and different cancers can be classified based on their miRNA profile (reviewed in Jansson and Lund, 2012). miRNA genes can be both targets for aberrant DNA methylation histone modification in cancer and regulator of it. Lujambio et al. (2008) showed, for instance, that hypermethylation of miR-148 resulted in its downregulation and the reactivation of miR-148 on treatment with a DNA-demethylating agent was associated with reduced tumor growth and inhibition of metastasis. Interestingly, miR-148 targets also de novo DNA methyltransferase DNMT3B (Duursma et al., 2008), and its deregulation could contribute to changes in DNA methylation. The same is true for the *miRNA-29* family, which is downregulated in cancers, and is complementary to DNMT3A and 3B. It was found that the enforced expression of *miR-29s* in lung cancer cell lines restored normal patterns of DNA methylation and induced reexpression of methylation-silenced tumor suppressor genes (Fabbri et al., 2007). There are also a number of miRNAs that are involved in cancer drug resistance. One is *miR-101*, which targets enhancer of zeste 2 polycomb repressive complex 2 subunit (EZH2), the enzyme responsible for trimethylating H3K27 (Sachdeva et al., 2011). The cartilage-specific miR-140 can target HDAC4 in mice (Tuddenham et al., 2006). It could therefore be that the interaction of miRNA with chromatin-modifying enzymes establishes stable feedback loops that are maintained during cell division since cytoplasmic miRNA will be transmitted to the daughter cells.

lncRNA are loosely defined as RNA molecules that are longer than 200 bp (Flippot et al., 2016) (sometimes restricted to >2 kb; Cao, 2014). About 25% of the total RNA in a cell represents lncRNA and their expression patterns allow for cancer subtype classification (Flippot et al., 2016). lncRNA can interact with proteins to guide them to DNA sequences; they can induce the correct folding of functional protein complexes and they can act as decoys for proteins with DNA-binding activity (reviewed in Flippot et al., 2016). lncRNA can thus interact with chromatin remodeling complexes at specific genomic targets. One of the first lncRNA that was described to reprogram chromatin structure in cancers was HOTAIR (Gupta et al., 2010). HOTAIR is highly expressed and overexpressed in various carcinomas and is a predictor of poor prognosis (Miao et al., 2016). HOTAIR recruits the polycomb repressive complex (PRC) 2 to the promotor regions of tumor suppressor genes leading to repression of transcription through methylation of H3K27. Other examples of cancer-promoting lncRNA are MALAT1 that influences the expression of numerous mobility-related genes (Tano et al., 2010) and ANRIL that can bind to protein regulator of cytokinesis 1 and 2 (PRC1 and PRC2) to regulate histone modification in the CDKN2A/B locus (reviewed in Congrains et al., 2013). Interestingly, lncRNA knockdown or overexpression experiment followed by microarray or RNA-seq showed that chromatin-modifying enzymes such as HDAC, EZH2 and DNMT1, and DNMT3A are themselves direct or indirect targets for lncRNA (Hou et al., 2014; Khalil et al., 2009). Consequently, and similar to miRNA, lncRNA could therefore establish stable feedback loops that allow to maintain their up- or downregulation through several cell generations.

CONCLUSIONS

Clearly, the different players of epigenetic regulations (DNA methylation, histone modification, nuclear topography, and noncoding RNAs) form a coherent interconnected system that contributes to cancer development and growth and underlies/determines the cancer phenotype.

Theoretical considerations and experimental evidence converge toward a scenario in which epigenetic changes are an early step in cancerogenesis, producing high epigenetic diversity in normal tissue that leads to the emergence of a precancerous phenotype and allows and/or favors the appearance of genetic mutations that fix the precancerous state and progress in a second step toward cancerous lesions. The initial epimutations can be provoked by exposure to epimutagens or infection.

As mentioned earlier, a second, slightly different scenario can be envisaged in which epigenetic changes do not necessarily precede genetic changes, but rather, both parts of the inheritance system work hand in hand. With the availability of massive sequencing techniques and standardized enrichment kits, in particular the Infinium Human Methylation BeadChip, it became clear that tumors show a high degree of spatial and temporal heterogeneity in genetic and epigenetic alterations, referred to as intratumor heterogeneity. Based on the idea that tumor progression corresponds to an evolutionary process, tree-building methods were used to compare genetic and epigenetic phylogenies. In at least three large studies, concordant phylogenetic relationships were established based on mutations and epimutations (Brocks et al., 2014; Mazor et al., 2015; Oakes et al., 2014). There was also strong correlation between genetic and epigenetic heterogeneity (Oakes et al., 2014). However, while it is clear that epigenetic and genetic evolutions are codependent, it was also found that the majority of mutations and epimutations occurred in different gene sets (Mazor et al., 2015). The independent findings that genetic modifications appear to be a continuous process while epimutations occur as response to environmental changes, for example (Feichtinger et al., 2016), and the observation that some genetic regions have the intrinsic property to attract epigenetic changes more easily than others, for example (Chen et al., 2016; Feinberg, 2014), have led to the idea of a "stochastic epigenetic codependency model" in which G×I are entirely interrelated. The model, put forward by Feinberg (2014), suggests that epigenetic heterogeneity occurs spontaneously in normal somatic tissue but only in specific genomic regions. In response to (repeated) changes in the microenvironment of the tissue, certain phenotypes, caused by specific combinations of genomic regions and epimutations, will be favored. As a result, specific environments (e.g., inflammation, toxins) will select specific genetic/epigenetic combinations that will have in the given environment a growth advantage and are characterized by typical genetic and epigenetic signatures. If this holds true, then diagnostic tools should not only be developed to identify cancer-type specific epimutations but also use changes in epigenetic diversity to detect early signs of cancer development (Feinberg, 2014). In addition, since epimutations are reversible, phenotypes should disappear (as long as they are not fixed by genetic changes) when the microenvironment changes. This could be important in the context of treatment-resistant cancers. Indeed, it was observed that cancer patients with treatment failure could be successfully retreated with the same anticancer drug after a "drug holiday." Sharma et al. (2010) showed that drug resistance can actually be a reversible drug-tolerant state that is based on chromatin structure changes that require histone demethylase KDM5 and that inhibition of histone deacetylases prevents drug "resistance."

Consequently, whatever scenario finally will be confirmed, integrating the epigenetic component of the inheritance system, diagnostics but also treatment of cancer can be made more efficient.

As a last note it should be mentioned that not only cancerology can profit from evolutionary and ecological thinking but also the latter from the former: clinicians are face to face with patients who wish to know "how long." In other words, they are badly in need for a prognosis. Naturally, they use any suitable source of information that is available to them and if epigenetics gives them an element of prognosis, they will use it. Time series that take into account the process character G×E→P are much more frequent in cancerology than in fundamental biology that focuses often simply on correlations (G ~ P). Learning from those approaches will be beneficial to ecologists as well as to oncologists who need to integrate epigenetics into their models.

References

Annunziato, A., 2008. DNA packaging: nucleosomes and chromatin. Nat. Educ. 1, 26.

Armaghany, T., et al., 2012. Genetic alterations in colorectal cancer. Gastrointest. Cancer Res. 5, 19–27.

Balch, C., et al., 2007. Epigenetic "bivalently marked" process of cancer stem cell-driven tumorigenesis. Bioessays 29, 842–845.

Bartel, D.P., 2009. MicroRNAs: target recognition and regulatory functions. Cell 136, 215–233.

Behbahani, T.E., et al., 2012. Alterations of global histone H4K20 methylation during prostate carcinogenesis. BMC Urol. 12, 5.

Bell, E.S., Lammerding, J., 2016. Causes and consequences of nuclear envelope alterations in tumour progression. Eur. J. Cell Biol., (Epub ahead of print).

Bernstein, B.E., et al., 2006. A bivalent chromatin structure marks key developmental genes in embryonic stem cells. Cell 125, 315–326.

Binda, O., et al., 2010. Trimethylation of histone H3 lysine 4 impairs methylation of histone H3 lysine 9: regulation of lysine methyltransferases by physical interaction with their substrates. Epigenetics 5, 767–775.

Bird, A., 2007. Perceptions of epigenetics. Nature 447, 396–398.

Bojang, P.J., Ramos, K.S., 2014. The promise and failures of epigenetic therapies for cancer treatment. Cancer Treat. Rev. 40, 153–169.

Bowman, J., 1972. Genotype × environment interactions. Ann. Genet. Sel. Anim. 4, 117–123.

Brocks, D., et al., 2014. Intratumor DNA methylation heterogeneity reflects clonal evolution in aggressive prostate cancer. Cell Rep. 8, 798–806.

Campos, E.I., Reinberg, D., 2009. Histones: annotating chromatin. Annu. Rev. Genet. 43, 559–599.

Cao, J., 2014. The functional role of long non-coding RNAs and epigenetics. Biol. Proced. Online 16, 11.

Cech, T.R., Steitz, J.A., 2014. The noncoding RNA revolution—trashing old rules to forge new ones. Cell 157, 77–94.

Chen, Y., et al., 2016. Tissue-independent and tissue-specific patterns of DNA methylation alteration in cancer. Epigenet. Chromatin 9, 10.

Christensen, B.C., et al., 2009. Aging and environmental exposures alter tissue-specific DNA methylation dependent upon CpG island context. PLoS Genet. 5, e1000602.

Congrains, A., et al., 2013. ANRIL: molecular mechanisms and implications in human health. Int. J. Mol. Sci. 14, 1278–1292.

Cooper, D.N., Krawczak, M., 1989. Cytosine methylation and the fate of CpG dinucleotides in vertebrate genomes. Hum. Genet. 83, 181–188.

Curtis, H., Barnes, N.S., 1989. Biology. Worth Publishers, New York, NY.

Del Monte, U., 2009. Does the cell number 10(9) still really fit one gram of tumor tissue? Cell Cycle 8, 505–506.

Delatte, B., Fuks, F., 2013. TET proteins: on the frenetic hunt for new cytosine modifications. Brief Funct. Genomics 12, 191–204.

Dodson, A.E., Rine, J., 2015. Heritable capture of heterochromatin dynamics in Saccharomyces cerevisiae. Elife 4, e05007.

Donley, N., Thayer, M.J., 2013. DNA replication timing, genome stability and cancer: late and/or delayed DNA replication timing is associated with increased genomic instability. Semin. Cancer Biol. 23, 80–89.

Duan, Q., et al., 2008. Phosphorylation of H3S10 blocks the access of H3K9 by specific antibodies and histone methyltransferase. Implication in regulating chromatin dynamics and epigenetic inheritance during mitosis. J. Biol. Chem. 283, 33585–33590.

Duursma, A.M., et al., 2008. miR-148 targets human DNMT3b protein coding region. RNA 14, 872–877.

Ehrlich, M., 2002. DNA methylation in cancer: too much, but also too little. Oncogene 21, 5400–5413.

Esteller, M., 2002. CpG island hypermethylation and tumor suppressor genes: a booming present, a brighter future. Oncogene 21, 5427–5440.

Etchegaray, J.P., et al., 2003. Rhythmic histone acetylation underlies transcription in the mammalian circadian clock. Nature 421, 177–182.

Fabbri, M., et al., 2007. MicroRNA-29 family reverts aberrant methylation in lung cancer by targeting DNA methyltransferases 3A and 3B. Proc. Natl. Acad. Sci. USA 104, 15805–15810.

Feichtinger, J., et al., 2016. Comprehensive genome and epigenome characterization of CHO cells in response to evolutionary pressures and over time. Biotechnol. Bioeng. 113, 2241–2253.

Feinberg, A.P., 2014. Epigenetic stochasticity, nuclear structure and cancer: the implications for medicine. J. Intern. Med. 276, 5–11.

Filion, G.J., et al., 2010. Systematic protein location mapping reveals five principal chromatin types in Drosophila cells. Cell 143, 212–224.

Flemming, W., 1882. Zellsubstanz, Kern und Zelltheilung. Vogel, Leipzig, Germany.

Flippot, R., et al., 2016. Cancer subtypes classification using long noncoding RNA. Oncotarget 7, 54082–54093.

Fraga, M.F., et al., 2005. Loss of acetylation at Lys16 and trimethylation at Lys20 of histone H4 is a common hallmark of human cancer. Nat. Genet. 37, 391–400.

Fryxell, K.J., Moon, W.J., 2005. CpG mutation rates in the human genome are highly dependent on local GC content. Mol. Biol. Evol. 22, 650–658.

Fullgrabe, J., Kavanagh, E., Joseph, B., 2011. Histone onco-modifications. Oncogene 30, 3391–3403.

Futuyma, D.J., 1998. Evolutionary Biology. Sinauer Associates, Sunderland, MA.

Gama-Sosa, M.A., et al., 1983. The 5-methylcytosine content of DNA from human tumors. Nucleic Acids Res. 11, 6883–6894.

Gatenby, R.A., Vincent, T.L., 2003. An evolutionary model of carcinogenesis. Cancer Res. 63, 6212–6220.

Greer, E.L., Shi, Y., 2012. Histone methylation: a dynamic mark in health, disease and inheritance. Nat. Rev. Genet. 13, 343–357.

Grunau, C., et al., 2005. Frequent DNA hypomethylation of human juxtacentromeric BAGE loci in cancer. Genes Chromosomes Cancer 43, 11–24.

Gupta, R.A., et al., 2010. Long non-coding RNA HOTAIR reprograms chromatin state to promote cancer metastasis. Nature 464, 1071–1076.

Haldane, J.B., 1946. The interaction of nature and nurture. Ann. Eugen. 13, 197–205.

Hasler, J., Samuelsson, T., Strub, K., 2007. Useful 'junk': Alu RNAs in the human transcriptome. Cell. Mol. Life Sci. 64, 1793–1800.

Hebert, P.D., Ward, R.D., 1972. Inheritance during parthenogenesis in Daphnia magna. Genetics 71, 639–642.

Hilton, I.B., et al., 2015. Epigenome editing by a CRISPR-Cas9-based acetyltransferase activates genes from promoters and enhancers. Nat. Biotechnol. 33, 510–517.

Horvath, S., 2013. DNA methylation age of human tissues and cell types. Genome Biol. 14, R115.

Horvath, S., 2015. Erratum to: DNA methylation age of human tissues and cell types. Genome Biol. 16, 96.

Hou, Z., et al., 2014. A long noncoding RNA Sox2ot regulates lung cancer cell proliferation and is a prognostic indicator of poor survival. Int. J. Biochem. Cell Biol. 53, 380–388.

Howe, C.G., Gamble, M.V., 2015. Enzymatic cleavage of histone H3: a new consideration when measuring histone modifications in human samples. Clin. Epigenet. 7, 7.

Issa, J.P., 2004. CpG island methylator phenotype in cancer. Nat. Rev. Cancer 4, 988–993.

Jacob, F., Monod, J., 1961. Genetic regulatory mechanisms in the synthesis of proteins. J. Mol. Biol. 3, 318–356.

Jansson, M.D., Lund, A.H., 2012. MicroRNA and cancer. Mol. Oncol. 6, 590–610.

Jenuwein, T., Allis, C.D., 2001. Translating the histone code. Science 293, 1074–1080.

Johannsen, W., 1911. The genotype conception of heredity. Am. Nat. 45, 129–159.

Kala, R., et al., 2013. MicroRNAs: an emerging science in cancer epigenetics. J. Clin. Bioinforma. 3, 6.

Keck, K.M., Pemberton, L.F., 2013. Histone chaperones link histone nuclear import and chromatin assembly. Biochim. Biophys. Acta 1819, 277–289.

Khalil, A.M., et al., 2009. Many human large intergenic noncoding RNAs associate with chromatin-modifying complexes and affect gene expression. Proc. Natl. Acad. Sci. USA 106, 11667–11672.

Kim, J.H., et al., 2011. Deep sequencing reveals distinct patterns of DNA methylation in prostate cancer. Genome Res. 21, 1028–1041.

Klironomos, F.D., Berg, J., Collins, S., 2013. How epigenetic mutations can affect genetic evolution: model and mechanism. Bioessays 35, 571–578.

Kornberg, R.D., 1974. Chromatin structure: a repeating unit of histones and DNA. Science 184, 868–871.

Kreuz, S., Fischle, W., 2016. Oxidative stress signaling to chromatin in health and disease. Epigenomics 8, 843–862.

Kundaje, A., et al., 2015. Integrative analysis of 111 reference human epigenomes. Nature 518, 317–330.

Lagios, M.D., 1993. Heterogeneity of ductal carcinoma in situ of the breast. J. Cell. Biochem Suppl. 17G, 49–52.

Laird, P.W., Jaenisch, R., 1996. The role of DNA methylation in cancer genetic and epigenetics. Annu. Rev. Genet. 30, 441–464.

Laird, C.D., et al., 2004. Hairpin-bisulfite PCR: assessing epigenetic methylation patterns on complementary strands of individual DNA molecules. Proc. Natl. Acad. Sci. USA 101, 204–209.

Lakshmaiah, K.C., et al., 2014. Epigenetic therapy of cancer with histone deacetylase inhibitors. J. Cancer Res. Ther. 10, 469–478.

Laland, K., et al., 2014. Does evolutionary theory need a rethink? Nature 514, 161–164.

Lander, E.S., et al., 2001. Initial sequencing and analysis of the human genome. Nature 409, 860–921.

Lee, Y.C., et al., 2011. Revisit of field cancerization in squamous cell carcinoma of upper aerodigestive tract: better risk assessment with epigenetic markers. Cancer Prev. Res. (Phila.) 4, 1982–1992.

Li, L., et al., 2013. Lamin B1 is a novel therapeutic target of betulinic acid in pancreatic cancer. Clin. Cancer Res. 19, 4651–4661.

Littau, V.C., et al., 1964. Active and inactive regions of nuclear chromatin as revealed by electron microscope autoradiography. Proc. Natl. Acad. Sci. USA 52, 93–100.

Lujambio, A., et al., 2008. A microRNA DNA methylation signature for human cancer metastasis. Proc. Natl. Acad. Sci. USA 105, 13556–13561.

Magnani, L., 2014. Epigenetic engineering and the art of epigenetic manipulation. Genome Biol. 15, 306.

Maynard Smith, J., 1990. Models of a dual inheritance system. J. Theor. Biol. 143, 41–53.

Mazor, T., et al., 2015. DNA methylation and somatic mutations converge on the cell cycle and define similar evolutionary histories in brain tumors. Cancer Cell 28, 307–317.

McBride, G., 1958. The environment and animal breeding problems. Anim. Breed. Abstr. 26, 340–358.

Miao, Z., et al., 2016. HOTAIR overexpression correlated with worse survival in patients with solid tumors. Minerva Med. 107, 392–400.

Minarovits, J., et al., 2016. Epigenetic dysregulation in virus-associated neoplasms. Adv. Exp. Med. Biol. 879, 71–90.

Miousse, I.R., Koturbash, I., 2015. The fine LINE: methylation drawing the cancer landscape. Biomed. Res. Int. 2015, 131547.

Nag, A., et al., 2013. Chromatin signature of widespread monoallelic expression. Elife 2, e01256.

Neguembor, M.V., Gabellini, D., 2010. In junk we trust: repetitive DNA, epigenetics and facioscapulohumeral muscular dystrophy. Epigenomics 2, 271–287.

Oakes, C.C., et al., 2014. Evolution of DNA methylation is linked to genetic aberrations in chronic lymphocytic leukemia. Cancer Discov. 4, 348–361.

Oltean, S., Bates, D.O., 2014. Hallmarks of alternative splicing in cancer. Oncogene 33, 5311–5318.

O'Neill, L.P., Turner, B.M., 2003. Immunoprecipitation of native chromatin: NChIP. Methods 31, 76–82.

Ooi, S.K., O'Donnell, A.H., Bestor, T.H., 2009. Mammalian cytosine methylation at a glance. J. Cell Sci. 122, 2787–2791.

Pan, B., et al., 2014. Alcohol consumption during gestation causes histone3 lysine9 hyperacetylation and an alternation of expression of heart development-related genes in mice. Alcohol Clin. Exp. Res. 38, 2396–2402.

Patai, A.V., et al., 2015. Comprehensive DNA methylation analysis reveals a common ten-gene methylation signature in colorectal adenomas and carcinomas. PLoS One 10, e0133836.

Pattamadilok, J., et al., 2008. LINE-1 hypomethylation level as a potential prognostic factor for epithelial ovarian cancer. Int. J. Gynecol. Cancer 18, 711–717.

Pavicic, W., et al., 2012. LINE-1 hypomethylation in familial and sporadic cancer. J. Mol. Med. 90, 827–835.

Rivenbark, A.G., et al., 2012. Epigenetic reprogramming of cancer cells via targeted DNA methylation. Epigenetics 7, 350–360.

Roquis, D., et al., 2016. Frequency and mitotic heritability of epimutations in Schistosoma mansoni. Mol. Ecol. 25, 1741–1758.

Roth, S.Y., Denu, J.M., Allis, C.D., 2001. Histone acetyltransferases. Annu. Rev. Biochem. 70, 81–120.

Russo, V.E.A., Martienssen, R.A., Riggs, A.D., 1996. Epigenetic Mechanisms of Gene Regulation. Cold Spring Harbor Laboratory Press, Plainview, NY.

Sachdeva, M., et al., 2011. MicroRNA-101-mediated Akt activation and estrogen-independent growth. Oncogene 30, 822–831.

Sadoni, N., et al., 1999. Nuclear organization of mammalian genomes. Polar chromosome territories build up functionally distinct higher order compartments. J. Cell Biol. 146, 1211–1226.

Santos-Rosa, H., et al., 2009. Histone H3 tail clipping regulates gene expression. Nat. Struct. Mol. Biol. 16, 17–22.

Sawan, C., Herceg, Z., 2010. Histone modifications and cancer. Adv. Genet. 70, 57–85.

Schones, D.E., et al., 2008. Dynamic regulation of nucleosome positioning in the human genome. Cell 132, 887–898.

Shaklai, S., et al., 2007. Gene silencing at the nuclear periphery. FEBS J. 274, 1383–1392.

Sharma, S.V., et al., 2010. A chromatin-mediated reversible drug-tolerant state in cancer cell subpopulations. Cell 141, 69–80.

Shay, J.W., Wright, W.E., 2011. Role of telomeres and telomerase in cancer. Semin. Cancer Biol. 21, 349–353.

Shiio, Y., Eisenman, R.N., 2003. Histone sumoylation is associated with transcriptional repression. Proc. Natl. Acad. Sci. USA 100, 13225–13230.

Slaughter, D.P., Southwick, H.W., Smejkal, W., 1953. Field cancerization in oral stratified squamous epithelium; clinical implications of multicentric origin. Cancer 6, 963–968.

Strahl, B.D., Allis, C.D., 2000. The language of covalent histone modifications. Nature 403, 41–45.

Sundar, I.K., et al., 2014. Cigarette smoke induces distinct histone modifications in lung cells: implications for the pathogenesis of COPD and lung cancer. J. Proteome Res. 13, 982–996.

Sveen, A., et al., 2016. Aberrant RNA splicing in cancer; expression changes and driver mutations of splicing factor genes. Oncogene 35, 2413–2427.

Tanabe, H., et al., 2002a. Non-random radial arrangements of inter-phase chromosome territories: evolutionary considerations and functional implications. Mutat. Res. 504, 37–45.

Tanabe, H., et al., 2002b. Evolutionary conservation of chromosome territory arrangements in cell nuclei from higher primates. Proc. Natl. Acad. Sci. USA 99, 4424–4429.

Tano, K., et al., 2010. MALAT-1 enhances cell motility of lung adeno-carcinoma cells by influencing the expression of motility-related genes. FEBS Lett. 584, 4575–4580.

Ther, M.L.D., 2014. A stool DNA test (Cologuard) for colorectal cancer screening. Med. Lett. Drugs Ther. 56, 100–101.

Tuddenham, L., et al., 2006. The cartilage specific microRNA-140 tar-gets histone deacetylase 4 in mouse cells. FEBS Lett. 580, 4214–4217.

Turker, M.S., 2002. Gene silencing in mammalian cells and the spread of DNA methylation. Oncogene 21, 5388–5393.

Undevia, S.D., et al., 2004. A phase I study of the oral combination of CI-994, a putative histone deacetylase inhibitor, and capecitabine. Ann. Oncol. 15, 1705–1711.

Ushijima, T., 2013. Epigenetic field for cancerization: its cause and clini-cal implications. BMC Proc. 7 (Suppl. 2), K22.

Vanyushin, B.F., et al., 1973. The content of 5-methylcytosine in animal DNA: the species and tissue specificity. Biochim. Biophys. Acta 299, 397–403.

Verdin, E., Dequiedt, F., Kasler, H.G., 2003. Class II histone deacety-lases: versatile regulators. Trends Genet. 19, 286–293.

Waddington, C.H., 1942. The epigenotype. Endeavour 1, 18–20.

Walker, E.J., et al., 2012. Monoallelic expression determines oncogenic progression and outcome in benign and malignant brain tumors. Cancer Res. 72, 636–644.

Wan, E.S., et al., 2012. Cigarette smoking behaviors and time since quit-ting are associated with differential DNA methylation across the human genome. Hum. Mol. Genet. 21, 3073–3082.

Yoo, S., Bieda, M.C., 2014. Differences among brain tumor stem cell types and fetal neural stem cells in focal regions of histone modifi-cations and DNA methylation, broad regions of modifications, and bivalent promoters. BMC Genomics 15, 724.

Zhou, H.L., et al., 2014. Regulation of alternative splicing by local his-tone modifications: potential roles for RNA-guided mechanisms. Nucleic Acids Res. 42, 701–713.

7

Evolution of Cancer Defense Mechanisms Across Species

Valerie K. Harris*,**, Joshua D. Schiffman†, Amy M. Boddy‡,§

*Cancer and Evolution Lab, Arizona State University, Tempe, AZ, United States
**School of Biological and Health Systems Engineering, Arizona State University, Tempe, AZ, United States
†Departments of Pediatrics and Oncological Sciences, Huntsman Cancer Institute, University of Utah, Salt Lake City, UT, United States
‡Department of Psychology, Arizona State University, Tempe, AZ, United States
§Center for Evolution & Cancer, University of California San Francisco, San Francisco, CA, United States

MECHANISMS OF CANCER SUPPRESSION: ORGANISMAL, MICROENVIRONMENTAL, AND MOLECULAR

Physical Mechanisms of Cancer Defense

Body Size

As body size increases over an evolutionary timescale, the need for cancer suppression mechanisms becomes much more apparent. After the origin of multicellularity, the selection pressures for cancer suppression remained high. Organisms that were able to suppress neoplastic growth while maintaining larger body size were evolutionarily favored. Overcoming the obstacles of balancing cell growth with the accumulation of oncogenic mutations is paramount to maintaining a robust organism with a larger body size that will not succumb to cancer prior to reproduction. As the number of cells increases in an organism, so does the probability of mutational events that lead to cancer. Additionally, the longer an organism's life, the more time for a cell to accumulate these genomic mutations in the DNA. However, it was observed that there is a surprising lack of cancer in large-bodied and long-lived species; this phenomenon is called "Peto's paradox" (Peto et al., 1975) (see also Chapter 1).

This paradox was reinforced by the recent exploration of cancer incidence across 36 different mammalian species. Results from these studies validated that cancer susceptibility and mortality does not always coincide with lifespan and body size (Abegglen et al., 2015). Perhaps the most notable confirmation of Peto's paradox is that of the African elephant (*Loxodonta africana*) and Asian elephant (*Elephas maximus*). Observations of deaths in over 600 African and Asian elephants revealed that the overall cancer incidence was less than 5% across species and age, compared to the 11–25% cancer risk estimated in humans; this low cancer incidence is notable because elephants are 100 times the size of humans and the expected cancer rate should be 100 times that of humans (Abegglen et al., 2015). Additionally, estimations of colon cancer in the blue whale (*Balaenoptera musculus*), based on mathematical calculations of somatic mutation rate and cellular number, suggest that 100% of blue whales would have colorectal cancer by age 90 (Caulin et al., 2015). Caulin and coworkers argue that while we currently do not know how often blue whales get colon cancer, it is improbable that all blue whales will have colon cancer by the age of 90, when the maximum lifespan of this animal is >100 years (de Magalhaes and Costa, 2009). These predicted cancer rates and limited observations also highlight an important gap within cancer research and the field of comparative oncology (see also Chapters 2 and 13). We currently have a limited understanding of how often other animals develop cancer and from what types of cancer these animals may suffer. Insights into accurate cancer incidence in other species of different sizes could play an important role in understanding cancer susceptibility among humans and may bring new perspectives on the prevention of cancer to light.

Ecology and Evolution of Cancer. http://dx.doi.org/10.1016/B978-0-12-804310-3.00007-7

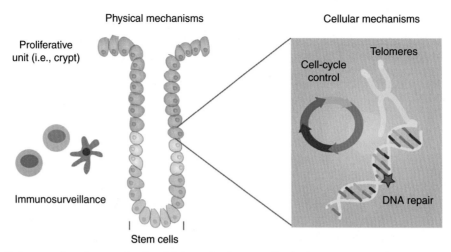

FIGURE 7.1 **Multicellular organisms have evolved multiple mechanisms to effectively suppress neoplastic growth and progression.** Here we illustrate examples of cancer defense mechanisms, including environmental (e.g., immunosurveillance), physical (e.g., proliferative units, tissue hierarchy, stem cell pools), and cellular (e.g., cell-cycle control, telomeres, and DNA repair) mechanisms.

Lifespan

It is well known that an increase in lifespan leads to a coincident increased risk for cancer (Frank, 2007). As an organism lives longer, damage accumulates over time within cells from broken chromatin and shortened telomeres, and DNA damage accumulates due to oxidative stress, ultraviolet, and ionizing radiation damage (Campisi, 2005). With the evolution of complex tissues, the ability to respond to damage and heal via self-renewal processes became paramount to increasing survival. It became necessary for the organism to control cellular divisions and have the ability to arrest those cells carrying DNA damage (mutations) before they could transform into cancer. One mechanism by which an organism can suppress cellular proliferation is senescence. Cells can withdraw from the cell-cycle process and become senescent as a response to increased age or DNA damage. This phenomenon is discussed in depth in Chapter 1.

While preventing progression and proliferation of a cell carrying DNA damage is an effective cancer defense mechanism, evidence has been found that the accumulation of senescent cells can contribute to late-in-life cancers (van Deursen, 2014). Selection for health and survival remains a weak evolutionary force in postreproductive years, and there is evidence that certain mechanisms that may benefit an organism early in life could lead to adverse health effects later in life (i.e., antagonistic pleiotropy) (Williams, 1957). Indeed, the role of senescent cells in both cancer prevention and progression could be an outcome of pleiotropic effects emphasizing the importance of lifespan when studying cancer risk across species.

Metabolic Rate

Metabolic rate is another physiological property that influences cancer risk. By-products of metabolism, such as reactive oxygen species (ROS), can cause DNA damage, ranging from point mutations, DNA breakage, and/

or large chromosomal rearrangements (Wiseman and Halliwell, 1996). While DNA damage produced from ROS is a consequence of normal cellular metabolism, it can also contribute to neoplastic initiation or fuel tumor progression; therefore, organisms may be under selection to reduce this DNA damage (Hoeijmakers, 2009; Wiseman and Halliwell, 1996). The production of ROS is positively associated with the basal metabolic rate (BMR) of an organism, whereas small-bodied mammals produced higher amounts of ROS (Adelman et al., 1988; Ku et al., 1993). Consequently, as species evolved larger bodies, they may have been subjected to selective pressure to lower their BMR to produce lower levels of ROS. This constitutional decrease in an organism's ROS production has been suggested as a cancer suppression mechanism, as lower BMR leads to a lower somatic mutation rate (Caulin and Maley, 2011; Totter, 1980). Interestingly, the cancer-resistant naked mole rats (*Heterocephalus glaber*) are reported to have lower BMR than predicted for an animal of that body mass (Caulin and Maley, 2011; de Magalhaes and Costa, 2009). This unexpectedly low BMR, combined with other physiological mechanisms of cancer suppression, may contribute to the storied cancer resistance of the naked mole rat.

Tissue Architecture

In addition to physiological means of cancer suppression, complex organisms possess several architectural mechanisms of cancer defense, which help to eliminate possibly pathogenic neoplasms by utilizing features inherent to the tissue itself. These characteristics of tissue organization function to prevent cancer at a physical level (Fig. 7.1), a crucial component of advanced multicellularity. John Cairns first described the importance of self-renewing hierarchical structures in 1975 with the discovery of compartmentalized stem cell niches for self-renewing epithelial tissues, such as the skin and gastrointestinal tract (Cairns, 1975).

The structure of organs as a whole can act as a powerful tumor suppressor. This is particularly evident in the skin and gastrointestinal tract, which are exposed to a significant amount of mutagens, microbiota, and cellular turnover throughout the organ's lifetime and still remain relatively cancer resistant (Creamer et al., 1961). These particular organs are structured to utilize a stem cell compartment, placing the most plastic and replication-competent cells at the base of the proliferative unit wherein cells undergo a differentiation process as they progress toward either the skin surface or the lumenal compartment of the gastrointestinal tract (Fig. 7.1) (Winton and Ponder, 1990). In this example, the gastrointestinal stem cells most susceptible to exposure to DNA-damaging agents and subsequent tumor transformation reside in the most distant and protected physical space in the organ itself. Furthermore, structures, such as the basement membrane provide additional walls that tumors must breach if they are to successfully invade and metastasize (Bischoff et al., 2014). All of these structural components serve to select for tumor cells when they develop that can disrupt the organ structure, either through the basement membrane or by invading neighboring proliferative units.

Stem cells and the hierarchical manner in which they divide are another key cancer suppression mechanism. Epithelial tissues undergo an astounding amount of cellular turnover within the life of an organism (Vermeulen and Snippert, 2014). These tissues need to have acquired exquisitely controlled mechanisms of cancer suppression in order to maintain a functioning organism. Accumulation of mutations in stem cells can create a mutation that is not only propagated in the stem cell itself but also exists in all of the cells that differentiated from that stem division (Jones, 2010). The consequences of accumulating mutations can be overcome by the manner in which the stem cell divides and cellular differentiation occurs. Stem cells can go through symmetric divisions in which the progeny cells from a stem cell division have equal chances of becoming either a stem cell or a differentiated cell. Each cell division on average produces one stem cell and one differentiated cell, but the chances of this occurring are completely random. By contrast, in asymmetric division, one cell is predetermined to remain a stem cell, whereas the other cell is fated to become a differentiated cell (Weinberg, 2007). Studies have shown that asymmetric division is able to reduce the impact of mutations caused by DNA synthesis by maintaining the older template strands and using those as the template by which new stem cells arise (Potten et al., 2002).

Molecular Mechanisms of Cancer Defense

There are many molecular mechanisms organisms have evolved to suppress cancer. As cancer is a disease of overproliferation, many of these mechanisms halt or put a limit on the proliferative capacity of a cell. In essence, a multicellular organism needs to suppress cell-level fitness for the fitness of the entire organisms (Grosberg and Strathmann, 2007; Michod, 2007). One example of suppressing cell-level fitness is to restrict the number of times a cell can divide (i.e., telomere length and function). Additionally, there are complex networks of signaling that suppress cellular growth (Evan and Vousden, 2001; Kastan and Bartek, 2004). In order to maintain the necessary level of proliferative control, an organism must protect itself from mutations that could lead to a fitness benefit at the cellular level (Brown and Aktipis, 2015). These protection mechanisms include halting the cell cycle at various checkpoints during proliferation to scan the genome for newly acquired mutations. If DNA damage is found, there are various cellular mechanisms in place to deal with these DNA aberrations, such as DNA repair functions. If the DNA damage cannot be repaired, other lines of defense, including cellular arrest and apoptosis (programmed cell death), can be used to reduce the probability of cancer subsequently developing from affected cells. Mutations in any of these cellular control mechanisms can enhance the probability of neoplastic progression, as these mutations can permit cells with DNA damage (and consequently genomic instability) to continue to grow and pass DNA alterations to daughter cells.

Telomeres and Telomerase Activity

Telomere length and telomerase activity can regulate cellular growth (Fig. 7.1). As described in Chapter 1, telomere dynamics across species has an important effect on cancer suppression, especially as it relates to body size and mass. In many cancers, there is a deregulation of telomere dynamics. Progressive shortening of telomeres due to replication can lead to chromosomal instability. In the case of normal cellular dynamics, critically short telomeres should induce cellular senescence. However, these shortened ends of the chromosome can lead to genome-destabilizing telomere fusions (Xu et al., 2013). Additionally, many tumors show reactivation of telomerase, which can then maintain the telomere length of the neoplastic cell, promoting the phenotype of overproliferation (Artandi and Depinho, 2010). Overexpression of telomerase in both human cell lines and mouse models triggers a variety of cancer-promoting phenotypes, including cellular transformation and cell survival (Artandi and Depinho, 2010).

Cell-Cycle Control

To reduce mutational events, cells in a multicellular body have many redundant and robust cell-cycle checkpoints (Evan and Vousden, 2001). The cell cycle is divided into four main phases: (1) G_1 phase (cell growth), (2) S phase (DNA synthesis), (3) G_2 phase (cell growth), and (4) M phase (mitosis). There are cell-cycle checkpoints at every phase of the cycle to ensure proper progression.

The majority of proteins involved in cell-cycle checkpoint are considered tumor suppressor proteins. The dominant cell-cycle checkpoint in response to DNA damage in mammalian cells is during the initial cell growth phase, G_1 (Kastan and Bartek, 2004). DNA damage in this phase can lead to G_1 arrest of the cell and initiation of DNA repair is signaled (Kastan and Bartek, 2004; Wahl and Carr, 2001). There are two critical tumor suppressor pathways that can be activated during the G_1 checkpoint, tumor protein P53 (p53) and retinoblastoma protein (pRB). These two tumor suppressor pathways may arguably be the most commonly deregulated pathways in human cancer (Sherr and McCormick, 2002; Vogelstein and Kinzler, 2004). Previous work has demonstrated functional p53 to be necessary for prolonged G_1 arrest, which can allow time for DNA repair or send signals for programmed cell death (Wahl and Carr, 2001).

While the body mass of an animal may drive the evolution of telomere regulation and replicative senescence as a cancer defense mechanism as discussed earlier, it seems that lifespan is associated with enhanced cell-cycle control mechanisms to prevent neoplastic growth. In 2008, Seluanov and colleagues reported a novel in vitro cellular phenotype of small, long-lived rodents. The fibroblasts of these rodents express telomerase and never enter replicative senescence, but their cells proliferate much slower than related shorter-lived cells. When compared to rodents of a variety of lifespans (ranging from a maximum lifespan of 4–28 years), the authors found fibroblast proliferation rates to have a negative correlation with lifespan, which was significant even after correcting for phylogeny (Seluanov et al., 2008). While there is no direct evidence that slower in vitro proliferation translates into more efficient cellular control mechanisms, it does suggest that alternative cancer defense mechanisms must be activated in these small, long-lived telomerase active fibroblasts. However, human cells do require more mutations to become immortalized than mouse cells and also require both pRB and p53 to be knocked out (Hahn and Weinberg, 2002; Rangarajan et al., 2004). These results suggest that humans have a more robust cell-cycle control (and hence slower proliferation) than mouse fibroblasts. This concept is supported by the observation that when human cells receive damage during S phase, replication is progressively slowed to coordinate with repair (Rothstein et al., 2000).

DNA Repair

DNA repair is a robust cancer defense mechanism. Mutational events are an inevitable part of cellular life, including internal metabolic damage and external sources of DNA damage (i.e., UV radiation, mutagens). Even after a somatic cell has been terminally differentiated, it can still accumulate DNA mutations in nonreplicating DNA (Lynch, 2010). Consequently, it becomes critical for an organism to be capable of repairing damaged DNA quickly and effectively in order to avoid a high burden of mutational accumulation. Just as many varied ways exist wherein DNA can be damaged, multiple ways have evolved that a cell can respond and repair damaged DNA. Responses other than DNA repair itself include halting cell-cycle progression or undergoing programmed cell death (discussed in sections "Cell-Cycle Control" and "Programmed Cell Death," respectively).

The DNA damage response pathway can be generally categorized into sensors, transducers, and effectors. DNA damage sensors recognize damage or aberrations to the genomic sequence, which can then activate signal transducers to signal the appropriate repair pathways (effectors) (Zhou and Elledge, 2000). Many of these DNA damage response genes are conserved from mammals to yeast (Zhou and Elledge, 2000). Two important kinases are considered central components to the DNA repair process, ataxia telangiectasia mutated (ATM) and ATM- and Rad3-related (ATR). These proteins are signal transducers and when activated, they begin rapid and extensive intermolecular phosphorylation of substrates that eventually leads to cell-cycle arrest (Kastan and Bartek, 2004). The loss of ATM in humans and mice increases the risk of many types of cancer, especially lymphoma (Shiloh, 2003). These signal transducers activate p53, which then requires more intracellular signaling to correctly respond to the DNA damage. The different types of DNA repair include direct repair, base excision repair, nucleotide excision repair (NER), double-strand break repair (including homologous recombination and nonhomologous end joining), and crosslink repair (Sancar et al., 2004). Deficiencies in DNA repair in humans can lead to conditions, such as xeroderma pigmentosum, wherein individuals are at a high risk of developing skin cancer. These patients have a mutation in the NER system, which removes DNA damage caused by UV radiation (Dworaczek and Xiao, 2007). Li–Fraumeni syndrome (LFS) is another cancer predisposition syndrome, where patients have p53 gene mutations leading to nearly 100% lifetime risk of cancer, multiple cancers, and early development of cancer often in childhood (McBride et al., 2014; Testa et al., 2013).

The rate and efficiency of DNA repair varies among organisms. Lower efficiency of repair rates may exert a fitness cost to the organisms and DNA repair mechanisms may have coevolved with lifespan or body size. Mammalian cell cultures exposed to ultraviolet light respond to damage at different rates and different extent to the repair; fibroblasts from short-lived species are worse at repairing DNA damage when compared to those from longer-lived organisms (Hanawalt, 2001). Additionally, there is a strong correlation between lifespan and DNA break recognition in mammals (Lorenzini et al., 2009). How robust different mechanisms of DNA repair have evolved in certain animals depends on the environmental pressures (e.g., sun exposure). Different repair mechanisms may be under stronger/weaker selection in depending on the organism's habitat (independent of body mass and lifespan).

Programmed Cell Death

An atavistic mechanism of cancer suppression is cell death by apoptosis, which simply removes the mutant cell from the population. In multicellular organisms, cell death can be more tolerated than inappropriate overproliferation. Accordingly, many physiological growth mechanisms that are responsible for tissue homeostasis are linked to apoptosis, as well as many mechanisms involved in DNA repair. Normal cells require survival signals (e.g., growth factors, cytokines, hormones) to stay alive and a lack of proper signaling can induce an apoptotic response (Elmore, 2007). Additionally, the immune system can trigger cellular death via apoptosis. Apoptotic pathways can be broken down into two main branches: intrinsic cellular signaling (via mitochondrial pathway) and extrinsic (death receptor pathway) signals (Elmore, 2007). Both pathways trigger a coordinated and energy-dependent cascade of events that lead to programmed cellular death. While apoptosis is a normal homeostatic mechanism to maintain cellular populations, other mechanisms of cellular death exist including necrosis (massive cell death) and autophagy (an orderly degradation and recycling process via the autophagosome) (Ouyang et al., 2012).

The blind mole rat (*Spalax galili*) is an excellent example of controlled cellular death as an efficient cancer suppression mechanism. This small, long-live rodent has been studied for over 40 years in a laboratory without incidence of spontaneous tumors (Gorbunova et al., 2012). In vitro manipulation of blind mole rat fibroblasts shows overproliferation leading to high expression of interferon beta (IFN-β). Cultures of blind mole rat fibroblasts that secreted IFN-β induced rapid cell death by necrosis, leaving no survivors. The mechanism of massive necrosis in these cultures is likely mediated through p53 and pRB pathways, as functional p53 and pRB were necessary for this rapid cell death mechanism (Gorbunova et al., 2012). While necrosis can be considered a less precise mechanism of cell death as the process additionally eliminates neighboring cells, it may be an efficient cancer suppression mechanism due to elimination of surrounding tumor stroma and neoplastic cells. Consequently, necrosis may actually be an appropriate cancer defense mechanism in small, long-lived species that constitutively express telomerase and have the regenerative capability to renew somatic tissue (unlike large, long-lived animals that suppress telomerase activity). Interestingly, elephant lymphocytes have a heightened response to DNA damage via apoptosis, when compared to healthy human lymphocytes and LFS lymphocytes (see subsequent discussion of p53) (Abegglen et al., 2015). Elephants are large, long-lived animals that repress telomerase in somatic tissues and, as such, may be under selection for a more controlled cellular death mechanism (i.e., apoptosis) to remove mutant cells.

Tumor Suppressor Genes

In most complex organisms, two classes of tumor suppressor genes play an important role as lifespan increases: the *gatekeepers* and the *caretakers* (Kinzler and Vogelstein, 1997). Gatekeepers serve to prevent cancer by directly inhibiting the growth of dysplastic cells through arrested mitosis or apoptosis. Gatekeeper genes can be found in many different tissue types and generally have specific functions directly related to the function of that tissue, for example, the adenomatous polyposis coli (APC) gene in the colon. Mutations within the APC gene lead to early onset development of multiple adenomatous polyps as compared to that of wild-type tissues (Markowitz and Bertagnolli, 2010; Nishisho et al., 1991). Caretaker genes function to increase the fidelity of DNA replication and DNA repair (Kinzler and Vogelstein, 1997). Inactivation of caretaker genes leads to the widespread accumulation of mutations that are more distributed throughout multiple tissue types (Kinzler and Vogelstein, 1998). A poignant example of this can be found in hereditary nonpolyposis colorectal cancer, wherein a class of DNA mismatch repair genes is mutated and causes large amounts of accumulated microsatellite instability, leading to the development of poorly differentiated neoplastic growths in the colon, endometrium, ovaries, small intestine, stomach, skin, and hepatobiliary and urinary tracts (Bellizzi and Frankel, 2009). Both gatekeeper and caretaker genes serve to guard the function and stability of complex, renewable tissues.

One of the most well-studied and crucial tumor suppressor genes is p53. p53 serves as the "guardian of the genome" and the "cellular gatekeeper" (Lane, 1992; Levine, 1997), although it also can be considered to be involved in caretaker functions in the cell. As described briefly earlier, p53 can be activated in response to DNA damage and halt the cell cycle. Subsequently, this activation can result in a variety of biological events dependent on the cue of activation, such as apoptosis, cell-cycle arrest, or senescence (Zilfou and Lowe, 2009). Mutations or loss of function of p53 have been found in the majority of human cancers (Hollstein et al., 1991; Muller and Vousden, 2013). Gene duplication events in tumor suppressor genes, such as p53, may be a potential mechanism for why some large, long-lived species develop less cancer than expected for their body size (Caulin and Maley, 2011). This prediction is supported from genetically engineered mouse models that demonstrate tumor resistance in mice with experimentally induced extra copies of p53 or cyclin dependent kinase inhibitor

2A (Cdkn2A) (Garcia-Cao et al., 2002; Matheu et al., 2004). Additionally, as discussed earlier, individuals with the inherited genetic disease LFS have only one functional allele of p53, whereas healthy individuals have two alleles. As discussed, patients with LFS have more than a 90% lifetime risk of getting cancer (Testa et al., 2013), demonstrating the crucial importance of functional p53 in cancer prevention. In 2015, Abegglen and colleagues examined the genomic sequence of the African and Asian elephants while investigating the phenomenon of less cancer in elephants than expected for a mammal of its body mass (Peto's paradox). Shockingly, the African elephant genome was found to contain 20 copies of p53 (40 alleles) and the Asian elephant was found to have 15–20 copies. The extra copies of p53 were found to lack introns, implying that they may actually be retrogenes reinserted throughout the evolution of the elephant species. The duplication likely occurred on the lineage leading to extant elephants as the hyrax, a close elephant relative (divergence of 50 million years), was found to have only one copy of p53 (two alleles). Elephant lymphocytes demonstrated higher rates of p53-mediated apoptosis in response to DNA-damaging assays when compared to healthy human lymphocytes, suggesting that these p53 retrogenes may help explain the functional mechanism of why elephants develop very little cancer (Abegglen et al., 2015).

Microenvironmental and Immune-Mediated Policing

Somatic mutations are an inevitable consequence to multicellularity and cell division. DNA repair and cell-cycle control mechanisms are not perfect (and may be vulnerable to trade-offs; see the section "A Life History Perspective: Organismal Level Trade-Offs in Cancer Suppression"), leading some cells vulnerable to the accumulation of mutations and progression to neoplastic phenotypes. However, organisms have evolved additional mechanisms at the microenvironmental level that can suppress neoplastic growth and progression if DNA repair and cell-cycle defense mechanisms should fail. One of the first lines of evidence to demonstrate the ability of the microenvironment to suppress malignancy is a study by Mintz in 1975 (Mintz and Illmensee, 1975). The researchers injected highly malignant mouse cells into a developing mouse embryo. These malignant cells, which typically would cause tumor formation when injected into an adult mouse, were found to lead to "normal" cell differentiation and development in the mouse embryo (Mintz and Illmensee, 1975). This work demonstrates that specific "antitumor" signaling from the microenvironment could override the cancerous phenotype in these cells, and that additional cancer defense mechanisms exist to deter cancer progression (Bissell and Hines, 2011).

Cell-to-Cell Contact

Cells within the multicellular body can suppress overproliferation through cell–cell contact. Normal cells can inhibit the growth of "transformed" neighboring cells simply by direct contact (Stoker et al., 1966). Direct cellular contact with normal epithelium or stroma is necessary for the survival and growth of normal cells. Tumor cells that lose direct contact can be inhibited (Altholland et al., 2005; Glick and Yuspa, 2005). Cell–cell interactions with the microenvironment may be a major reason why certain neoplastic cells never progress to malignancy (Klein, 2009). There has been specific focus in the literature on adherens junctions as an important mechanism of cell–cell contact and tumor suppression (Martin-Belmonte and Perez-Moreno, 2012). Additionally, many tumors lose the expression of E-cadherin, a structural component of these junctions, and reexpression in an in vitro 3D model leads to a reversion of "morphologically normal structures" (Wang et al., 2002).

Immune Surveillance

One of the most complex and potent forms of cancer suppression comes from tumor interactions with the host immune system (see also Chapter 9). The immune system is divided into two different branches: the *adaptive* and the *innate*. In addition to protecting us from bacterial and viral invasion, both branches of the immune system serve to eliminate neoplastic growths in unique ways. Tumoral control can also be divided into two branches: *elimination* and *equilibrium* (Koebel et al., 2007). In the course of the elimination branch, the adaptive and innate immune systems work in concert to seek out and destroy cancer cells that may express aberrant surface receptors. In the equilibrium phase, the population size of cells within a tumor is kept at a near-constant level due to consistent immune predation. Work by Koebel et al. (2007) has shown that the immune system is able to keep a large number of tumors within a small, near-undetectable population level by maintaining this equilibrium phase (Koebel et al., 2007). These small tumors below a detection level are known as "occult tumors" and can help to explain why tumor prevalence may increase after treatment with immunosuppressant drugs or transplantation of an organ into a new host naive to the presented antigen on the surface of the tumor cell (Bongartz et al., 2006; Kouklakis et al., 2013; MacKie et al., 2003).

The immune system is known to decrease in function as organismal age increases, and this may help to explain the surge in cancer rates among aging populations. This process of decreased immune function with age has been coined "immunosenescence" (Pawelec et al., 2010). The inherent immunogenicity of cancer cells can assist the immune system in controlling the tumor population, as described earlier, but this control mechanism may fail during organismal aging and immunosenescence, allowing the tumor to escape immune control and establish a clinically detectable neoplasm. Humans with immunodeficiencies are also shown to have higher incidence of cancer compared to humans with normal immune systems, serving as further evidence of the importance of immune surveillance in controlling cancer risk (Kersey et al., 1973).

Tumors may also be able to establish an immunosuppressive network as they establish malignancy by recruiting immune-suppressing cells, such as T-regulator cells and immature tumor-associated dendritic cells that can abrogate the activity of cytotoxic immune cells that previously were able to destroy tumor cells (Kim et al., 2006). This immunosuppressive process can extend outside of the primary tumor site and spread systemically, creating the ideal environment for the promotion of metastatic seeding. Understanding the complex interplay and coevolution between the host immune system of different species and tumor cells can help to direct future treatments that harness the cytotoxic abilities of the immune system to seek and destroy neoplastic growths. Already, we are seeing a surge in clinical immunotherapy for cancer with great promise (Allison, 2015; Economopoulou et al., 2016; Jacob, 2015; Ribas, 2015; Sznol and Longo, 2015).

A LIFE HISTORY PERSPECTIVE: ORGANISMAL LEVEL TRADE-OFFS IN CANCER SUPPRESSION

In the previous sections we reviewed potential cancer defense mechanisms; in the subsequent text we will discuss that despite these defenses, multicellular animals are still susceptible to cancer. However, some species are better at suppressing cancer than others. From an evolutionary perspective, there are likely costs and benefits to defenses against cancer (Boddy et al., 2015; Casás-Selves and DeGregori, 2011; Crespi and Summers, 2006). Benefits early in life that increase an organism's fitness may lead to disease susceptibility later in life, an evolutionary term we previously introduced called "antagonistic pleiotropy" (Williams, 1957). Additionally, cancer defense mechanisms, such as DNA repair, immune function, and cell-cycle control can be energetically costly. Investing in such mechanisms can lead to trade-offs in other areas, such as reproduction and growth. This is one potential explanation for why organisms vary in degree of cancer susceptibility across the tree of life and within mammals (Abegglen et al., 2015; Aktipis et al., 2015). In the subsequent text we will provide a brief review of the current knowledge of evolutionary forces potentially driving cancer risk in organisms and also provide support for relevant human implications in disease susceptibility.

Sexually Selected Traits May Increase Cancer Risk
Body Size

Fitness is a function of increasing body size in animals and, thus, sexual dimorphism in body size is a common sexually selected feature in animals (Weckerly, 1998). An example of such a trade-off exists in the species of freshwater fish, *Xiphophorus maculatus*, where males with melanoma are significantly longer and have more opportunities to mate (Fernandez and Morris, 2008; Summers and Crespi, 2010). Additional support for the link between body size and cancers can be found in the case of dwarfism. Dwarf mice with mutations in genes that encode for insulin-like growth factor (IGF1) and growth hormone (GH) have a reduced body size and reduced fecundity, but are longer-lived than wild-type mice and are cancer resistant (Bartke et al., 2013). These examples provide clues to the mating advantage of being large despite cancer risk, but also suggest weaker selection for cancer defense mechanisms.

Does this same framework apply within humans, wherein larger humans get more cancer? Although risk factors for cancer are often complex with a combination of genetic and environmental factors, there are patterns to suggest that tall stature/height is associated with certain types of cancer. A recent prospective study on the influence of body size and height in Australian men and women demonstrated that height was significantly associated with squamous cell carcinoma in men and basal cell carcinoma in women (Lahmann et al., 2016). Additionally, height was shown to elevate the risk of melanoma in men (but not women) and adult height was shown to be associated with testicular cancer and prostate cancer in men, while short stature showed protective effects for testicular cancer (Dieckmann et al., 2008; Rasmussen et al., 2003; Shors et al., 2001; Zuccolo et al., 2008). In women, there is strong evidence that adult height may be linked to breast cancer risk (Ahlgren et al., 2004; van den Brandt et al., 2000; Zhang et al., 2015). Moreover, dwarf families in Ecuador and Israel have Laron syndrome due to GH–IGF1 signaling defects and these very short individuals appear to be protected from cancer (Janecka et al., 2016; Lapkina-Gendler et al., 2016; Laron, 2015; Leslie, 2011; Printz, 2011). Also, increased size in dogs has been noted to associate with increased cancer risk, particularly for the development of canine osteosarcoma (Grüntzig et al., 2016; Ru et al., 1998; Song et al., 2013). These data support the idea that while Peto's paradox holds true across species (e.g., bigger species may get less cancer than smaller species), the same paradox does not apply *within* a species (e.g., bigger humans may get more cancer).

Rapid Growth

Large body size is important for an individual's reproductive success, but the rate at which an organism reaches an adult can influence both the fitness and the disease susceptibility of the organism. Increased rate of development (i.e., early puberty or rapid increase in body size) is important for competition in mating or could be advantageous for predator avoidance. However, there is an energetic cost to rapid growth. From a life history framework, there are likely trade-offs between growth and reproduction and somatic maintenance. Selection experiments with fruit flies (*Drosophila melanogaster*) demonstrate a fitness trade-off with rapid growth and fecundity in females when the flies were selected for rapid larval development (Nunney, 1996). Some support can be found in the literature stating that rapid growth prior to reproductive maturity may influence the risk of cancer in humans, including increased osteosarcoma risk in rapidly growing teenagers (Ahlgren et al., 2004; Arora et al., 2011; Gelberg et al., 1997; Giles et al., 2003). While the mechanisms underlying these trade-offs in humans are unclear, high levels of circulating hormones and growth factors may contribute to disease susceptibility. Additionally, faster growth may lead to shorter generation times at the cellular level, leading to faster rates of somatic evolution.

Ornamentation/Coloring

Ornamentation and coloring are two common secondary sexual characteristics in animals. Such body structures are costly displays to attract mates, sometimes through lure of the color or weapons in a battle. While these courting displays may increase an individual's reproductive success, some male-mediated exaggerated morphologies can impose costs on survival (Kotiaho, 2000; Moore et al., 2015). These effects on survivorship are likely due to energetic trade-offs between reproduction and survival. Color signals are a sign of health and resources, but can be energetically costly in animals (Doutrelant et al., 2012; Kemp et al., 2012). Resources invested in these costly traits may reduce the organism's ability to defend against cancer (Boddy et al., 2015). An interesting example of this includes the massive growths found on the antlers of free-ranging deer, which occur frequently enough to have coined the term "antleromas" (Goss, 1990; Munk et al., 2015).

Physiological Mechanisms That May Influence Cancer Susceptibility

Wound Healing

Wound healing is physiological mechanism that ties in closely to cancer susceptibility. Both neoplastic growth and wound healing require concerted communications and signaling between multiple different cell types in order to restore (or alter, in the case of neoplastic growth) tissue function (Arwert et al., 2012). For example, signal transducer and activator of transcription 3 (Stat3) is an important regulator of wound healing, but evidence suggests that Stat3 activity can also promote neoplastic

invasion (Dauer et al., 2005). In tissues with chronic damage, this wound healing can be disrupted and lead to a higher incidence of cancer in a number of different tissue types, suggesting to some that cancer may be a "deregulation of normal wound healing process" (Coussens and Werb, 2002; Dauer et al., 2005). Additionally, wound healing may require higher telomerase activity as the need for rapid cellular proliferation is increased, but as discussed earlier, high expression of telomerase can be associated with increased cancer risk (Osanai et al., 2002).

An extreme case of wound healing is the capacity for full regeneration of a tissue. One of the most notable examples of superior wound healing exists among freshwater planarians (*Planaria torva*). Planarians serve as a convenient experimental model for wound healing and cancer susceptibility due to the fact that they maintain bilateral symmetry as well as organ systems derived from all three germ layers (Newmark and Alvarado, 2002). They also possess the ability to generate entirely new planarians from very small body fragments after injury (Reddien and Alvarado, 2004). In addition to their regenerative capacity, planarians also carry many of the candidate genes that are required for carcinogenesis (Schaeffer, 1993). Another useful model for studying the relationship between wound healing in vertebrates is the axolotl (*Ambystoma mexicanum*), an amphibious animal capable of healing full-thickness excisional wounds without scarring. In the case of the axolotl, scar-free healing is a trade-off with rapid healing: axolotl wounds heal at a considerably slower rate with less immunocyte infiltration as compared to mammalian wounds (Seifert et al., 2012). While little is known on the cancer incidence of these animals, there are very limited literature reports on these species getting cancer (Aktipis et al., 2015). Wound healing may be a trade-off in some organisms, but it seems that animals with the capacity of full regeneration have additional proliferative control mechanisms to reduce the chances of neoplastic growth. Regeneration and cancer suppression may be a novel and fruitful avenue for comparative oncologists.

Placentation

Placentation and cancer invasion share similar biological processes. In placental mammals, the implantation of the embryo requires placental cells to become invasive, degrade the extracellular matrix, promote angiogenesis, migrate, and evade the maternal immune system (D'Souza and Wagner, 2014; Murray and Lessey, 1999). These characteristics of placentation lead some to hypothesize that cellular processes used for implantation may be "reused" by cancer cells during metastatic progression (Murray and Lessey, 1999). For example, the placental-specific 8 (PLAC8) protein is highly expressed in the placenta but high PLAC8 expression in tumors is also associated with a more invasive phenotype in neoplastic

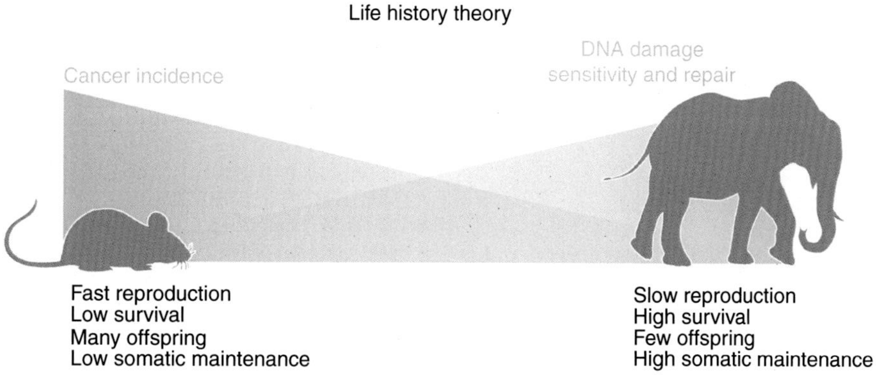

FIGURE 7.2 **Life history theory (LHT) is important for comparative oncology.** Life history strategies fall on a continuum from fast to slow, and organisms with a fast life history strategy allocate less energy to somatic maintenance (i.e., cancer defense), while slow-life-history organisms have higher investment in cancer defense mechanisms, such as DNA damage sensitivity and repair. For this evolutionary perspective, we can predict that large, long-lived animals (fast life history strategists) will invest more in cancer defenses.

murine models (Galaviz-Hernandez et al., 2003; Kaistha et al., 2016; Li et al., 2014). Indeed, surveys of cancer across various complex life stages report that mammals have higher cancer rates than birds or reptiles, although many other factors may contribute to this finding including higher BMRs, immune function, and even sampling bias, as mammals are more often studied in relation to human diseases (Aktipis et al., 2015; Effron et al., 1977) (see also Chapter 2).

Additionally, the depth of placentation varies across placental mammals (Wildman, 2016). The degree of placental invasiveness may be correlated with susceptibility to metastatic disease, where species with the least invasive placenta type have been reported to have lower rates of metastatic cancer (D'Souza and Wagner, 2014). To translate these findings into humans, we may predict that a woman's cancer risk may be higher with greater depth of placentation. There is accumulating evidence that women diagnosed with preeclampsia (characterized by abnormally shallow placentation during pregnancy) are less likely to develop breast cancer; however, these results do not hold for all populations studied (Calderon-Margalit et al., 2009; Kim et al., 2013; Pacheco et al., 2015; Vatten et al., 2002). Interestingly, a protein encoded by the KiSS-1 metastasis-suppressor (KISS1) gene is important for regulating trophoblast invasion and migration during placentation (Ohtaki et al., 2001); this protein also has been implicated in a variety of cancers, and is considered a metastasis suppressor gene (Cartwright and Williams, 2012; Hiden et al., 2007). The enhanced benefits of placental morphology, including increased nutrient transfer and faster fetal growth rates, may come at potential costs to less regulation on invasive cellular phenotypes (Boddy et al., 2015; Capellini et al., 2011).

Extrinsic Mortality

Extrinsic mortality itself can be an important driving force in the energetic investment of cancer defense

mechanisms. Environments with high extrinsic mortality (i.e., predation, starvation, accidents) and unpredictable resources tend to favor the evolution of organisms that mature early and invest in reproduction, at the cost of growth or somatic maintenance in order to successfully reproduce during their lifetime (Drenos and Kirkwood, 2005). Tumor suppression is a major component of somatic maintenance (Boddy et al., 2015; Casás-Selves and DeGregori, 2011). However, large long-lived organisms tend to occupy predictable environments with low extrinsic mortality and these species can afford to invest in growth and maintenance and delay investment in reproduction (Fig. 7.2). Investment in somatic maintenance (i.e., cancer defenses) has allowed species to extend their lifespans (Campisi, 2005; Gorbunova et al., 2014). Species that live in an environment with high extrinsic mortality are likely to die of causes other than cancer and thus have weak selection for cancer defense mechanisms (Boddy et al., 2015). More energetic investment into cancer defense mechanisms may explain why short-lived mice are cancer prone, while the long-lived naked mole rat and blind mole rat are seemingly cancer resistant, as both the naked mole rat and the blind mole rat evolved in a subterranean environment with no predators and very low levels of extrinsic mortality (Gorbunova et al., 2014).

These findings on comparative oncology and extrinsic mortality across organisms have implications in human health and disease.

There is plasticity in how an organism divides effort into reproduction, growth, and maintenance (i.e., cancer defense). Environmental cues help guide these resource allocations. Similar to life history strategies among organisms, humans who live in environments of high extrinsic mortality [low socioeconomic status (SES) environment] may invest more in reproduction than in survival (Ellis, 2004; Hidaka and Boddy, 2016). Indeed, if this is the case, we may expect to see individuals in low SES environments investing less in cancer defense mechanisms (Hidaka and Boddy, 2016). This life history framework of cancer susceptibility could prove to be very beneficial in understanding lifestyle and environmental risk factors in humans.

CONCLUSIONS

Cancer defense mechanisms vary across species. As organisms grow bigger and more complex in tissue type and cell numbers, they likely need more cancer suppression. Additionally, as species live longer, they must acquire biological ways to slow down and/or regulate somatic evolution (as the increasing probability of mutations arises). Certain species, especially the long-lived or large-bodied, have evolved different cancer defense mechanisms, such as high-molecular-mass hyaluronan in the naked mole rat and p53 amplification in elephants. Why these differences in cancer suppression mechanisms evolved is likely due to the ecological environment and the phylogenetic context of the organism. Using an evolutionary and comparative approach to study cancer defense mechanisms has implications in human health and disease. It can provide new insights to cancer treatment (e.g., p53, immunotherapy) and prevention (e.g., reduction in extrinsic mortality conditions, such as low SES). However, more information is needed on cancer incidence across species, as well as continued research into the underlying molecular, physical, and environmental mechanisms of cancer defense.

References

Abegglen, L.M., et al., 2015. Potential mechanisms for cancer resistance in elephants and comparative cellular response to DNA damage in humans. JAMA 314 (17), 1850–1860.

Adelman, R., Saul, R.L., Ames, B.N., 1988. Oxidative damage to DNA—relation to species metabolic-rate and life-span. Proc. Natl. Acad. Sci. USA 85 (8), 2706–2708.

Ahlgren, M., et al., 2004. Growth patterns and the risk of breast cancer in women. N. Engl. J. Med. 351 (16), 1619–1626.

Aktipis, C.A., et al., 2015. Cancer across the tree of life: cooperation and cheating in multicellularity. Philos. Trans. R. Soc. Lond. B Biol. Sci. 370 (1673), 20140219.

Allison, J.P., 2015. Immune checkpoint blockade in cancer therapy: the 2015 Lasker–DeBakey Clinical Medical Research Award. JAMA 314, 1113–1114.

Altholland, A., 2005. Microenvironmental control of premalignant disease: the role of intercellular adhesion in the progression of squamous cell carcinoma. Semin. Cancer Biol. 15 (2), 84–96.

Arora, R.S., et al., 2011. Relationship between height at diagnosis and bone tumours in young people: a meta-analysis. Cancer Causes Control 22, 681–688.

Artandi, S.E., Depinho, R.A., 2010. Telomeres and telomerase in cancer. Carcinogenesis 31 (1), 9–18.

Arwert, E.N., Hoste, E., Watt, F.M., 2012. Epithelial stem cells, wound healing and cancer. Nat. Rev. Cancer 12 (3), 170–180.

Bartke, A., Sun, L.Y., Longo, V., 2013. Somatotropic signaling: trade-offs between growth, reproductive development, and longevity. Physiol. Rev. 93 (2), 571–598.

Bellizzi, A.M., Frankel, W.L., 2009. Colorectal cancer due to deficiency in DNA mismatch repair function: a review. Adv. Anat. Pathol. 16 (6), 405–417.

Bischoff, S.C., et al., 2014. Intestinal permeability—a new target for disease prevention and therapy. BMC Gastroenterol. 14 (1), 189.

Bissell, M.J., Hines, W.C., 2011. Why don't we get more cancer? A proposed role of the microenvironment in restraining cancer progression. Nat. Med. 17 (3), 320–329.

Boddy, A.M., et al., 2015. Cancer susceptibility and reproductive trade-offs: a model of the evolution of cancer defences. Philos. Trans. R. Soc. Lond. B Biol. Sci. 370 (1673), 20140220.

Bongartz, T., et al., 2006. Anti-TNF antibody therapy in rheumatoid arthritis and the risk of serious infections and malignancies: systematic review and meta-analysis of rare harmful effects in randomized controlled trials. JAMA 295, 2275–2285.

Brown, J.S., Aktipis, C.A., 2015. Inclusive fitness effects can select for cancer suppression into old age. Philos. Trans. R. Soc. Lond. B Biol. Sci. 370 (1673), 20150160.

Cairns, J., 1975. Mutation selection and the natural history of cancer. Nature 255, 197–200.

Calderon-Margalit, R., et al., 2009. Preeclampsia and subsequent risk of cancer: update from the Jerusalem Perinatal Study. Am. J. Obstet. Gynecol. 200 (1), 5.

Campisi, J., 2005. Senescent cells, tumor suppression, and organismal aging: good citizens, bad neighbors. Cell 120 (4), 513–522.

Capellini, I., Venditti, C., Barton, R.A., 2011. Placentation and maternal investment in mammals. Am. Nat. 177 (1), 86–98.

Cartwright, J.E., Williams, P.J., 2012. Altered placental expression of kisspeptin and its receptor in pre-eclampsia. J. Endocrinol. 214 (1), 79–85.

Casás-Selves, M., DeGregori, J., 2011. How cancer shapes evolution and how evolution shapes cancer. Evolution (N. Y.) 4 (4), 624–634.

Caulin, A.F., Maley, C.C., 2011. Peto's paradox: evolution's prescription for cancer prevention. Trends Ecol. Evol. 26 (4), 175–182.

Caulin, A.F., et al., 2015. Solutions to Peto's paradox revealed by mathematical modelling and cross-species cancer gene analysis. Philos. Trans. R. Soc. Lond. B Biol. Sci. 370 (1673), 20140222.

Coussens, L.M., Werb, Z., 2002. Inflammation and cancer. Nature 420 (6917), 860–867.

Creamer, B., Shorter, R.G., Bamforth, J., 1961. The turnover and shedding of epithelial cells. Part I. The turnover in the gastro-intestinal tract. Gut 2 (2), 110–116.

Crespi, B.J., Summers, K., 2006. Positive selection in the evolution of cancer. Biol. Rev. 81 (3), 407–424.

D'Souza, A.W., Wagner, G.P., 2014. Malignant cancer and invasive placentation: a case for positive pleiotropy between endometrial and malignancy phenotypes. Evol. Med. Public Health 2014 (1), 136–145.

Dauer, D.J., et al., 2005. Stat3 regulates genes common to both wound healing and cancer. Oncogene 24 (21), 3397–3408.

de Magalhaes, J.P., Costa, J., 2009. A database of vertebrate longevity records and their relation to other life-history traits. J. Evol. Biol. 22 (8), 1770–1774.

Dieckmann, K.-P., et al., 2008. Tallness is associated with risk of testicular cancer: evidence for the nutrition hypothesis. Br. J. Cancer 99 (9), 1517–1521.

Doutrelant, C., et al., 2012. Female plumage coloration is sensitive to the cost of reproduction. An experiment in blue tits. J. Anim. Ecol. 81 (1), 87–96.

Drenos, F., Kirkwood, T., 2005. Modelling the disposable soma theory of ageing. Mech. Ageing Dev. 126 (1), 99–103.

Dworaczek, H., Xiao, W., 2007. Xeroderma pigmentosum: a glimpse into nucleotide excision repair, genetic instability, and cancer. Crit. Rev. Oncog. 13 (2), 159–177.

Economopoulou, P., et al., 2016. The promise of immunotherapy in head and neck squamous cell carcinoma. Ann. Oncol. 27, 1675–1685.

Effron, M., Griner, L., Benirschike, K., 1977. Nature and rate of neoplasia found in captive wild mammals, birds, and reptiles at necropsy. J. Natl. Cancer Inst. 59 (1), 185–198.

Ellis, B.J., 2004. Timing of pubertal maturation in girls: an integrated life history approach. Psychol. Bull. 130 (6), 920–958.

Elmore, S., 2007. Apoptosis: a review of programmed cell death. Toxicol. Pathol. 35 (4), 495–516.

Evan, G.I., Vousden, K.H., 2001. Proliferation, cell cycle and apoptosis in cancer. Nature 411 (6835), 342–348.

Fernandez, A.A., Morris, M.R., 2008. Mate choice for more melanin as a mechanism to maintain a functional oncogene. Proc. Natl. Acad. Sci. USA 105 (36), 13503–13507.

Frank, S.A., 2007. In: Allen Orr, H. (Ed.), Dynamics of Cancer. Princeton University Press, Princeton.

Galaviz-Hernandez, C., et al., 2003. Plac8 and Plac9, novel placental-enriched genes identified through microarray analysis. Gene 309 (2), 81–89.

Garcia-Cao, I., et al., 2002. "Super p53" mice exhibit enhanced DNA damage response, are tumor resistant and age normally. EMBO J. 21 (22), 6225–6235.

Gelberg, K.H., et al., 1997. Growth and development and other risk factors for osteosarcoma in children and young adults. Int. J. Epidemiol. 26, 272–278.

Giles, G.G., et al., 2003. Early growth, adult body size and prostate cancer risk. Int. J. Cancer 103 (2), 241–245.

Glick, A., Yuspa, S., 2005. Tissue homeostasis and the control of the neoplastic phenotype in epithelial cancers. Semin. Cancer Biol. 15 (2), 75–83.

Gorbunova, V., et al., 2012. Cancer resistance in the blind mole rat is mediated by concerted necrotic cell death mechanism. Proc. Natl. Acad. Sci. USA 109 (47), 19392–19396.

Gorbunova, V., et al., 2014. Comparative genetics of longevity and cancer: insights from long-lived rodents. Nat. Rev. Genet. 15 (8), 531–540.

Goss, R.J., 1990. Tumor-like growth of antlers in castrated fallow deer—an electron-microscopic study. Scanning Microsc. 4 (3), 715–721.

Grosberg, R.K., Strathmann, R.R., 2007. The evolution of multicellularity: a minor major transition? Annu. Rev. Ecol. Evol. Syst. 38 (1), 621–654.

Grüntzig, K., et al., 2016. Swiss Canine Cancer Registry 1955–2008: occurrence of the most common tumour diagnoses and influence of age, breed, body size, sex and neutering status on tumour development. J. Comp. Pathol. 155, 156–170.

Hahn, W.C., Weinberg, R.A., 2002. Mechanisms of disease: rules for making human tumor cells. N. Engl. J. Med. 347 (20), 1593–1603.

Hanawalt, P.C., 2001. Revisiting the rodent repairadox. Environ. Mol. Mutagen. 38 (2–3), 89–96.

Hidaka, B.H., Boddy, A.M., 2016. Is estrogen receptor negative breast cancer risk associated with a fast life history strategy? Evol. Med. Public Health 2016 (1), 17–20.

Hiden, U., et al., 2007. Kisspeptins and the placenta: regulation of trophoblast invasion. Rev. Endocr. Metab. Disord. 8 (1), 31–39.

Hoeijmakers, J.H.J., 2009. DNA damage, aging, and cancer. N. Engl. J. Med. 361 (15), 1475–1485.

Hollstein, M., et al., 1991. P53 mutations in human cancers. Science 253 (5015), 49–53.

Jacob, J.A., 2015. Cancer immunotherapy researchers focus on refining checkpoint blockade therapies. JAMA 314, 2117–2119.

Janecka, A., Kołodziej-Rzepa, M., Biesaga, B., 2016. Clinical and molecular features of Laron syndrome, a genetic disorder protecting from cancer. In Vivo 30, 375–381.

Jones, P.H., 2010. Stem cell fate in proliferating tissues: equal odds in a game of chance. Dev. Cell 19 (4), 489–490.

Kaistha, B.P., et al., 2016. PLAC8 localizes to the inner plasma membrane of pancreatic cancer cells and regulates cell growth and disease progression through critical cell-cycle regulatory pathways. Cancer Res. 76 (1), 96–107.

Kastan, M.B., Bartek, J., 2004. Cell-cycle checkpoints and cancer. Nature 432 (7015), 316–323.

Kemp, D.J., Herberstein, M.E., Grether, G.F., 2012. Unraveling the true complexity of costly color signaling. Behav. Ecol. 23 (2), 233–236.

Kersey, J.H., Spector, B.D., Good, R.A., 1973. Immunodeficiency and cancer. Adv. Cancer Res. 18, 211–230.

Kim, R., et al., 2006. Tumor-driven evolution of immunosuppressive networks during malignant progression. Cancer Res. 66 (11), 5527–5536.

Kim, J.S., et al., 2013. The relationship between preeclampsia, pregnancy-induced hypertension and maternal risk of breast cancer: a meta-analysis. Acta Oncol. 52 (8), 1643–1648.

Kinzler, K.W., Vogelstein, B., 1997. Cancer-susceptibility genes—gatekeepers and caretakers. Nature 386 (6627), 761, 763.

Kinzler, K.W., Vogelstein, B., 1998. Landscaping the cancer terrain. Science 280 (5366), 1036–1037.

Klein, G., 2009. Toward a genetics of cancer resistance. Proc. Natl. Acad. Sci. USA 106 (3), 859–863.

Koebel, C.M., et al., 2007. Adaptive immunity maintains occult cancer in an equilibrium state. Nature 450 (7171), 903–907.

Kotiaho, J.S., 2000. Testing the assumptions of conditional handicap theory: costs and condition dependence of a sexually selected trait. Behav. Ecol. Sociobiol. 48 (3), 188–194.

Kouklakis, G., Efremidou, E.I., Pitiakoudis, M., 2013. Development of primary malignant melanoma during treatment with a TNF-α antagonist for severe Crohn's disease: a case report and review of the hypothetical association between TNF-α blockers and cancer. Drug Des. Dev. 7, 195–199.

Ku, H.H., Brunk, U.T., Sohal, R.S., 1993. Relationship between mitochondrial superoxide and hydrogen-peroxide production and longevity of mammalian-species. Free Radic. Biol. Med. 15 (6), 621–627.

Lahmann, P.H., et al., 2016. A prospective study of measured body size and height and risk of keratinocyte cancers and melanoma. Cancer Epidemiol. 40, 119–125.

Lane, D.P., 1992. p53, guardian of the genome. Nature 358 (6381), 15–16.

Lapkina-Gendler, L., et al., 2016. Identification of signaling pathways associated with cancer protection in Laron syndrome. Endocr. Relat. Cancer 23 (5), 399–410.

Laron, Z., 2015. Lessons from 50 years of study of Laron syndrome. Endocr. Pract. 21 (12), 1395–1402.

Leslie, M., 2011. Genetics and disease. Growth defect blocks cancer and diabetes. Science 331 (6019), 837.

Levine, A.J., 1997. p53, the cellular gatekeeper for growth and division. Cell 88 (3), 323–331.

Li, C., et al., 2014. Excess PLAC8 promotes an unconventional ERK2-dependent EMT in colon cancer. J. Clin. Invest. 124 (5), 2172–2187.

Lorenzini, A., et al., 2009. Significant correlation of species longevity with DNA double strand break recognition but not with telomere length. Mech. Ageing Dev. 130 (11–12), 784–792.

Lynch, M., 2010. Rate, molecular spectrum, and consequences of human mutation. Proc. Natl. Acad. Sci. USA 107 (3), 961–968.

MacKie, R.M., Reid, R., Junor, B., 2003. Fatal melanoma transferred in a donated kidney 16 years after melanoma surgery. N. Engl. J. Med. 348 (6), 567–568.

Markowitz, S.D., Bertagnolli, M.M., 2010. Molecular origins of cancer: molecular basis of colorectal cancer. N. Engl. J. Med. 361 (25), 2449–2460.

Martin-Belmonte, F., Perez-Moreno, M., 2012. Epithelial cell polarity, stem cells and cancer. Nat. Rev. Cancer 12 (1), 23–38.

Matheu, A., et al., 2004. Increased gene dosage of Ink4a/Arf results in cancer resistance and normal aging. Genes Dev. 18 (22), 2736–2746.

McBride, K.A., et al., 2014. Li–Fraumeni syndrome: cancer risk assessment and clinical management. Nat. Rev. Clin. Oncol. 11, 260–271.

Michod, R.E., 2007. Evolution of individuality during the transition from unicellular to multicellular life. Proc. Natl. Acad. Sci. USA 104 (Suppl. 1), 8613–8618.

Mintz, B., Illmensee, K., 1975. Normal genetically mosaic mice produced from malignant teratocarcinoma cells. Proc. Natl. Acad. Sci. USA 72 (9), 3585–3589.

Moore, F.R., et al., 2015. Investment in a sexual signal results in reduced survival under extreme conditions in the male great tit (*Parus major*). Behav. Ecol. Sociobiol. 69 (1), 151–158.

Muller, P.A.J., Vousden, K.H., 2013. p53 mutations in cancer. Nat. Cell Biol. 15 (1), 2–8.

Munk, B.A., et al., 2015. Antleroma in a free-ranging white-tailed deer (*Odocoileus virginianus*). Vet. Pathol. Online 52 (1), 213–216.

Murray, M.J., Lessey, B.A., 1999. Embryo implantation and tumor metastasis: common pathways of invasion and angiogenesis. Semin. Reprod. Endocrinol. 17 (3), 275–290.

Newmark, P.A., Alvarado, A.S., 2002. Not your father's planarian: a classic model enters the era of functional genomics. Nat. Rev. Genet. 3 (3), 210–219.

Nishisho, I., et al., 1991. Mutations of chromosome-5q21 genes in Fap and colorectal-cancer patients. Science 253 (5020), 665–669.

Nunney, L., 1996. The response to selection for fast larval development in *Drosophila melanogaster* and its effect on adult weight: an example of a fitness trade-off. Evolution 50 (3), 1193–1204.

Ohtaki, T., et al., 2001. Metastasis suppressor gene KiSS-1 encodes peptide ligand of a G-protein-coupled receptor. Nature 411 (6837), 613–617.

Osanai, M., et al., 2002. Transient increase in telomerase activity of proliferating fibroblasts and endothelial cells in granulation tissue of the human skin. Wound Repair Regen. 10 (1), 59–66.

Ouyang, L., et al., 2012. Programmed cell death pathways in cancer: a review of apoptosis, autophagy and programmed necrosis. Cell Prolif. 45 (6), 487–498.

Pacheco, N.L.P., Andersen, A.-M.N., Kamper-Jorgensen, M., 2015. Preeclampsia and breast cancer: the influence of birth characteristics. Breast 24 (5), 613–617.

Pawelec, G., Derhovanessian, E., Larbi, A., 2010. Immunosenescence and cancer. J. Geriatr. Oncol. 1 (1), 20–26.

Peto, R., et al., 1975. Cancer and ageing in mice and men. Br. J. Cancer 32 (4), 411–426.

Potten, C.S., Owen, G., Booth, D., 2002. Intestinal stem cells protect their genome by selective segregation of template DNA strands. J. Cell Sci. 115 (11), 2381–2388.

Printz, C., 2011. Study shows link between dwarfism and cancer/diabetes protection. Cancer 117 (11), 2356.

Rangarajan, A., et al., 2004. Species- and cell type-specific requirements for cellular transformation. Cancer Cell 6 (2), 171–183.

Rasmussen, F., et al., 2003. Birth weight, adult height, and testicular cancer: cohort study of 337,249 Swedish young men. Cancer Causes Control 14 (6), 595–598.

Reddien, P.W., Alvarado, A.S., 2004. Fundamentals of planarian regeneration. Annu. Rev. Cell Dev. Biol. 20 (1), 725–757.

Ribas, A., 2015. Releasing the brakes on cancer immunotherapy. N. Engl. J. Med. 373, 1490–1492.

Rothstein, R., Michel, B., Gangloff, S., 2000. Replication fork pausing and recombination or "gimme a break". Genes Dev. 14 (1), 1–10.

Ru, G., Terracini, B., Glickman, L.T., 1998. Host related risk factors for canine osteosarcoma. Vet. J. 156, 31–39.

Sancar, A., et al., 2004. Molecular mechanisms of mammalian DNA repair and the DNA damage checkpoints. Annu. Rev. Biochem. 73 (1), 39–85.

Schaeffer, D.J., 1993. Planarians as a model system for in vivo tumorigenesis studies. Ecotoxicol. Environ. Saf. 25 (1), 1–18.

Seifert, A.W., et al., 2012. Skin regeneration in adult axolotls: a blueprint for scar-free healing in vertebrates. PLoS One 7 (4), e32875.

Seluanov, A., et al., 2008. Distinct tumor suppressor mechanisms evolve in rodent species that differ in size and lifespan. Aging Cell 7 (6), 813–823.

Sherr, C.J., McCormick, F., 2002. The RB and p53 pathways in cancer. Cancer Cell 2 (2), 103–112.

Shiloh, Y., 2003. ATM and related protein kinases: safeguarding genome integrity. Nat. Rev. Cancer 3 (3), 155–168.

Shors, A.R., et al., 2001. Melanoma risk in relation to height, weight, and exercise (United States). Cancer Causes Control 12 (7), 599–606.

Song, R.B., et al., 2013. Postmortem evaluation of 435 cases of intracranial neoplasia in dogs and relationship of neoplasm with breed, age, and body weight. J. Vet. Intern. Med. 27 (5), 1143–1152.

Stoker, M.G., Shearer, M., O'Neill, C., 1966. Growth inhibition of polyoma-transformed cells by contact with static normal fibroblasts. J. Cell Sci. 1 (3), 297–310.

Summers, K., Crespi, B.J., 2010. Xmrks the spot: life history tradeoffs, sexual selection and the evolutionary ecology of oncogenesis. Mol. Ecol. 19 (15), 3022–3024.

Sznol, M., Longo, D.L., 2015. Release the hounds! Activating the T-cell response to cancer. N. Engl. J. Med. 372, 374–375.

Testa, J.R., Malkin, D., Schiffman, J.D., 2013. Connecting molecular pathways to hereditary cancer risk syndromes. Am. Soc. Clin. Oncol. Educ. Book 33, 81–90.

Totter, J.R., 1980. Spontaneous cancer and its possible relationship to oxygen-metabolism. Proc. Natl. Acad. Sci. USA 77 (4), 1763–1767.

van den Brandt, P.A., et al., 2000. Pooled analysis of prospective cohort studies on height, weight, and breast cancer risk. Am. J. Epidemiol. 152 (6), 514–527.

van Deursen, J.M., 2014. The role of senescent cells in ageing. Nature 509 (7501), 439–446.

Vatten, L.J., et al., 2002. Pre-eclampsia in pregnancy and subsequent risk for breast cancer. Br. J. Cancer 87 (9), 971–973.

Vermeulen, L., Snippert, H.J., 2014. Stem cell dynamics in homeostasis and cancer of the intestine. Nat. Rev. Cancer 14, 468–480.

Vogelstein, B., Kinzler, K.W., 2004. Cancer genes and the pathways they control. Nat. Med. 10 (8), 789–799.

Wahl, G.M., Carr, A.M., 2001. The evolution of diverse biological responses to DNA damage: insights from yeast and p53. Nat. Cell Biol. 3 (12), E277–E286.

Wang, F., et al., 2002. Phenotypic reversion or death of cancer cells by altering signaling pathways in three-dimensional contexts. J. Natl. Cancer Inst. 94 (19), 1494–1503.

Weckerly, F.W., 1998. Sexual-size dimorphism: influence of mass and mating systems in the most dimorphic mammals. J. Mammal. 79 (1), 33–52.

Weinberg, R.A., 2007. Dynamics of cancer: incidence, inheritance, and evolution. Nature 449 (7165), 978.

Wildman, D.E., 2016. IFPA award in placentology lecture: phylogenomic origins and evolution of the mammalian placenta. Placenta 48 (Suppl. 1), S31–S39.

Williams, G.C., 1957. Pleiotropy, natural-selection, and the evolution of senescence. Evolution 11 (4), 398–411.

Winton, D.J., Ponder, B., 1990. Stem-cell organization in mouse small intestine. Proc. Biol. Sci. 241, 13–18.

Wiseman, H., Halliwell, B., 1996. Damage to DNA by reactive oxygen and nitrogen species: role in inflammatory disease and progression to cancer. Biochem. J. 313 (Pt 1), 17–29.

Xu, L., Li, S., Stohr, B.A., 2013. The role of telomere biology in cancer. Annu. Rev. Pathol. 8 (1), 49–78.

Zhang, B., et al., 2015. Height and breast cancer risk: evidence from prospective studies and Mendelian randomization. J. Natl. Cancer Inst. 107 (11).

Zhou, B.-B.S., Elledge, S.J., 2000. The DNA damage response: putting checkpoints in perspective. Nature 408 (6811), 433–439.

Zilfou, J.T., Lowe, S.W., 2009. Tumor suppressive functions of p53. Cold Spring Harb. Perspect. Biol. 1 (5), a001883.

Zuccolo, L., et al., 2008. Height and prostate cancer risk: a large nested case–control study (ProtecT) and meta-analysis. Cancer Epidemiol. Biomarkers Prev. 17 (9), 2325–2336.

Coevolution of Tumor Cells and Their Microenvironment: "Niche Construction in Cancer"

Arig Ibrahim-Hashim*, Robert J. Gillies*,**, Joel S. Brown†, Robert A. Gatenby**,‡

*Department of Cancer Imaging and Metabolism, H. Lee Moffitt Cancer Center and Research Institute, Tampa, FL, United States
**Department of Radiology, H. Lee Moffitt Cancer Center and Research Institute, Tampa, FL, United States
†Department of Evolutionary Biology, University of Illinois, Chicago, IL, United States
‡Department of Integrated Mathematical Oncology, H. Lee Moffitt Cancer Center and Research Institute, Tampa, FL, United States

INTRODUCTION

Cancer as an Evolutionary and Ecological Process

The conceptual model of cancer as an evolutionary problem is not new. Nowell (1976) and others described the genetic events leading to cancer as "somatic evolution" more than 60 years ago. However, the application of evolutionary biology to cancer is often confined to the molecular changes (genetic and epigenetic events) that are associated with the origin and progression of cancer. But evolution by natural selection includes a wide range of other interactions at nonmolecular spatial and temporal scales that collectively comprise the environmental selection forces and adaptive strategies governing the Darwinian dynamics of cancer. We propose that investigation of these forces in cancer progression and therapy resistance can provide new insights and potential strategies to improve cancer prevention and control (see also the Introductory Chapter).

Natural selection is the result of complex interactions among living organisms and their environment. Species are continuously shaped through environmental alterations that act on heritable variations in phenotypic properties (Polyak et al., 2009). Proliferation of each species is determined by its fitness within the context of the local ecosystem including the effects of competing populations. Importantly, however, the species also affect the ecosystem in complex ways. Often these actions are passive but some species evolve strategies in which they engineer their environment in ways that benefit them often at the expense of competitors. This "niche construction" strategy is probably best recognized as dam building by the North American beaver (*Castor canadensis*), but can be more subtle as we will see later.

In general, we assume that cancer cells are subject, like all other living systems, to Darwinian first principles so that cells with highest level of fitness within the context of local environmental selection forces will have proliferative advantages and, hence, will be continuously selected for success. Thus, cancer development within an individual is an evolutionary process that, in many ways, resembles species evolution.

The biological forces that drive cancer include accumulating mutations, a topic that is extensively discussed. However, we note that cancer cells also have access to the vast informational content of the human genome so that many adaptive strategies simply require up- or downregulation of genes already present. Finally, we note that Darwinian dynamics fundamentally are governed by interactions of environmental selection

Ecology and Evolution of Cancer. http://dx.doi.org/10.1016/B978-0-12-804310-3.00008-9

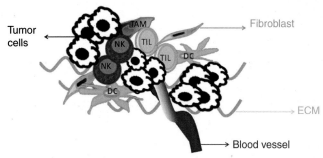

FIGURE 8.1 **Cellular compartments of tumor microenvironment.** Schematic illustration of cells that typically exist in the tumor microenvironment. These environmental cells cooperatively interact with tumor cells and contribute to tumor progression. *DC*, Dendritic cell; *ECM*, extracellular matrix; *NK*, natural killer; *TAM*, tumor-associated macrophage; *TIL*, tumor infiltrating lymphocyte.

TABLE 8.1 Noncellular Compartments of Tumor Microenvironment

Big molecules	Small molecules
Cytokines [interleukins 4 and 10 (IL-4 and IL-10), TGF-β, epidermal growth factor (EGF)]	Nitric oxide (NO)
Receptors	Water
Hydrolases (proteases, glycosidases, glucogenases)	Hydrogen ions
Collagens	Oxygen
Vimentins	Glucose
Fibronectins	Metabolites

The table shows big and small molecules that typically exist in the tumor microenvironment.

forces and the phenotypic (not genotypic) properties of organisms within an ecosystem. Those phenotypic properties are, of course, governed by the molecular properties of the cells but also reflect the reaction norm of that genotype. That is, each tumor cell has some degree of plasticity and can exhibit a wide range of phenotypic properties to acclimate to changes in environmental conditions.

Furthermore, the microenvironment can be modulated by the tumor cells in ways to favor their growth (an evolutionary strategy termed "niche construction"), thus establishing a distinctive cancer ecosystem (Merlo et al., 2006). Niche construction was originally defined in evolutionary biology, as the interaction between natural selection via environmental condition and the modifications of these conditions by the organism itself. In metastatic progression, cells originating from tumor and traveling via blood or lymphatic circulation need to modify to their distal organ environment in order to survive and proliferate, a process known as a "premetastatic niche" construction, mediated through release of paracrine factors and exosomes. Specific bone marrow (BM)–derived cell populations, in particular BM progenitor's derived cells, are targeted and recruited by tumor cell's secreted factors, such as vascular endothelial growth factor (VEGF)-A, placenta derived growth factor (PlGF), transforming growth factor beta 1 (TGF-β), tumor necrosis factor (TNF-α), and lysyl oxidase (LOX), to generate the premetastatic niche to which tumor will metastasize (Kaplan et al., 2005, 2006; Peinado et al., 2011; Steeg, 2005).

Although the conceptual model of cancer as an ecological and evolutionary process is perhaps 40 years old, our true understanding of this process is surprising rudimentary. In approaching cancers like any ecosystem, evolutionary biologists would begin with very basic questions: How many niches and tumor species exist in each cancer? What are the birth and death rates of each

population? What governs the cycles of key nutrients, such as carbon, nitrogen, phosphate, and iron? What are the diurnal and other temporal changes in the environment? It is in many ways astonishing that these questions remain unasked, much less unanswered.

Cancer as a Complex Dynamic System

Most current conceptual models of cancer are both cancer-cell-centric and gene-centric. However, some investigators have consistently noted that in situ cancer cells do not exist in isolation (Bissell et al., 1999). In fact, clinical tumors are spatially complex ecosystems with rich cancer–stroma interactions that include cell–cell contacts, as well as communication through small molecules, such as cytokines and metabolites (Fig. 8.1; Table 8.1). Under pathological conditions, such as wound healing, the stroma exerts homeostatic controls to initiate repair and return to a steady state. However, under some conditions, the stroma remains fibrotic or chronically inflamed, and these conditions are pathological and are known to predispose to cancer (Coussens and Werb, 2002). Remarkably, several human cancers have been shown to induce a stromal reaction or desmoplasia as a component of carcinoma progression (Noel and Foidart, 1998; Tuxhorn et al., 2002). This permanently reactive stroma can be characterized as a "wound that does not heal" (Dvorak, 1986).

Thus, the interaction between cells and their microenvironment, although governed by Darwinian first principles, is generally nonlinear and bidirectional, so that environmental conditions select for adapted tumor cell phenotype and the tumor cell properties can affect the environment. These complex dynamics result in multiscale temporal and spatial heterogeneity in virtually every component of a tumor (Anderson and Quaranta, 2008; Gillies et al., 2010).

THE CONSEQUENCES OF TUMOR-STROMA METABOLIC COMPARTMENTATION AND COLLABORATION

Niche Construction Strategy

Throughout the biome, organisms routinely evolve adaptations to engineer their environmental circumstances in order to favor their own fitness and undermine competitors (Laland et al., 1999; Odling-Smee et al., 2013; Silver and Di Paolo, 2006). Similarly, cancer phenotypes include traits that can alter the environment by, for example, altering the blood supply, generating an acidic environment, or coopting elements of the host mesenchyma and immune system. We propose that a key dynamic in these tumor–host interactions is an ecoevolutionary link between tumor invasion and metabolism through an evolutionary strategy termed "niche construction." Similar to the prairie dog communities, cancer cells can change their local habitat to benefit their survival and proliferation. Perhaps the most apparent niche construction is angiogenesis. That is, tumor cells excrete factors, such as VEGF that promotes ingrowth of blood vessels to provide the necessary nutrients and remove metabolites (such as H^+). In an alternative and probably competing strategy (termed "perturbatory niche construction), tumor cells generate an acidic local environment through upregulation of glycolytic pathways even in the present of oxygen. Tumor cell that are adapted to an acidic environment gain a proliferative advantage over nonadapted competitors. Furthermore, the acidic environment promotes invasion through degradation of extracellular matrix (ECM) and reduces the immune response to tumor-associated antigens. Clinical studies using positron emission tomography (PET) scans have demonstrated that the vast majority of human cancers have greatly increased glucose uptake compared to normal cells, suggesting that this strategy is common. Furthermore, theoretical studies have shown that even a few cancer cells (<16) can acidify the environment so that this mechanism can also promote the growth at metastatic sites. Understanding and targeting these strategies can potentially steer the evolutionary trajectory of tumor cells into a less invasive phenotype through selective application of relatively small microenvironmental perturbations.

An Altered Blood Flow in Establishing Cancer Habitats

Carcinomas typically evolve and exist in a nonphysiological microenvironment, characterized by acidosis, hypoxia, and reactive oxygen's (ROS) and nutrient deprivation (Gillies and Gatenby, 2007; Wykoff et al., 2000). It is tempting to assume that tumors grow despite the negative effects of such conditions. However, a likely alternative view is that tumors progress *because* of these conditions, within which cancer cells are more fit relative to nonneoplastic cells. Such environments can promote a malignant phenotype through induction of genomic instability and concomitant selection for aggressive clones (clades) of cells, leading to genomic diversity, which is evident in intratumoral genetic heterogeneity (Baird and Caldas, 2013; Gerlinger et al., 2012; Sottoriva et al., 2013; Yachida and Iacobuzio-Donahue, 2013).

Perhaps the most well-recognized and well-studied link between the tumor environment and the tumor cell evolution is intratumoral hypoxia. Hypoxia is frequently observed in tumors as a consequence of diffusion-limited O_2 delivery, wherein high rates of cancer cell proliferation lead to cells that are distant from blood vessels, and thus encounter perfusion-limited delivery of O_2. Furthermore, the majority of tumor vessels are structurally and functionally abnormal, and do not maintain steady blood flow, which can lead to volumes of profound hypoxia, even in cells adjacent to blood vessels (Vaupel et al., 2004). Reduced oxygen concentrations in tumors are typically observed in both radiological and pathological tumor images. Importantly, hypoxia acts as a stimulating signal that initiates changes in gene expression leading to modification in cell signaling similar to those caused by somatic mutations or epigenetic modifications.

Cells react to hypoxia via changes in gene expression, generally mediated by the hypoxia-inducible transcription factors (HIFs). These factors are heterodimers composed of HIF-1β subunits that are constitutively present and α subunits (HIF-1 α, HIF-2 α, or HIF-3 α) that are constitutively synthesized yet are degraded in an O_2-dependent manner, such that they accumulate rapidly under hypoxic conditions (Semenza, 2012). Increased HIF-1 α accumulation results in induction of a wide range of client proteins that increase survival pathways as well as glycolysis, acid production, and motility. While these adaptations increase survival and proliferation in hypoxic regions, they are also linked to tumor progression and metastasis. Furthermore, increased accumulation of HIFs allows the tumor cells to "engineer" their environment through induction of VEGF expression that increases vascularization and microvessel density (Bos et al., 2001). The evolutionary value of HIF-1α is evident by observations that tumors can stimulate hypoxic responses in *normoxic conditions* [known as *pseudohypoxia* (see also Chapter 4)] through neutralizing the HIF-1α proteasome recognition site (von Hippel–Lindau protein), promoting areas of normoxic acidosis (Mekhail et al., 2004). In other words, while increased expression of HIF-1α was likely stimulated initially as a response to local hypoxia, tumor cells can constitutively accumulate HIF-1α under normoxic conditions, indicating that its client proteins confer an adaptive advantage beyond just a response to decreased local oxygen concentrations.

Thus, it is not surprising that the observed consequences of HIF-1 activation include suppression of *CD8+* T-cell activation, thus reducing immune response to tumor antigens and remodeling of the ECM that permits tumor growth and invasion (Gabrilovich and Nagaraj, 2009).

The Consequences of Low pH on Establishing Cancer Habitat

An important consequence of increased HIF-1α under normoxic conditions is aerobic glycolysis—the fermentative metabolism of glucose even in the presence of oxygen. First observed by Warburg (and often termed "the Warburg effect"), aerobic glycolysis appears to be an evolutionary conundrum because it results in lower efficiency of ATP production from glucose (aerobic glucose metabolism produces 36ATP/glucose molecule compared to 2ATP/glucose in the fermentative pathways) and higher rates of acid production. Despite this apparent negative selection pressure, the vast majority of human cancers exhibit aerobic glycolysis (with increased glucose uptake on fludeoxyglucose (FdG) PET imaging studies) and the extracellular pH (pHe) of tumors is typically acidic. We propose that the Warburg effect represents a niche construction strategy that promotes tumor cell survival and proliferation by generating an acidic microenvironment. These benefits include reduced proliferation in competitors that are not acid-adapted, suppression of immune response to tumor antigens, and promotion of tumor invasion by ECM degradation. Importantly, however, generating an acidic environment also imposes significant costs on the tumor cells, including the loss of efficiency in ATP production by aerobic glycolysis and the energy required to operate membrane pumps to maintain a steep transmembrane pH gradient. The observation that most clinical cancers exhibit aerobic glycolysis indicates that this phenotype promotes a favorable cost/benefit ratio. Because of this, it presents potential opportunities for altering the course of tumor growth, as we will see later.

One benefit to the tumor is that an acidic pHe significantly induces invasive tumor growth. In one system, it has been shown that the effect of pH on local invasion can involve an acid-stimulated release of cathepsin B, which leads to degradation of the ECM (Cuvier et al., 1997; Montcourrier et al., 1994). However, there are many alternative mechanisms whereby acid can stimulate metastasis, such as selecting for cells that are resistant to programmed cell death, or apoptosis (Ohtsubo et al., 1997). We have interpreted these effects to

generate an "acid-mediated invasion" hypothesis, which proposed that H+ flows along concentration gradients from the tumor into peritumoral normal tissue, causing normal cell death and ECM degradation. Cancer cells, which are acid-adapted, can then invade into the damaged adjacent normal tissue. It has been shown that exposure of tumor cells to acidic media prior to intravascular injection substantially increases their ability to metastasize (Rofstad et al., 2006).

Cancer cells that evolve enhanced glucose uptake and glycolysis appear to be engaging in ecological engineering in which the acidification of their environment permits more aggressive proliferation and opens space for invasion by suppressing other cancer cells, inhibiting immune efficacy, damaging normal cells, and inducing remodeling of the ECM. Prior studies have demonstrated that increased acid production is intimately coupled to invasion both in vitro and in vivo (Estrella et al., 2013) and a recent genome-wide computational analysis has shown high degree of correlation between aerobic glycolysis and cell motility (Yizhak et al., 2014). In addition, tumor acidosis confers a survival advantage to tumor cells that have adapted to this environment and can cause chemoresistance via a phenomenon known as ion trapping (Wojtkowiak et al., 2011). Even intraductal hyperplasia is expected to be acidic, as cells grow hundreds of micrometers away from their blood supply (Gatenby and Gillies, 2004). Cells able to survive this harsh acidic condition display aggressive phenotypes with survival benefits, including augmented autophagy (Wojtkowiak et al., 2011) and increased invasion (Estrella et al., 2013; Gatenby et al., 2006). An important advantage of the niche construction strategy is suppression of host immune response to tumor antigens (Lardner, 2001). More recent studies demonstrated that acidosis can alter the functions of immune cells including T cells, neutrophils, macrophages, and dendritic cells (DCs) (Draghiciu et al., 2011; Pardoll, 2012). For example, one recent study demonstrates that tumor acidosis can cause T-cell anergy that can be reversed by inhibiting acidosis (Calcinotto et al., 2012). Acid pH has also been shown to activate the inhibitory receptor, T-cell death-associated gene-8 (TDAG-8), a specific acid-sensing G-protein (Ishii et al., 2005), and ROS production in neutrophils (Mogi et al., 2009). Proton pump inhibition on tumor cells can hamper tumor-induced suppression of macrophages in vitro and in vivo (Vishvakarma and Singh, 2010, 2011).

We view this microenvironmental acidification as a niche construction evolutionary strategy in which acid-producing/acid-adapted cancer cells benefit by decreasing the fitness of nonadapted competitors, reducing efficacy of immune response, and promoting local invasion resulting in transition from in situ to invasive cancer and aggressive growth of primary or

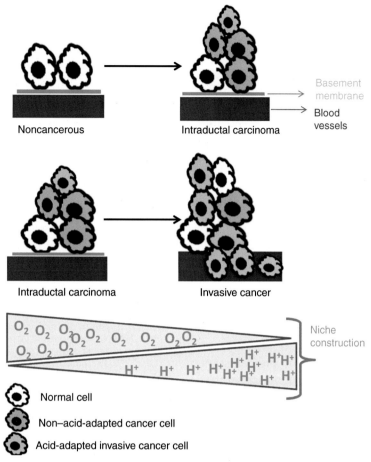

FIGURE 8.2 **Schematic illustration of niche engineering strategy.** This model explains from an ecological point of view the role of low oxygen and high proton production in promoting cancer progression. The change from noncancerous to intraductal cancer is achieved when the environment becomes hypoxic and acidic, which will generate more fit cells to survive this harsh environment. Niche modifications by further increase in hypoxia and acidity will progress tumor from in situ to invasive cancer type by generating the invader cells. Those cells are able to invade through basement membrane to bloodstream, avoid immune cells, and travel to distal organs.

metastatic tumors (Fig. 8.2). However, this strategy incurs a significant, and potentially exploitable, phenotypic "cost" due to the decreased efficiency in energy production using fermentative versus oxidative glucose metabolic pathways (2ATP/glucose vs. ~36ATP/glucose) and the energy necessary for adaptations to an acidic environment (e.g., extrusion of H^+ against a transmembrane gradient). The vast majority of human cancers have increased glucose flux on FdG PET imaging (Hawkins and Phelps, 1988), suggesting that the cost/benefit ratio of the niche engineering strategy is generally favorable.

Niche Construction Strategy and Immune Response

The immune system is made up of various cellular and molecular components including T cells, B cells, macrophages, DCs, natural killer (NK) T cells, basophils, neutrophils, eosinophils, mast cells, cytokines, antibodies,

opsonins, complement, etc. In contrast to their normal course of action, in the tumor microenvironment, immune cells contribute to various hallmarks of cancer including sustaining tumor proliferative ability; resisting cell death, angiogenesis, invasion, and metastasis; evading growth suppressors; and avoiding immune destruction (Hanahan and Coussens, 2012).

Cells of the immune system massively infiltrate tissue sites containing malignant cells. The dual roles of immune cells in both tumor surveillance and tumor promotion are now well accepted but poorly understood (see also Chapter 9). Due to immune surveillance of host tissues, cells of the immune system are dominant at sites of hyperplasia and, hence, it is accepted that malignant progression must contain an immune component (Hussein et al., 2009; Kramer et al., 2007). The vast majority of studies characterizing tumor-induced alteration of the immune response have been focused on macromolecules secreted or expressed on cell surface by cells residing in the tumor microenvironment. Moreover, the hyperplastic niche, which can be devoid of immunosuppressive macromolecules, somehow shifts the balance of immune responses in favor of tumor progression (DeNardo and Coussens, 2007). Moreover, premalignant tissue is mostly characterized with inflammation (Mantovani et al., 2008). Immune cells secrete growth factors as they fight inflammation, which sequentially can be utilized by the premalignant cells thus generating a new growth-factor-rich microenvironment (Johansson et al., 2008; Ruffell et al., 2010). In cases wherein inflammation does not resolve quickly, new cancerous clones will be generated that further exploit the factor-rich milieu. Once the process has been initiated, inhibition of inflammation may not be sufficient to reverse tumor growth, as tumor cells eventually secrete their own growth factors, or manipulate other, stromal, cells, such as tumor-associated macrophages (TAMs) to do so (Grimshaw and Balkwill, 2001; Mantovani et al., 2004).

Can We Reverse Cancer Niche?

Most cancer treatment and prevention strategies directly target proliferating malignant or premalignant cells, but this imposes Darwinian selection forces that consistently promote resistance (Gatenby, 2009; Gatenby et al., 2009; Jansen et al., 2015). In contrast, the model of cancer niche construction potentially allows alternative Darwinian strategies that use small but selective perturbations of complex tumor dynamics to steer the cancer system into a less invasive evolutionary trajectory by perturbing the niche construction dynamics. For example, reversing the acidosis of the niche construction strategy while simultaneously increasing the immune response to tumor with immune checkpoint inhibitors (anti–PD-1 or anti–CTLA-4) and/or adoptive T-cell therapy leads to rarely observed cures in the B16 melanoma model (Pilon-Thomas et al., 2016). While this is promising, and while there is evidence that long-term consumption of alkaline buffers can be tolerated (Mann and Stuart, 1974), there is significant concern that the current clinical regimen is unsustainable, hence the need to investigate alternative pharmacological interventions that are designed to accomplish the same objective, that is, to neutralize the pH of tumors and remove this immunosuppressive effect. One class of approaches will be to induce compensated metabolic alkalosis (as bicarbonate does) either through direct ingestion of alkali ($NaHCO_3$ or Na-citrate) or through the use of loop diuretics, such as furosemide, and another will inhibit the production of tumor-derived acid through pharmacological interventions, such as inhibition of the following: (1) lactate dehydrogenase inhibitor (LDH-A); (2) carbonic anhydrases; (3) monocarboxylate transporters; or (4) proton pumps.

CONCLUSIONS

Cancers, like other complex dynamic systems, can be steered into a less aggressive course through application of relatively small but highly selective biological "force." However, we believe that the cost of the niche construction strategy may be exploited to perturb the underlying ecoevolutionary dynamics during cancer development and growth. Specifically, if the benefits of the niche construction phenotype can be reduced, the Darwinian selection pressures may produce a phase transition in which metabolically normal, noninvasive phenotypes can be selected.

References

Anderson, A.R., Quaranta, V., 2008. Integrative mathematical oncology. Nat. Rev. Cancer 8, 227–234.

Baird, R.D., Caldas, C., 2013. Genetic heterogeneity in breast cancer: the road to personalized medicine? BMC Med. 11, 151.

Bissell, M.J., Weaver, V.M., Lelievre, S.A., Wang, F., Petersen, O.W., Schmeichel, K.L., 1999. Tissue structure, nuclear organization, and gene expression in normal and malignant breast. Cancer Res. 59, 1757–1763 (discussion 1763s–1764s).

Bos, R., Zhong, H., Hanrahan, C.F., et al., 2001. Levels of hypoxia-inducible factor-1 alpha during breast carcinogenesis. J. Natl. Cancer Inst. 93, 309–314.

Calcinotto, A., Filipazzi, P., Grioni, M., et al., 2012. Modulation of microenvironment acidity reverses anergy in human and murine tumor-infiltrating T lymphocytes. Cancer Res. 72, 2746–2756.

Coussens, L.M., Werb, Z., 2002. Inflammation and cancer. Nature 420, 860–867.

Cuvier, C., Jang, A., Hill, R.P., 1997. Exposure to hypoxia, glucose starvation and acidosis: effect on invasive capacity of murine tumor cells and correlation with cathepsin (L + B) secretion. Clin. Exp. Metastasis 15, 19–25.

DeNardo, D.G., Coussens, L.M., 2007. Inflammation and breast cancer. Balancing immune response: crosstalk between adaptive and innate immune cells during breast cancer progression. Breast Cancer Res. 9, 212.

Draghiciu, O., Nijman, H.W., Daemen, T., 2011. From tumor immunosuppression to eradication: targeting homing and activity of immune effector cells to tumors. Clin. Dev. Immunol. 2011, 439053.

Dvorak, H.F., 1986. Tumors: wounds that do not heal. Similarities between tumor stroma generation and wound healing. N. Engl. J. Med. 315, 1650–1659.

Estrella, V., Chen, T., Lloyd, M., et al., 2013. Acidity generated by the tumor microenvironment drives local invasion. Cancer Res. 73, 1524–1535.

Gabrilovich, D.I., Nagaraj, S., 2009. Myeloid-derived suppressor cells as regulators of the immune system. Nat. Rev. Immunol. 9, 162–174.

Gatenby, R.A., 2009. A change of strategy in the war on cancer. Nature 459, 508–509.

Gatenby, R.A., Gillies, R.J., 2004. Why do cancers have high aerobic glycolysis? Nat. Rev. Cancer 4, 891–899.

Gatenby, R.A., Gawlinski, E.T., Gmitro, A.F., Kaylor, B., Gillies, R.J., 2006. Acid-mediated tumor invasion: a multidisciplinary study. Cancer Res. 66, 5216–5223.

Gatenby, R.A., Silva, A.S., Gillies, R.J., Frieden, B.R., 2009. Adaptive therapy. Cancer Res. 69, 4894–4903.

Gerlinger, M., Rowan, A.J., Horswell, S., et al., 2012. Intratumor heterogeneity and branched evolution revealed by multiregion sequencing. N. Engl. J. Med. 366, 883–892.

Gillies, R.J., Gatenby, R.A., 2007. Hypoxia and adaptive landscapes in the evolution of carcinogenesis. Cancer Metastasis Rev. 26, 311–317.

Gillies, R.J., Anderson, A.R., Gatenby, R.A., Morse, D.L., 2010. The biology underlying molecular imaging in oncology: from genome to anatome and back again. Clin. Radiol. 65, 517–521.

Grimshaw, M.J., Balkwill, F.R., 2001. Inhibition of monocyte and macrophage chemotaxis by hypoxia and inflammation—a potential mechanism. Eur. J. Immunol. 31, 480–489.

Hanahan, D., Coussens, L.M., 2012. Accessories to the crime: functions of cells recruited to the tumor microenvironment. Cancer Cell. 21, 309–322.

Hawkins, R.A., Phelps, M.E., 1988. PET in clinical oncology. Cancer Metastasis Rev. 7, 119–142.

Hussein, M.R., Al-Assiri, M., Musalam, A.O., 2009. Phenotypic characterization of the infiltrating immune cells in normal prostate, benign nodular prostatic hyperplasia and prostatic adenocarcinoma. Exp. Mol. Pathol. 86, 108–113.

Ishii, S., Kihara, Y., Shimizu, T., 2005. Identification of T cell death-associated gene 8 (TDAG8) as a novel acid sensing G-protein-coupled receptor. J. Biol. Chem. 280, 9083–9087.

Jansen, G., Gatenby, R., Aktipis, C.A., Opinion:, 2015. Control vs. eradication: applying infectious disease treatment strategies to cancer. Proc. Natl. Acad. Sci. USA 112, 937–938.

Johansson, M., Denardo, D.G., Coussens, L.M., 2008. Polarized immune responses differentially regulate cancer development. Immunol. Rev. 222, 145–154.

Kaplan, R.N., Riba, R.D., Zacharoulis, S., et al., 2005. VEGFR1-positive haematopoietic bone marrow progenitors initiate the pre-metastatic niche. Nature 438, 820–827.

Kaplan, R.N., Rafii, S., Lyden, D., 2006. Preparing the "soil": the pre-metastatic niche. Cancer Res. 66, 11089–11093.

Kramer, G., Mitteregger, D., Marberger, M., 2007. Is benign prostatic hyperplasia (BPH) an immune inflammatory disease? Eur. Urol. 51, 1202–1216.

Laland, K.N., Odling-Smee, F.J., Feldman, M.W., 1999. Evolutionary consequences of niche construction and their implications for ecology. Proc. Natl. Acad. Sci. USA 96, 10242–10247.

Lardner, A., 2001. The effects of extracellular pH on immune function. J. Leukoc. Biol. 69, 522–530.

Mann, J.R., Stuart, J., 1974. Sodium bicarbonate prophylaxis of sickle cell crisis. Pediatrics 53, 414–416.

Mantovani, A., Allavena, P., Sica, A., 2004. Tumour-associated macrophages as a prototypic type II polarised phagocyte population: role in tumour progression. Eur. J. Cancer 40, 1660–1667.

Mantovani, A., Allavena, P., Sica, A., Balkwill, F., 2008. Cancer-related inflammation. Nature 454, 436–444.

Mekhail, K., Gunaratnam, L., Bonicalzi, M.E., Lee, S., 2004. HIF activation by pH-dependent nucleolar sequestration of VHL. Nat. Cell Biol. 6, 642–647.

Merlo, L.M., Pepper, J.W., Reid, B.J., Maley, C.C., 2006. Cancer as an evolutionary and ecological process. Nat. Rev. Cancer 6, 924–935.

Mogi, C., Tobo, M., Tomura, H., et al., 2009. Involvement of proton-sensing TDAG8 in extracellular acidification-induced inhibition of proinflammatory cytokine production in peritoneal macrophages. J. Immunol. 182, 3243–3251.

Montcourrier, P., Mangeat, P.H., Valembois, C., et al., 1994. Characterization of very acidic phagosomes in breast cancer cells and their association with invasion. J. Cell Sci. 107 (Pt 9), 2381–2391.

Noel, A., Foidart, J.M., 1998. The role of stroma in breast carcinoma growth in vivo. J. Mammary Gland Biol. Neoplasia 3, 215–225.

Nowell, P.C., 1976. The clonal evolution of tumor cell populations. Science 194, 23–28.

Odling-Smee, J., Erwin, D.H., Palkovacs, E.P., Feldman, M.W., Laland, K.N., 2013. Niche construction theory: a practical guide for ecologists. Q. Rev. Biol. 88, 4–28.

Ohtsubo, T., Wang, X., Takahashi, A., et al., 1997. p53-dependent induction of WAF1 by a low-pH culture condition in human glioblastoma cells. Cancer Res. 57, 3910–3913.

Pardoll, D.M., 2012. The blockade of immune checkpoints in cancer immunotherapy. Nat. Rev. Cancer 12, 252–264.

Peinado, H., Lavotshkin, S., Lyden, D., 2011. The secreted factors responsible for pre-metastatic niche formation: old sayings and new thoughts. Semin. Cancer Biol. 21, 139–146.

Pilon-Thomas, S., Kodumudi, K.N., El-Kenawi, A.E., et al., 2016. Neutralization of tumor acidity improves antitumor responses to immunotherapy. Cancer Res. 76, 1381–1390.

Polyak, K., Haviv, I., Campbell, I.G., 2009. Co-evolution of tumor cells and their microenvironment. Trends Genet. 25, 30–38.

Rofstad, E.K., Mathiesen, B., Kindem, K., Galappathi, K., 2006. Acidic extracellular pH promotes experimental metastasis of human melanoma cells in athymic nude mice. Cancer Res. 66, 6699–6707.

Ruffell, B., DeNardo, D.G., Affara, N.I., Coussens, L.M., 2010. Lymphocytes in cancer development: polarization towards pro-tumor immunity. Cytokine Growth Factor Rev. 21, 3–10.

Semenza, G.L., 2012. Hypoxia-inducible factors in physiology and medicine. Cell 148, 399–408.

Silver, M., Di Paolo, E., 2006. Spatial effects favour the evolution of niche construction. Theor. Popul. Biol. 70, 387–400.

Sottoriva, A., Spiteri, I., Piccirillo, S.G., et al., 2013. Intratumor heterogeneity in human glioblastoma reflects cancer evolutionary dynamics. Proc. Natl. Acad. Sci. USA 110, 4009–4014.

Steeg, P.S., 2005. Cancer biology: emissaries set up new sites. Nature 438, 750–751.

Tuxhorn, J.A., McAlhany, S.J., Dang, T.D., Ayala, G.E., Rowley, D.R., 2002. Stromal cells promote angiogenesis and growth of human prostate tumors in a differential reactive stroma (DRS) xenograft model. Cancer Res. 62, 3298–3307.

Vaupel, P., Mayer, A., Hockel, M., 2004. Tumor hypoxia and malignant progression. Methods Enzymol. 381, 335–354.

Vishvakarma, N.K., Singh, S.M., 2010. Immunopotentiating effect of proton pump inhibitor pantoprazole in a lymphoma-bearing murine host: implication in antitumor activation of tumor-associated macrophages. Immunol. Lett. 134, 83–92.

Vishvakarma, N.K., Singh, S.M., 2011. Augmentation of myelopoiesis in a murine host bearing a T cell lymphoma following in vivo administration of proton pump inhibitor pantoprazole. Biochimie 93, 1786–1796.

Wojtkowiak, J.W., Verduzco, D., Schramm, K.J., Gillies, R.J., 2011. Drug resistance and cellular adaptation to tumor acidic pH microenvironment. Mol. Pharm. 8, 2032–2038.

Wykoff, C.C., Beasley, N.J.P., Watson, P.H., et al., 2000. Hypoxia-inducible expression of tumor-associated carbonic anhydrases. Cancer Res. 60, 7075–7083.

Yachida, S., Iacobuzio-Donahue, C.A., 2013. Evolution and dynamics of pancreatic cancer progression. Oncogene 18, 29.

Yizhak, K., Le Devedec, S.E., Rogkoti, V.M., et al., 2014. A computational study of the Warburg effect identifies metabolic targets inhibiting cancer migration. Mol. Syst. Biol. 10, 744.

9

Evolutionary Perspective of Tumorigenesis and Antitumor Immunity: A Comparative Approach

Jacques Robert, Francisco De Jesús Andino, Maureen Banach, Kun Hyoe Rhoo, Eva-Stina Edholm

Department of Microbiology and Immunology, University of Rochester Medical Center, Rochester, NY, United States

INTRODUCTION

Although the recent successes of cancer immunotherapy have renewed interest in cancer immunity, clinical applications of these therapies are still challenging (Koster et al., 2015). One potential limitation of the current approach is the heavy reliance on murine models, which may not always fully recapitulate human characteristics (Dranoff, 2012). The use of a comparative approach focusing on nonmammalian animal models may broaden our understanding of interactions between tumors and immune systems and lead to novel avenues in cancer immunotherapy (Aktipis et al., 2015; Robert, 2010). Furthermore, an evolutionary perspective of the struggle between cancers and the host immune system may help to better delineate antitumor immune functions and tumor escape mechanisms that have remained conserved during evolution.

IMMUNOLOGY OF CANCER

Although tumor cells are derived from normal nontumoral tissues of an organism, and thus are not typical foreign entities as are pathogens, it is now well established that tumors can be recognized and specifically attacked by the host immune system (reviewed in Gajewski et al., 2013; Table 9.1). In mammals, the immune repertoire of cancer-bearing animals (e.g., mice) or human patients contains B and T cells that recognize antigens expressed by autologous cancer cells (Coulie et al., 2014). Among those, cytotoxic CD8 T cells have emerged as key antitumor cell

effectors that are able to recognize tumor antigens bound to Major Histocompatibility Complex (MHC) class I molecules and kill tumor cells through perforin- and granzyme-mediated apoptosis. CD8 T cells also produce interferon gamma (IFN-γ) that has antitumor properties. Importantly, CD8 T-cell activation and expansion is determined by antigen-presenting cells (APCs), such as dendritic cells (DCs) and macrophages, which present tumor antigens, express critical stimulatory surface molecules (e.g., B7), and release cytokines (Pardoll, 2012). Some tumor antigens triggering immune responses, termed mutation-derived tumor antigens, neoantigens, or tumor-specific antigens, are derived from somatic mutations accumulating during tumorigenesis (Robbins et al., 2013). Others known as tumor-associated antigens are products from deregulated or overexpressed genes in malignant cells. Unlike neoantigens that are unique for each individual tumor, these tumor-associated antigens are generally shared among multiple individual organisms with the same type of cancer (Coulie et al., 2014; van der Bruggen et al., 1991).

The finding of specific antitumor B- and T-cell responses in mammals is consistent with the concept of tumor immune surveillance first proposed by Ehrlich (1957) in the early 20th century and then further developed by Burnet and Thomas (Burnet, 1970), which states that in multicellular organisms a fundamental function of the immune system is to detect and eliminate transformed cells. Evidence supporting this hypothesis has been obtained in mouse by impairing immune function with drugs or sublethal irradiation (reviewed in Dunn et al., 2004; Ostrand-Rosenberg, 2008), as well

Ecology and Evolution of Cancer. http://dx.doi.org/10.1016/B978-0-12-804310-3.00009-0

TABLE 9.1   Main Adaptive and Innate Immune Cell Effectors, Surface Receptors, and Soluble Factors Promoting or Preventing Tumor Growth

	Antitumorigenic	Protumorigenic
Adaptive cell effectors	CD8 cytotoxic T cells T-helper 1 CD4 T cells Type 1 iNKT cells	T-regulatory cells T-helper 2 CD4 T cells Type II NKT cells
Innate cell effectors	NK cells M1 proinflammatory macrophages	MDSC M2 TAM
Receptors	Classical MHC class Ia CD1d[a] MICA, MICB, NKG2D[b] CD40	CD1d[a] HLA E[b] CTLA-4 PD-1/PD-1L
Soluble factors	IFN-γ Granzyme B, perforin IL-2 IL-12 iNOS IFN-α/β	IL-10, IL-6 IL-4, IL-13 TGF-β CCL22 CSF-1

[a]The nonclassical MHC class I CD1d can activate both antitumor type I iNKT cells and tumor-promoting type II NKT cells depending on lipid ligand it presents.
[b]MHC class I–like MICA and MICB, as well as NK receptor NKG2D expressed at the tumor cell surface are NK cell activators, whereas nonclassical MHC molecules, such as HLA E function as inhibitory receptor for NK cells.
IFN-γ, Interferon gamma; IL, interleukin; iNKT, invariant natural killer T; MDSC, myeloid-derived suppressor cell; NK, natural killer; NKT, natural killer T; PD-1, programmed T-cell death 1; TAM, tumor-associated macrophage; TGF, transforming growth factor.

as by generating mice genetically deficient for specific genes controlling adaptive immunity, such as RAG2 (lacking B and T cells), IFN-γ (deficient effector T-cell response), or Baft3 (deficient antigen presentation; Hildner et al., 2008). These immunodeficient mice have an increased incidence of spontaneous tumors and are more susceptible to the induction of tumors by chemical carcinogens (Ostrand-Rosenberg, 2008).

While tumor immune surveillance is usually focused on adaptive T and B cells, the differentiation and function of these cells depend on a highly integrated immune system that also includes innate immune cells. For example, natural killer (NK) cells are able to detect and kill tumor cells that avoid CD8 T-cell recognition by downregulating surface MHC class I expression, and thus are critical for tumor immune surveillance (Fuertes et al., 2011; Liu et al., 2012).

STRATEGIES OF TUMOR IMMUNE EVASION

Despite immune surveillance and active antitumor T cells, tumors are often not eradicated. Indeed, under the selective pressure imposed by the immune system, cancers evolve multiple ways to resist and ultimately overcome immune responses. Tumor factors counteracting host immunity range in two main categories: the first one is evading immune recognition to become undetectable to NK and T cells; the second one is to actively suppress antitumor immune responses (reviewed in Gabrilovich et al., 2012).

A major mechanism of tumors to prevent immune recognition focuses on MHC class I presentation. Given the importance of MHC class I molecules for presentation and recognition of tumor antigen by cytotoxic CD8 T cells, it is not surprising that tumors often downregulate MHC class Ia expression to escape CD8 T-cell–mediated immune recognition and killing (Zitvogel et al., 2006). This is achieved by various ways including the downregulation of genes encoding the MHC class Ia chain itself or the associated β2-microglobulin that is required for the complex to be expressed at the cell surface (Garcia-Lora et al., 2003). Alternatively, some tumors impair the transfer and/or processing of antigens in the endoplasmic reticulum (Garcia-Lora et al., 2003). The importance of MHC class I expression for detecting and preventing tumor growth is particularly illustrated by so-called transmissible or contagious tumors (reviewed in Ostrander et al., 2016; Siddle and Kaufman, 2013; see also Chapter 12). The canine transmissible venereal tumors (CTVTs) are clonal tumor cells that pass from one donor dog to another during coitus (Murgia et al., 2006). During the progressive phase, the tumor cells do not express MHC class I or class II, and the tumor secretes transforming growth factor-β1 (TGF-β1), a cytokine that inhibits tumor-infiltrating lymphocyte (including NK cells) cytotoxicity. Only at later stages when rejection is observed do these tumors express MHC class I. Another notable transmissible tumor is the devil facial tumor disease (DFTD), which is a fatal monophyletic clonally transmissible tumor, or basically an allograft transmitted between devils by biting during fights for mating (Pearse and Swift, 2006; Siddle et al., 2013). Similar to CTVTs,

DFTD cancer cells avoid allogeneic immune recognition by downregulating cell surface MHC class I expression. Whether the NK cell response is insufficient or inhibited by these tumors is currently unclear (Brown et al., 2011).

A second general way tumors antagonize host immune responses is to actively inhibit or suppress immune cell effectors by expressing various inhibitory surface receptors and producing soluble inhibitory factors. A notable inhibitory receptor expressed by many tumors is PD-L1, the ligand for programmed T-cell death 1 (PD-1), which is a transmembrane protein found on the surface of T cells and that on ligation induces apoptosis of activated T cells. Soluble immunosuppressive factors produced by tumors include TGF-β, interleukin (IL)-10, IL-6, reactive oxygen species, and nitric oxide (Gabrilovich et al., 2012; Lippitz, 2013). In addition, when tumors become established, they exhibit the remarkable ability to control their microenvironment and promote a potent suppressive state by recruiting a variety of suppressor leukocytes, such as myeloid-derived suppressor cells (MDSCs), macrophages, as well as tolerogenic DCs, and regulatory T cells (Gabrilovich et al., 2012; Quail and Joyce, 2013).

The complex interactions of tumor with the immune system are summarized in Table 9.1, as a continual competition between factors preventing and promoting tumor growth. At the cellular level, besides cytotoxic CD8 T cells and NK cells, proinflammatory Th1 CD4 T cells and invariant natural killer T (iNKT) cells are critical for preventing tumor growth by assisting CD8 T cells and producing antitumor cytokines, such as IFNγ, which are critical players preventing tumor growth. In contrast, tumor growth is promoted by Th2 CD4 T and type II natural killer T (NKT) cells that produce factor-inhibiting antitumor T-cell responses, as well as by T-regulatory cells that suppress immune surveillance. In addition, different innate immune cells are critical components of the tumor microenvironment that significantly influence the tumor outcome. For example, proinflammatory macrophages (M1) are antitumorigenic by secreting Th1 cytokines and thus promoting Th1 T-cell differentiation. Conversely, macrophage polarized into M2 tumor-associated macrophages (TAMs) and leukocytes induced into MDSCs by the tumor microenvironment suppress antitumor activity. The anti- and protumor antagonism is also reflected at the level of cell surface receptors. On the one hand, receptors activating CD8 T cells (MHC class Ia), iNKT (MHC class I–like), Th1 cells (CD40), and NK cells (NKG2D, MICA, MICB) contribute to antitumor responses. On the other hand, checkpoint inhibitory receptors, such as CTLA-4 and PD-L1 that suppress Th1 and CD8 T-cell activation, as well as MHC class I–like molecules bound to certain ligands activating type II NKT cells are promoting tumor growth. Finally, secreted factors preventing tumor growth include the following: IFN-γ, granzyme B, and perforin produced by cytotoxic CD8 T cells; the major

T-cell growth factor IL-2 and IL-12 (an important cytokine driving the differentiation of naive T cells into Th1 cells and promoting cytotoxic activities of T and NK cells) (Liao et al., 2013); as well as IFNs and iNOs associated with inflammatory macrophage functions. These factors are counterbalanced by antiinflammatory cytokines IL-10 and IL-6, Th2 cytokines (IL-4, IL-13), the T-cell inhibitory factor TGF-β (Lippitz, 2013), as well as the M2-promoting factor colony stimulating factor 1 (CSF-1) and the chemokine CCL22 that recruit Tregs.

IMMUNOEDITING

According to the immune surveillance theory, the immune system patrols the body to recognize and destroy cancer cells. However, the occurrence of malignancies indicates that this protection is not perfect. Moreover, the many mechanisms exhibited to counteract immune responses imply that malignant tumors undergo a selection process established by the immune system. Accordingly, a modified theory, cancer immunoediting theory, has been proposed taking into account that the immune system not only can protect the host against tumors but also, by selecting for tumors of lower immunogenicity, has the capacity to shape and even promote the generation of new variants displaying increased ability to escape and overcome immune control (Dunn et al., 2004; Mittal et al., 2014). The immunoediting comprises three phases (Dunn et al., 2004): (1) an elimination phase where the immune system detects and destroys many precancerous and malignant cells; (2) an equilibrium phase where editing and selection occur; and (3) an escape phase where immunological modified or sculpted tumors progressively grow and overcome immune responses. Immunoediting is thought to continue throughout the life of the tumor so that the phenotype of an established tumor has been directed by the host's immune response.

This evolutionary view of the role of the immune system in tumorigenesis has received some experimental support in mice by comparing the properties of tumors induced by chemical carcinogens in animals that are either immunocompetent or immunodeficient (e.g., RAG2 or perforin-deficient). Tumors obtained from immunocompetent mice were more tumorigenic and, therefore, more modified than those that have developed under low or absent immune pressure (Matsushita et al., 2012).

TUMORS IN INVERTEBRATES AND ECTOTHERMIC VERTEBRATES

Another way to assess the dual involvement of the immune system in preventing, as well as shaping cancer is to extend studies to other evolutionary distant organisms. In addition, a wider more inclusive comparison of tumorigenesis not limited to mammals, and mouse model especially, may better underscore conserved features in the development of malignancies. Studies over the years have revealed that cancerous diseases are found not only in vertebrates but also in invertebrates (reviewed in Aktipis et al., 2015; Robert, 2010; see also Chapter 2). For example, calicoblastic neoplasms affect several species of corals, although these tumors remain localized and are not invasive like a typical malignant cancer (Peters et al., 1986; Yamashiro et al., 2000). In the nematode, *Caenorhabditis elegans*, tumors from the germinal layer are induced by a member of the Notch family (Berry et al., 1997). In mollusks, the bivalve *Mya arenaria* is affected by a type of cancer similar to leukemia. Notably, these putative malignant leukemia cells are polyploid and transplantable, can be grown in vitro, and are horizontally transmitted, even across species (Metzger et al., 2015, 2016; Walker et al., 2009). In insects, multiple types of tumors are known in the fly, *Drosophila melanogaster*, including neuroblastoma and ovarian and imaginal disk tumors (Beaucher et al., 2007; Herranz et al., 2016). Moreover, similar to humans the incidence of tumors of the testis and gut in *D. melanogaster* increases with age (Salomon and Jackson, 2008). In crustaceans, described tumors include a lymphoma-like neoplasm possibly caused by a viral infection in the white shrimp (*Penaeus vannamei*; Lightner and Brock, 1987)

and a putative carcinoma-like neoplasm in the hindgut of the red king crab, *Paralithodes camtschatica* (Sparks and Morado, 1987). Similarly, species among all jawless and jawed vertebrate classes develop tumors and cancer. Malignancies in jawless hagfish and lampreys include a high incidence of hepatomas and other tumors in the gut, pancreas, and kidneys (Falkmer et al., 1978; Hardisty, 1976).

Among jawed vertebrates, sharks have been mistakenly claimed to be tumor-free (Finkelstein, 2005), whereas they are in fact as susceptible as any other vertebrates to cancer, as a large variety of neoplasms and malignancies have been reported from neuroblastomas and osteomas to various carcinomas, fibrosarcomas, melanomas, and lymphomas (Ostrander et al., 2004). There is a large body of data about cancer in bony fish for more than 200 different species due to their economical importance (Groff, 2004; Smith, 2000), and although not as extensive, a large variety of cancerous diseases are also known in amphibians (see next section) and reptiles (Garner et al., 2004; Sykes and Trupkiewicz, 2006). Moreover, there is evidence of tumors in dinosaur fossil records from the Cretaceous (Rothschild et al., 1999, 2003). Thus, there is ample evidence that neoplasia and malignancies are widespread in the whole animal kingdom and thus of ancient evolutionary origin, which goes well with the idea of an ancestral origin of tumor immune surveillance (Table 9.2).

TABLE 9.2 Summary of Known Tumors and Antitumor Defenses in Different Animal Taxa

Animal taxa	Tumors		Immunity				
	Neoplasia	Invasive malignancy	Innate			Adaptive	
			Antitumor molecules	NK-like cells	Adaptive-like	T cells	B cells (Ab)
Sponges (porifers)	+	NF	+	+	−	−	−
Corals, hydra (cnidarians)	+	NF	+	+	+	−	−
C. elegans (nematodes)	+	NF	?	?	−	−	−
Earthworms (annelids)	+	NF	?	+	−	−	−
Snails, bivalves (mollusks)	+	(±)	+	+	+	−	−
Drosophila (insects)	+	(±)	+	+	+	−	−
Shrimps, crabs (crustaceans)	+	(±)	+	+	+	−	−
Lamprey, hagfish (agnathans)	+	+	?	?	?	T-like	B-like
Sharks, rays (Chondrichthyes)	+	+	+	+	+	+	+
Zebrafish, trout (teleosts)	+	+	+	+	+	+	+
Xenopus (amphibians)	+	+	+	+	+*	+*	+
Chicken (avians)	+	+	+	+	+*	+*	+*
Mouse, human (mammals)	+	+	+	+	+*	+*	+*

Adaptive-like refers to systems of immune receptors' somatic diversification not mediated by RAG1 and 2. T- and B-like refers to cells using nonimmunoglobulin variable lymphocyte receptors (VLR) ?, Unknown; +, definitely found; ±, some evidence; +*, cognate antitumor activity; NF: not found; NK, natural killer.

EVOLUTION OF IMMUNITY AND TUMOR IMMUNITY

The advances in genomics and comparative immunological studies in a wide range of invertebrate and vertebrates species have provided many important insights into the evolution and diversification of immune defenses (reviewed in Buchon et al., 2014; Flajnik and Kasahara, 2010; Litman et al., 2010). From an evolutionary point of view, the adaptive immune system relying on the generation of large repertoires of immunoglobulin domain-based T-cell receptors (TCRs) and B-cell receptors is shared among all jawed vertebrates and is thought to have emerged approximately 500 million years ago in a common ancestor of the now extinct placoderms. This emergence was coincident with two rounds of genome-wide duplication that distinguish invertebrates from vertebrates. In addition, the advent of adaptive immunity is postulated to have occurred in a relatively short period of evolutionary time with the acquisition by horizontal transfer (e.g., from viral infection) of the RAG1 and 2 genes to produce somatic diversification of lymphocyte receptors and by a yet unknown origin of the MHC selection/tolerance system.

However, the evolutionary history of the vertebrate immune system has become more complicated by the identification of another adaptive system that arose independently in basal vertebrate lineage giving rise to jawless vertebrates (Cooper and Alder, 2006). The jawless adaptive immune system uses a non–RAG-mediated gene conversion mechanism to generate diversity of leucin-rich repeats containing variable lymphocyte receptors (Boehm et al., 2012; Pancer et al., 2004). Furthermore, there is increasing evidence of adaptive feature in invertebrate immune systems. First, RAG1- and RAG2-like gene clusters, as well as sequence elements with similarities to RAG1 have been identified in the echinoderm purple sea urchin *Strongylocentrotus purpuratus* and the amphioxus *Branchiostoma floridae* (Fugmann et al., 2006; Huang et al., 2016). This raises the possibility of a more ancient origin of the genetic mechanisms associated with V(D)J recombination that may have generated repertoire of immune receptors in invertebrates and prochordate species extinct or still existing. Second, there is multiple evidence in invertebrates of expansion of immune receptors including Toll-like receptors (TLRs) and the Sp185/333 system in sea urchin (Ghosh et al., 2011), fibrinogen-related proteins (FREP) in snails (Hanington et al., 2010), and Down syndrome cell adhesion molecules (DSCAM) in insects (Brites and Du Pasquier, 2015). These receptors are in some cases somatically diversified by mechanisms including RNA processing and hypermutations, although there is no evidence to date of clonal selection and expansion as it is encountered in jawless and jawed vertebrates. The dichotomy adaptive/innate immune system has also become more blurred in jawed vertebrates by the expansion and diversification of some family of innate-like immune receptors including Fc-like receptors and other NK receptors, and nonclassical MHC. This is paralleled with the finding of innate-like immune cell effectors, such as iNKT cells and γδT cells recognizing typical innate-like pathogen-associated molecular patterns (PAMPs), as well as adaptive-like immunological memory of some NK cells. Overall, the comparative and evolutionary view of the immune system is of a highly redundant, integrated, and plastic system that can rapidly adapt to new external pressures from pathogens and/or internal pressure from malignancies (Fig. 9.1).

Despite the progress in comparative and evolutionary studies of immune functions, the role of the immune system in detecting and controlling neoplasia is still relatively poorly investigated in species other than mammals, chickens, and the amphibian *Xenopus* (Robert, 2010; Table 9.2). In invertebrates, soluble factors produced by various marine crustaceans and insect species have shown to exhibit activity against mammalian tumor cells in vitro, but it is not known whether these factors are also antitumorigenic in the species of origin (Chernysh et al., 2002; O'Hanlon, 2006). Also invertebrate species including sponges, annelids, and insects have NK-like cells that can kill allogeneic mammalian tumor cells, but again it remains to be determined whether these NK-like cells can kill cognate tumors (De Tomaso, 2009; Franceschi et al., 1991; Nappi and Ottaviani, 2000). In jawless (lamprey and hagfish), cartilaginous (sharks), and bony (trout, zebrafish) fish, very little is known about tumor immunity despite the substantial record of reported malignancies. All the fundamental elements of adaptive immunity with B and T cells, as well as genes involved in antigen processing and presentation (e.g., MHC class I, class II, immunoproteasome) and generation of somatic repertoires of lymphocyte receptors (e.g., RAG, Ig, TCRα, β, γ, δ) are present from sharks and bony fish (Flajnik and Kasahara, 2010; Litman et al., 2010). In addition, cell-mediated CD8 T and NK cell–mediated killing has been characterized in fish species including carp, zebrafish, and catfish, although their involvement in cognate antitumor response is still unclear (Moss et al., 2009; Nakanishi et al., 2011; Shen et al., 2002).

IMMUNOTHERAPIES

In parallel with progress in delineating tumor immune surveillance, as well as immune evasion and suppression by tumors, cancer immunotherapy approaches have encountered encouraging success recently. Based on the antagonism between antitumor immunity and tumor-mediated suppression, immunotherapy of cancer can be either active by stimulating the activities of specific components of the immune system or passive by counteracting signals from cancer cells that suppress

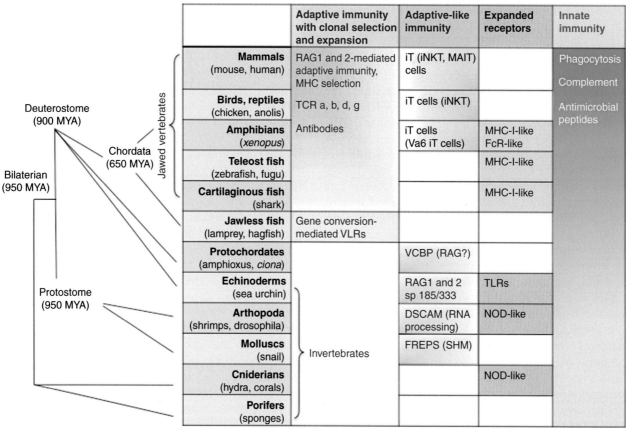

FIGURE 9.1 **Comparative and evolutionary overview of innate and adaptive features of immune systems in major invertebrate and vertebrate taxa.** Typical adaptive immunity based on RAG mediated somatic diversification of immune receptors followed by selection, expansion, and long-term immunological memory is limited to jawless and jawed vertebrates. Innate immunity, typically based on immune receptors germline-encoded recognizing pathogen-associated molecular patterns that are expressed by immune cells serving in a first line of defense and that are not expending, is present to some extent in all animals. Additional features in between these two categories have evolved in invertebrate and vertebrate species including somatic diversifications in invertebrates and expansion of different families of immune cell receptors. *DSCAM*, Down syndrome cell adhesion molecule; *iNKT*, invariant natural killer T; *iT*, innate T; *MAIT*, mucosal-associated invariant T; *MYA*, millions of years ago; *TCR*, T-cell receptor.

immune responses. Among active immunotherapy methods, one can mention the various attempts to immunize against tumor antigens and elicit specific antitumor T-cell responses using, for example, the patient DCs pulsed with tumor antigens as vaccine (Palucka and Banchereau, 2012). Adoptive cell transfer with engineered T cells transformed in vitro to express chimeric antigen receptors (CAR) is another type of active treatment for cancers, such as leukemia (Maude et al., 2015).

The development of passive type of immunotherapy is more recent and has led to remarkable clinical effects. This approach uses antibodies that block the checkpoint inhibitors CTLA-4 and PD-1, of two important immune pathways (Farkona et al., 2016; Pardoll, 2012; Sharma et al., 2011). The blockade of these receptors enhances antitumor T-cell response by preventing or abrogating suppressive signals. Effects of targeting additional inhibitory ligands, such as LAG3 or TIM are currently in preclinical development (Perez-Gracia et al., 2014). However, not all patients respond to these immunotherapies, highlighting a need for novel strategies that could

help transform the nonresponders to responders' category. Besides finding new targets and treatments, efforts also involve combinations of immune monotherapies, as well as integration of these immunotherapies with other targeted cancer therapies (e.g., radiation and chemotherapies) into a personalized therapy. A real and not fully resolved concern for developing and optimizing these novel and combined immunotherapies is the reliability of the animal model systems.

OVERVIEW OF ANIMAL MODEL SYSTEM USED TO INVESTIGATE THE ROLE OF IMMUNE SYSTEM IN CANCER

Immunology and tumor immunology has and still is heavily relying almost exclusively on the mouse as animal model. The extensive resources in inbred and genetically modified mouse strains, as well as tools and reagents including many transplantable tumors types have been key for elucidating the mechanisms of antitumor immunity

TABLE 9.3 Summary of the Main Animal Models Used for Studying Immunology

Mammals	Mouse	Main model for immunology and tumor immunity
	Rat	Immunology (transplantation)
	Rabbit	Immunology (B cell, mucosal immunity)
	Goat	Immunology (mucosal immunity)
Birds	Chicken	Immunology (virally induced cancer)
Amphibians	*Xenopus*	Immunology (ontogeny, tumor immunity)
	Axolotl	Immunology (ontogeny, regeneration)
Fish	Zebrafish	Immunology (innate immunity, some tumor immunity)
	Fugu	Immunology (immunogenetics)
	Trout	Immunology (mucosal and microbial immunity)
	Catfish	Immunology (lymphoid cell lines)
Invertebrates	*Ciona intestinalis*	Immunogenetics
	Botryllus schlosseri	Allorecognition
	Sea urchin	Innate immunology
	Snail (*B. glabrata*)	Innate immunology
	Bivalve (*Mya arenaria*)	Leukemia
	Drosophila	Innate immunology, cancer
	Nematode (*C. elegans*)	Oncogenes, immune defenses, cancer

and tumor escape. The mouse model has and still is valuable to design and explore immunotherapeutic approaches (reviewed in Budhu et al., 2014; Dranoff, 2012).

The power and success of the mouse model has overshadowed other comparative models. However, a generalization based on a single animal model may reveal to be limited and problematic. Indeed, application from mouse to human has been challenging, especially with regards to immunotherapy. Different solutions are explored to obtain a better translation, such as humanized mouse systems where both tumor and immune cells are of human origin (Dranoff, 2012). Nevertheless, it seems clear that species-specific differences and even small nuances in the immune system are critical in how well an immunotherapy will translate from any given animal model into successful clinical use. As such alternative comparative models remain an unappreciated option that would merit to better develop. Indeed, there are numerous nonmurine animal models currently used for the comparative study of different aspects of immunology (Table 9.3). Each model presents attractive features that could be useful for investigating various aspects of tumor immunity.

AMPHIBIANS AS MODEL SYSTEMS TO EXPLORE THE EVOLUTIONARY CONSERVATION OF TUMOR IMMUNITY

Amphibians, because of their evolutionary distance to mammals, and their position as a taxon connecting cartilaginous and bony fish with tetrapods, provide an interesting animal model to explore the evolutionary conservation of tumor immunity. The *Xenopus* genus has diverged from common ancestors with mammals approximately 350 million years ago (Evans et al., 2015).

As such studies in *Xenopus* allow investigators to distinguish interesting species-specific adaptations, as well as conserved fundamental features.

Notably, *Xenopus* has proven instrumental to exploring the evolutionary perspective of interactions between the host immune system and cancerous cells (reviewed in Goyos and Robert, 2009; Robert and Ohta, 2009). The high degree of functional conservation of the *Xenopus* immune system with human, as well as the amenability of *Xenopus* to in vivo experimentation makes it a highly relevant and attractive nonmammalian comparative model (reviewed in Robert and Cohen, 2011).

THE *XENOPUS* IMMUNE SYSTEM

The amphibian *X. laevis* has an immune system with the typical elements shared across jawed vertebrates' immunity including the primary immune organs thymus and spleen, T and B lymphocytes expressing a wide repertoire of rearranged TCRs and B-cell receptors, respectively, hallmark cytokines, and MHC class I and class II genes (reviewed in Du Pasquier et al., 1989; Robert and Ohta, 2009). The thymus dependency, T-cell differentiation, and T-cell–mediated responses are relatively well characterized in *Xenopus*, using, for example, permanent T-cell–deficient frogs obtained by thymectomy (Horton et al., 1992, 1998) or by reverse genetics (e.g., disruption of the *Foxn1* gene; Nakai et al., 2016). MHC-restricted cytotoxic and helper T-cell responses have been identified, as well as a prominence of innate-like T cells (Edholm et al., 2013, 2015). Similar to mammals, *Xenopus* also possess the whole range of innate cell effectors, such as neutrophils, eosinophils, basophils, DCs, and macrophages (Du Pasquier et al., 1989; Grayfer and Robert, 2016;

Mescher et al., 2007). The fundamental conservation of the immune system between *Xenopus* and mammals is further supported by comparative genomics of the fully sequenced and annotated genomes of two different *Xenopus* species, *X. tropicalis* and *X. laevis* (Hellsten et al., 2010; Session, 2016).

While the *Xenopus* immune system is fundamentally conserved, it also presents distinctive features not encountered in mammals. Indeed, in contrast to mammals, the immune system develops twice in *Xenopus*: first during embryogenesis and again during metamorphosis (Flajnik and Du Pasquier, 1990; Flajnik et al., 1986). The *Xenopus* thymus is first colonized by embryonic stem cells a few days after fertilization (Kau and Turpen, 1983). During metamorphosis, the thymus loses about 90% of its lymphocytes (Du Pasquier and Weiss, 1973). This loss is followed by a second wave of stem cell immigration (Bechtold et al., 1992; Turpen and Smith, 1989). Another salient aspect of *Xenopus* immunology is the absence of classical MHC class Ia protein expression in tadpoles until the onset of metamorphosis (Flajnik and Du Pasquier, 1988; Flajnik et al., 1986; Rollins-Smith et al., 1997). Although tadpoles are immunocompetent and have CD8 T cells, the larval thymus lacks significant expression of class Ia and *LMP7* genes until metamorphosis, which suggests an inefficient class Ia–restricted T-cell education during larval life (Flajnik et al., 1986; Salter-Cid et al., 1998). Conversely, multiple *XNC* genes are expressed by thymocytes at the onset of thymic organogenesis consistent with a role of class I–like molecules in early T-cell development.

Presumably because of the small number of lymphocytes, the tadpole's immune system appears to rely on a prominence of *XNC*-restricted invariant or innate (i) T cells. Similar to mammalian iT cells, such as iNKT and mucosal-associated invariant T (MAIT) cells, these *Xenopus* iT-cell subsets express limited or invariant TCR rearrangements that presumably recognize distinct conserved antigens or common pathogen patterns (Edholm et al., 2013; Robert and Edholm, 2014). One particular iT-cell subset is restricted by *XNC10* and strictly depends on *XNC10* function for its development as shown by reverse genetic approach combining transgenesis with RNA interference (Edholm et al., 2013, 2014b).

TRANSPLANTABLE THYMIC LYMPHOID TUMOR CELL LINES

Although *X. laevis* and *X. tropicalis* are widely used as experimental animal models, there have been relatively few reports of spontaneous tumors in *Xenopus* species (reviewed in Anver, 1992; Stacy and Parker, 2004; Table 9.4). Additionally, one difficulty with studying tumor immunity in nontraditional model animals is that

TABLE 9.4 Spontaneous Malignancies Reported in *Xenopus*

Organs/tissues	Type	References
Intestine	Carcinoma	Elkan (1970)
Kidney	Carcinoma, nephroblastoma	Meyer-Rochow et al. (1991)
Liver	Adenomas	Reichenbach-Klinke and Elkan (1965)
Pancreas	Carcinoma	Stern et al. (2014)
Ovary	Dysgerminoma	Goyos and Robert (2009)
Pigment cells	Melanophoroma	Reichenbach-Klinke and Elkan (1965)
Spleen	Leukocytic	Du Pasquier et al. (2009)
Thymus	Lymphoid	Du Pasquier and Robert (1992), Earley et al. (1995), Robert et al. (1994)
Thyroid	Carcinomas	Cheong et al. (2000)

tumors cannot be transplanted because of the lack of a MHC-compatible inbred system. Thus, the occurrence of several spontaneous thymic tumors in the MHC-defined partially inbred F strain and *X. laevis*/*X. gilli* isogenetic clones (LG clones) provided an opportunity to derive lymphoid tumor lines growing in vitro and in vivo following transplantation in compatible amphibian host (Robert et al., 1994, 1995).

The derived four thymic lymphoid tumor cell lines (B3B7, ff-2, 15/40, and 15/0) share a mixed immature T/B-cell phenotype and highly express several *Xenopus* nonclassical MHC class Ib (*XNC*) genes, including *XNC1, 4, 10,* and *11*, as well as β_2-microglobulin (*b2m*; Goyos et al., 2009). In contrast, classical MHC class Ia protein is not (15/0, 15/40, B3B7) or barely (ff-2) expressed (Robert et al., 1994, 1995). The ff-2 tumor, derived from the *Xenopus* inbred F strain, has remained transplantable and tumorigenic in compatible tadpoles but is rejected by F adult frogs. This rejection is dependent on the differentiation of new T cells in adults (Robert et al., 1995, 1997). The 15/0 tumor is highly tumorigenic when transplanted into both MHC-defined isogenic cloned frogs LG-6 and LG-15 tadpoles and adults (Goyos et al., 2004; Robert et al., 1994, 1995). These clones share the same MHC haplotype (a/c) but differ at multiple minor histocompatibility (H) loci (Kobel and Du Pasquier, 1975, 1977).

TUMOR IMMUNITY IN *X. LAEVIS*

The transplantable ff-2 and LG-15 tumors and their respective compatible *X. laevis* hosts have permitted exploration of tumor immunity outside mammals. Several studies, both in vivo by antibody treatment and in vitro by cytotoxic assays, have demonstrated that in *X. laevis* as in mammals NK and CD8 T cells are critical antitumor cell effectors able to kill tumor cells (Goyos et al., 2004; Horton et al., 2000; Rau et al., 2002). The requirement of conventional T cells, and particularly CD8 T cells, for controlling malignancy was also shown with the ff-2 tumor line by thymectomy and by sublethal γ-irradiation (Horton et al., 2003; Robert et al., 1995, 1997).

Importantly, in the case of the 15/0 tumor line the absence of MHC class Ia expression has unveiled the unappreciated but evolutionarily conserved roles of MHC class Ia–unrestricted NK and unconventional T cells in vertebrate tumor immunity.

CONSERVATION OF ANTITUMOR PROPERTIES OF HEAT SHOCK PROTEINS

The comparative tumor immunity model in *X. laevis* has provided valuable evolutionary support for the postulated roles of certain heat shock proteins (hsps) in immunity. In mouse and human, hsps, such as the endoplasmic resident gp96 and the cytosolic hsp70, generate potent antitumor protective responses by inducing proinflammatory cytokines, stimulating NK cells, and eliciting potent cytotoxic CD8 T-cell responses against the antigenic peptides they chaperone (Calderwood and Gong, 2016; Srivastava, 2002a,b). The cross-presentation of antigens chaperoned by these hsps in the context of MHC class Ia by APCs critically involves endocytic receptors including CD91 (Basu et al., 2001; Binder et al., 2000) and other scavenger receptors (Delneste et al., 2002; Facciponte et al., 2007; Murshid et al., 2015). The additional interaction of these hsps with various signaling receptors, such as TLRs is associated with their ability to stimulate inflammation (Asea et al., 2002; Warger et al., 2006).

Given the high degree of evolutionary conservation of gp96 and hsp70 genes, it was of interest to determine how conserved their role was in tumor immunity. Notably, *X. laevis* gp96 binds to synthetic antigenic peptides and facilitates their cross-presentation by mouse APCs as efficiently as mouse gp96 (Robert et al., 2001). Moreover, *Xenopus* gp96 and hsp70 can cross-present chaperoned LG-6 or LG-15 minor-H-Ags in a MHC-restricted fashion and generate efficient antiminor H-Ag specific cytotoxic CD8 T cells (Robert et al., 2002). Hsp-mediated antigen cross-presentation within *Xenopus* was further demonstrated by the memory T-cell response generated against skin minor H-Ags after adoptive transfer of peritoneal *Xenopus* macrophages pulsed with hsp70/gp96-minor H-Ag complexes (Nedelkovska and Robert, 2013; Robert et al., 2002). As in mouse, this cross-presentation required the endocytic receptor CD91 (Robert et al., 2008). These findings suggest that the involvement of certain hsps (gp96, hsp70) and hsp receptors (CD91) in immune surveillance is evolutionarily conserved between amphibians and mammals.

In absence of MHC class Ia expression, the antitumor response generated by gp96 or hsp70 is intriguing and suggests that these hsps can also stimulate unconventional T cells. Indeed, anti-15/0 tumor response elicited by hsps critically involves unconventional CD8 cytotoxic T cells that are not restricted by classical MHC class Ia but require chaperoned Ags (Nedelkovska and Robert, 2013; Robert et al., 2001). Adoptive transfer of *X. laevis* peritoneal macrophages pulsed with 15/0 tumor-derived hsps further showed that the antitumor unconventional T-cell response was dependent on CD91-mediated cross-presentation (Nedelkovska and Robert, 2013; Robert et al., 2008). For hsp70, the respective role of the inducible hsp72 and the cognate or constitutively expressed hsc73 was further investigated in *Xenopus*. Although these two types of cytosolic hsp70 share extensive molecular structure, they have significant differences in their ligand binding domains, subcellular localization, and some of their functions (Callahan et al., 2002). Interestingly, recombinant hsp72 and hsc73-Ag complexes produced by 15/0 tumor transfectants were as efficient in mediating cross-presentation and conventional T-cell responses against minor H-Ags. However, hsp72 was markedly more potent than hsc73 in generating protective immune responses against the class Ia–negative 15/0 tumors in an Ag-dependent but MHC class Ia–independent manner, which suggested the involvement of class Ib molecules and class Ia–unrestricted effector T-cell populations (Nedelkovska and Robert, 2013).

CONSERVED ROLES OF NONCLASSICAL MHC AND INNATE T CELLS IN TUMOR IMMUNITY

The importance of MHC class Ia for education and function of CD8 T cells is well established (Klein et al., 2014). Thus, tumors often downregulate class Ia expression to escape conventional T-cell–mediated immune recognition and killing (Zitvogel et al., 2006). However, by losing the expression of class Ia, these tumors become susceptible to NK cell–mediated cytotoxicity. Therefore, as an additional evasion mechanism from immune cell-mediated killing, tumors are also frequently found to induce or upregulate the expression of MHC class I–like genes (also called HLA in human literature). Class I–like molecules can serve as indicators of malignancy and/or intracellular stress (Gleimer and Parham, 2003). Still, the functional relevance of class I–like molecules in the cancer field is not fully established. A few clinical studies have confirmed class I–like deregulated expression by certain tumors, and found correlation between increased class I–like expression and unfavorable prognostics. For example, HLA-E and HLA-G are indicators of poor clinical outcome in several different types of cancer (Benevolo et al., 2011; de Kruijf et al., 2010; He et al., 2010; Ye et al., 2007; Yie et al., 2007). Contrasting with these findings, several other class I–like proteins, in human and mouse, can mediate protective immunity against different cancers. Notably, CD1d-restricted NKT cells appear to affect the outcome of tumor immunity (Fox et al., 2013; Hix et al., 2011; Hong and Park, 2007), and several ongoing clinical trials are evaluating the effect of the class Ib CD1d-mediated stimulation of iNKT cells with α-galactosylceramide (α-GalCer) on cancer patients (reviewed in Altman et al., 2015).

From an evolutionary perspective, both class Ia– and class I–like genes have been found in all jawed vertebrates studied to date (reviewed in Flajnik and Kasahara, 2001). Although relationships between evolutionarily distant class I–like molecules are difficult to establish, functional analogs, such as the primate HLA-E and the mouse Qa-1b, have been identified (Yeager et al., 1997). CD1 is found in mammals (Brossay et al., 1998; Dascher, 2007), birds (Miller et al., 2005; Salomonsen et al., 2005), and reptiles (Yang et al., 2015) but neither in fish nor in amphibians. In *X. laevis* there are at least 23 class I–like nonclassical (*XNCs*) genes that, like other vertebrate class I–like genes, are heterogeneous, less polymorphic, and less ubiquitously expressed than class Ia (Edholm et al., 2014a; Flajnik et al., 1993; Goyos et al., 2009, 2011).

In this context, the high expression levels of several *XNC* genes by genetically distinct *X. laevis* lymphoid thymic tumors (ff-2, 15/0) take a particular relevance. The possible involvement of some of these *XNCs* in tumorigenesis, as well as their requirement for the development and function of different subsets of iT cells expressing semiinvariant TCR repertoires has opened new avenues of investigation in *Xenopus*. For example, 15/0 tumor transfectants functionally deficient for multiple *XNCs* were more susceptible to NK cells, whereas they were more resistant to CD8 T cells and grew faster in vivo when transplanted into LG-15 adult recipients (Goyos et al., 2007). However, when only *XNC10*, which is among the highest *XNC* expressed in 15/0 tumor, was silenced, the tumor transfectants were acutely rejected by both syngeneic LG-15 adults and naturally class Ia–deficient LG-15 tadpoles (Haynes-Gilmore et al., 2014). Furthermore, this rejection was due to a cell-mediated killing that could be enhanced by priming. These unexpected findings hint to a more complex role of *XNCs* in tumor immunity. On the one hand, some *XNCs* appear to be critical for immune recognition and antitumor response against 15/0 tumors, as suggested by the overall silencing of *XNCs* or *b2m*. On the other hand, the immune evasion of 15/0 tumors is promoted by *XNC10* that appears to have a dominant effect over other *XNCs*, as indicated by its requirement for growing in vivo avoiding immune rejection and class Ia–independent cytotoxicity. It is also

possible that the different *XNCs* expressed by 15/0 tumor cells interact with distinct effector cells resulting in a balance between inhibitory and activating signals leading to either increased or decreased tumorigenicity.

Interestingly, *XNC10* is required for the development and function of a distinct iT cell subset (Edholm et al., 2013). These *XNC10*-restricted iT cells are early responders following transplantation of 15/0 tumors into LG-15 tadpoles (Haynes-Gilmore et al., 2014) and ff-2 tumor into inbred F tadpoles (Fig. 9.2). Intriguingly, knockdown of *XNC10* in 15/0 tumor results in increased infiltration of *XNC10*-iT cells, which is again consistent with the use of *XNC10* as an immune evasion strategy by the 15/0 tumors.

XENOPUS TADPOLE TUMOR MODEL

Early developmental stages of *Xenopus* are attractive for inducing tumors and studying links between development and cancer (reviewed in Wallingford, 1999). The accessibility of one-cell stage embryos for experimentation has allowed a loss-of-function investigation by injection of antibody to determine the role of factors involved in pigmentation and melanoma (von Strandmann et al., 2001). Gene disruption by reverse genetics has been also developed using Transcription activator-like effector nucleases (TALEN) (and now CRISPR/Cas9). For example, tadpoles derived from embryos injected with TALENs targeting the tumor suppressor gene *adenomatous polyposis coli* rapidly developed intestinal hyperplasia and other neoplasms typically observed in patients presenting with familial adenomatous polyposis (FAP), including desmoid tumors and medulloblastomas (Van Nieuwenhuysen et al., 2015).

With the development of modern microscopic technology, the *Xenopus* tadpole has become a valuable model for gathering real-time insights into neovascularization and lymphangiogenesis during tumorigenesis, as well as tumor microenvironment (Haynes-Gimore et al., 2015; Kalin et al., 2009; Ny et al., 2005). Free-swimming tadpoles are amenable to surgical (e.g., thymectomy, transplantation) and nonsurgical (e.g., adoptive transfer of leukocytes, injection of hormones, antibodies) manipulations, as well as intravital microscopic observations because of their transparent skin and a small size (10- to 20-mm body length). As such, *Xenopus* tadpoles are useful for collecting information during early development, which is more difficult to gather in utero in mammals (e.g., development of self-tolerance to adult-specific antigens, ontogeny of T-cell subsets, and early antitumor immunity in a natural setting).

Hence, the *Xenopus* tadpole is a powerful model for regeneration and wound healing research (Slack et al., 2008), as well as for blood and lymphatic vascular

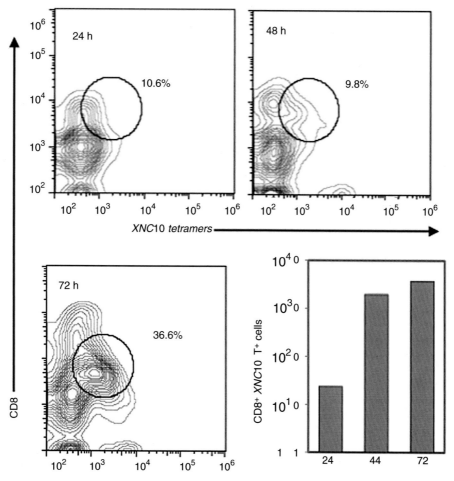

FIGURE 9.2 Rapid infiltration of *XNC10*-restricted iT cells following tumor transplantation. At 24, 48, and 72 h following intraperitoneal transplantation of 100,000 ff-2 cells, peritoneal exudate was collected and cells were analyzed by flow cytometry with *XNC10* tetramers + APC-conjugated streptavidin and anti-CD8 mAb + Fluorescein isothiocyanate (FITC)-conjugated goat antimouse Ab. *APC*, antigen-presenting cell; *iT*, innate T.

development and angiogenesis (Levine et al., 2003; Ny et al., 2005). Molecular and cellular mechanisms involved in these processes are conserved between *Xenopus* and mammals (Ny et al., 2005, 2008). These include the key role of macrophages and endothelial precursor cells (EPCs), as well as critical genes in cancer cell signaling (Hardwick and Philpott, 2015) and cellular interactions (e.g., VGEF-R, CD34; Ny et al., 2008; Saharinen et al., 2010). An exciting area of investigation using *Xenopus* tadpoles concerns modulation of membrane potential during tumor development and metastasis (Chernet and Levin, 2014; Lobikin et al., 2012). Recently, optogenetics has been implemented in *Xenopus* tadpoles to manipulate ion flux–mediated regulation of membrane potential specifically to prevent and cause regression of oncogene-induced tumors (Chernet et al., 2016). Least but not last, a semisolid tumor system has been developed in *X. laevis* by embedding 15/0 tumor cells into a collagen matrix before tadpole implantation dorsally and bilaterally onto single LG-15 tadpoles under their transparent skin (Fig. 9.3). This approach permits convenient comparisons between a wild-type tumor control and a tumor genetically modified (e.g., a stable *XNC10*-deficient transfectant abrogating its tumorigenicity) within the same animal recipient. This system recapitulates many facets of mammalian tumorigenesis and permits real-time visualization of the active formation of the tumor microenvironment induced by 15/0 tumor cells including neovascularization, collagen rearrangements, as well as infiltration of immune cells and melanophores (Fig. 9.3).

CONCLUSIONS AND PERSPECTIVES

The use of animal models alternative to mouse, such as *Xenopus*, is important not only for exploring innovative approaches for cancer immunotherapy but also for better understanding the fundamental tumor and immune system interactions. With the development of extensive genomic and reverse genetic resources, *Xenopus* has become a powerful model for understanding various aspects of human health and diseases, including cancer

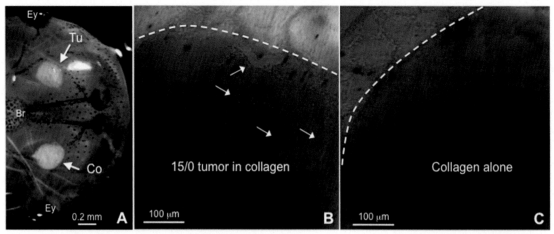

FIGURE 9.3 **Neovascularization initiated by tumor engraftment in tadpoles.** An LG-6 tadpole recipient was grafted with 1×10^5, 15/0 lymphoid tumor cells embedded in 10 μL of rat tail collagen on one side and 10 μL of rat collagen alone on the other side. (A) Dorsal view of the tadpole with 15/0 tumor (*Tu*) and empty collagen (*Co*) grafts transplanted bilaterally under the dorsal skin of the tadpole in the vicinity of the eyes (*Ey*). At 6 days posttransplantation, the anesthetized tadpole was microinjected with 10 μL of Texas-Red dextran into the pulmonary artery. One percent agarose gel was used to stabilize the tadpole throughout imaging. Images of the 15/0 tumor (B) and empty collagen (C) were taken with an Olympus FV1000 laser scanning confocal microscope at the URMC Confocal and Conventional Microscopy Core. Note the presence of newly formed blood vessels inside the tumor graft (*white arrows*) but not in the empty collagen graft.

(Hardwick and Philpott, 2015; LaBonne and Zorn, 2015). Furthermore, the availability of MHC-defined inbred strains and clones with transplantable lymphoid tumors, the attractive biology (accessible external development, small size, and transparency of immunocompetent tadpoles), naturally MHC-deficient tadpoles, and the rapid advance in genome editing and transgenesis, all make *Xenopus* a useful comparative model for research in oncogenesis and cancer immunity.

More generally, pathology, etiology, and particularly immunity to cancer remain to date insufficiently investigated in ectothermic vertebrate and invertebrate species. Indeed despite recent successes of cancer immunotherapy, clinical applications of these therapies still remain challenging. For example, not all patients are responding to immunothereapy. In addition, not all tumor types can be targeted by immunotherapy. While current cancer research efforts are focused on improving current immunotherapeutic approaches and developing other methodologies, one potential limitation often underappreciated is the heavy reliance on murine models, which may not always fully recapitulate human characteristics. Extending comparative studies to evolutionary distant organisms warrants fruitful potential to discover unappreciated or unknown mechanisms of cancer immune surveillance and immunoediting that can be exploited for immunotherapies. A broader comparative approach would also improve our understanding of the coevolution or arms race between tumors and host immune systems. This in turn would permit finding ways to minimize the selection of more malignant variants during cancer treatments, especially when using immunotherapies.

Acknowledgments

We would like to thank Dr. Edith Lord for critical reading of the manuscript. This work was supported by an R24-AI-059830 grant from the National Institute of Allergy and Infectious Diseases (NIH/NIAID) and from the Kesel Fund of Rochester Area Community Foundation, Rochester, NY. M.B. was supported by a predoctoral fellowship Ruth L. Kirschstein Predoctoral F31 (F31CA192664) from the National Cancer Institute (NIH/NCI). E.-S.E. was supported by the National Science Foundation IOS-1456213 and a 2015 Career in Immunology Fellowship from the American Association of Immunologists.

References

Aktipis, C.A., Boddy, A.M., Jansen, G., Hibner, U., Hochberg, M.E., Maley, C.C., Wilkinson, G.S., 2015. Cancer across the tree of life: cooperation and cheating in multicellularity. Philos. Trans. R. Soc. Lond. B Biol. Sci. 370(1673). pii: 20140219..

Altman, J.B., Benavides, A.D., Das, R., Bassiri, H., 2015. Antitumor responses of invariant natural killer T cells. J. Immunol. Res. 2015, 652875.

Anver, M.R., 1992. Amphibian tumors: a comparison of anurans and urodeles. In Vivo 6, 435–437.

Asea, A., Rehli, M., Kabingu, E., Boch, J.A., Bare, O., Auron, P.E., Stevenson, M.A., Calderwood, S.K., 2002. Novel signal transduction pathway utilized by extracellular HSP70: role of toll-like receptor (TLR) 2 and TLR4. J. Biol. Chem. 277, 15028–15034.

Basu, S., Binder, R.J., Ramalingam, T., Srivastava, P.K., 2001. CD91 is a common receptor for heat shock proteins gp96, hsp90, hsp70, and calreticulin. Immunity 14, 303–313.

Beaucher, M., Goodliffe, J., Hersperger, E., Trunova, S., Frydman, H., Shearn, A., 2007. *Drosophila* brain tumor metastases express both neuronal and glial cell type markers. Dev. Biol. 301, 287–297.

Bechtold, T.E., Smith, P.B., Turpen, J.B., 1992. Differential stem cell contributions to thymocyte succession during development of *Xenopus laevis*. J. Immunol. 148, 2975–2982.

Benevolo, M., Mottolese, M., Tremante, E., Rollo, F., Diodoro, M.G., Ercolani, C., Sperduti, I., Lo Monaco, E., Cosimelli, M., Giacomini, P., 2011. High expression of HLA-E in colorectal carcinoma is associated with a favorable prognosis. J. Transl. Med. 9, 184. doi: 10.1186/1479-5876-9-184.

Berry, L.W., Westlund, B., Schedl, T., 1997. Germ-line tumor formation caused by activation of glp-1, a *Caenorhabditis elegans* member of the Notch family of receptors. Development 124, 925–936.

Binder, R.J., Han, D.K., Srivastava, P.K., 2000. CD91: a receptor for heat shock protein gp96. Nat. Immunol. 1, 151–155.

Boehm, T., McCurley, N., Sutoh, Y., Schorpp, M., Kasahara, M., Cooper, M.D., 2012. VLR-based adaptive immunity. Annu. Rev. Immunol. 30, 203–220.

Brites, D., Du Pasquier, L., 2015. Somatic and germline diversification of a putative immunoreceptor within one phylum: Dscam in arthropods. Results Probl. Cell Differ. 57, 131–158.

Brossay, L., Chioda, M., Burdin, N., Koezuka, Y., Casorati, G., Dellabona, P., Kronenberg, M., 1998. CD1d-mediated recognition of an alpha-galactosylceramide by natural killer T cells is highly conserved through mammalian evolution. J. Exp. Med. 188, 1521–1528.

Brown, G.K., Kreiss, A., Lyons, A.B., Woods, G.M., 2011. Natural killer cell mediated cytotoxic responses in the Tasmanian devil. PLoS One 6, e24475.

Buchon, N., Silverman, N., Cherry, S., 2014. Immunity in *Drosophila melanogaster*—from microbial recognition to whole-organism physiology. Nat. Rev. Immunol. 14, 796–810.

Budhu, S., Wolchok, J., Merghoub, T., 2014. The importance of animal models in tumor immunity and immunotherapy. Curr. Opin. Genet. Dev. 24, 46–51.

Burnet, F.M., 1970. The concept of immunological surveillance. Prog. Exp. Tumor Res. 13, 1–27.

Calderwood, S.K., Gong, J., 2016. Heat shock proteins promote cancer: it's a protection racket. Trends Biochem. Sci. 41, 311–323.

Callahan, M.K., Chaillot, D., Jacquin, C., Clark, P.R., Menoret, A., 2002. Differential acquisition of antigenic peptides by Hsp70 and Hsc70 under oxidative conditions. J. Biol. Chem. 277, 33604–33609.

Cheong, S.W., et al., 2000. Spontaneous thyroid-containing teratoma associated with impaired development in the African clawed frog, *Xenopus laevis*. J. Comp. Pathol. 123 (2–3), 110–118.

Chernet, B.T., Levin, M., 2014. Transmembrane voltage potential of somatic cells controls oncogene-mediated tumorigenesis at long-range. Oncotarget 5, 3287–3306.

Chernet, B.T., Adams, D.S., Lobikin, M., Levin, M., 2016. Use of genetically encoded, light-gated ion translocators to control tumorigenesis. Oncotarget 7, 19575–19588.

Chernysh, S., Kim, S.I., Bekker, G., Pleskach, V.A., Filatova, N.A., Anikin, V.B., Platonov, V.G., Bulet, P., 2002. Antiviral and antitumor peptides from insects. Proc. Natl. Acad. Sci. USA 99, 12628–12632.

Cooper, M.D., Alder, M.N., 2006. The evolution of adaptive immune systems. Cell 124, 815–822.

Coulie, P.G., Van den Eynde, B.J., van der Bruggen, P., Boon, T., 2014. Tumour antigens recognized by T lymphocytes: at the core of cancer immunotherapy. Nat. Rev. Cancer 14, 135–146.

Dascher, C.C., 2007. Evolutionary biology of CD1. Curr. Top. Microbiol. Immunol. 314, 3–26.

de Kruijf, E.M., Sajet, A., van Nes, J.G., Natanov, R., Putter, H., Smit, V.T., Liefers, G.J., van den Elsen, P.J., van de Velde, C.J., Kuppen, P.J., 2010. HLA-E and HLA-G expression in classical HLA class I-negative tumors is of prognostic value for clinical outcome of early breast cancer patients. J. Immunol. 185, 7452–7459.

De Tomaso, A.W., 2009. Sea squirts and immune tolerance. Dis. Model Mech. 2, 440–445.

Delneste, Y., Magistrelli, G., Gauchat, J., Haeuw, J., Aubry, J., Nakamura, K., Kawakami-Honda, N., Goetsch, L., Sawamura, T., Bonnefoy, J., Jeannin, P., 2002. Involvement of LOX-1 in dendritic cell-mediated antigen cross-presentation. Immunity 17, 353–362.

Dranoff, G., 2012. Experimental mouse tumour models: what can be learnt about human cancer immunology? Nat. Rev. Immunol. 12, 61–66.

Du Pasquier, L., Weiss, N., 1973. The thymus during the ontogeny of the toad *Xenopus laevis*: growth, membrane-bound immunoglobulins and mixed lymphocyte reaction. Eur. J. Immunol. 3, 773–777.

Du Pasquier, L., Schwager, J., Flajnik, M.F., 1989. The immune system of *Xenopus*. Annu. Rev. Immunol. 7, 251–275.

Du Pasquier, L., Robert, J., 1992. In vitro growth of thymic tumor cell lines from *Xenopus*. Dev. Immunol. 2 (4), 295–307.

Du Pasquier, L., et al., 2009. The fate of duplicated immunity genes in the dodecaploid *Xenopus ruwenzoriensis*. Front. Biosci. 14, 177–191.

Dunn, G.P., Old, L.J., Schreiber, R.D., 2004. The three Es of cancer immunoediting. Annu. Rev. Immunol. 22, 329–360.

Earley, E.M., et al., 1995. Tissue culture of a mixed cell thymic tumor from *Xenopus laevis*. In Vitro Cell Dev. Biol. Anim. 31 (4), 255–257.

Edholm, E.S., Albertorio Saez, L.M., Gill, A.L., Gill, S.R., Grayfer, L., Haynes, N., Myers, J.R., Robert, J., 2013. Nonclassical MHC class I-dependent invariant T cells are evolutionarily conserved and prominent from early development in amphibians. Proc. Natl. Acad. Sci. USA 110, 14342–14347.

Edholm, E.S., Goyos, A., Taran, J., De Jesus Andino, F., Ohta, Y., Robert, J., 2014a. Unusual evolutionary conservation and further species-specific adaptations of a large family of nonclassical MHC class Ib genes across different degrees of genome ploidy in the amphibian subfamily Xenopodinae. Immunogenetics 66, 411–426.

Edholm, E.S., Grayfer, L., Robert, J., 2014b. Evolution of nonclassical MHC-dependent invariant T cells. Cell. Mol. Life Sci. 71, 4763–4780.

Edholm, E.S., Grayfer, L., De Jesus Andino, F., Robert, J., 2015. Nonclassical MHC-restricted invariant Valpha6 T cells are critical for efficient early innate antiviral immunity in the amphibian *Xenopus laevis*. J. Immunol. 195, 576–586.

Ehrlich, P., 1957. The Collected Papers of Paul Ehrlich. Pergamon Press, London.

Elkan, E., 1970. A spontaneous anaplastic intestinal metastasising carcinoma in a South African clawed toad (*Xenopus laevis* Daudin). J. Pathol. 100 (3), 205–207.

Evans, B.J., Carter, T.F., Greenbaum, E., Gvozdik, V., Kelley, D.B., McLaughlin, P.J., Pauwels, O.S., Portik, D.M., Stanley, E.L., Tinsley, R.C., Tobias, M.L., Blackburn, D.C., 2015. Genetics, morphology, advertisement calls, and historical records distinguish six new polyploid species of African clawed frog (*Xenopus*, Pipidae) from West and Central Africa. PLoS One 10, e0142823.

Facciponte, J.G., Wang, X.Y., Subjeck, J.R., 2007. Hsp110 and Grp170, members of the Hsp70 superfamily, bind to scavenger receptor-A and scavenger receptor expressed by endothelial cells-I. Eur. J. Immunol. 37, 2268–2279.

Falkmer, S., Marklund, S., Mattsson, P.E., Rappe, C., 1978. Hepatomas and other neoplasms in the Atlantic hagfish (*Myxine glutinosa*): a histopathologic and chemical study. Ann. N. Y. Acad. Sci. 298, 342–355.

Farkona, S., Diamandis, E.P., Blasutig, I.M., 2016. Cancer immunotherapy: the beginning of the end of cancer? BMC Med. 14, 73.

Finkelstein, J.B., 2005. Sharks do get cancer: few surprises in cartilage research. J. Natl. Cancer Inst. 97, 1562–1563.

Flajnik, M.F., Du Pasquier, L., 1988. MHC class I antigens as surface markers of adult erythrocytes during the metamorphosis of *Xenopus*. Dev. Biol. 128, 198–206.

Flajnik, M.F., Du Pasquier, L., 1990. The major histocompatibility complex of frogs. Immunol. Rev. 113, 47–63.

Flajnik, M.F., Kasahara, M., 2001. Comparative genomics of the MHC: glimpses into the evolution of the adaptive immune system. Immunity 15, 351–362.

Flajnik, M.F., Kasahara, M., 2010. Origin and evolution of the adaptive immune system: genetic events and selective pressures. Nat. Rev. Genet. 11, 47–59.

Flajnik, M.F., Kaufman, J.F., Hsu, E., Manes, M., Parisot, R., Du Pasquier, L., 1986. Major histocompatibility complex-encoded class I molecules are absent in immunologically competent *Xenopus* before metamorphosis. J. Immunol. 137, 3891–3899.

Flajnik, M.F., Kasahara, M., Shum, B.P., Salter-Cid, L., Taylor, E., Du Pasquier, L., 1993. A novel type of class I gene organization in vertebrates: a large family of non-MHC-linked class I genes is expressed at the RNA level in the amphibian *Xenopus*. EMBO J. 12, 4385–4396.

Fox, L.M., Miksanek, J., May, N.A., Scharf, L., Lockridge, J.L., Veerapen, N., Besra, G.S., Adams, E.J., Hudson, A.W., Gumperz, J.E., 2013. Expression of CD1c enhances human invariant NKT cell activation by alpha-GalCer. Cancer Immun. 13, 9.

Franceschi, C., Cossarizza, A., Monti, D., Ottaviani, E., 1991. Cytotoxicity and immunocyte markers in cells from the freshwater snail *Planorbarius corneus* (L.) (Gastropoda pulmonata): implications for the evolution of natural killer cells. Eur. J. Immunol. 21, 489–493.

Fuertes, M.B., Kacha, A.K., Kline, J., Woo, S.R., Kranz, D.M., Murphy, K.M., Gajewski, T.F., 2011. Host type I IFN signals are required for antitumor CD8+ T cell responses through CD8{alpha}+ dendritic cells. J. Exp. Med. 208, 2005–2016.

Fugmann, S.D., Messier, C., Novack, L.A., Cameron, R.A., Rast, J.P., 2006. An ancient evolutionary origin of the Rag1/2 gene locus. Proc. Natl. Acad. Sci. USA 103, 3728–3733.

Gabrilovich, D.I., Ostrand-Rosenberg, S., Bronte, V., 2012. Coordinated regulation of myeloid cells by tumours. Nat. Rev. Immunol. 12, 253–268.

Gajewski, T.F., Schreiber, H., Fu, Y.X., 2013. Innate and adaptive immune cells in the tumor microenvironment. Nat. Immunol. 14, 1014–1022.

Garcia-Lora, A., Algarra, I., Garrido, F., 2003. MHC class I antigens, immune surveillance, and tumor immune escape. J. Cell. Physiol. 195, 346–355.

Garner, M.M., Hernandez-Divers, S.M., Raymond, J.T., 2004. Reptile neoplasia: a retrospective study of case submissions to a specialty diagnostic service. Veterinary Clin. North Am. Exot. Anim. Pract. 7, 653–671, vi.

Ghosh, J., Lun, C.M., Majeske, A.J., Sacchi, S., Schrankel, C.S., Smith, L.C., 2011. Invertebrate immune diversity. Dev. Comp. Immunol. 35, 959–974.

Gleimer, M., Parham, P., 2003. Stress management: MHC class I and class I-like molecules as reporters of cellular stress. Immunity 19, 469–477.

Goyos, A., Robert, J., 2009. Tumorigenesis and anti-tumor immune responses in *Xenopus*. Front. Biosci. 14, 167–176.

Goyos, A., Cohen, N., Gantress, J., Robert, J., 2004. Anti-tumor MHC class Ia-unrestricted CD8 T cell cytotoxicity elicited by the heat shock protein gp96. Eur. J. Immunol. 34, 2449–2458.

Goyos, A., Guselnikov, S., Chida, A.S., Sniderhan, L.F., Maggirwar, S.B., Nedelkovska, H., Robert, J., 2007. Involvement of nonclassical MHC class Ib molecules in heat shock protein-mediated anti-tumor responses. Eur. J. Immunol. 37, 1494–1501.

Goyos, A., Ohta, Y., Guselnikov, S., Robert, J., 2009. Novel nonclassical MHC class Ib genes associated with CD8 T cell development and thymic tumors. Mol. Immunol. 46, 1775–1786.

Goyos, A., Sowa, J., Ohta, Y., Robert, J., 2011. Remarkable conservation of distinct nonclassical MHC class I lineages in divergent amphibian species. J. Immunol. 186, 372–381.

Grayfer, L., Robert, J., 2016. Amphibian macrophage development and antiviral defenses. Dev. Comp. Immunol. 58, 60–67.

Groff, J.M., 2004. Neoplasia in fishes. Veterinary Clin. North Am. Exot. Anim. Pract. 7, 705–756, vii.

Hanington, P.C., Forys, M.A., Dragoo, J.W., Zhang, S.M., Adema, C.M., Loker, E.S., 2010. Role for a somatically diversified lectin in resistance of an invertebrate to parasite infection. Proc. Natl. Acad. Sci. USA 107, 21087–21092.

Hardisty, M.W., 1976. Cysts and tumour-like lesions in the endocrine pancreas of the lamprey (*Lampetra fluviatilis*). J. Zool. 178, 305–317.

Hardwick, L.J., Philpott, A., 2015. An oncologists friend: how *Xenopus* contributes to cancer research. Dev. Biol. 408, 180–187.

Haynes-Gilmore, N., Banach, M., Edholm, E.S., Lord, E., Robert, J., 2014. A critical role of non-classical MHC in tumor immune evasion in the amphibian *Xenopus* model. Carcinogenesis 35, 1807–1813.

Haynes-Gimore, N., Banach, M., Brown, E., Dawes, R., Edholm, E.S., Kim, M., Robert, J., 2015. Semi-solid tumor model in *Xenopus laevis*/gilli cloned tadpoles for intravital study of neovascularization, immune cells and melanophore infiltration. Dev. Biol. 408, 205–212.

He, X., Dong, D.D., Yie, S.M., Yang, H., Cao, M., Ye, S.R., Li, K., Liu, J., Chen, J., 2010. HLA-G expression in human breast cancer: implications for diagnosis and prognosis, and effect on allocytotoxic lymphocyte response after hormone treatment in vitro. Ann. Surg. Oncol. 17, 1459–1469.

Hellsten, U., Harland, R.M., Gilchrist, M.J., Hendrix, D., Jurka, J., Kapitonov, V., Ovcharenko, I., Putnam, N.H., Shu, S., Taher, L., Blitz, I.L., Blumberg, B., Dichmann, D.S., Dubchak, I., Amaya, E., Detter, J.C., Fletcher, R., Gerhard, D.S., Goodstein, D., Graves, T., Grigoriev, I.V., Grimwood, J., Kawashima, T., Lindquist, E., Lucas, S.M., Mead, P.E., Mitros, T., Ogino, H., Ohta, Y., Poliakov, A.V., Pollet, N., Robert, J., Salamov, A., Sater, A.K., Schmutz, J., Terry, A., Vize, P.D., Warren, W.C., Wells, D., Wills, A., Wilson, R.K., Zimmerman, L.B., Zorn, A.M., Grainger, R., Grammer, T., Khokha, M.K., Richardson, P.M., Rokhsar, D.S., 2010. The genome of the Western clawed frog *Xenopus tropicalis*. Science 328, 633–636.

Herranz, H., Eichenlaub, T., Cohen, S.M., 2016. Cancer in *Drosophila*: imaginal discs as a model for epithelial tumor formation. Curr. Top. Dev. Biol. 116, 181–199.

Hildner, K., Edelson, B.T., Purtha, W.E., Diamond, M., Matsushita, H., Kohyama, M., Calderon, B., Schraml, B.U., Unanue, E.R., Diamond, M.S., Schreiber, R.D., Murphy, T.L., Murphy, K.M., 2008. Batf3 deficiency reveals a critical role for CD8alpha+ dendritic cells in cytotoxic T cell immunity. Science 322, 1097–1100.

Hix, L.M., Shi, Y.H., Brutkiewicz, R.R., Stein, P.L., Wang, C.R., Zhang, M., 2011. CD1d-expressing breast cancer cells modulate NKT cell-mediated antitumor immunity in a murine model of breast cancer metastasis. PLoS One 6, e20702.

Hong, C., Park, S.H., 2007. Application of natural killer T cells in antitumor immunotherapy. Crit. Rev. Immunol. 27, 511–525.

Horton, J.D., Horton, T.L., Ritchie, P., Varley, C.A., 1992. Skin xenograft rejection in *Xenopus*—immunohistology and effect of thymectomy. Transplantation 53, 473–476.

Horton, J.D., Horton, T.L., Dzialo, R., Gravenor, I., Minter, R., Ritchie, P., Gartland, L., Watson, M.D., Cooper, M.D., 1998. T-cell and natural killer cell development in thymectomized *Xenopus*. Immunol. Rev. 166, 245–258.

Horton, T.L., Minter, R., Stewart, R., Ritchie, P., Watson, M.D., Horton, J.D., 2000. *Xenopus* NK cells identified by novel monoclonal antibodies. Eur. J. Immunol. 30, 604–613.

Horton, T.L., Stewart, R., Cohen, N., Rau, L., Ritchie, P., Watson, M.D., Robert, J., Horton, J.D., 2003. Ontogeny of *Xenopus* NK cells in the absence of MHC class I antigens. Dev. Comp. Immunol. 27, 715–726.

Huang, S., Tao, X., Yuan, S., Zhang, Y., Li, P., Beilinson, H.A., Zhang, Y., Yu, W., Pontarotti, P., Escriva, H., Le Petillon, Y., Liu, X., Chen, S., Schatz, D.G., Xu, A., 2016. Discovery of an active RAG transposon illuminates the origins of V(D)J recombination. Cell 166, 102–114.

Kalin, R.E., Banziger-Tobler, N.E., Detmar, M., Brandli, A.W., 2009. An in vivo chemical library screen in *Xenopus* tadpoles reveals novel pathways involved in angiogenesis and lymphangiogenesis. Blood 114, 1110–1122.

Kau, C.L., Turpen, J.B., 1983. Dual contribution of embryonic ventral blood island and dorsal lateral plate mesoderm during ontogeny of hemopoietic cells in *Xenopus laevis*. J. Immunol. 131, 2262–2266.

Klein, L., Kyewski, B., Allen, P.M., Hogquist, K.A., 2014. Positive and negative selection of the T cell repertoire: what thymocytes see (and don't see). Nat. Rev. Immunol. 14, 377–391.

Kobel, H.R., Du Pasquier, L., 1975. Production of large clones of histocompatible, fully identical clawed toads (Xenopus). Immunogenetics 2, 87–91.

Kobel, H.R., Du Pasquier, L., 1977. Strains and species of Xenopus for immunological research. In: Solomon, J.B., Horton, J.D. (Eds.), Developmental Immunobiology. Elsevier/North Holland Publishing Co., Amsterdam.

Koster, B.D., de Gruijl, T.D., van den Eertwegh, A.J., 2015. Recent developments and future challenges in immune checkpoint inhibitory cancer treatment. Curr. Opin. Oncol. 27, 482–488.

LaBonne, C., Zorn, A.M., 2015. Modeling human development and disease in Xenopus. Dev. Biol. 408, 179.

Levine, A.J., Munoz-Sanjuan, I., Bell, E., North, A.J., Brivanlou, A.H., 2003. Fluorescent labeling of endothelial cells allows in vivo, continuous characterization of the vascular development of Xenopus laevis. Dev. Biol. 254, 50–67.

Liao, W., Lin, J.X., Leonard, W.J., 2013. Interleukin-2 at the crossroads of effector responses, tolerance, and immunotherapy. Immunity 38, 13–25.

Lightner, D.V., Brock, J.A., 1987. A lymphoma-like neoplasm arising from hematopoietic tissue in the white shrimp, Penaeus vannamei Boone (Crustacea: Decapoda). J. Invertebr. Pathol. 49, 188–193.

Lippitz, B.E., 2013. Cytokine patterns in patients with cancer: a systematic review. Lancet Oncol. 14, e218–e228.

Litman, G.W., Rast, J.P., Fugmann, S.D., 2010. The origins of vertebrate adaptive immunity. Nat. Rev. Immunol. 10, 543–553.

Liu, R.B., Engels, B., Arina, A., Schreiber, K., Hyjek, E., Schietinger, A., Binder, D.C., Butz, E., Krausz, T., Rowley, D.A., Jabri, B., Schreiber, H., 2012. Densely granulated murine NK cells eradicate large solid tumors. Cancer Res. 72, 1964–1974.

Lobikin, M., Chernet, B., Lobo, D., Levin, M., 2012. Resting potential, oncogene-induced tumorigenesis, and metastasis: the bioelectric basis of cancer in vivo. Phys. Biol. 9, 065002.

Matsushita, H., Vesely, M.D., Koboldt, D.C., Rickert, C.G., Uppaluri, R., Magrini, V.J., Arthur, C.D., White, J.M., Chen, Y.S., Shea, L.K., Hundal, J., Wendl, M.C., Demeter, R., Wylie, T., Allison, J.P., Smyth, M.J., Old, L.J., Mardis, E.R., Schreiber, R.D., 2012. Cancer exome analysis reveals a T-cell-dependent mechanism of cancer immunoediting. Nature 482, 400–404.

Maude, S.L., Teachey, D.T., Porter, D.L., Grupp, S.A., 2015. CD19-targeted chimeric antigen receptor T-cell therapy for acute lymphoblastic leukemia. Blood 125, 4017–4023.

Mescher, A.L., Wolf, W.L., Moseman, E.A., Hartman, B., Harrison, C., Nguyen, E., Neff, A.W., 2007. Cells of cutaneous immunity in Xenopus: studies during larval development and limb regeneration. Dev. Comp. Immunol. 31, 383–393.

Metzger, M.J., Reinisch, C., Sherry, J., Goff, S.P., 2015. Horizontal transmission of clonal cancer cells causes leukemia in soft-shell clams. Cell 161, 255–263.

Metzger, M.J., Villalba, A., Carballal, M.J., Iglesias, D., Sherry, J., Reinisch, C., Muttray, A.F., Baldwin, S.A., Goff, S.P., 2016. Widespread transmission of independent cancer lineages within multiple bivalve species. Nature 534, 705–709.

Meyer-Rochow, V.B., et al., 1991. Nephroblastoma in the clawed frog Xenopus laevis. J. Exp. Anim. Sci. 34 (5–6), 225–228.

Miller, M.M., Wang, C., Parisini, E., Coletta, R.D., Goto, R.M., Lee, S.Y., Barral, D.C., Townes, M., Roura-Mir, C., Ford, H.L., Brenner, M.B., Dascher, C.C., 2005. Characterization of two avian MHC-like genes reveals an ancient origin of the CD1 family. Proc. Natl. Acad. Sci. USA 102, 8674–8679.

Mittal, D., Gubin, M.M., Schreiber, R.D., Smyth, M.J., 2014. New insights into cancer immunoediting and its three component phases—elimination, equilibrium and escape. Curr. Opin. Immunol. 27, 16–25.

Moss, L.D., Monette, M.M., Jaso-Friedmann, L., Leary, III, J.H., Dougan, S.T., Krunkosky, T., Evans, D.L., 2009. Identification of phagocytic cells, NK-like cytotoxic cell activity and the production of cellular exudates in the coelomic cavity of adult zebrafish. Dev. Comp. Immunol. 33, 1077–1087.

Murgia, C., Pritchard, J.K., Kim, S.Y., Fassati, A., Weiss, R.A., 2006. Clonal origin and evolution of a transmissible cancer. Cell 126, 477–487.

Murshid, A., Borges, T.J., Calderwood, S.K., 2015. Emerging roles for scavenger receptor SREC-I in immunity. Cytokine 75, 256–260.

Nakai, Y., Nakajima, K., Robert, J., Yaoita, Y., 2016. Ouro proteins are not essential to tail regression during Xenopus tropicalis metamorphosis. Genes Cells 21, 275–286.

Nakanishi, T., Toda, H., Shibasaki, Y., Somamoto, T., 2011. Cytotoxic T cells in teleost fish. Dev. Comp. Immunol. 35, 1317–1323.

Nappi, A.J., Ottaviani, E., 2000. Cytotoxicity and cytotoxic molecules in invertebrates. Bioessays 22, 469–480.

Nedelkovska, H., Robert, J., 2013. Hsp72 mediates stronger antigen-dependent non-classical MHC class Ib anti-tumor responses than hsc73 in Xenopus laevis. Cancer Immun. 13, 4.

Ny, A., Koch, M., Schneider, M., Neven, E., Tong, R.T., Maity, S., Fischer, C., Plaisance, S., Lambrechts, D., Heligon, C., Terclavers, S., Ciesiolka, M., Kalin, R., Man, W.Y., Senn, I., Wyns, S., Lupu, F., Brandli, A., Vleminckx, K., Collen, D., Dewerchin, M., Conway, E.M., Moons, L., Jain, R.K., Carmeliet, P., 2005. A genetic Xenopus laevis tadpole model to study lymphangiogenesis. Nat. Med. 11, 998–1004.

Ny, A., Koch, M., Vandevelde, W., Schneider, M., Fischer, C., Diez-Juan, A., Neven, E., Geudens, I., Maity, S., Moons, L., Plaisance, S., Lambrechts, D., Carmeliet, P., Dewerchin, M., 2008. Role of VEGF-D and VEGFR-3 in developmental lymphangiogenesis, a chemicogenetic study in Xenopus tadpoles. Blood 112, 1740–1749.

O'Hanlon, L.H., 2006. Scientists are searching the seas for cancer drugs. J. Natl. Cancer Inst. 98, 662–663.

Ostrander, G.K., Cheng, K.C., Wolf, J.C., Wolfe, M.J., 2004. Shark cartilage, cancer and the growing threat of pseudoscience. Cancer Res. 64, 8485–8491.

Ostrander, E.A., Davis, B.W., Ostrander, G.K., 2016. Transmissible tumors: breaking the cancer paradigm. Trends Genet. 32, 1–15.

Ostrand-Rosenberg, S., 2008. Immune surveillance: a balance between protumor and antitumor immunity. Curr. Opin. Genet. Dev. 18, 11–18.

Palucka, K., Banchereau, J., 2012. Cancer immunotherapy via dendritic cells. Nat. Rev. Cancer 12, 265–277.

Pancer, Z., Amemiya, C.T., Ehrhardt, G.R., Ceitlin, J., Gartland, G.L., Cooper, M.D., 2004. Somatic diversification of variable lymphocyte receptors in the agnathan sea lamprey. Nature 430, 174–180.

Pardoll, D.M., 2012. The blockade of immune checkpoints in cancer immunotherapy. Nat. Rev. Cancer 12, 252–264.

Pearse, A.M., Swift, K., 2006. Allograft theory: transmission of devil facial-tumour disease. Nature 439, 549.

Perez-Gracia, J.L., Labiano, S., Rodriguez-Ruiz, M.E., Sanmamed, M.F., Melero, I., 2014. Orchestrating immune check-point blockade for cancer immunotherapy in combinations. Curr. Opin. Immunol. 27, 89–97.

Peters, E.C., Halas, J.C., McCarty, H.B., 1986. Calicoblastic neoplasms in Acropora palmata, with a review of reports on anomalies of growth and form in corals. J. Natl. Cancer Inst. 76, 895–912.

Quail, D.F., Joyce, J.A., 2013. Microenvironmental regulation of tumor progression and metastasis. Nat. Med. 19, 1423–1437.

Rau, L., Gantress, J., Bell, A., Stewart, R., Horton, T., Cohen, N., Horton, J., Robert, J., 2002. Identification and characterization of Xenopus CD8+ T cells expressing an NK cell-associated molecule. Eur. J. Immunol. 32, 1574–1583.

Reichenbach-Klinke, H., Elkan, E., 1965. The Principal Diseases of Lower Vertebrates. Academic Press, New York, New York.

Robbins, P.F., Lu, Y.C., El-Gamil, M., Li, Y.F., Gross, C., Gartner, J., Lin, J.C., Teer, J.K., Cliften, P., Tycksen, E., Samuels, Y., Rosenberg, S.A.,

2013. Mining exomic sequencing data to identify mutated antigens recognized by adoptively transferred tumor-reactive T cells. Nat. Med. 19, 747–752.

Robert, J., 2010. Comparative study of tumorigenesis and tumor immunity in invertebrates and nonmammalian vertebrates. Dev. Comp. Immunol. 34, 915–925.

Robert, J., Cohen, N., 2011. The genus *Xenopus* as a multispecies model for evolutionary and comparative immunobiology of the 21st century. Dev. Comp. Immunol. 35, 916–923.

Robert, J., Edholm, E.S., 2014. A prominent role for invariant T cells in the amphibian *Xenopus laevis* tadpoles. Immunogenetics 66, 513–523.

Robert, J., Ohta, Y., 2009. Comparative and developmental study of the immune system in *Xenopus*. Dev. Dyn. 238, 1249–1270.

Robert, J., Guiet, C., Du Pasquier, L., 1994. Lymphoid tumors of *Xenopus laevis* with different capacities for growth in larvae and adults. Dev. Immunol. 3, 297–307.

Robert, J., Guiet, C., Du Pasquier, L., 1995. Ontogeny of the alloimmune response against a transplanted tumor in *Xenopus laevis*. Differentiation 59, 135–144.

Robert, J., Guiet, C., Cohen, N., Du Pasquier, L., 1997. Effects of thymectomy and tolerance induction on tumor immunity in adult *Xenopus laevis*. Int. J. Cancer 70, 330–334.

Robert, J., Menoret, A., Basu, S., Cohen, N., Srivastava, P.R., 2001. Phylogenetic conservation of the molecular and immunological properties of the chaperones gp96 and hsp70. Eur. J. Immunol. 31, 186–195.

Robert, J., Gantress, J., Rau, L., Bell, A., Cohen, N., 2002. Minor histocompatibility antigen-specific MHC-restricted CD8 T cell responses elicited by heat shock proteins. J. Immunol. 168, 1697–1703.

Robert, J., Ramanayake, T., Maniero, G.D., Morales, H., Chida, A.S., 2008. Phylogenetic conservation of glycoprotein 96 ability to interact with CD91 and facilitate antigen cross-presentation. J. Immunol. 180, 3176–3182.

Rollins-Smith, L.A., Flajnik, M.F., Blair, P.J., Davis, A.T., Green, W.F., 1997. Involvement of thyroid hormones in the expression of MHC class I antigens during ontogeny in *Xenopus*. Dev. Immunol. 5, 133–144.

Rothschild, B.M., Witzke, B.J., Hershkovitz, I., 1999. Metastatic cancer in the Jurassic. Lancet 354, 398.

Rothschild, B.M., Tanke, D.H., Helbling, II, M., Martin, L.D., 2003. Epidemiologic study of tumors in dinosaurs. Naturwissenschaften 90, 495–500.

Saharinen, P., Helotera, H., Miettinen, J., Norrmen, C., D'Amico, G., Jeltsch, M., Langenberg, T., Vandevelde, W., Ny, A., Dewerchin, M., Carmeliet, P., Alitalo, K., 2010. Claudin-like protein 24 interacts with the VEGFR-2 and VEGFR-3 pathways and regulates lymphatic vessel development. Genes Dev. 24, 875–880.

Salomon, R.N., Jackson, F.R., 2008. Tumors of testis and midgut in aging flies. Fly (Austin) 2, 265–268.

Salomonsen, J., Sorensen, M.R., Marston, D.A., Rogers, S.L., Collen, T., van Hateren, A., Smith, A.L., Beal, R.K., Skjodt, K., Kaufman, J., 2005. Two CD1 genes map to the chicken MHC, indicating that CD1 genes are ancient and likely to have been present in the primordial MHC. Proc. Natl. Acad. Sci. USA 102, 8668–8673.

Salter-Cid, L., Nonaka, M., Flajnik, M.F., 1998. Expression of MHC class Ia and class Ib during ontogeny: high expression in epithelia and coregulation of class Ia and lmp7 genes. J. Immunol. 160, 2853–2861.

Session, A.M., Uno, Y., Kwon, T., Chapman, J.A., Toyoda, A., Takahashi, S., Fukui, A., Hikosaka, A., Suzuki, A., Kondo, M., van Heeringen, S.J., Quigley, I., Heinz, S., Ogino, H., Ochi, H., Hellsten, U., Lyons, J.B., Simakov, O., Putnam, N., Stites, J., Kuroki, Y., Tanaka, T., Michiue, T., Watanabe, M., Bogdanovic, O., Lister, R., Georgiou, G., Paranjpe, S.S., van Kruijsbergen, I., Shu, S., Carlson, J., Kinoshita, T., Ohta, Y., Mawaribuchi, S., Jenkins, J., Grimwood, J., Schmutz, J., Mitros, T., Mozaffari, S.V., Suzuki, Y., Haramoto, Y., Yamamoto,

T.S., Takagi, C., Heald, R., Miller, K., Haudenschild, C., Kitzman, J., Nakayama, T., Izutsu, Y., Robert, J., Fortriede, J., Burns, K., Lotay, V., Karimi, K., Yasuoka, Y., Dichmann, D.S., Flajnik, M.F., Houston, D.W., Shendure, J., DuPasquier, L., Vize, P.D., Zorn, A.M., Ito, M., Marcotte, E.M., Wallingford, J.B., Ito, Y., Asashima, M., Ueno, N., Matsuda, Y., Veenstra, G.J., Fujiyama, A., Harland, R.M., Taira, M., Rokhsar, D.S., 2016. Genome evolution in the allotetraploid frog *Xenopus laevis*. Nature 538 (7625), 336–343.

Sharma, P., Wagner, K., Wolchok, J.D., Allison, J.P., 2011. Novel cancer immunotherapy agents with survival benefit: recent successes and next steps. Nat. Rev. Cancer 11, 805–812.

Shen, L., Stuge, T.B., Zhou, H., Khayat, M., Barker, K.S., Quiniou, S.M., Wilson, M., Bengten, E., Chinchar, V.G., Clem, L.W., Miller, N.W., 2002. Channel catfish cytotoxic cells: a mini-review. Dev. Comp. Immunol. 26, 141–149.

Siddle, H.V., Kaufman, J., 2013. A tale of two tumours: comparison of the immune escape strategies of contagious cancers. Mol. Immunol. 55, 190–193.

Siddle, H.V., Kreiss, A., Tovar, C., Yuen, C.K., Cheng, Y., Belov, K., Swift, K., Pearse, A.M., Hamede, R., Jones, M.E., Skjodt, K., Woods, G.M., Kaufman, J., 2013. Reversible epigenetic downregulation of MHC molecules by devil facial tumour disease illustrates immune escape by a contagious cancer. Proc. Natl. Acad. Sci. USA 110, 5103–5108.

Slack, J.M., Lin, G., Chen, Y., 2008. The *Xenopus* tadpole: a new model for regeneration research. Cell. Mol. Life Sci. 65, 54–63.

Smith, A.C., 2000. Comparative pathology: human disease counterparts in marine animals. Arch. Pathol. Lab. Med. 124, 348–352.

Sparks, A.K., Morado, J.F., 1987. A putative carcinoma-like neoplasm in the hindgut of a red king crab, *Paralithodes camtschatica*. J. Invertebr. Pathol. 50, 45–52.

Srivastava, P., 2002a. Interaction of heat shock proteins with peptides and antigen presenting cells: chaperoning of the innate and adaptive immune responses. Annu. Rev. Immunol. 20, 395–425.

Srivastava, P., 2002b. Roles of heat-shock proteins in innate and adaptive immunity. Nat. Rev. Immunol. 2, 185–194.

Stacy, B.A., Parker, J.M., 2004. Amphibian oncology. Veterinary Clin. North Am. Exot. Anim. Pract. 7, 673–695, vi–vii.

Stern, A.W., et al., 2014. Pancreatic carcinoma in an African clawed frog (*Xenopus laevis*). Comp. Med. 64 (6), 421–423.

Sykes, J.M.t., Trupkiewicz, J.G., 2006. Reptile neoplasia at the Philadelphia Zoological Garden, 1901–2002. J. Zoo Wildl. Med. 37, 11–19.

Turpen, J.B., Smith, P.B., 1989. Precursor immigration and thymocyte succession during larval development and metamorphosis in *Xenopus*. J. Immunol. 142, 41–47.

van der Bruggen, P., Traversari, C., Chomez, P., Lurquin, C., De Plaen, E., Van den Eynde, B., Knuth, A., Boon, T., 1991. A gene encoding an antigen recognized by cytolytic T lymphocytes on a human melanoma. Science 254, 1643–1647.

Van Nieuwenhuysen, T., Naert, T., Tran, H.T., Van Imschoot, G., Geurs, S., Sanders, E., Creytens, D., Van Roy, F., Vleminckx, K., 2015. TALEN-mediated apc mutation in *Xenopus tropicalis* phenocopies familial adenomatous polyposis. Oncoscience 2, 555–566.

von Strandmann, E.P., Senkel, S., Ryffel, G., Hengge, U.R., 2001. Dimerization co-factor of hepatocyte nuclear factor 1/pterin-4alpha-carbinolamine dehydratase is necessary for pigmentation in *Xenopus* and overexpressed in primary human melanoma lesions. Am. J. Pathol. 158, 2021–2029.

Walker, C., Bottger, S.A., Mulkern, J., Jerszyk, E., Litvaitis, M., Lesser, M., 2009. Mass culture and characterization of tumor cells from a naturally occurring invertebrate cancer model: applications for human and animal disease and environmental health. Biol. Bull. 216, 23–39.

Wallingford, J.B., 1999. Tumors in tadpoles: the *Xenopus* embryo as a model system for the study of tumorigenesis. Trends Genet. 15, 385–388.

Warger, T., Hilf, N., Rechtsteiner, G., Haselmayer, P., Carrick, D.M., Jonuleit, H., von Landenberg, P., Rammensee, H.G., Nicchitta, C.V., Radsak, M.P., Schild, H., 2006. Interaction of TLR2 and TLR4 ligands with the N-terminal domain of Gp96 amplifies innate and adaptive immune responses. J. Biol. Chem. 281, 22545–22553.

Yamashiro, H., Yamamoto, M., van Woesik, R., 2000. Tumor formation on the coral *Montipora informis*. Dis. Aquat. Organ. 41, 211–217.

Yang, Z., Wang, C., Wang, T., Bai, J., Zhao, Y., Liu, X., Ma, Q., Wu, X., Guo, Y., Zhao, Y., Ren, L., 2015. Analysis of the reptile CD1 genes: evolutionary implications. Immunogenetics 67, 337–346.

Ye, S.R., Yang, H., Li, K., Dong, D.D., Lin, X.M., Yie, S.M., 2007. Human leukocyte antigen G expression: as a significant prognostic indicator for patients with colorectal cancer. Mod. Pathol. 20, 375–383.

Yeager, M., Kumar, S., Hughes, A.L., 1997. Sequence convergence in the peptide-binding region of primate and rodent MHC class Ib molecules. Mol. Biol. Evol. 14, 1035–1041.

Yie, S.M., Yang, H., Ye, S.R., Li, K., Dong, D.D., Lin, X.M., 2007. Expression of HLA-G is associated with prognosis in esophageal squamous cell carcinoma. Am. J. Clin. Pathol. 128, 1002–1009.

Zitvogel, L., Tesniere, A., Kroemer, G., 2006. Cancer despite immunosurveillance: immunoselection and immunosubversion. Nat. Rev. Immunol. 6, 715–727.

10

The Response of Cancer Cell Populations to Therapies

Danika Lindsay*, Colleen M. Garvey**, Shannon M. Mumenthaler**, Jasmine Foo*

*School of Mathematics, University of Minnesota, Minneapolis, MN, United States
**Lawrence J. Ellison Institute for Transformative Medicine, University of Southern California, Los Angeles, CA, United States

INTRODUCTION

In this chapter we discuss, from an evolutionary perspective, the response of cancer cell populations to therapeutic intervention. There are many factors that influence therapeutic response. Here we focus on examples of cell-intrinsic and -extrinsic mechanisms of current interest to the cancer biology and evolution communities: the roles of (1) phenotypic plasticity and (2) the tumor microenvironment.

PHENOTYPIC PLASTICITY

Cellular plasticity refers to the intrinsic ability of cells to take on biological properties, such as gene expression profiles or phenotypes, of other cell types for undefined periods of time. For example, the three main sources of phenotypic plasticity that have been identified to be important in cancer are epigenetic modifications, epithelial-to-mesenchymal transitions (EMTs), and plasticity in differentiation status. In the following we discuss these sources of phenotypic plasticity and their impact on cancer treatment response.

Epigenetics

Epigenetics is defined as heritable changes in gene expression that occur without changes in DNA sequence and are thus more easily reversible (Wright, 2013). Two of the most studied mechanisms of epigenetic modifications include DNA methylation and histone modification (see Chapter 6). DNA methylation refers to the addition or removal of a methyl group to DNA (Jones, 2012). The addition of a methyl group can occur either in the promoter region or within the gene body, which typically results in either gene silencing (Baylin and Jones, 2011) or overexpression (Yang et al., 2014), respectively. Histones can undergo multiple kinds of post-translational modifications (e.g., methylations, acetylations, phosphorylations), leading to more "open" or "closed" states of chromatin structure that can modulate gene expression levels (Rodriguez-Paredes and Esteller, 2011). The degree to which these modifications are heritable or transient depends on the exact mechanism and cellular environment.

In recent years, epigenetic alterations have been found in many cancers, and evidence suggests that they play an important role in all aspects of cancer initiation, progression, and response to therapy. Epigenetic disruptions triggering cancer initiation have primarily focused on methylation of CpG sites within the promoter regions of tumor suppressor genes, such as breast-cancer-associated-1 (BRCA1) and cyclin-dependent kinase inhibitor 2A (CDKN2A) (Esteller, 2008; Jones and Baylin, 2007), resulting in gene silencing. More recently, it has been shown that histone modifications play a crucial role in promoting and maintaining key features of a malignant phenotype (Cohen et al., 2011). For example, alterations in expression of genes encoding histone deacetylases have been linked to tumor development since they impact transcription of genes [e.g., cyclin-dependent kinase inhibitors (CDKI), cell-cycle activator cyclin D1, transforming growth factor-beta (TGF-β)] regulating the cell cycle, proliferation, and apoptosis (Glozak and Seto, 2007; Ropero and Esteller, 2007).

Ecology and Evolution of Cancer. http://dx.doi.org/10.1016/B978-0-12-804310-3.00010-7

To date, several excellent reviews have been written that provide ample support for the idea that cancer is as much driven by epigenetic events as genetic alterations (Iacobuzio-Donahue, 2009). In the next sections we discuss the role of epigenetic modifications on cancer treatment response from an evolutionary perspective. In particular, we focus on two main areas in which epigenetic programming has been shown to play a major role: (1) promotion of intratumoral heterogeneity, which has been linked to more aggressive tumors and poorer initial clinical response, and (2) driving reversible drug-resistant states in cancer cell populations.

Epigenetic Modifications Promote Tumor Heterogeneity and Lead to Poor Initial Therapeutic Response

Various studies have demonstrated that epigenetic inactivation of DNA mismatch repair genes can occur through promoter hypermethylation. In healthy tissues, these genes encode for proteins that repair DNA damage to prevent the accumulation of mutations. Methylation in the promoter regions of such genes can lead to a "mutator phenotype" in which cells are unable to repair DNA damage occurring stochastically or as a result of environmental/therapeutic stress, and consequently experience an elevated rate of mutation. For example, methylation of MLH1 in colorectal cancer can result in microsatellite instability (Goodfellow et al., 2015; Herman et al., 1998; Kane et al., 1997; Toyota et al., 1999a,b). Epigenetic silencing of the MGMT DNA repair gene has been associated with mutations in p53 and KRAS (Esteller et al., 2000) and leads to a mutator phenotype in a variety of human cancers (Esteller and Herman, 2004).

Epigenetic modifications have also been found to directly contribute to intratumor heterogeneity through disordered methylation. A recent study on chronic lymphocytic leukemia (CLL) by Landau et al. (2014) reported that a high level of intrasample heterogeneity was due to DNA methylation. Specifically, the authors found locally disordered, stochastic methylation patterns within the samples that contributed to a more "noisy" transcriptional landscape, with a decoupling of the relationship between promoter methylation and gene expression. They postulate that (possibly transient) effects on transcription may result in a subpopulation of cells capable of surviving environmental stresses and propagating new genotypes in the population. Indeed, CLLs with a higher number of subclonal mutations also exhibit higher intrasample methylation heterogeneity. Epigenetic alterations have also been implicated in cancer stem cell (CSC) reprogramming, fueling heterogeneity in another dimension: differentiation status (Munoz et al., 2012).

These studies demonstrate that epigenetic changes can collaborate with genetic changes to drive the genetic diversification of a tumor, either indirectly or directly. What are the evolutionary and treatment consequences of this phenomenon? While normally occurring mutations can also induce a mutator phenotype and drive the diversification of tumor populations, epigenetic modifications occur at a much higher rate as compared to mutations in somatic cells (Rando and Verstrepen, 2007). One study reported that the spontaneous mutation rate in normal and cancer cells is about 10^{-10} mutations per nucleotide base pair per cell division (Jones et al., 2008), whereas the error rate for gaining or losing methylation is estimated to be about 2×10^{-5} per CpG site per cell division (Easwaran et al., 2014; Yatabe et al., 2001). Thus, epigenetic induction of a mutator phenotype may indirectly provide a more rapid mechanism for intratumor genomic diversification than solely provided by genetic processes. This diversification leads to a lack of effectiveness of cancer therapy by better equipping tumor cells to explore the evolutionary fitness space and optimize survival strategies under conditions of environmental or treatment stress. For example, epigenetic silencing of estrogen and progesterone receptors has been implicated in poor initial response to drugs, such as tamoxifen and raloxifene in hormone-related malignancies (Rodriguez-Paredes and Esteller, 2011). In addition, intratumor methylation heterogeneity has been found to be predictive of time to tumor relapse in diffuse large B-cell lymphoma (Pan et al., 2015). Since CSCs have been shown to exhibit resistance to antitumor treatments, epigenetic programs driving heterogeneity in differentiation hierarchy also contribute to poor initial treatment response. In general, the phenotypic heterogeneity conferred by epigenetic processes may confer an evolutionary "robustness" to the overall tumor population, by allowing individual subpopulations to explore different evolutionary strategies to deal with environmental or therapy-mediated stresses.

On the other hand, some have proposed that tumors in which DNA repair mechanisms are epigenetically silenced are in fact more sensitive to the killing effects of alkylating drugs used in chemotherapy, due to the inability of cells to repair damage induced by treatment (Esteller and Herman, 2004). Such observations provide a more nuanced and context-dependent view of the role of epigenetic modifications in tumor evolution, and demonstrate the possibility of new treatment strategies incorporating epigenetic targeting or modification in combination with existing therapies.

A Reversible Nature of Epigenetically Induced Drug Resistance

As noted earlier, epigenetic modifications occur at substantially higher rates (up to five orders of magnitude) than genetic alterations, but these changes are more easily reversed. Several studies have sought to quantify the generation length of transient epigenetic phenomena, and to understand the mechanisms driving these changes. For example, recent work by Sharma et al. (2010) observed a transiently drug-tolerant state in epidermal growth factor receptor (EGFR)-driven non-small cell lung cancer (NSCLC) lines. In particular, a small percentage of an initially untreated population was found to be two orders of magnitude less sensitive to an EGFR tyrosine kinase inhibitor (TKI) than the bulk of the population. The mechanism underlying this reversible drug tolerance was found to be a chromatin state established by the histone demethylase KDM5A. Drug sensitivity was restored after a limited number of passages cultured in the absence of drug. In another study, DNA methylation was found to mediate transient resistance to sunitinib in renal clear-cell carcinoma; interestingly, drug sensitivity in this case was restored by escalation of the dose (Adelaiye et al., 2015).

As touched upon previously, it has been postulated that transient epigenetic phenomena provide short-term adaptations that allow lineages to survive therapy (perhaps undetected at low frequency), until more permanent genetic resistance mechanisms can be established. Depending on the timescales of epigenetic processes occurring in individual tumors, the outcome of these phenomena can be more aggressively growing tumors prior to diagnosis and treatment, poorer initial tumor response to therapy, and/or tumor recurrence after initially successful therapy. Indeed, a very recent work reported on how coevolution of the epigenome with the genome drives clinical recurrence in glioblastoma (Mazor et al., 2015). The reversibility of the drug-tolerant state may also explain why these cells are able to persist at low frequencies in the population.

More quantitative evolutionary models are required to fully understand the impact of reversible epigenetic phenomena on the long-term evolution of tumors. This is especially necessary for the design of therapeutic strategies aimed at targeting or modulating epigenetic processes. Most of the current literature in mathematical and evolutionary modeling of tumor growth and response to therapy focuses on stochastically arising genetic alterations. However, some technical challenges arise when trying to incorporate epigenetic phenomena into standard stochastic evolutionary models used in cancer (e.g., multitype branching processes). In particular the transient nature of the phenomena suggests that there may be a limited-length "memory" to the system, which violates common mathematical assumptions that make standard analyses of these processes feasible.

Rethinking Therapeutic Strategies in Light of Epigenetic Phenomena

It has been proposed that drugs altering the methylation status of cancer cells may have potential in combination strategies to enhance the effectiveness of existing anticancer therapies. For example, Montenegro et al. (2016) found that an adjuvant hypomethylating treatment sensitized cancer cells to radiotherapy. More generally, the use of DNA-demethylating agents and histone deacetylase inhibitors could sensitize cancer cells to any existing therapy that acts via DNA damage.

In addition, activity of histone demethylase KDM5A has been implicated in resistance to erlotinib in NSCLC as well as resistance to temozolomide in glioblastoma (Banelli et al., 2015; Sharma et al., 2010), suggesting that histone demethylase inhibitors may be useful in combination with these drugs. Several clinical trials have been undertaken studying the effectiveness of chemotherapies in combination with DNA methyltransferase inhibitors and histone deacetalyse inhibitors (Rodriguez-Paredes and Esteller, 2011). However, determining the optimal temporal sequence, doses, and timing of these combination therapies remains an open question. Another major challenge is the relative nonspecificity of many of the epigenetically targeted drugs—which may lead to undesirable side effects. It is clear that the benefits of epigenetically targeted drugs as part of an anticancer therapeutic strategy are extremely dependent on the context of the particular cancer type, drug, and driving oncogenic processes. Nevertheless, due to the emerging importance of epigenetic phenomena in cancer progression and response to treatment, it remains an important task to further develop quantitative evolutionary theory of epigenetic phenomena in cancer.

Epithelial-to-Mesenchymal Transitions

EMT is a complex molecular and cellular program by which cells lose their polarity and adhesion and gain migratory and invasive properties. It is often characterized by a change in gene expression markers (e.g., epithelial markers, such as e-cadherin and cytokeratins as well as mesenchymal markers, such as vimentin, n-cadherin, and fibronectin) and cell morphology features (e.g., epithelial morphology—cobblestone; mesenchymal morphology—elongated, spindle-like). EMT is essential for embryonic development as well as wound healing and tissue regeneration. However, EMT has also been shown to play a crucial role in the evolution of cancer by driving invasion, metastasis, and the emergence of drug resistance.

Here we discuss the role of EMT in mediating cancer cell response to therapy from an evolutionary perspective. We start by providing evidence that anticancer therapy drives cellular adaptation via the EMT process, and consider the role of the tumor microenvironment (TME) in this process. We then discuss the relationship between EMT and the CSC population as well as the evolutionary consequences of EMT, namely, invasion, metastasis, and treatment resistance. We conclude this discussion with a section on how viewing the role of EMT in therapeutic response and cancer progression through an evolutionary lens can lend new treatment strategies with the potential to improve cancer therapy.

Anticancer Therapy Drives Cellular Adaptation of Cancer Cells via EMT

Many studies have shown that treatment with anticancer therapy drives cellular adaptation by inducing EMT in the cancer cell population. For example, there is compelling evidence suggesting that cancer cells evolve under the stress imposed by radiation therapy by undergoing EMT. This type of adaptation of cancer cells to harsh conditions has been observed in cell lines from a variety of cancers, including colorectal, breast, and lung (Zhou et al., 2011).

Several studies have been aimed at investigating the role of chemotherapy or targeted therapies in mediating cancer cell evolution via EMT in specific cancers. Vazquez-Martin et al. (2013) showed that short-term exposure to erlotinib, a first-line targeted therapy used to treat NSCLC, is sufficient to induce EMT in some NSCLC cells (containing mutations in EGFR). Chronic exposure to cisplatin, paclitaxel, or radiation has also been shown to induce the EMT phenotype in NSCLC (Shintani et al., 2011). In addition, treatment has been shown to induce phenotype switching in melanoma cells from a proliferative phenotype to an invasive phenotype with EMT properties (Shintani et al., 2011; Zipser et al., 2011).

In addition to therapy, TME stressors promote tumor plasticity and heterogeneity through the induction of EMT, which allows cancer cells to adapt to microenvironmental pressures. For example, the diffusion limitation of oxygen in tumor tissue leads to local regions of hypoxia (areas with a low oxygen concentration) in tumors. The resulting competition among cancer cells for limited resources promotes cellular adaptation, which is driven by molecular signaling pathways that allow cells to survive and thrive in a hostile oxygen-deprived TME and can lead to the induction of EMT (Cannito et al., 2008; Imai et al., 2003; Marie-Egyptienne et al., 2013; Theys et al., 2011). The immune system has also been shown to function as an EMT inducer via production and secretion of TGF-β (Donkor et al., 2011; Santisteban et al., 2009). The interdependence of tumor cells and immune cells in the TME results in adaptive phenotypic plasticity during tumor progression (Chouaib et al., 2014). In addition to driving cellular adaptation via EMT, the TME also produces signals to induce the reverse process, mesenchymal-to-epithelial transitions (METs) (Drasin et al., 2011). A more in-depth discussion of the effect of the TME on therapeutic response can be found in a later section of this chapter.

EMT Drives Therapeutic Resistance

Possibly the most significant evolutionary consequence of EMT in cancer cell populations is the emergence of therapeutic resistance. Numerous studies have suggested that cellular adaptation via EMT is responsible for conferring resistance to anticancer therapeutics in various cancers, including head and neck, bladder, pancreatic, breast, lung, and prostate cancers (Black et al., 2008; Buck et al., 2007; Engl et al., 2015; Frederick et al., 2007; Oliveras-Ferraros et al., 2012).

The mechanisms by which EMT confers resistance to anticancer therapies remain largely unknown, and much research effort has been focused on answering this question. Lim et al. (2013) concluded that EMT confers chemoresistance and cellular plasticity by regulating genes involved in cell death and stem cell maintenance. They showed that EMT induction confers resistance to drug-induced apoptosis as well as apoptosis due to withdrawal of survival factors. On the other hand, Robson et al. (2006) postulated that EMT may drive resistance to therapy by promoting a cell's ability to heal at a faster rate after wounding by therapy. Under this theory, EMT-induced populations do not display resistance to apoptosis; instead, these cells continue to proliferate and replace cells that apoptose.

Many studies have shown that EGFR kinase inhibitor-resistant cells have adapted mesenchymal properties in response to drugs (Thomson et al., 2008; Witta et al., 2006). In EGFR-mutant NSCLC, reversible epigenetic changes emerging during acquired resistance likely reflect EMT and the emergence of chemorefractory cells with stem cell-like features (Sharma et al., 2010; Voulgari and Pintzas, 2009). EMT has been shown to drive resistance of NSCLC cell lines to EGFR inhibitors by eliminating the erlotinib-sensitizing effect of EGFR mutations (Vazquez-Martin et al., 2013; Yao et al., 2010; Yauch et al., 2005). This confers a relative fitness advantage to those cells that have undergone EMT during treatment with EGFR inhibitors. In fact, the EMT process is sufficient to induce full refractoriness to erlotinib (Vazquez-Martin et al., 2013). This is largely due to the fact that NSCLC cells that have adapted by undergoing EMT become less dependent on EGFR signaling for cell proliferation and survival, and are thus less responsive to erlotinib (Thomson et al., 2005). In addition to inducing resistance to EGFR inhibitors in NSCLC, the induction of EMT has also been shown to promote cellular adaptations that reduce sensitivity of NSCLC cell lines to cisplatin and paclitaxel (Shintani et al., 2011).

EMT has been shown to drive the evolution of resistance in not only lung cancer but many other cancers as well. For example, experimental and clinical studies have found that EMT in breast cancer cell populations promotes resistance to cytotoxic chemotherapy and radiation therapy (Koh et al., 2008; Luo et al., 2015).

One clinical trial showed that residual tumor cells from two different types of breast cancer express the EMT phenotype following therapy (Creighton et al., 2009). Additionally, melanoma cells with the invasive phenotype display resistance to therapy (Anastas et al., 2014; Konieczkowski et al., 2014; O'Connell et al., 2013; Tap et al., 2010). This phenotype switching may offer a crucial escape from treatment for melanoma cells: the switch from the proliferative phenotype to the invasive phenotype buys cells time to accumulate additional (epi)genetic resistance mutations, which would then allow them to revert back to the proliferative phenotype and survive in the presence of treatment (Kemper et al., 2014). Acquisition of oxaliplatin resistance in colorectal cancer also corresponds with morphological changes consistent with EMT and allows cells to switch from a proliferative to an invasive phenotype. Since proliferation is required for oxaliplatin-induced chemosensitivity, a decrease in proliferation may be one means by which cells escape the effect of chemotherapy (Yang et al., 2004).

There is compelling evidence supporting a causal relationship between the EMT process and the emergence of the CSC phenotype in cancer cell populations (Ben-Porath et al., 2008; Lim et al., 2013; Mani et al., 2008; Marie-Egyptienne et al., 2013; Morel et al., 2008; Santisteban et al., 2009). For example, breast CSCs with the EMT phenotype have been found at the invasive edges of tumors, and the mesenchymal features of these cells allow them to quickly invade surrounding tissue. The presence of CSCs has been implicated in the increased resistance to anticancer therapies (Hollier et al., 2009; Marie-Egyptienne et al., 2013; Phillips et al., 2006). Studies have shown that exposure to hypoxia or TGF-β causes changes in cancer cells consistent with the EMT process, which then leads to the evolution of radioresistance (Theys et al., 2011).

An Evolutionary Perspective on Potential Therapeutic Strategies by Exploiting EMT

It may be possible to exploit the role of EMT in mediating drug resistance by identifying biomarkers that predict drug response. In particular, EMT plays a role in defining sensitivity to EGFR inhibitors and provides a molecular and morphological signature to define NSCLC tumors most likely to respond to treatment. EGFR-mutant human NSCLC cell lines grown both in vitro and in xenografts show a range of sensitivities to EGFR inhibition depending on the degree to which they have undergone EMT. Specifically, NSCLC lines that have not undergone EMT show greater sensitivity to EGFR inhibition, while NSCLC lines that have undergone EMT are insensitive to EGFR inhibition, and cell lines having undergone partial EMT are intermediately sensitive (Thomson et al., 2005). Researchers have also found that a microRNA gene expression signature targeting EMT can be used to predict response to EGFR inhibitors (Bryant

et al., 2012). This novel discovery could provide clinicians with the tools necessary to make better-informed decisions regarding optimal therapeutic strategies for individual patients.

Since both EMT and MET play a significant role in driving tumor invasion and metastasis as well as the evolution of drug resistance, it is crucial to devise methods for targeting these processes in cancer cells. In the following, we use evolutionary theory to discuss several ideas for therapeutic strategies taking into account the role of EMT in cancer progression:

1. *Exploit the evolution and adaptation of a tumor to its microenvironment.* As discussed previously, hypoxia has been shown to drive cellular adaptation via EMT and contribute to the emergence of drug resistance in cancer cells. Because of this, one beneficial strategy may be to target cells in hypoxic regions of the tumor. Several drugs that aim to accomplish this goal (called hypoxia-activated prodrugs) are currently in clinical development. By preferentially killing cells in hypoxic regions of the TME, the number of cells that undergo EMT in order to adapt to hypoxic conditions will be minimized, and hence so will the fraction of cells that are resistant to therapy.

2. *Use combination therapy to simultaneously target multiple phenotypes within the tumor.* The standard approach to treat cancer is the use of cytotoxic chemotherapeutic agents. These drugs work by targeting rapidly proliferating cells. However, cancer cells that undergo EMT and display a more mesenchymal-like phenotype become resistant to chemotherapy because they are no longer rapidly proliferating. One strategy may be to induce differentiation in these mesenchymal-like cells via MET in order to resensitize them to therapy. Another standard approach to treat cancer is to target the CSC population, since this group of cells is thought to be the catalyst behind tumor growth. However, these therapies can fail due to the fact that CSCs can reemerge from the non-CSC population via EMT. To combat this, one promising strategy is to simultaneously target rapidly proliferating cells in the epithelial state as well as mesenchymal cells with the CSC phenotype by alternating between different drugs that control each of these populations individually. This therapeutic approach has the potential to mitigate the evolution of resistance due to cellular adaptation and plasticity via EMT and MET processes.

Cancer Stem Cells

CSCs are cancer cells that possess characteristics associated with normal adult stem cells. They have unlimited proliferative potential, the ability to renew indefinitely in an undifferentiated state, a high rate of resistance to therapies, a high DNA repair capacity, the ability to resist apoptosis, and the ability to drive tumor expansion using differentiated cells (Ahmed et al., 2013; Gil et al., 2008; Stassi et al., 2003; Wodarz and Komarova, 2007). CSCs are capable of differentiating into heterogeneous cancer subpopulations (Fornari et al., 2011). It is because of these properties that if a given treatment fails to kill the population of CSCs, there is a large probability of tumor regeneration and recurrence (Enderling and Hahnfeldt, 2011; Fornari et al., 2011). Hence, CSCs play a crucial role in not only cancer initiation but also its evolution, metastasis, and recurrence.

The CSC hypothesis refers to the idea that tumors arise from a small number of CSCs with the ability of self-renewal and differentiation into non–stem cancer cells (Clarke et al., 2006; Jordan et al., 2006; Leder et al., 2010; Mimeault and Batra, 2008; Monteagudo and Santos, 2015a; Reya et al., 2001; Sehl et al., 2015; Wodarz and Komarova, 2007). According to this hypothesis, tumors have a cellular hierarchy in which CSCs are at the top (Dalerba et al., 2007a; Fornari et al., 2011; Meacham and Morrison, 2013). It remains unclear what fraction of cancers actually follow the CSC hypothesis. However, this hypothesis is consistent with several different types of cancer, including chronic myeloid leukemia (CML) (Wang et al., 1998), acute myeloid leukemia (AML) (Bonnet and Dick, 1997; Hope et al., 2004; Lapidot et al., 1994), breast cancer (Al-Hajj et al., 2003), colorectal cancer (Dalerba et al., 2007b; O'Brien et al., 2007; Ricci-Vitiani et al., 2007), head and neck cancers (Prince et al., 2007), glioma (Bao et al., 2006; Galli et al., 2004; Ignatova et al., 2002; Singh et al., 2004), ovarian cancer (Alvero et al., 2009; Bapat et al., 2005; Burgos-Ojeda et al., 2012; Curley et al., 2009; Ferrandina et al., 2008; Gao et al., 2010; Pan and Huang, 2008; Stewart et al., 2011; Szotek et al., 2006; Wang et al., 2012; Zhang et al., 2008), and pancreatic cancer (Li et al., 2007, 2009).

Many studies have shown that stemness is not a cellular phenotype, but rather a transient state that cells may acquire (Meacham and Morrison, 2013; Poulsom et al., 2002; Rapp et al., 2008; Wagers and Weissman, 2004; Wang et al., 2009), and this plasticity of cancer cells plays an essential role in maintaining phenotypic diversity (Zhou et al., 2014). For example, Chen et al. (2016) compared a hierarchical model of tumor progression with a model incorporating cellular plasticity and found that the plasticity model provided far more accurate predictions of short-term transient dynamics as well as long-term stability. CSCs hold the potential for multidirectional differentiation (Gil et al., 2008). Specifically, differentiated cancer cells are capable of undergoing dedifferentiation and reverting to CSCs (Barker et al., 2007; Chaffer et al., 2011; Charles et al., 2010; Gupta et al., 2011; Meyer et al., 2009; Quintana et al., 2010; Scaffidi and Misteli, 2011; Taussig et al., 2010; Yang et al., 2012).

Cancer Stem Cells are Associated with Therapeutic Resistance

It is widely accepted that CSCs drive drug resistance (Ahmed et al., 2010, 2013; Bao et al., 2006; Diehn et al., 2009; Hollier et al., 2009; Latifi et al., 2011; Meacham and Morrison, 2013; Oravecz-Wilson et al., 2009; Vathipadiekal et al., 2012) and tumor recurrence (Sehl et al., 2015). Many studies have found recurrent tumors to be enriched with CSCs following therapy (Bao et al., 2006; Diehn et al., 2009; Dylla et al., 2008; Latifi et al., 2012; Steg et al., 2011, 2012), which supports this hypothesis. Some CSCs are inherently resistant to current therapies that target terminally differentiated cells, including chemotherapy and radiation therapy (Li et al., 2008; Woodward et al., 2007). This is largely due to the fact that cytotoxic therapies preferentially target rapidly proliferating cells and only a small fraction of CSCs actively proliferate (Monteagudo and Santos, 2015b; Vainstein et al., 2012). This may also be because CSCs display resistance to apoptosis (Gil et al., 2008; Jagani and Khosravi-Far, 2008), increased DNA repair (Desai et al., 2014), and enhanced drug-efflux ability through ABC transporter expression (Dean et al., 2005). These resistant CSCs are able to repopulate the original tumor (Ahmed et al., 2013; Aguilar-Gallardo et al., 2012; Curley et al., 2011), thus leading to tumor recurrence. In fact, evidence suggests that some tumor cells may evade chemotherapy by adapting and reverting to a CSC state, resulting in tumor progression and recurrence (Ahmed et al., 2013). For example, studies have shown that glioma CSCs are preferentially resistant to radiation and temozolomide, which results in disease recurrence (Bao et al., 2006; Chen et al., 2012; Liu et al., 2006). CSCs have also been shown to be resistant to particular targeted therapies (Meacham and Morrison, 2013). For example, CML stem cells are inherently resistant to imatinib (Chaudhary and Roninson, 1991; Corbin et al., 2011; Graham et al., 2002; Jiang et al., 2007; Mahon et al., 2003; Neering et al., 2007). Imatinib targets a specific genetic alteration found within differentiated CML cells, and hence significantly reduces tumor burden. But upon discontinuation of therapy, the leukemic cell count rises to the same level as or higher than the pretreatment count (Michor, 2008). This rapid increase in the number of leukemic cells after imatinib indicates that the CSC population is not depleted by imatinib and thus drives recurrence since CSCs are able to produce more differentiated cells (Michor, 2008; Rousselot et al., 2007; Savona and Talpaz, 2008).

Incorporating Cancer Stem Cell Dynamics in Treatment Design

Directly Targeting CSCs

Although chemotherapy is efficient at removing the tumor bulk, it has been shown to leave behind the CSC core (Latifi et al., 2011; Oliver et al., 2010). As a result, combination treatments involving drugs that directly target the CSC population in addition to drugs reducing tumor bulk have been the focus of some recent studies. Abubaker et al. (2014) studied the effect of using CYT387, a small molecule inhibitor designed to target CSCs, in combination with paclitaxel to treat ovarian cancer. When these drugs were given in combination, a much smaller tumor burden was observed as well as a loss of CSC characteristics (Abubaker et al., 2014). Other studies have sought to induce terminal differentiation in the CSC population so that the CSCs lose their self-renewal properties (Aguilar-Gallardo et al., 2012; Meacham and Morrison, 2013; Sell, 2004).

Mathematical modeling efforts have also been conducted to evaluate the theoretical potential of directly targeting CSCs, which currently remains a challenging task in the clinic. Fornari et al. (2011) showed that therapy targeting the CSC population leads to a reduction in CSCs followed by a reduction in terminally differentiated cells. Another theoretical study concluded that higher levels of dedifferentiation substantially reduce the effectiveness of therapy directed at CSCs (Leder et al., 2010).

Therapeutic Strategies Exploiting (De)differentiation Dynamics

A variety of anticancer therapies have been shown to induce cellular adaptation via dedifferentiation of non–stem cancer cells and hence increase the CSC population within the tumor. For example, in order to escape treatment, differentiated cancer cells may revert to stem-like cancer cells after exposure to radiation (Bleau et al., 2009; Charles et al., 2010; Chen et al., 2012; Li et al., 2009; Pistollato et al., 2010). Chemotherapy has also been shown to prompt a cascade of events that enhances the expression of CSC markers (Ahmed et al., 2013). This observation poses an interesting challenge for the design of treatment schedules, since the temporal dynamics of both the treatment response and the dedifferentiation process producing resistant populations are closely intertwined.

Instead of directly targeting CSCs, several promising modeling efforts have studied or tried to exploit the impact of therapeutic scheduling on shifting the population dynamics of the CSC and differentiated cancer cell populations. Monteagudo and Santos (2015b) found that under a strong pulsative therapy, the differentiated cancer cells are quickly eliminated, which allows room for the CSC population to expand and drive aggressive tumor growth, whereas a weak pulsative therapy leads to a slower growth of the tumor bulk. In a theoretical study, Vainstein et al. (2012) found that simultaneously promoting CSC differentiation and inhibiting CSC proliferation can substantially reduce the tumor population size. A combined experimental and mathematical modeling study by Leder et al. (2014) revealed that by taking into account dedifferentiation induced by radiotherapy, treatment schedules can be optimized to improve outcomes in mouse models of glioblastoma. Perhaps counterintuitively, the improvement in survival outcome in this study was attributed to optimized radiation scheduling that enriched for the slower-growing CSC population.

Here we briefly summarize a few main conclusions from these studies on rethinking future treatment strategies. First, in designing combination strategies using CSC-targeting drugs with tumor debulking agents, early inhibition of CSC proliferation is vital for preventing the emergence of resistant CSCs. It may also be beneficial to utilize low doses of therapies aimed at reducing the differentiated cell population, since a rapid increase in CSC population has been sometimes observed in response to debulking, leading to rapid tumor growth. Therefore, it seems that combinations must be effectively designed and dosed at the correct level to maintain the two subpopulations in a constrained competition with one another. Perhaps even more intriguing are the studies utilizing mathematical modeling to optimize the scheduling of existing standard therapies to exploit the dynamics of the CSC differentiation and dedifferentiation processes. It may be possible to leverage the timescale of these cellular processes; however, much more effort is needed to determine whether such refined control of these temporal processes can be achieved in the clinic, and the magnitude of the patient-to-patient variability that will almost surely impact the outcomes.

TUMOR MICROENVIRONMENT

Cancer is a highly complex, rapidly evolving ecosystem that relies on more than the underlying genetics of a cell to fuel the disease. Just as organisms have evolved over time to adapt to their surroundings, tumor evolution is the result of accumulation of mutations leading to a survival advantage within particular microenvironmental landscapes (also see Chapter 14). The TME plays an active role in tumor progression and treatment response. It is heterogeneous in nature, consisting of naturally occurring gradients of oxygen, nutrients, cytokines, and drug, other cell types (i.e., fibroblast, immune, and endothelial cells), and extracellular matrix factors, which create physical environmental niches in the tumor that drive cellular adaption. In response to the selective pressures imposed by the microenvironment, cells may: (1) produce offspring (proliferate) at a faster or slower rate, (2) die (apoptosis, necrosis), (3) enter a stressed state (autophagy), or (4) migrate toward or away from stimuli. This dynamic process results in the propagation of tumor clones with a relative fitness advantage under the surrounding selective pressures (Greaves and Maley, 2012).

Tumor cells often take advantage of the protective and supportive features of the microenvironment. The importance of a tumor's surroundings on its response to therapy has been appreciated for some time. Over 20 years ago, cancer cells were implanted either subcutaneously or into different visceral organs prone to metastasis. While the tumors were genetically identical on implantation, they varied in their response to therapy. The subcutaneous tumors were sensitive to the chemotherapeutic agent doxorubicin, whereas tumors implanted in the lung or liver did not respond (Fidler et al., 1994). This result, among others, highlights the importance of "context" in cancer where tumor cells require a receptive environment to prosper. More recently, researchers have begun to use evolutionary theory to outsmart evolution (Willyard, 2016). Understanding how the selective pressures of a tumor drive therapeutic resistance can ultimately be used to overcome or at least delay this process.

In the subsequent text we discuss the evolutionary impact of various microenvironmental factors on therapeutic response. We have subclassified the environmental selective pressures into (1) public goods and (2) stromal cell types to account for both biochemical and heterocellular interactions in a tumor. The heterogeneous microenvironment can either select for tumor cells with a relative fitness advantage or contribute to phenotypic adaptations of these tumor cells. Both processes can result in clonal populations that are resistant to particular therapies and ultimately dominate the tumor.

Availability of Public Goods

In solid tumors, the vasculature is notoriously disorganized, with regions of increased interstitial fluid pressure and extracellular matrix resulting in irregular blood flow (Tan et al., 2015). Irrespective of phenotype, the location of a tumor cell in the context of the TME can provide an evolutionary advantage. Due to limitations in drug penetration in solid tumors, a proportion of tumor cells are not exposed to lethal concentrations of drug, allowing for cellular adaptation and resistance outgrowth to occur. Furthermore, the location of cells in relation to blood vessels or other stromal cell subtypes may dictate the amount of resources, including oxygen, nutrients, and cytokines, that are available.

Oxygen

Tumor hypoxia, or low-oxygen conditions, can lead to cellular adaptions or select for clones that have decreased drug sensitivity through various mechanisms. One such mechanism is the result of reduced cell proliferation in oxygen-deprived regions of a tumor, which limits the cells' responsiveness to chemotherapy agents whose mechanisms of action target rapidly dividing cells (Sullivan et al., 2008). In addition, the effectiveness of radiation

therapy to kill cells is highly dependent on oxygen availability and thus regions of low oxygen are considered radiation resistant (Brown and Wilson, 2004). Tumor cells also actively upregulate expression of specific markers in low-oxygen conditions that lead to drug resistance including P-glycoprotein [also known as multidrug resistance protein (MDR1)], which is an ATP-dependent efflux pump involved in actively exporting drugs out of the cell (Wartenberg et al., 2003). Higher expression of this protein results in lower efficacy of many anticancer therapies. In addition, hypoxia-inducible factor-1alpha (HIF-1α), which is stable only in low-oxygen conditions, has been shown to increase pyruvate dehydrogenase kinase-3 expression (an enzyme involved in the regulation of glucose metabolism) and results in metabolic rewiring that confers chemoresistance in cervical (Lu et al., 2008) and colon (Lu et al., 2011) cancer models.

Hypoxia has been known to contribute to genomic diversification in tumors, resulting in production of more aggressive phenotypes including enhanced motility and invasiveness (Hongo et al., 2013; Zeng et al., 2010). In particular, genomic instability (i.e., point mutations, gene amplifications, etc.) can be induced by low oxygen levels through a variety of mechanisms, including deregulation of DNA damage checkpoint signaling (Luoto et al., 2013) and selection for clonal populations with loss of DNA mismatch repair (Kondo et al., 2001). Therefore, the likelihood of acquiring advantageous mutations in oxygen scarce regions is higher (Luoto et al., 2013; Reynolds et al., 1996).

Nutrients

It has been postulated that tumor cells are not as adept as normal host cells at adapting to low-nutrient conditions. Most of the energy production in tumor cells is produced via aerobic glycolysis rather than through oxidative phosphorylation utilized by normal cells (Vander Heiden et al., 2009). This breakdown of glucose is not as efficient as oxidative phosphorylation and the adaptive advantages of this "Warburg effect" are still not entirely clear, although hypotheses, such as the acidity induced by glycolysis confers survival advantages for tumor cells (Vander Heiden et al., 2009), exist (see also Chapters 4 and 8). Since tumor cells require more glucose for survival, low-nutrient conditions appear to impact their survival more profoundly than normal cells. Indeed, recent evidence has suggested that fasting-based methods may protect normal cells from chemotherapy while still targeting malignant cells (Lee et al., 2012a,b; Raffaghello et al., 2008).

On the other hand, it also been suggested that altered nutrient levels in tumors can drive resistance to targeted cancer treatments. Masui et al. (2015) observed that elevated glucose levels promoted oncogenic signaling (EGFRvIII-dependent signaling through activation of mTORC2) in glioblastoma cells, which rendered them resistant to EGFR, phosphoinositide 3-kinase (PI3K), or serine/threonine kinase 1 (AKT)-targeted therapies. In contrast, low glucose conditions in colorectal cancer have been observed to select for KRAS mutant cells over cells with wild-type KRAS alleles. KRAS mutants are resistant to EGFR-targeted therapy (cetuximab) and therefore these low glucose conditions help to foster the drug resistant population (Yun et al., 2009).

Cytokines

A common means of resistance to therapies found across many cancer types is the deregulated expression and secretion of cytokines. Cytokines refer to a number of secreted proteins that serve as signals for integral physiological processes, such as inflammation, migration, and apoptosis. Several studies have found that high levels of cytokines, such as interleukin 6 (IL-6), insulin-like growth factor 1 (IGF-1), transforming growth factor beta (TGFB1), and vascular endothelial growth factor (VEGF), are correlated with poorer overall survival in many cancer types (Janes et al., 2005; Jones et al., 2016; Hansson, 2005; Moser et al., 2004). While cytokines are essential for long-term cellular homeostasis, high expression can alter drug sensitivity through a variety of mechanisms. For example, Conze et al. (2001) provided evidence that IL-6 expression increased breast cancer cell resistance to chemotherapy by inducing expression of the efflux pump protein MDR1. In neuroblastoma, IL-6 was shown to activate signal transducer and activator of transcription 3 (STAT3), a prosurvival protein that contributes to drug resistance (Ara et al., 2013) by upregulating several antiapoptotic proteins.

TGF-β is a regulatory protein that inhibits cell growth under normal conditions. However, under some stressed conditions, such as addition of drug, high levels of this protein provide an evolutionary advantage to cells. In squamous cell carcinoma, TGF-β signaling increased protection of CSCs against the DNA-damaging agent cisplatin through metabolic reprogramming and increased glutathione metabolism (Oshimori et al., 2015). Additional work by Lopez-Diaz et al. (2013) found that TGF-β represses the tumor suppressor gene TP53 transcription and translation, thereby impairing stress response and increasing drug resistance. These cells also experienced slower rates of proliferation, so it was only on the addition of drug that this phenotype provided a fitness advantage.

The most common way cytokines act to confer resistance to molecular targeted therapies is by activating compensatory pathways that allow cells to bypass the imposed inhibitory blockade. In response to AKT inhibition, ER+ breast cancer cells upregulated IGF and its receptor, which provide compensatory signaling for survival and limit the effectiveness of the treatment (Fox et al., 2013). TGF-β signaling has been shown to cause activation of the mitogen-activated protein kinase/ extracellular receptor kinase (MAPK/ERK) pathway, which provides an alternative pathway for many cells to confer resistance to TKIs (Huang et al., 2012). Increased hepatocyte growth factor (HGF) expression activates the MET receptor tyrosine kinase and its downstream signaling cascade to confer resistance to gefitinib (an EGFR TKI inhibitor) in lung adenocarcinoma (Yano et al., 2008). This was due to the fitness advantage garnered by MET-amplified lung cancer clones in the presence of HGF (Turke et al., 2010).

Tumor–Stromal Interactions

Cancer cells are in physical and biochemical contact with many different cells native to the host environment (termed stromal cells), including fibroblasts, immune, vascular, fat, glial, and smooth muscle cells. A balance exists between these different cell types, with each cellular species competing for public goods and uniquely responding to their availability. The interplay between tumor and stromal cells drives many processes in cancer progression, including response to therapeutics (see also Chapters 4, 8, and 14). Often times it is thought that therapeutic resistance is due to tumor cell-intrinsic mechanisms. However, more recent evidence points to the bidirectional feedback between tumor and stromal cells leading to the TME as a source of acquired resistance.

Cancer-Associated Fibroblasts

Fibroblasts are the dominant cellular component in tumor stroma. Fibroblasts found in normal tissue are responsible for the deposition of noncellular scaffolding; on appropriate stimuli, they become "activated" to orchestrate the wound healing process. In cancer, the associated fibroblasts are found in this activated state even in the absence of stimuli. It has become apparent that these cells have an integral role in cancer progression and response to therapy. Indeed, fibroblast composition in patient tumors is prognostic of overall survival in many cancer types, including colorectal cancer, with fibroblast percentages over 50% associated with decreased survival rate (Huijbers et al., 2013).

Various factors secreted by fibroblasts are known to inhibit therapeutic response in cancer (Rasanen and Vaheri, 2010). Stroma-mediated resistance, especially to targeted therapies, has been shown to be a common occurrence in in vitro studies (Straussman et al., 2012). Straussman et al. (2012) observed that HGF secreted by fibroblasts was responsible for innate resistance to RAF inhibitors in melanoma cells. In ovarian cancer, fibroblast-mediated resistance to platinum-based chemotherapies has been shown to occur through reducing the intracellular content of platinum in cancer cells.

However, CD8+ T cells abolish this resistance by altering fibroblast metabolism through the interferon gamma (IFN-γ) pathway (Wang et al., 2016). Cancer-associated fibroblasts also secrete more extracellular matrix components than normal fibroblasts. This deposition of matrix protein (i.e., hyaluronan, collagen) has been found to affect drug penetration, which provides another mechanism of drug resistance (Flach et al., 2011; Loeffler et al., 2006; Misra et al., 2003).

Immune Cells

A delicate balance exists between the host's immune system and tumor cells (see also Chapters 8 and 9). The strength of a host's immune component has been correlated to therapeutic response. In breast cancer, a strong immune component is correlated with sensitivity to chemotherapy (Fridman et al., 2012; Iwamoto et al., 2011; Sotiriou and Pusztai, 2009). In metastatic colorectal cancer, high levels of CD8+ cells in the invasive margin predict better response to chemotherapy (Halama et al., 2011). The underlying mechanisms behind these clinical observations have yet to be determined, but they suggest that the immune presence is protecting the patient from expansion of the malignancy and therefore prolonging life. Under the right conditions, the immune response may be capable of eradicating the host of foreign malignant cells, as is suggested by the recent successes of immunotherapy in the clinic (Hodi et al., 2010; Pardoll, 2012; Rizvi et al., 2015; Topalian et al., 2012).

However, the impact of the immune system on cancer is complex and not yet fully understood. Evidence from other studies suggests that some immune components can actually have a protumorigenic effect (Pribluda et al., 2013). For example, macrophages, a type of immune cell, have been observed to promote tumor progression. In glioblastoma multiforme (GBM), colony-stimulating factor-1 receptor (CSF-1R) is currently being investigated as a therapy aimed at depleting tumor-associated macrophages (Quail et al., 2016). Further complicating matters, distinct immune cell types are often found in different regions of the tumor. For example, natural killer (NK) cells are found in the stroma surrounding the tumor and are not in direct contact with tumor cells (Fridman et al., 2012), while B cells are mostly found in the invasive margins of growing tumors, and T cells are found in both the margins and the tumor core (Halama et al., 2011). This suggests that each immune component has specific roles within the cancer ecosystem and that the spatial dynamics may play an important role in response to immune therapy. Furthermore, in some cancers there is very little immune infiltrate, whereas in others, immune cells are recruited to actively influence tumorigenesis (Pages et al., 2010). This implies that the coevolution of cancer and immune cells is a heterogeneous and possibly patient-specific process. The use of mathematical modeling is helping to shed light on this process—by enlisting models of tumor–immune interactions, one may be able to predict which patients will respond to immunotherapy (Babbs, 2012; de Pillis et al., 2005).

The complexity of the TME can make it difficult to study experimentally. Existing biological models capture certain aspects of the TME, but rarely are they comprehensive. Therefore, mathematical modeling can help bridge the gap between biological models and clinical workflows and was the topic of a recent review in *Nature Reviews Cancer* (Altrock et al., 2015).

Therapeutic Strategies Targeting the Tumor Microenvironment

For many years, the cancer field has focused on developing therapies that directly target tumor cells. However, most treatments currently used in the clinic today are plagued with the emergence of drug resistance. The key is to use evolutionary principles to identify novel ways of delaying or preventing therapeutic resistance. Targeting the TME may be one way of achieving this goal. If one thinks of a tumor as a "seed," and the microenvironment as the "soil," as proposed by Stephen Paget in 1889, then the challenge is to find ways to disrupt the "soil" to make it difficult for the tumor cells to grow. Therapeutically targeting the tumor cells' interactions with the microenvironment may suppress a benefit from this niche and restore control over the tumor cell burden. In addition, targeting the microenvironment rather than the tumor cells should be less prone to evolve resistance. Several TME-directed therapies are clinically approved or in the development pipeline, including angiogenesis inhibitors and immunotherapies (Giantonio et al., 2007; Junttila and de Sauvage, 2013; Quail and Joyce, 2013). Here we will highlight a few treatment strategies whereby understanding the symbiotic relationship between the TME and cancer cells may be beneficial for improving patient outcome.

Recent experimental (Hata et al., 2016) and mathematical modeling (Bhang et al., 2015; Diaz et al., 2012) work has shown that often small resistant subpopulations exist prior to receiving therapy and it is the addition of drug that results in the competitive expansion of these resistant subclones. One proposed evolutionary strategy is to alter the competitive dynamics of clonal populations within a tumor by increasing the fitness advantage of chemosensitive cells prior to drug treatment (Maley et al., 2004). Within tumors, there is a limited supply of public goods resulting in competition (or cooperation) for resources to survive. Therapeutically altering public goods to shift the clonal balance toward chemosensitive cells would result in improved treatment response (Driscoll and Pepper, 2010). However, in order for this approach to be successful, one must have knowledge as to the TME conditions that selectively and positively impact the fitness of chemosensitive cells.

As touched upon earlier, cancer cells are highly addicted to glycolysis. Because of this dependency, it was postulated that therapeutic targeting of glycolysis would effectively kill tumor cells. However, clinically this approach has not been successful, in part due to the metabolic reprogramming, or "plasticity," of tumor cells to escape glycolysis addiction by upregulating alternative fuel-source pathways. This suggests that future strategies will require simultaneous targeting of multiple metabolic networks to result in improved outcomes, which is the focus of several recent articles (Pusapati et al., 2016). In addition, therapeutic modulation of tumor stroma metabolism may be just as important as we have seen that stromal glucose and amino acid metabolism can directly stimulate the growth of cancer cells through cytokine production (Valencia et al., 2014).

While most research focuses on the role of the TME in contributing to cancer progression, it is also important to consider the role of the TME in restraining tumorigenesis. Pancreatic cancer is notorious for its large presence of stromal cells, which was believed to enhance cancer cell proliferation and invasion phenotypes (Hwang et al., 2008; Ikenaga et al., 2010). However, it was found that the deletion of a subset of stromal elements, namely, those that are driven by sonic hedgehog (Shh) signaling, resulted in more aggressive tumors (Rhim et al., 2014). One cannot ignore that within solid tumors, resources are limited and external goods are shared among tumor and stromal cells. Consequently, if the stroma is targeted, it may shift the resource balance in favor of the tumor cells leading to tumor growth. This emphasizes the importance of considering the coevolution that occurs between the tumor cells and their surrounding microenvironment and the impact treatment will have on this balance.

SUMMARY AND OUTLOOK ON THERAPEUTIC STRATEGIES

In this chapter we have discussed the evolutionary impacts of (1) phenotypic plasticity (e.g., epigenetic programming, EMTs, and fluidity of the "CSC" phenotype) and (2) the TME (e.g., availability of public goods, tumor–stromal interactions) on treatment response in cancer. These discussions lead naturally to the exploration of novel directions for treating cancer cell populations in light of these evolutionary processes. One consistent theme emerges—more nuanced treatment strategies aimed at controlling the overall evolution of a (likely genotypically and phenotypically heterogeneous) tumor cell population may be much more promising than more traditional strategies aimed at debulking the majority

component of a tumor. Many such proposed strategies currently being considered involve combinations of agents targeting processes within the TME or controlling plasticity alongside more standard cytotoxic or targeted therapies (see also Chapter 14). Some examples we have already noted include the following: (1) DNA-demethylating agents and histone deacetylase inhibitors can be used in combinations to sensitize cancer cells to any existing therapy that acts via DNA damage (e.g., radiation, chemotherapy), (2) CSC-targeting drugs can be used in combination with tumor debulking agents to prevent the rapid increase in the CSC population, which often drives the resistance that is observed with debulking, and (3) altering public good availability to select for drug/chemosensitive cells followed by treatment with drug/chemotherapy can reduce the likelihood of intrinsic resistance. However, the design of effective combination therapies is more than twice as difficult as designing single-agent therapies, due to the myriad of possibilities of dose combination and scheduling choices as well as the challenge of delineating a combination toxicity constraint space (the nature of which also varies with the scheduling protocol). It has been demonstrated that quantitative models can be of tremendous use in exploring the space of possible combination strategies and identifying the most promising schedules for further experimental and clinical testing. However, in order to realize the potential of quantitative models for the design of such combination therapies, much more work is needed to develop a suite of experimentally driven quantitative models describing the complex evolutionary processes driving treatment responses to the novel agents described here (e.g., CSC-targeting therapies, environmentally activated agents, therapies targeting epigenetic processes) as well as the role of factors, such as plasticity and the TME on response to standard therapies.

References

Abubaker, K., et al., 2014. Inhibition of the JAK2/STAT3 pathway in ovarian cancer results in the loss of cancer stem cell-like characteristics and a reduced tumor burden. BMC Cancer 14, 317.

Adelaiye, R., et al., 2015. Sunitinib dose escalation overcomes transient resistance in clear cell renal cell carcinoma and is associated with epigenetic modifications. Mol. Cancer Ther. 14, 513–522.

Aguilar-Gallardo, C., et al., 2012. Overcoming challenges of ovarian cancer stem cells: novel therapeutic approaches. Stem Cell Rev. 8, 994–1010.

Ahmed, N., Abubaker, K., Findlay, J., Quinn, M., 2010. Epithelial mesenchymal transition and cancer stem cell-like phenotypes facilitate chemoresistance in recurrent ovarian cancer. Curr. Cancer Drug Targets 10, 268–278.

Ahmed, N., Abubaker, K., Findlay, J., Quinn, M., 2013. Cancerous ovarian stem cells: obscure targets for therapy but relevant to chemoresistance. J. Cell. Biochem. 114, 21–34.

Al-Hajj, M., Wicha, M.S., Benito-Hernandez, A., Morrison, S.J., Clarke, M.F., 2003. Prospective identification of tumorigenic breast cancer cells. Proc. Natl. Acad. Sci. USA 100, 3983–3988.

Altrock, P.M., Liu, L.L., Michor, F., 2015. The mathematics of cancer: integrating quantitative models. Nat. Rev. Cancer 15, 730–745.

Alvero, A.B., et al., 2009. Molecular phenotyping of human ovarian cancer stem cells unravels the mechanisms for repair and chemoresistance. Cell Cycle 8, 158–166.

Anastas, J.N., et al., 2014. WNT5A enhances resistance of melanoma cells to targeted BRAF inhibitors. J. Clin. Invest. 124, 2877–2890.

Ara, T., et al., 2013. Critical role of STAT3 in IL-6-mediated drug resistance in human neuroblastoma. Cancer Res. 73, 3852–3864.

Babbs, C.F., 2012. Predicting success or failure of immunotherapy for cancer: insights from a clinically applicable mathematical model. Am. J. Cancer Res. 2, 204–213.

Banelli, B., et al., 2015. The histone demethylase KDM5A is a key factor for the resistance to temozolomide in glioblastoma. Cell Cycle 14, 3418–3429.

Bao, S., et al., 2006. Glioma stem cells promote radioresistance by preferential activation of the DNA damage response. Nature 444, 756–760.

Bapat, S.A., Mali, A.M., Koppikar, C.B., Kurrey, N.K., 2005. Stem and progenitor-like cells contribute to the aggressive behavior of human epithelial ovarian cancer. Cancer Res. 65, 3025–3029.

Barker, N., et al., 2007. Identification of stem cells in small intestine and colon by marker gene Lgr5. Nature 449, 1003–1007.

Baylin, S.B., Jones, P.A., 2011. A decade of exploring the cancer epigenome—biological and translational implications. Nat. Rev. Cancer 11, 726–734.

Ben-Porath, I., et al., 2008. An embryonic stem cell-like gene expression signature in poorly differentiated aggressive human tumors. Nat. Genet. 40, 499–507.

Bhang, H.E., et al., 2015. Studying clonal dynamics in response to cancer therapy using high-complexity barcoding. Nat. Med. 21, 440–448.

Black, P.C., et al., 2008. Sensitivity to epidermal growth factor receptor inhibitor requires E-cadherin expression in urothelial carcinoma cells. Clin. Cancer Res. 14, 1478–1486.

Bleau, A.M., et al., 2009. PTEN/PI3K/Akt pathway regulates the side population phenotype and ABCG2 activity in glioma tumor stem-like cells. Cell Stem Cell 4, 226–235.

Bonnet, D., Dick, J.E., 1997. Human acute myeloid leukemia is organized as a hierarchy that originates from a primitive hematopoietic cell. Nat. Med. 3, 730–737.

Brown, J.M., Wilson, W.R., 2004. Exploiting tumour hypoxia in cancer treatment. Nat. Rev. Cancer 4, 437–447.

Bryant, J.L., et al., 2012. A microRNA gene expression signature predicts response to erlotinib in epithelial cancer cell lines and targets EMT. Br. J. Cancer 106, 148–156.

Buck, E., et al., 2007. Loss of homotypic cell adhesion by epithelial–mesenchymal transition or mutation limits sensitivity to epidermal growth factor receptor inhibition. Mol. Cancer Ther. 6, 532–541.

Burgos-Ojeda, D., Rueda, B.R., Buckanovich, R.J., 2012. Ovarian cancer stem cell markers: prognostic and therapeutic implications. Cancer Lett. 322, 1–7.

Cannito, S., et al., 2008. Redox mechanisms switch on hypoxia-dependent epithelial–mesenchymal transition in cancer cells. Carcinogenesis 29, 2267–2278.

Chaffer, C.L., et al., 2011. Normal and neoplastic nonstem cells can spontaneously convert to a stem-like state. Proc. Natl. Acad. Sci. USA 108, 7950–7955.

Charles, N., et al., 2010. Perivascular nitric oxide activates notch signaling and promotes stem-like character in PDGF-induced glioma cells. Cell Stem Cell 6, 141–152.

Chaudhary, P.M., Roninson, I.B., 1991. Expression and activity of P-glycoprotein, a multidrug efflux pump, in human hematopoietic stem cells. Cell 66, 85–94.

Chen, J., et al., 2012. A restricted cell population propagates glioblastoma growth after chemotherapy. Nature 488, 522–526.

Chen, X., et al., 2016. The overshoot and phenotypic equilibrium in characterizing cancer dynamics of reversible phenotypic plasticity. J. Theor. Biol. 390, 40–49.

Chouaib, S., Janji, B., Tittarelli, A., Eggermont, A., Thiery, J.P., 2014. Tumor plasticity interferes with anti-tumor immunity. Crit. Rev. Immunol. 34, 91–102.

Clarke, M.F., et al., 2006. Cancer stem cells—perspectives on current status and future directions: AACR Workshop on cancer stem cells. Cancer Res. 66, 9339–9344.

Cohen, I., Poreba, E., Kamieniarz, K., Schneider, R., 2011. Histone modifiers in cancer: friends or foes? Genes Cancer 2, 631–647.

Conze, D., et al., 2001. Autocrine production of interleukin 6 causes multidrug resistance in breast cancer cells. Cancer Res. 61, 8851–8858.

Corbin, A.S., et al., 2011. Human chronic myeloid leukemia stem cells are insensitive to imatinib despite inhibition of BCR-ABL activity. J. Clin. Invest. 121, 396–409.

Creighton, C.J., et al., 2009. Residual breast cancers after conventional therapy display mesenchymal as well as tumor-initiating features. Proc. Natl. Acad. Sci. USA 106, 13820–13825.

Curley, M.D., et al., 2009. CD133 expression defines a tumor initiating cell population in primary human ovarian cancer. Stem Cells 27, 2875–2883.

Curley, M.D., Garrett, L.A., Schorge, J.O., Foster, R., Rueda, B.R., 2011. Evidence for cancer stem cells contributing to the pathogenesis of ovarian cancer. Front. Biosci. (Landmark Ed.) 16, 368–392.

Dalerba, P., Cho, R.W., Clarke, M.F., 2007a. Cancer stem cells: models and concepts. Annu. Rev. Med. 58, 267–284.

Dalerba, P., et al., 2007b. Phenotypic characterization of human colorectal cancer stem cells. Proc. Natl. Acad. Sci. USA 104, 10158–10163.

de Pillis, L.G., Radunskaya, A.E., Wiseman, C.L., 2005. A validated mathematical model of cell-mediated immune response to tumor growth. Cancer Res. 65, 7950–7958.

Dean, M., Fojo, T., Bates, S., 2005. Tumour stem cells and drug resistance. Nat. Rev. Cancer 5, 275–284.

Desai, A., Webb, B., Gerson, S.L., 2014. CD133+ cells contribute to radioresistance via altered regulation of DNA repair genes in human lung cancer cells. Radiother. Oncol. 110, 538–545.

Diaz, Jr., L.A., et al., 2012. The molecular evolution of acquired resistance to targeted EGFR blockade in colorectal cancers. Nature 486, 537–540.

Diehn, M., et al., 2009. Association of reactive oxygen species levels and radioresistance in cancer stem cells. Nature 458, 780–783.

Donkor, M.K., et al., 2011. T cell surveillance of oncogene-induced prostate cancer is impeded by T cell-derived TGF-beta1 cytokine. Immunity 35, 123–134.

Drasin, D.J., Robin, T.P., Ford, H.L., 2011. Breast cancer epithelial-to-mesenchymal transition: examining the functional consequences of plasticity. Breast Cancer Res. 13, 226.

Driscoll, W.W., Pepper, J.W., 2010. Theory for the evolution of diffusible external goods. Evolution Int. J. Org. Evolution 64, 2682–2687.

Dylla, S.J., et al., 2008. Colorectal cancer stem cells are enriched in xenogeneic tumors following chemotherapy. PLoS One 3, e2428.

Easwaran, H., Tsai, H.C., Baylin, S.B., 2014. Cancer epigenetics: tumor heterogeneity, plasticity of stem-like states, and drug resistance. Mol. Cell 54, 716–727.

Enderling, H., Hahnfeldt, P., 2011. Cancer stem cells in solid tumors: is 'evading apoptosis' a hallmark of cancer? Prog. Biophys. Mol. Biol. 106, 391–399.

Engl, W., Viasnoff, V., Thiery, J.P., 2015. Epithelial mesenchymal transition influence on CTL activity. Resist. Target Anti-C 7, 267–284.

Esteller, M., 2008. Epigenetics in cancer. N. Engl. J. Med. 358, 1148–1159.

Esteller, M., Herman, J.G., 2004. Generating mutations but providing chemosensitivity: the role of O6-methylguanine DNA methyltransferase in human cancer. Oncogene 23, 1–8.

Esteller, M., et al., 2000. Inactivation of the DNA repair gene O6-methylguanine-DNA methyltransferase by promoter hypermethylation is associated with G to A mutations in K-ras in colorectal tumorigenesis. Cancer Res. 60, 2368–2371.

Ferrandina, G., et al., 2008. Expression of CD133-1 and CD133-2 in ovarian cancer. Int. J. Gynecol. Cancer 18, 506–514.

Fidler, I.J., et al., 1994. Modulation of tumor cell response to chemotherapy by the organ environment. Cancer Metastasis Rev. 13, 209–222.

Flach, E.H., Rebecca, V.W., Herlyn, M., Smalley, K.S., Anderson, A.R., 2011. Fibroblasts contribute to melanoma tumor growth and drug resistance. Mol. Pharm. 8, 2039–2049.

Fornari, C., Cordero, F., Manini, D., Balbo, G., Calogero, R., 2011. Mathematical approach to predict the drug effects on cancer stem cell models. Electronic Notes Theor. Comput. Sci. 277, 29–39.

Fox, E.M., Kuba, M.G., Miller, T.W., Davies, B.R., Arteaga, C.L., 2013. Autocrine IGF-I/insulin receptor axis compensates for inhibition of AKT in ER-positive breast cancer cells with resistance to estrogen deprivation. Breast Cancer Res. 15, R55.

Frederick, B.A., et al., 2007. Epithelial to mesenchymal transition predicts gefitinib resistance in cell lines of head and neck squamous cell carcinoma and non-small cell lung carcinoma. Mol. Cancer Ther. 6, 1683–1691.

Fridman, W.H., Pages, F., Sautes-Fridman, C., Galon, J., 2012. The immune contexture in human tumours: impact on clinical outcome. Nat. Rev. Cancer 12, 298–306.

Galli, R., et al., 2004. Isolation and characterization of tumorigenic, stem-like neural precursors from human glioblastoma. Cancer Res. 64, 7011–7021.

Gao, M.Q., Choi, Y.P., Kang, S., Youn, J.H., Cho, N.H., 2010. CD24+ cells from hierarchically organized ovarian cancer are enriched in cancer stem cells. Oncogene 29, 2672–2680.

Giantonio, B.J., et al., 2007. Bevacizumab in combination with oxaliplatin, fluorouracil, and leucovorin (FOLFOX4) for previously treated metastatic colorectal cancer: results from the Eastern Cooperative Oncology Group Study E3200. J. Clin. Oncol. 25, 1539–1544.

Gil, J., Stembalska, A., Pesz, K.A., Sasiadek, M.M., 2008. Cancer stem cells: the theory and perspectives in cancer therapy. J. Appl. Genet. 49, 193–199.

Glozak, M.A., Seto, E., 2007. Histone deacetylases and cancer. Oncogene 26, 5420–5432.

Goodfellow, P.J., et al., 2015. Combined microsatellite instability, MLH1 methylation analysis, and immunohistochemistry for Lynch syndrome screening in endometrial cancers from GOG210: an NRG Oncology and Gynecologic Oncology Group Study. J. Clin. Oncol. 33, 4301–4308.

Graham, S.M., et al., 2002. Primitive, quiescent, Philadelphia-positive stem cells from patients with chronic myeloid leukemia are insensitive to STI571 in vitro. Blood 99, 319–325.

Greaves, M., Maley, C.C., 2012. Clonal evolution in cancer. Nature 481, 306–313.

Gupta, P.B., et al., 2011. Stochastic state transitions give rise to phenotypic equilibrium in populations of cancer cells. Cell 146, 633–644.

Halama, N., et al., 2011. Localization and density of immune cells in the invasive margin of human colorectal cancer liver metastases are prognostic for response to chemotherapy. Cancer Res. 71, 5670–5677.

Hansson, G.K., 2005. Inflammation, atherosclerosis, and coronary artery disease. N. Engl. J. Med. 352, 1685–1695.

Hata, A.N., et al., 2016. Tumor cells can follow distinct evolutionary paths to become resistant to epidermal growth factor receptor inhibition. Nat. Med. 22, 262–269.

Herman, J.G., et al., 1998. Incidence and functional consequences of hMLH1 promoter hypermethylation in colorectal carcinoma. Proc. Natl. Acad. Sci. USA 95, 6870–6875.

Hodi, F.S., et al., 2010. Improved survival with ipilimumab in patients with metastatic melanoma. N. Engl. J. Med. 363, 711–723.

Hollier, B.G., Evans, K., Mani, S.A., 2009. The epithelial-to-mesenchymal transition and cancer stem cells: a coalition against cancer therapies. J. Mammary Gland Biol. Neoplasia 14, 29–43.

Hongo, K., et al., 2013. Hypoxia enhances colon cancer migration and invasion through promotion of epithelial–mesenchymal transition. J. Surg. Res. 182, 75–84.

Hope, K.J., Jin, L., Dick, J.E., 2004. Acute myeloid leukemia originates from a hierarchy of leukemic stem cell classes that differ in self-renewal capacity. Nat. Immunol. 5, 738–743.

Huang, S., et al., 2012. MED12 controls the response to multiple cancer drugs through regulation of TGF-beta receptor signaling. Cell 151, 937–950.

Huijbers, A., et al., 2013. The proportion of tumor-stroma as a strong prognosticator for stage II and III colon cancer patients: validation in the VICTOR trial. Ann. Oncol. 24, 179–185.

Hwang, R.F., et al., 2008. Cancer-associated stromal fibroblasts promote pancreatic tumor progression. Cancer Res. 68, 918–926.

Iacobuzio-Donahue, C.A., 2009. Epigenetic changes in cancer. Annu. Rev. Pathol. 4, 229–249.

Ignatova, T.N., et al., 2002. Human cortical glial tumors contain neural stem-like cells expressing astroglial and neuronal markers in vitro. Glia 39, 193–206.

Ikenaga, N., et al., 2010. CD10+ pancreatic stellate cells enhance the progression of pancreatic cancer. Gastroenterology 139, 1041–1051, 1051.e1–1051.e8.

Imai, T., et al., 2003. Hypoxia attenuates the expression of E-cadherin via up-regulation of SNAIL in ovarian carcinoma cells. Am. J. Pathol. 163, 1437–1447.

Iwamoto, T., et al., 2011. Gene pathways associated with prognosis and chemotherapy sensitivity in molecular subtypes of breast cancer. J. Natl. Cancer Inst. 103, 264–272.

Jagani, Z., Khosravi-Far, R., 2008. Cancer stem cells and impaired apoptosis. Adv. Exp. Med. Biol. 615, 331–344.

Janes, K.A., et al., 2005. A systems model of signaling identifies a molecular basis set for cytokine-induced apoptosis. Science 310, 1646–1653.

Jiang, X., et al., 2007. Chronic myeloid leukemia stem cells possess multiple unique features of resistance to BCR-ABL targeted therapies. Leukemia 21, 926–935.

Jones, P.A., 2012. Functions of DNA methylation: islands, start sites, gene bodies and beyond. Nat. Rev. Genet. 13, 484–492.

Jones, P.A., Baylin, S.B., 2007. The epigenomics of cancer. Cell 128, 683–692.

Jones, S., et al., 2008. Comparative lesion sequencing provides insights into tumor evolution. Proc. Natl. Acad. Sci. USA 105, 4283–4288.

Jones, V.S., et al., 2016. Cytokines in cancer drug resistance: cues to new therapeutic strategies. Biochim. Biophys. Acta 1865, 255–265.

Jordan, C.T., Guzman, M.L., Noble, M., 2006. Cancer stem cells. N. Engl. J. Med. 355, 1253–1261.

Junttila, M.R., de Sauvage, F.J., 2013. Influence of tumour micro-environment heterogeneity on therapeutic response. Nature 501, 346–354.

Kane, M.F., et al., 1997. Methylation of the hMLH1 promoter correlates with lack of expression of hMLH1 in sporadic colon tumors and mismatch repair-defective human tumor cell lines. Cancer Res. 57, 808–811.

Kemper, K., de Goeje, P.L., Peeper, D.S., van Amerongen, R., 2014. Phenotype switching: tumor cell plasticity as a resistance mechanism and target for therapy. Cancer Res. 74, 5937–5941.

Koh, M.Y., et al., 2008. Molecular mechanisms for the activity of PX-478, an antitumor inhibitor of the hypoxia-inducible factor-1alpha. Mol. Cancer Ther. 7, 90–100.

Kondo, A., et al., 2001. Hypoxia-induced enrichment and mutagenesis of cells that have lost DNA mismatch repair. Cancer Res. 61, 7603–7607.

Konieczkowski, D.J., et al., 2014. A melanoma cell state distinction influences sensitivity to MAPK pathway inhibitors. Cancer Discov. 4, 816–827.

Landau, D.A., et al., 2014. Locally disordered methylation forms the basis of intratumor methylome variation in chronic lymphocytic leukemia. Cancer Cell 26, 813–825.

Lapidot, T., et al., 1994. A cell initiating human acute myeloid leukaemia after transplantation into SCID mice. Nature 367, 645–648.

Latifi, A., et al., 2011. Cisplatin treatment of primary and metastatic epithelial ovarian carcinomas generates residual cells with mesenchymal stem cell-like profile. J. Cell. Biochem. 112, 2850–2864.

Latifi, A., et al., 2012. Isolation and characterization of tumor cells from the ascites of ovarian cancer patients: molecular phenotype of chemoresistant ovarian tumors. PLoS One 7, e46858.

Leder, K., Holland, E.C., Michor, F., 2010. The therapeutic implications of plasticity of the cancer stem cell phenotype. PLoS One 5, e14366.

Leder, K., et al., 2014. Mathematical modeling of PDGF-driven glioblastoma reveals optimized radiation dosing schedules. Cell 156, 603–616.

Lee, C., et al., 2012a. Fasting cycles retard growth of tumors and sensitize a range of cancer cell types to chemotherapy. Sci. Transl. Med. 4, 124ra127.

Lee, C., Raffaghello, L., Longo, V.D., 2012b. Starvation, detoxification, and multidrug resistance in cancer therapy. Drug Resist. Updat. 15, 114–122.

Li, C., et al., 2007. Identification of pancreatic cancer stem cells. Cancer Res. 67, 1030–1037.

Li, X., et al., 2008. Intrinsic resistance of tumorigenic breast cancer cells to chemotherapy. J. Natl. Cancer Inst. 100, 672–679.

Li, C., Lee, C.J., Simeone, D.M., 2009. Identification of human pancreatic cancer stem cells. Methods Mol. Biol. 568, 161–173.

Lim, S., et al., 2013. SNAI1-mediated epithelial–mesenchymal transition confers chemoresistance and cellular plasticity by regulating genes involved in cell death and stem cell maintenance. PLoS One 8, e66558.

Liu, G., et al., 2006. Analysis of gene expression and chemoresistance of CD133+ cancer stem cells in glioblastoma. Mol. Cancer 5, 67.

Loeffler, M., Kruger, J.A., Niethammer, A.G., Reisfeld, R.A., 2006. Targeting tumor-associated fibroblasts improves cancer chemotherapy by increasing intratumoral drug uptake. J. Clin. Invest. 116, 1955–1962.

Lopez-Diaz, F.J., et al., 2013. Coordinate transcriptional and translational repression of p53 by TGF-beta1 impairs the stress response. Mol. Cell 50, 552–564.

Lu, C.W., Lin, S.C., Chen, K.F., Lai, Y.Y., Tsai, S.J., 2008. Induction of pyruvate dehydrogenase kinase-3 by hypoxia-inducible factor-1 promotes metabolic switch and drug resistance. J. Biol. Chem. 283, 28106–28114.

Lu, C.W., et al., 2011. Overexpression of pyruvate dehydrogenase kinase 3 increases drug resistance and early recurrence in colon cancer. Am. J. Pathol. 179, 1405–1414.

Luo, M., Brooks, M., Wicha, M.S., 2015. Epithelial–mesenchymal plasticity of breast cancer stem cells: implications for metastasis and therapeutic resistance. Curr. Pharm. Des. 21, 1301–1310.

Luoto, K.R., Kumareswaran, R., Bristow, R.G., 2013. Tumor hypoxia as a driving force in genetic instability. Genome Integrity 4, 5.

Mahon, F.X., et al., 2003. MDR1 gene overexpression confers resistance to imatinib mesylate in leukemia cell line models. Blood 101, 2368–2373.

Maley, C.C., Reid, B.J., Forrest, S., 2004. Cancer prevention strategies that address the evolutionary dynamics of neoplastic cells: simulating benign cell boosters and selection for chemosensitivity. Cancer Epidemiol. Biomarkers Prev. 13, 1375–1384.

Mani, S.A., et al., 2008. The epithelial–mesenchymal transition generates cells with properties of stem cells. Cell 133, 704–715.

Marie-Egyptienne, D.T., Lohse, I., Hill, R.P., 2013. Cancer stem cells, the epithelial to mesenchymal transition (EMT) and radioresistance: potential role of hypoxia. Cancer Lett. 341, 63–72.

Masui, K., et al., 2015. Glucose-dependent acetylation of Rictor promotes targeted cancer therapy resistance. Proc. Natl. Acad. Sci. USA 112, 9406–9411.

Mazor, T., et al., 2015. DNA methylation and somatic mutations converge on the cell cycle and define similar evolutionary histories in brain tumors. Cancer Cell 28, 307–317.

Meacham, C.E., Morrison, S.J., 2013. Tumour heterogeneity and cancer cell plasticity. Nature 501, 328–337.

Meyer, M.J., et al., 2009. Dynamic regulation of CD24 and the invasive, CD44posCD24neg phenotype in breast cancer cell lines. Breast Cancer Res. 11, R82.

Michor, F., 2008. Mathematical models of cancer stem cells. J. Clin. Oncol. 26, 2854–2861.

Mimeault, M., Batra, S.K., 2008. Targeting of cancer stem/progenitor cells plus stem cell-based therapies: the ultimate hope for treating and curing aggressive and recurrent cancers. Panminerva Med. 50, 3–18.

Misra, S., Ghatak, S., Zoltan-Jones, A., Toole, B.P., 2003. Regulation of multidrug resistance in cancer cells by hyaluronan. J. Biol. Chem. 278, 25285–25288.

Monteagudo, Á., Santos, J., 2015a. Treatment analysis in a cancer stem cell context using a tumor growth model based on cellular automata. PLoS One 10, e0132306.

Monteagudo, A., Santos, J., 2015. Genetic and Evolutionary Computation Conference. ACM, New York, pp. 233–240.

Montenegro, M.F., et al., 2016. Targeting the epigenetics of the DNA damage response in breast cancer. Cell Death Dis. 7, e2180.

Morel, A.P., et al., 2008. Generation of breast cancer stem cells through epithelial–mesenchymal transition. PLoS One 3, e2888.

Moser, B., Wolf, M., Walz, A., Loetscher, P., 2004. Chemokines: multiple levels of leukocyte migration control. Trends Immunol. 25, 75–84.

Munoz, P., Iliou, M.S., Esteller, M., 2012. Epigenetic alterations involved in cancer stem cell reprogramming. Mol. Oncol. 6, 620–636.

Neering, S.J., et al., 2007. Leukemia stem cells in a genetically defined murine model of blast-crisis CML. Blood 110, 2578–2585.

O'Brien, C.A., Pollett, A., Gallinger, S., Dick, J.E., 2007. A human colon cancer cell capable of initiating tumour growth in immunodeficient mice. Nature 445, 106–110.

O'Connell, M.P., et al., 2013. Hypoxia induces phenotypic plasticity and therapy resistance in melanoma via the tyrosine kinase receptors ROR1 and ROR2. Cancer Discov. 3, 1378–1393.

Oliver, T.G., et al., 2010. Chronic cisplatin treatment promotes enhanced damage repair and tumor progression in a mouse model of lung cancer. Genes Dev. 24, 837–852.

Oliveras-Ferraros, C., et al., 2012. Epithelial-to-mesenchymal transition (EMT) confers primary resistance to trastuzumab (Herceptin). Cell Cycle 11, 4020–4032.

Oravecz-Wilson, K.I., et al., 2009. Persistence of leukemia-initiating cells in a conditional knockin model of an imatinib-responsive myeloproliferative disorder. Cancer Cell 16, 137–148.

Oshimori, N., Oristian, D., Fuchs, E., 2015. TGF-beta promotes heterogeneity and drug resistance in squamous cell carcinoma. Cell 160, 963–976.

Pages, F., et al., 2010. Immune infiltration in human tumors: a prognostic factor that should not be ignored. Oncogene 29, 1093–1102.

Pan, Y., Huang, X., 2008. Epithelial ovarian cancer stem cells—a review. Int. J. Clin. Exp. Med. 1, 260–266.

Pan, H., et al., 2015. Epigenomic evolution in diffuse large B-cell lymphomas. Nat. Commun. 6, 6921.

Pardoll, D.M., 2012. The blockade of immune checkpoints in cancer immunotherapy. Nat. Rev. Cancer 12, 252–264.

Phillips, T.M., McBride, W.H., Pajonk, F., 2006. The response of CD24(-/low)/CD44+ breast cancer-initiating cells to radiation. J. Natl. Cancer Inst. 98, 1777–1785.

Pistollato, F., et al., 2010. Intratumoral hypoxic gradient drives stem cells distribution and MGMT expression in glioblastoma. Stem Cells 28, 851–862.

Poulsom, R., Alison, M.R., Forbes, S.J., Wright, N.A., 2002. Adult stem cell plasticity. J. Pathol. 197, 441–456.

Pribluda, A., et al., 2013. A senescence-inflammatory switch from cancer-inhibitory to cancer-promoting mechanism. Cancer Cell 24, 242–256.

Prince, M.E., et al., 2007. Identification of a subpopulation of cells with cancer stem cell properties in head and neck squamous cell carcinoma. Proc. Natl. Acad. Sci. USA 104, 973–978.

Pusapati, R.V., et al., 2016. mTORC1-dependent metabolic reprogramming underlies escape from glycolysis addiction in cancer cells. Cancer Cell 29, 548–562.

Quail, D.F., Joyce, J.A., 2013. Microenvironmental regulation of tumor progression and metastasis. Nat. Med. 19, 1423–1437.

Quail, D.F., et al., 2016. The tumor microenvironment underlies acquired resistance to CSF-1R inhibition in gliomas. Science 352, aad3018.

Quintana, E., et al., 2010. Phenotypic heterogeneity among tumorigenic melanoma cells from patients that is reversible and not hierarchically organized. Cancer Cell 18, 510–523.

Raffaghello, L., et al., 2008. Starvation-dependent differential stress resistance protects normal but not cancer cells against high-dose chemotherapy. Proc. Natl. Acad. Sci. USA 105, 8215–8220.

Rando, O.J., Verstrepen, K.J., 2007. Timescales of genetic and epigenetic inheritance. Cell 128, 655–668.

Rapp, U.R., Ceteci, F., Schreck, R., 2008. Oncogene-induced plasticity and cancer stem cells. Cell Cycle 7, 45–51.

Rasanen, K., Vaheri, A., 2010. Activation of fibroblasts in cancer stroma. Exp. Cell Res. 316, 2713–2722.

Reya, T., Morrison, S.J., Clarke, M.F., Weissman, I.L., 2001. Stem cells, cancer, and cancer stem cells. Nature 414, 105–111.

Reynolds, T.Y., Rockwell, S., Glazer, P.M., 1996. Genetic instability induced by the tumor microenvironment. Cancer Res. 56, 5754–5757.

Rhim, A.D., et al., 2014. Stromal elements act to restrain, rather than support, pancreatic ductal adenocarcinoma. Cancer Cell 25, 735–747.

Ricci-Vitiani, L., et al., 2007. Identification and expansion of human colon-cancer-initiating cells. Nature 445, 111–115.

Rizvi, N.A., et al., 2015. Activity and safety of nivolumab, an anti-PD-1 immune checkpoint inhibitor, for patients with advanced, refractory squamous non-small-cell lung cancer (CheckMate 063): a phase 2, single-arm trial. Lancet Oncol. 16, 257–265.

Robson, E.J., Khaled, W.T., Abell, K., Watson, C.J., 2006. Epithelial-to-mesenchymal transition confers resistance to apoptosis in three murine mammary epithelial cell lines. Differentiation 74, 254–264.

Rodriguez-Paredes, M., Esteller, M., 2011. Cancer epigenetics reaches mainstream oncology. Nat. Med. 17, 330–339.

Ropero, S., Esteller, M., 2007. The role of histone deacetylases (HDACs) in human cancer. Mol. Oncol. 1, 19–25.

Rousselot, P., et al., 2007. Imatinib mesylate discontinuation in patients with chronic myelogenous leukemia in complete molecular remission for more than 2 years. Blood 109, 58–60.

Santisteban, M., et al., 2009. Immune-induced epithelial to mesenchymal transition in vivo generates breast cancer stem cells. Cancer Res. 69, 2887–2895.

Savona, M., Talpaz, M., 2008. Getting to the stem of chronic myeloid leukaemia. Nat. Rev. Cancer 8, 341–350.

Scaffidi, P., Misteli, T., 2011. In vitro generation of human cells with cancer stem cell properties. Nat. Cell Biol. 13, 1051–1061.

Sehl, M.E., Shimada, M., Landeros, A., Lange, K., Wicha, M.S., 2015. Modeling of cancer stem cell state transitions predicts therapeutic response. PLoS One 10, e0135797.

Sell, S., 2004. Stem cell origin of cancer and differentiation therapy. Crit. Rev. Oncol. Hematol. 51, 1–28.

Sharma, S.V., et al., 2010. A chromatin-mediated reversible drug-tolerant state in cancer cell subpopulations. Cell 141, 69–80.

Shintani, Y., et al., 2011. Epithelial to mesenchymal transition is a determinant of sensitivity to chemoradiotherapy in non-small cell lung cancer. Ann. Thorac. Surg. 92, 1794–1804, (discussion 1804).

Singh, S.K., et al., 2004. Identification of human brain tumour initiating cells. Nature 432, 396–401.

Sotiriou, C., Pusztai, L., 2009. Gene-expression signatures in breast cancer. N. Engl. J. Med. 360, 790–800.

Stassi, G., et al., 2003. Thyroid cancer resistance to chemotherapeutic drugs via autocrine production of interleukin-4 and interleukin-10. Cancer Res. 63, 6784–6790.

Steg, A.D., et al., 2011. Targeting the notch ligand JAGGED1 in both tumor cells and stroma in ovarian cancer. Clin. Cancer Res. 17, 5674–5685.

Steg, A.D., et al., 2012. Stem cell pathways contribute to clinical chemoresistance in ovarian cancer. Clin. Cancer Res. 18, 869–881.

Stewart, J.M., et al., 2011. Phenotypic heterogeneity and instability of human ovarian tumor-initiating cells. Proc. Natl. Acad. Sci. USA 108, 6468–6473.

Straussman, R., et al., 2012. Tumour micro-environment elicits innate resistance to RAF inhibitors through HGF secretion. Nature 487, 500–504.

Sullivan, R., Pare, G.C., Frederiksen, L.J., Semenza, G.L., Graham, C.H., 2008. Hypoxia-induced resistance to anticancer drugs is associated with decreased senescence and requires hypoxia-inducible factor-1 activity. Mol. Cancer Ther. 7, 1961–1973.

Szotek, P.P., et al., 2006. Ovarian cancer side population defines cells with stem cell-like characteristics and Mullerian inhibiting substance responsiveness. Proc. Natl. Acad. Sci. USA 103, 11154–11159.

Tan, Q., Saggar, J.K., Yu, M., Wang, M., Tannock, I.F., 2015. Mechanisms of drug resistance related to the microenvironment of solid tumors and possible strategies to inhibit them. Cancer J. 21, 254–262.

Tap, W.D., et al., 2010. Pharmacodynamic characterization of the efficacy signals due to selective BRAF inhibition with PLX4032 in malignant melanoma. Neoplasia 12, 637–649.

Taussig, D.C., et al., 2010. Leukemia-initiating cells from some acute myeloid leukemia patients with mutated nucleophosmin reside in the CD34(−) fraction. Blood 115, 1976–1984.

Theys, J., et al., 2011. E-cadherin loss associated with EMT promotes radioresistance in human tumor cells. Radiother. Oncol. 99, 392–397.

Thomson, S., et al., 2005. Epithelial to mesenchymal transition is a determinant of sensitivity of non-small-cell lung carcinoma cell lines and xenografts to epidermal growth factor receptor inhibition. Cancer Res. 65, 9455–9462.

Thomson, S., Petti, F., Sujka-Kwok, I., Epstein, D., Haley, J.D., 2008. Kinase switching in mesenchymal-like non-small cell lung cancer lines contributes to EGFR inhibitor resistance through pathway redundancy. Clin. Exp. Metastasis 25, 843–854.

Topalian, S.L., et al., 2012. Safety, activity, and immune correlates of anti-PD-1 antibody in cancer. N. Engl. J. Med. 366, 2443–2454.

Toyota, M., et al., 1999a. CpG island methylator phenotype in colorectal cancer. Proc. Natl. Acad. Sci. USA 96, 8681–8686.

Toyota, M., et al., 1999b. Aberrant methylation in gastric cancer associated with the CpG island methylator phenotype. Cancer Res. 59, 5438–5442.

Turke, A.B., et al., 2010. Preexistence and clonal selection of MET amplification in EGFR mutant NSCLC. Cancer Cell 17, 77–88.

Vainstein, V., Kirnasovsky, O.U., Kogan, Y., Agur, Z., 2012. Strategies for cancer stem cell elimination: insights from mathematical modeling. J. Theor. Biol. 298, 32–41.

Valencia, T., et al., 2014. Metabolic reprogramming of stromal fibroblasts through p62-mTORC1 signaling promotes inflammation and tumorigenesis. Cancer Cell 26, 121–135.

Vander Heiden, M.G., Cantley, L.C., Thompson, C.B., 2009. Understanding the Warburg effect: the metabolic requirements of cell proliferation. Science 324, 1029–1033.

Vathipadiekal, V., et al., 2012. Identification of a potential ovarian cancer stem cell gene expression profile from advanced stage papillary serous ovarian cancer. PLoS One 7, e29079.

Vazquez-Martin, A., et al., 2013. IGF-1R/epithelial-to-mesenchymal transition (EMT) crosstalk suppresses the erlotinib-sensitizing effect of EGFR exon 19 deletion mutations. Sci. Rep. 3, 2560.

Voulgari, A., Pintzas, A., 2009. Epithelial–mesenchymal transition in cancer metastasis: mechanisms, markers and strategies to overcome drug resistance in the clinic. Biochim. Biophys. Acta 1796, 75–90.

Wagers, A.J., Weissman, I.L., 2004. Plasticity of adult stem cells. Cell 116, 639–648.

Wang, J.C., et al., 1998. High level engraftment of NOD/SCID mice by primitive normal and leukemic hematopoietic cells from patients with chronic myeloid leukemia in chronic phase. Blood 91, 2406–2414.

Wang, L., et al., 2009. Bistable switches control memory and plasticity in cellular differentiation. Proc. Natl. Acad. Sci. USA 106, 6638–6643.

Wang, Y.C., et al., 2012. ALDH1-bright epithelial ovarian cancer cells are associated with CD44 expression, drug resistance, and poor clinical outcome. Am. J. Pathol. 180, 1159–1169.

Wang, W., et al., 2016. Effector T cells abrogate stroma-mediated chemoresistance in ovarian cancer. Cell 165, 1092–1105.

Wartenberg, M., et al., 2003. Regulation of the multidrug resistance transporter P-glycoprotein in multicellular tumor spheroids by hypoxia-inducible factor (HIF-1) and reactive oxygen species. FASEB J. 17, 503–505.

Willyard, C., 2016. Cancer therapy: an evolved approach. Nature 532, 166–168.

Witta, S.E., et al., 2006. Restoring E-cadherin expression increases sensitivity to epidermal growth factor receptor inhibitors in lung cancer cell lines. Cancer Res. 66, 944–950.

Wodarz, D., Komarova, N., 2007. Can loss of apoptosis protect against cancer? Trends Genet. 23, 232–237.

Woodward, W.A., et al., 2007. WNT/beta-catenin mediates radiation resistance of mouse mammary progenitor cells. Proc. Natl. Acad. Sci. USA 104, 618–623.

Wright, J., 2013. Epigenetics: reversible tags. Nature 498, S10–S11.

Yang, J., et al., 2004. Twist, a master regulator of morphogenesis, plays an essential role in tumor metastasis. Cell 117, 927–939.

Yang, G., et al., 2012. Dynamic equilibrium between cancer stem cells and non-stem cancer cells in human SW620 and MCF-7 cancer cell populations. Br. J. Cancer 106, 1512–1519.

Yang, X., et al., 2014. Gene body methylation can alter gene expression and is a therapeutic target in cancer. Cancer Cell 26, 577–590.

Yano, S., et al., 2008. Hepatocyte growth factor induces gefitinib resistance of lung adenocarcinoma with epidermal growth factor receptor-activating mutations. Cancer Res. 68, 9479–9487.

Yao, Z., et al., 2010. TGF-beta IL-6 axis mediates selective and adaptive mechanisms of resistance to molecular targeted therapy in lung cancer. Proc. Natl. Acad. Sci. USA 107, 15535–15540.

Yatabe, Y., Tavare, S., Shibata, D., 2001. Investigating stem cells in human colon by using methylation patterns. Proc. Natl. Acad. Sci. USA 98, 10839–10844.

Yauch, R.L., et al., 2005. Epithelial versus mesenchymal phenotype determines in vitro sensitivity and predicts clinical activity of erlotinib in lung cancer patients. Clin. Cancer Res. 11, 8686–8698.

Yun, J., et al., 2009. Glucose deprivation contributes to the development of KRAS pathway mutations in tumor cells. Science 325, 1555–1559.

Zeng, M., Kikuchi, H., Pino, M.S., Chung, D.C., 2010. Hypoxia activates the K-ras proto-oncogene to stimulate angiogenesis and inhibit apoptosis in colon cancer cells. PLoS One 5, e10966.

Zhang, S., et al., 2008. Identification and characterization of ovarian cancer-initiating cells from primary human tumors. Cancer Res. 68, 4311–4320.

Zhou, Y.C., et al., 2011. Ionizing radiation promotes migration and invasion of cancer cells through transforming growth factor-beta-mediated epithelial–mesenchymal transition. Int. J. Radiat. Oncol. Biol. Phys. 81, 1530–1537.

Zhou, D., Wang, Y., Wu, B., 2014. A multi-phenotypic cancer model with cell plasticity. J. Theor. Biol. 357, 35–45.

Zipser, M.C., et al., 2011. A proliferative melanoma cell phenotype is responsive to RAF/MEK inhibition independent of BRAF mutation status. Pigment Cell Melanoma Res. 24, 326–333.

11

Ecology of the Metastatic Process

Mark C. Lloyd*,†, Robert A. Gatenby**, Joel S. Brown**,†

*Inspirata, Inc., Tampa, FL, United States
**Cancer Biology and Evolution Program, H. Lee Moffitt Cancer Center
and Research Institute, Tampa, FL, United States
†Department of Biological Sciences, University of Illinois at Chicago, Chicago, IL, United States

INTRODUCTION

"Your cancer has metastasized" are disheartening words for a physician to speak or a patient to hear. It means the cancer has spread from its primary site to new sites and organs. Such a diagnosis nearly always portends more toxic, debilitating, and onerous therapies. The diagnosis is generally poor often with life expectation of only months or a few years (Greenlee et al., 2001). Over 90% of cancer deaths are attributed to disseminated tumor spread (Chaffer and Weinberg, 2011). While the prognosis for locally confined cancer is excellent and steadily improving for most tumor types, the diagnosis of metastases is the equivalent of a death sentence. And, this has changed little in the last several decades (Siegel et al., 2015). Hence, there is a compelling clinical need for understanding, preventing, and advancing better therapies for metastatic cancer. Here we focus on understanding how and why metastases happen with some mention of prevention and treatment.

We shall take an ecoevolutionary approach to understand the metastatic process. Why? Cancer has variously been defined as a disease of the genes—and indeed cancer initiation involves a series of unfortunate mutations. Cancer has also been defined as a disease of unregulated proliferation—and indeed a hallmark of cancer is the ability of cancer cells to proliferate indefinitely if conditions permit. But, by the time cancer is at a stage where metastases are a possibility, it has become a disease of Darwinian evolution, a disease propelled by natural selection (Greaves and Maley, 2012) (see also the introductory chapter).

Natural selection has three ingredients. The first is heritable variation. Cancer cells exhibit considerable heritable variation often due to an elevated mutation rate. Indeed, the "mutator phenotype" is often considered a necessary condition for carcinogenesis and cancer cells frequently exhibit elevated rates of mutations as a result of poorer DNA repair mechanisms and chromosomal instability. One might think this is cancer's potential "Achilles' heel." Might not the cancer cells mutate themselves into oblivion—an "error catastrophe" as predicted theoretically by Eigen and Shuster (Eigen, 1971). Since so many mutations can be deleterious and even fatal, this hope seems plausible. Sadly, elevated mutation rates appear to be adaptive and evolvable so that loss of fitness due to accumulating mutations in an asexually reproducing organism (i.e., "Mueller's ratchet") seems to provide no relief from the onset and progression of cancer (Rosenberg and Queitsch, 2014). Recent research suggests that mutations are not randomly distributed across the whole genome, but rather associated with genes that do useful things or that can be improved, at least improved from the cancer cell's perspective (Galhardo et al., 2007; Rosenberg, 2001). These apparently nonrandom mutations throughout the genome appear to reflect a phenomenon termed "evolutionary triage" in which the gene's linkage to fitness determines its observable prevalence in a population. Thus, while deleterious mutations do and will occur, the mutation rate may evolve to facilitate evolution's march of "creative destruction." The capacity to generate heritable variation likely plays a major role in the early and late stages of the metastatic process.

A second essential component of natural selection requires a struggle for existence. For an organism, creature or cell to be subjected to natural selection it must have the capacity to produce many more offspring than can possibly survive. The environment can only support so many individuals. Under ideal conditions, organisms and cancer cells can grow exponentially; one cell begets two, two beget four, four beget eight and so on. But at some point the space and nutrients required to promote

Ecology and Evolution of Cancer. http://dx.doi.org/10.1016/B978-0-12-804310-3.00011-9

survival and proliferation run out or become depleted. The struggle for existence means that all organisms experience limits to growth. Cancer cells are no exception. As the density of tumor cells grow their proliferation rates decline and their mortality rates increase. Carrying capacity is reached when proliferation and mortality rates balance. This can occur at quite high densities—hence the observation that breast cancer can be first noticed as a lump, as a very dense mass of cells. General tumors have densities of cancer cells that are 10–100 times that of the original, normal tissue. It seems that the limiting factors to cancer cells within their tumor ecosystem include glucose, glutamine, oxygen, acidic conditions, safety from the immune system, and a myriad of trace organic and inorganic molecules most of which are supplied by the blood. To the tumor, blood brings nourishment and takes away potentially toxic metabolites. As the blood leaves the tumor it also takes away living cancer cells—the potential colonizers of distant organs.

Third, the heritable variation influences the struggle for existence. Heritable variation is the "fuel" of evolution. The struggle for existence describes ecology, which includes the factors influencing births, deaths, and population growth. The genetic variability leads to phenotypic variability. For cancer cells, certain phenotypes will be better than others at harvesting nutrients either quickly or efficiently, others may be better at evading the immune system, some phenotypes may allow the cell to tolerate and even thrive under acidic conditions or conditions of low oxygen. Some phenotypes may be better than others at moving from areas of overcrowding or scarcity into areas of plenty. In this way the tumor environment becomes the selective force that determines which phenotypes survive and which die-off. We will use the term *strategies* when referring to these different heritable phenotypes. In the end, the combination of *strategy* variation and its influence on births and deaths generate the Darwinian dynamics. Darwinian dynamics include evolutionary dynamics, changes in the *strategies* of the cancer cells with time as more successful *strategies* replace less successful ones. Darwinian dynamics also include ecological dynamics, changes in the population size, or density of cancer cells with time via survival and proliferation. In the end, natural selection equips cancer cells with adaptations. Such adaptations are *strategies* that maximize the fitness of the cancer cell given its circumstances. And these circumstances include its tumor environment, the density of other cells, and the *strategies* of the other cancer cells (see also Chapter 17).

A given tumor will grow according to ecological dynamics, and the characteristics and *strategies* of the cancer cells will be shaped by the evolutionary dynamics. The size and characteristics of the tumor, as well as the different adaptive *strategies* of the cancer cells must be central to understanding metastatic progression. Furthermore, significant spatial and temporal variation in circumstances occur within the same tumor due the, for example, regional variations in vascular density and blood flow. It is likely that some regions may contribute disproportionately to the cells that successfully metastasize to different sites. For the aforementioned reasons we see metastatic progression as an ecological process that begins with a primary tumor shedding cancer cells into the blood or lymphatics. Some of these cells colonize a distant organ of the body. To the tumor cell, this new site is a foreign land, inhabited by normal cells but uninhabited by cancer cells. It is a land for which natural selection has not directly prepared the cancer cell. Thus, the cancer cell's ability to survive and proliferate is based on preadaptations, *strategies* that were honed by natural selection in the primary tumor that coincidentally serve the cancer cell well at its new site.

Metastases are defined in the NCI Dictionary of terms (https://www.cancer.gov/publications/dictionaries/cancer-terms) as "The spread of cancer cells from the place where they first formed to another part of the body." Thus, the metastasizing cancer cell can be viewed, analogous to exotic species found in nature, as a long range disperser that successfully invades and proliferates in a foreign site. In this way, the metastasizing cancer cell is like an invasive species that succeeds in dispersing from some original place or continent to a novel place from which the species had been absent. Throughout, we shall see metastatic progression as something that can and should be viewed through the lens of invasion ecology, the study of how invasive species disperse, colonize, and spread (Chen and Pienta, 2011).

We begin by providing some historical background on metastases and the metastatic progression. We then introduce a framework for metastases based on invasion ecology (Colautti et al., 2006; Keane and Crawley, 2002; Williamson, 1996). We shall then, in turn, discuss the ecology and evolution of the dispersers, their colonization, establishment, spread, and continued evolution within the novel organ and secondary tumors (Fidler, 2003). From this we shall draw the conclusion that it is unlikely that successful metastatic cancer cells are a random subset of the cancer cells within the tumor—certain *strategies* should preadapt a cancer cell to metastasize (Amend et al., 2015). We shall then seek observations and conclusions regarding the pairings of primary sites and secondary sites—why do brain cancers (glioblastomas) almost never metastasize, yet skin cancers (melanomas) and breast cancers frequently metastasize to the brain? We shall see that there is an emergent pattern where tissues with high blood flow and low cell density tend to be recipients of metastases whereas cancers from organs with low blood flow to normal tissue density ratios seem to produce metastases. We shall conclude with how knowledge of the evolutionary ecology of metastases may inform prevention and therapy. Throughout we shall apply the evolutionary and ecological principles discussed in this introduction.

INTRODUCTION TO METASTASIS

Metastatic cancer is an invasive process in which cancer cells from the primary tumor travel long distances through the body and compete with the body's native cells for survival at a distant organ (Poste and Fidler, 1980). Continuing with the NCI's dictionary of terms: "In metastasis, cancer cells break away from the original (primary) tumor, travel through the blood or lymph system, and form a new tumor in other organs or tissues of the body. The new, metastatic tumor is the same type of cancer as the primary tumor. For example, if breast cancer spreads to the lung, the cancer cells in the lung are breast cancer cells, not lung cancer cells." To be complete, we note that some investigators propose that cells from a tumor can circulate and metastasize to another site within the same tumor (Norton, 2005). While an interesting subject, we will not pursue this topic further.

To understand a term in detail it is helpful to understand the etymology of the word. The word metastasis comes from the Greek root word *methistanai*, which may be translated to "change" which was then redefined in the late 16th century as a rhetorical term meaning a "rapid transition from one point to another." Only one century later, in the mid-1660s the term was first used to describe a "shift of disease from one part of the body to another" (Random House Dictionary, 2016). By the late 1800s it was further used specifically to describe an entire manuscript titled "The distribution of secondary growths in cancer of the breast" (Paget, 1889).

The Metastatic Cascade

Consistently, sites of metastatic cancer share a well-defined and finite number of patterns of invasion—often termed the "metastatic cascade." Here, a general outline is presented for the four well-established stages for metastasis based on the invasion process following the shedding of cancer cells from the primary tumor: (1) intravasation (invasion into the blood stream or lymphatics within the primary tumor), (2) circulation and evasion of immune system, (3) extravasation (withdrawal from the blood stream), and (4) establishment and angiogenesis (Paterlini-Brechot and Benali, 2007).

Fortunately, it appears that successful completion of the metastatic cascade is quite rare. In experimental systems (Chambers et al., 2002; Gupta and Massagué, 2006) in which large numbers of tumor cells are injected, only about 2% of circulating tumor cells (CTC) are still found alive at some foreign tissue site. Only about 1/10th of these form micrometastases and about 1/10th of those form macroscopic tumors. The rate of success for the metastatic cascade in humans is not known and may vary from one individual to another. However, observations that tumor cells can be found in the bone marrow of patients with even stage 1 primary tumors suggests that tumor cell arrival at a distant site is far more common than clinically significant colonization events (i.e., metastases formation).

Furthermore, the microenvironment of tumor cells may dramatically affect the growth of the primary tumor and the progression toward metastases (Kenny and Bissell, 2003). In disseminated cancers, the "seed and soil hypothesis" originally described by Stephen Paget in 1889 suggests that metastasis is analogous to the colonization of a new habitat (Fidler, 2003). Successful colonization depends on the conditions of the new environment including resource availability, competition from normal cells, and predation from the immune system. From an ecological and evolutionary perspective, a tumor can be viewed as a large population of heterogeneous cells competing for similar resources. Selection for dispersal and colonization likely depends on the heterogeneity of these tumor cells and a series of filters. As each cancer patient is a novel evolutionary event, tumor cells cannot anticipate or be selected to metastasize. Rather, metastasis must be an unintended consequence and emergent property of the Darwinian dynamics driving tumor cell evolution and diversification within the primary tumor.

Heterogeneity within a tumor produces a variety of different *strategies* among the cancer cells. But, who among these has the greatest capacity to metastasize? Perhaps they are the cancer cells that can survive under extreme conditions (Amend and Pienta, 2015). Or, cancer cells that are more mobile may have increased chances of encountering vasculature and moving into and eventually out of the bloodstream (Amend et al., 2015). In prostate and breast cancers, an epithelial to mesenchymal shift in cancer cell morphology (Barriere et al., 2015) may generate motility and a phenotype conducive for metastasizing to the bone or elsewhere (Shiraishi et al., 2014).

When a cancer cell establishes and develops as a tumor in the secondary site, it may dominate that organ and disrupt the normal physiological processes of that organ, which places the native cells of the organ at a greater risk for survival. As noted earlier, the general process of the metastatic cascade seems quite predictable and many primary tumors spread to other organs in a reasonably consistent pattern. For example, prostate cancers commonly metastasize to the bones and retroperitoneal lymph nodes. Other sites, such as the liver and brain are occasionally observed but generally quite rare. On the other hand, many specifics of the metastatic disease can be frustratingly unpredictable. For example, metastatic sites may remain quiescent for years or even decades before suddenly progressing. In some individuals, many metastatic sites are present and grow at more or less the same rate. In other patients, only one or a few metastatic sites emerge and in some the tumors that grow at each site are highly variable.

What are the Factors Determining/Defying the Success of a Metastatic Event?

Having described how the stages of metastasis are known to occur at a high level, we shall raise a number of questions related to what contributes to successful metastatic events. Cancer biologists increasingly appreciate that heterogeneous tumors are comprised of subpopulations of cancer cells with different traits, morphologies, mutations and survival strategies. These various tumor cell populations are likely the result of temporal and regional variations in environmental selection forces. Thus, each population may have variable capacities to migrate from the primary tumor and then survive and thrive at a distant location. There is potential, then, for cells within heterogeneous tumors to be described by at least three general adaptive *strategies* that influence their capacity to form metastases: (1) motile tumor cells that engage in short-distance dispersal/or get shoved into the blood system; (2) generalists capable of surviving in a myriad of diverse potential habitats including the lymphatic or vascular system as well as the secondary site in which it settles and/or; (3) highly proliferative *strategies* which lead to 100,000s if not millions of cancer cell shedding events that eventually, by sheer numbers, result in a successful metastasis.

Of course, as described earlier by the seed and soil hypothesis, apart from the cancer cells' *strategies*, the host and environment at the distant secondary site(s) are also defining factors in successful metastatic events. The soil may be viewed as a habitat that has properties which provide a spectrum of rich or sparse resources for arriving tumor cells. Such resources may include oxygen and glucose to enable active cellular functions, the presence or absence of an aggressive immune cell defense and a normal cell density which facilitates or compromises the availability of resources and the space to migrate or proliferate. The success of a metastatic event is currently controversial and is a prominent scientific question evaluated rigorously by numerous scientific groups around the world. In addition to the seed and soil hypothesis, researchers have suggested a role for the architecture of the circulatory system (Weiss et al., 1986), the extra-cellular matrix structure of the donor and recipient organs (Barney et al., 2016), and a phenomenon known as "tropism" where a recipient organ, such as the bone may have structures, chemical cues or signals that attract and/or promote the success of an arriving cancer cell (Hess et al., 2006).

INTRODUCTION TO INVASION ECOLOGY

Invasion ecology is a rapidly expanding ecological discipline, with the number of articles on the subject having grown exponentially since the early 1900s (Lockwood et al., 2009). An invasive species in ecology is an organism/species that is nonnative to an environment, but nevertheless capable of surviving and flourishing in that environment. If introduced, it is able to successfully colonize and expand into other sites sometimes far away from its initial establishment site (Lockwood et al., 2007). Not all invasion events are successful, much like the invasion events of a circulating tumor cell described previously. Akin to metastasis, the stages of ecological invasion are generally: (1) transport, (2) colonization, (3) establishment, and (4) landscape spread (Theoharides and Dukes, 2007). These stages are analogous to the steps in the metastatic cascade described earlier (Gatenby et al., 2009) (Fig. 11.1).

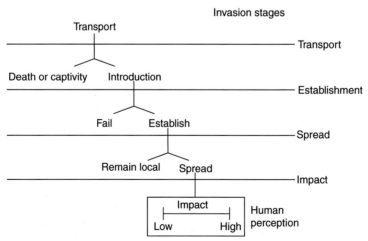

FIGURE 11.1 **The stages of ecological invasions.** *Source: Recreated from Lockwood, J.L., Martha, F.H., Michael, P.M., 2007. Invasion Ecology. Blackwell Publishing.*

The transport step requires species whose propagules can *survive* and be *transported* over potentially long distances to a new region. The *colonization* step contains abiotic filters that determine whether propagules survive in the new habitat, and affect growth rates (Theoharides and Dukes, 2007). The establishment step has several processes including *biotic resistance* from the native species, and the suppression of the noninvasive species reproductive rates. At the *landscape-spread* step, the expansion rates of invasive species depend on *establishment*, *dispersal ability*, and *habitat connectivity* (Theoharides and Dukes, 2007).

There are failures within each of these steps that serve as filters for the invasive species. During transport, the invasive species might simply die from starvation or depredation. During establishment, the invasive species might be outcompeted by the native species, may have limited access to nutrients, or may be immediately killed off by a predator. The result of any of these challenges is a failure to establish at the secondary site. During landscape spread, the invasive species may not have the opportunity to disperse effectively; therefore it remains in the local region. Often invasive species are able to succeed because of a vacancy within the new niche, or the absence of predators and competitors, providing for effective establishment. Furthermore, a change in the donor region can lead to a population increase of preexisting resident species, such that more individuals would be available to interface with a transport mechanism, such as ballast water (Rothlisberger et al., 2010), aircraft transport of fire ants to Florida (King et al., 2009), banana box transport of multiple spider species (Pospischil, 2014) or range expansion of local species into previously uninhabitable areas of the donor region, making these species available for transport (Carlton, 1996).

Likewise, changes in the recipient region can lead to altered ecological, biological, chemical, or physical states, thus increasing the susceptibility of the recipient region to invasion (Carlton, 1996). For example, altered water quality conditions lead to increased ability of pollution-intolerant or pollution-tolerant species to invade. The release of a very large number of invasive species to the recipient region increases the number of introduction events, thus increasing the probability of a successful colonization and establishment. Alternatively, the environment of the donor or recipient regions may not change and an invasion event could still occur if the invasive species can survive the transport step.

All sites that receive new invasions thus become new potential donor regions. For example, zebra mussels (*Dreissena polymorpha*) from Europe invaded the Laurentian Great Lakes in the 1980s. By hitchhiking on shipping moving from the Great Lakes, the zebra mussel has now spread throughout much of the Mississippi drainage.

An analysis of in-ballast vessel traffic patterns departing these regions could provide insight into where these species may next appear (Carlton, 1996). A similar analysis could be applied to the study of metastatic cancer in the human body. Following the vasculature and blood flow patterns departing the infected organs could provide insight into where the next establishment may occur, given that CTCs occur in the blood.

A striking contrast between invasive species and metastatic cancer concerns the damage done. Once the metastatic burden of the cancer becomes sufficiently large, the patient dies. Invasive species cannot completely destroy an ecosystem. They generally augment, replace or alter the abundances of the other species. Yet, the effect of an invasive species can achieve catastrophic proportions as seen in the spiny water flea (*Bythotrephes longimanus*), a zooplankton that like the zebra mussel invaded the Great Lakes via ships' ballast water. This large species of zooplankton feeds upon and eliminates native species, competes with some fish species and denies other parts of the fish community a food source (Barbiero and Rockwell, 2008).

METASTASIS AND INVASION ECOLOGY—A NOVEL FRAMEWORK FOR CLINICAL TREATMENT?

As described earlier, clear parallels exist between the invasion ecology at the organism level (metastatic cancers) and at the species/ecosystem level (invasive species). The congruence between the two biological phenomena invites the question: Can the principles and concepts from invasion ecology meaningfully contribute to a novel framework for understanding the metastatic cancer process and introduce new ideas for clinical treatment? Can we characterize the properties of tumor cells that may impede or facilitate migration and establishment in distant organs? And, are there discernable patterns and associations between the organs of the primary tumor (donor community) and the organs subject to metastasis (recipient community)?

The Invasive Geno-/Phenotype/Strategy

A neoplasm can be viewed from an evolutionary perspective as a large, morphologically, genetically and epigenetically heterogeneous population of individual cells (Lloyd et al., 2015; Merlo et al., 2006). Comparative analyses of cancer cell phenotypes, genotypes, and phylogenies could be utilized to explain the molecular changes occurring during tumor progression to distinguish which tumor cell types give rise to the development of invasive, metastatic and resistant cell *strategies*. As stated in the article by Gatenby et al. (2009), "invasive populations are more likely to be successful when there are multiple introductions and if the originating population is genetically and phenotypically diverse" (Lavergne and Molofsky, 2007; Suarez and Tsutsui, 2008). Populations that do invade may pass through a genetic bottleneck in which the population is initially relatively genetically homogeneous (Chakraborty and Nei, 1977). Following such a bottleneck, successful invasive species typically undergo rapid evolution (Nei et al., 1975). Natural selection may favor an increased diversity of phenotypes and subpopulations with adaptations to exploit the opportunities and avoid the hazards of their novel environments (Lee, 2002; Sakai et al., 2003). Thus, treatment of metastatic sites will more likely be successful early in the local evolutionary process when the tumor population is small and relatively homogeneous thus reducing the probability of an extant phenotype that is resistant to treatment. The relative success of adjuvant therapy (see also Chapters 10 and 20) in reducing the subsequent formation of clinically-evident metastases may be the result of these favorable dynamics.

The Necessity to Spread—the Ecological Drivers of Invasions

It is important to recognize that there are several general hypotheses in invasion ecology which may be used to elucidate the ecology of the metastatic process. Here, we briefly review four which were identified by the authors as relevant to oncology and metastasis. Perhaps the most poignant connection between ecological invasion concepts and distantly metastasizing cancer cells is the "enemy release" hypothesis (Colautti et al., 2004). As early as Thellung (1915), this concept is described as invasive success because of release from its natural enemies. It is plausible that cancer cells contend not only with the immune system but also with other cancer cells (Moreno-Smith et al., 2010). For this reason, successfully metastasizing to a distant site may permit a release from competition from other cancer cell populations and/or lowered exposure to any elevated immune response at the primary tumor. Paradoxically, many metastatic tumor cells leave a tumor region via the lymphatic system and are first successful in colonizing the lymph nodes, a structure whose sole purpose is the identification and eradication of foreign bodies and threats to the normal system.

Next, one might consider the propagule pressure hypothesis in which an increased number and frequency of invasion events (in cancer, metastatic events) increases the likelihood of success (Colautti et al., 2006; Lockwood et al., 2005). Namely, higher rates of release of CTCs from the primary tumor may increase the probability that a cancer cell will take hold and multiply in a distant organ.

The Tens Rule is similar in that approximately 10% of a species' population takes a given step in the four-step cascade of becoming an invasive species. Thus if 10% of a species are transported, 10% of those colonize, 10% of those become established, and 10% of those spread in the landscape (Jeschke et al., 2014; Lockwood et al., 2001), then at most 1 in 10,000 individuals will give rise to a successful invasion. In cancer one might suggest proposing at least one order of magnitude less at each stage, and metastatic cancer cells may not support the general linearity of this "rule of thumb." A 1% rule poses interesting estimates, tumors are often made up of billions of cells, if 100 million cells intravasated and 1 million could successfully circulate then 10,000 might extravasate and 100 might establish. This could be reasonable. This is purely conjecture but a 1% rule provides an interesting thought-exercise and hypothesis to investigate further.

Finally, the last hypothesis for invasion ecology that we will discuss is the Biotic Resistance hypothesis (Rius and Darling, 2014). Here, the greater the biodiversity of the recipient ecosystem, the greater its ability to resist invasion. This is a well-studied hypothesis in ecology, yet it is unclear that this hypothesis readily applies to cancer. Yet, perhaps this is why metastasis is such a difficult (yet deadly) feat for cancer cells to achieve. There is at least some evidence that one thousand to one million CTCs are present in about 20% of all breast cancer patients, yet very, very few actually succeed (Rack et al., 2014). Perhaps this is due in part to the body's diverse cell types. This is an area in which a great deal more investigation is merited.

In summary, the success of an individual tumor cell resulting in a metastatic event is dependent on the primary tumor type and is generally poorly understood (Merlo et al., 2006). The success of invasive species, such as tumor cells, heavily depends on a series of filters. The important filters that accompany the stages of metastases may be dispersal-limitation, recipient-limitation, organ-specific signaling, number of circulatory rounds before arrest, or nutrient-limitation (Fidler, 2003).

WHAT IS SPECIAL ABOUT THE INVASION ECOLOGY OF CANCER CELLS?

Size Range and Dispersal Dimensions

While metastatic progression is likely a complex interplay of deterministic and random processes, it has some features that make it quite distinct from and in some ways simpler than the broader field of invasion ecology. In nature the donor community may have plants, invertebrates, birds, reptiles, mammals, algae, protists, and bacteria. The size range of potential invasive species can be up to 10 orders of magnitude (Simberloff, 2013;

Woodward and Quinn, 2011). North America has seen the invasion of the Dutch elm disease (a fungus, *Ophiostoma* sp.) and west Nile virus (*Flavivirus*), one decimating the native American elm tree (*Ulmus americana*), the other numerous groups of birds, such as crows (*Corvus brachyrhynchos*). Some of the most ecologically damaging invasive species include the brown tree snake (*Boiga irregularis*), originally from Australia and New Guinea, that has spread to oceanic islands. Within a few years of WWII, it arrived as a stowaway (either ship or airplane wheel well) on the island of Guam where it has decimated the native bird populations. Kudzu (*Pueraria* sp.) is an invasive plant. In the southeastern United States, the vines of this plant can carpet and smother whole stands of trees. Known as the "mile-a-minute" vine because of its ability to grow up to 30 cm per day, kudzu was introduced from Japan with the thought that it might help control soil erosion. The black rat (*Rattus rattus*) has traveled with humans as a stowaway on ships, wagons and baggage trains for centuries. The Asian tiger mosquito (*Aedes albopictus*), a carrier of dengue fever and west Nile virus, has spread from Southeast Asia. Water buildup in automobile tires stacked outdoors and then shipped may have provided entry to over 28 nonnative countries. Cane toads (*Rhinella marina*) (native to South and Central America) and the European rabbit (*Oryctolagus cuniculus*) have spread widely due to intentional introductions either as a form of pest control (cane toads) or for nostalgia and food (European rabbit).

These examples of biological invasions illustrate three features that make metastatic cancer a narrower subset of invasion ecology. First, all of the cancer cells and the normal cells are relatively of the same size and characteristics. This is unlike invasions of macro environments, e.g. the ecosystems of Florida in the United States that have seen invasive plants, reptiles, fish, insects, amphibians, birds, and mammals. Second, while biological invasions occur by very diverse pathways, metastases result from dispersal through the lymph or blood systems—both of which are relatively two dimensional pathways. Third, the donor and recipient communities offer diverse food webs with multiple trophic levels. An invasive species may find itself in a novel environment offering completely new prey species. Furthermore, the invasive species may be free of its traditional predator species or find itself encountering completely new predators. Not so for metastasizing cancer cells. The resources they require and the suite of normal cells they encounter in both the donor and recipient organs will be relatively the same. While the structure of the different organs, say the bone, lung, liver or colon may be somewhat different, the normal cells of these different organs are genetically the same. Differences only arise between normal cells via the cell and organ differentiation that occurred during the ontogeny of the person or multicellular organism. In terms of predation

risk from the immune system, the immune system differs only by degree across the different organs of the body. A cancer cell is likely to encounter very similar immune threats across the organs of an individual's body. While organ-to-organ differences in immune response may be important and sometimes critical to the metastatic process, it offers far fewer complexities and differences than those faced by biological invasions in nature.

Primary Succession and Conquering Over Natives

Biological invasions, colonizations or long distance dispersals generally involve two sorts (Simberloff, 2013). The first, termed primary succession, involves the colonization of otherwise novel and sterile habitats. A volcanic explosion in 1883 left the islands of Krakatau in present day Indonesia essentially devoid of terrestrial life. A spider was the only thing found in May 1884, yet by October various grasses, mosses and lichens were evident. Such founder populations came from seeds and spores brought by sea, wind and birds. At other times colonization occurs in communities already occupied by similar or quite different species. Canada thistle (*Cirsium arvense*) is quite a misnomer. Native to Europe, its spread throughout much of North America began as early as the 1600s. Its seeds likely occurred as "impurities" in crop seeds brought to the New World. It spread faster than the Europeans that brought it. Despite competition, it has been able to successfully establish in communities already occupied by native thistles of the same genus, even going so far as to exclude some, such as the Dune thistle (*Cirsium pitcher*), from their own indigenous communities.

Metastasizing cancer cells appear to exist in a twilight between these extremes of biological invasions. When a cancer cell establishes in a distant organ it is sterile in the sense that it is unoccupied by any other cancer cells. The cancer cell experiences no competition from other eco-evolutionarily dynamic cancer cell subpopulations. Yet, the new organ is not devoid of "life" in the sense of Krakatau, there is competition, predation and opportunities from the normal cells. Unlike the thistles, the normal cells are not units of natural selection—they do not engage in a struggle for existence nor evolve adaptations in the same manner as the cancer cells. The normal cells are part of not separate from the whole organism program of homeostasis. But like the thistle example, the normal cells are remarkably similar to the cancer cells in their overall structure and metabolic functions. They also represent dynamically active subpopulations of cells that die and proliferate. As an ecological phenomenon, the founding cancer cell or small aggregation of cancer cells enters a space within an organ free of other cancer cells, but still full of normal cells that operate at a similar scale and in a similar manner to the newly arrived cancer cells.

Obligate Invaders and Bet-Hedging Strategies

Some biological invasions are literally "intentional." Species must and do evolve adaptations for dispersal. Dispersal provides a means of reducing competition among offspring. Some plants, such as those in the genus *Heterotheca* produce two types of seeds, one falls near the mother and another type that disperses far away (Gibson and Tomlinson, 2002). Dispersal provides a means of avoiding overcrowding or lack of resources. Various species of spiny lobsters (*Palinuridae*) are famous for lengthy sojourns to exit overcrowded rocky areas in search of less crowded habitats (Phillips and Kittaka, 2008). Presumably for safety from predators, such as triggerfish (*Balistidae*), dispersing lobsters travel as a group in single file. One such line was seen to be 50 m long. Finally, long range dispersal may evolve as a bet-hedging *strategy* in response to the catastrophic collapses of widely spaced subpopulations (e.g. Childs et al., 2010; Starrfelt and Kokko, 2012). Disease, predation or habitat destruction might result in the local extinction of a population. While the individual prospects of a long range disperser may be lower than an individual that stays closer to home, occasionally the far ranging disperser may find an empty or near empty patch of suitable habitat thus reestablishing an extirpated population. However, for such long-range dispersal to evolve, there must be an ongoing and repeatable process of local extinction followed by recolonization. Over ecological and evolutionary time, the source population must at some point become the recipient population and vice-versa.

For cancer cells it is easy to imagine the evolution of short-range dispersal, perhaps over distances of microns to millimeters. Indeed, many tumor populations appear to expand by invasion into and proliferation within adjacent normal tissue. We predict that the ecological dynamics at the tumor–host interface may well represent tumor cell *strategies* to optimize local dispersal in a hostile environment. Reducing competition between daughter cells of a clade, avoiding overcrowding, and seeking better opportunities elsewhere could all represent evolutionary pressures that promote this *strategy*. All of these selection forces may also increase the likelihood of a successful metastasis and propel the metastatic process by enabling local invasion of blood vessels and lymphatics. *But, cancer cells cannot evolve an adaptation whose specific function is to metastasize.* This is because successful metastases are rare, near one off events. The cells of the primary tumor have no collective ecological or evolutionary experience with the distant organ. Natural selection can never select for an adaption to a circumstance that has not yet happened.

Theories and ideas regarding the evolution of a metastatic phenotype or cancer cell clades driven by evolution to metastasize are perhaps fanciful, unless seen as *strategies* or evolutionary trajectories that preadapt cancer cells for metastasis (Amend et al., 2015). What of theories and research that postulate and identify the invasive phenotype? Cancer cells should evolve adaptations for invading the normal tissue surrounding a growing tumor—a simple means for avoiding over-crowding. The traits needed to expand into normal tissue may increase the chances for metastasizing into similar circumstances of distant organs.

The Metastatic Process as a Matter of Weighted Probabilities

Not all, but many of the species that have the capacity for long-range dispersal and colonization possess a distinct life history stage known as the dispersal morph. For plants, it is generally the seed that has the great dispersal capacity, either through wind, animal dispersal or buoyancy. For many flying insects and birds the adults provide the capacity for long distance travel, migration and dispersal. For tropical reef fish and marine bivalves, a larval stage provides the opportunity for long-range dispersal via ocean currents. The vast majority of these dispersers likely die while a sufficiently large number settle on suitable reefs or substrates. In the case of seeds and larvae, success is insured by the sheer number of dispersers. For species with adult dispersal, it is usually the capacity of the individual to manage and direct its dispersal that permits success. Formation of a dispersal morph may well explain some properties that are available within the reaction norm of cancer cells. For example, epithelial to mesenchymal transformation is a well-described process and the resulting phenotype has properties that likely foster local invasion. Similarly, "stemness" is a tumor property that may be acquired and seems to be a bet hedging *strategy* that permits population regrowth in the event of a catastrophic, destructive event.

The metastatic process then may be a matter of weighted probabilities—weighted by the preadaptations of the cancer cells, weighted by the characteristics of the primary tumor, and weighted by the recipient organs. A single cancer cell circulating within the blood stream likely has a negligible chance of metastasizing. But, a primary tumor of billions of cells may generate a steady flow of such CTCs, many that may possess preadaptations permitting at least a chance at metastasizing. At any given moment metastatic progression may be small, but over months and years the probability must rise until at some point it becomes very high or a near certainty within some organs while remaining vanishingly small in others. As an invasion process and as an ecological process the metastatic process must be understood as a series of filters.

FILTERING CONCEPTS AS LIMITS TO METASTATIC SUCCESS

The success of invasion depends heavily upon a series of filters. Ecologically, metastases, particularly in traveling from the organ of the primary tumor to another organ, represent long distance dispersal events. There are at least three ways in which dispersal can be important in cancer: the movement of cells between the partially isolated subpopulations of proliferative units, local invasion of neighboring tissues, and emigration of metastatic cells from the primary tumor. Whether dispersal distance is declared to be short or long will largely depend on the system and particular colonizer under scrutiny (Showalter et al., 2008). Nonetheless, short-distance dispersal is considered to be primarily between adjacent, or nearly adjacent, environments. On the other hand, long-distance dispersal can be viewed as movement typically between widely distant environments, usually separated by a barrier of some sort, a process termed saltation dispersal or punctuated dispersal (Showalter et al., 2008). Note that subsequent short-distance dispersal originating from within the newly colonized environment may often follow an initial colonization episode precipitated by punctuated dispersal.

Recruitment of Long Range Dispersers—the Seeds of Metastasis

The first filter toward a successful metastasis requires invasion into blood vessel or lymphatics and survival in the primary tumor to allow viable CTCs. Selection forces, thus, promote the ability of certain tumor cell types to intravasate the blood vasculature and survive in circulation. Thomas Ashworth (1869) was the first to note the presence of cancer cells within the blood and he correctly associated these cells with causing metastases. Since then, and particularly in the last 10 years, great strides have been made in detecting, isolating and measuring the characteristics of cancer cells found in the blood. But, important questions remain. Just how many CTCs are there? How many times on average can a CTC make a complete round and pass through the heart and pulmonary system? How is the number of CTCs influenced by tumor volume, density and shape? How is the number of CTCs influenced by the frequency and diversity of adaptations found among the different cancer cell types within the primary tumor? Much of this is poorly understood or known only inexactly. But, these are the important issues in determining the available pool of long distance migrants.

In natural systems there are the adaptations that encourage and facilitate long distance migration. However, with cancer cells, none can have been specifically selected to become a CTC because, by definition, these cells are lost to the population so that "CTC traits" cannot be expanded through proliferation. Thus, each CTC is an "accidental" disperser. Nature is full of examples in which individuals of species become accidental and incidental dispersers. Storms, floods, and falling vegetation create rafts of logs, branches, and plant material that float down rivers. Particularly in great rivers, such as the Amazon and Congo, these rafts can find themselves moving downstream 1000 km or more, and even well into the ocean. Such mats from the Amazon, can harbor a diverse array of insects, spiders, snails, reptiles, amphibians, and even mammals. Storms can send birds hundreds of kilometers from their intended path or habitat. To the delight of birdwatchers such birds become "accidentals." A tropical trogon (Trogonidae) from Mexico arrives in the southwestern United States, or a bird belonging to the Atlantic seaboard of North America arrives in Europe. Until fumigation and other quarantine measures, crates of bananas from Central America would include all matter of insects and even snakes. Each of these individuals unintentionally and quite by accident becomes a long range disperser. But, some characteristics of these species and their habits may increase the likelihood that they find themselves traveling far from home and across the globe or seas.

So it must be with cancer cells. Where are the most likely entry points in the vasculature? The edge of the tumor offers the normal array of blood vessels and capillary beds. Any cancer cell entering such vessels can likely become swept away into major portions of the circulatory system. The flow properties of vessels close to but external to the tumor would offer less variance in blood flow and a greater pressure gradient between the interior of the vessel and the surrounding tissue. The other entry point may lie deep in the tumor in association with the vessels produced via angiogenesis. Such vessels are known to be poorly structured, more "swamp-like" in their organization, and highly leaky (Amend and Pienta, 2015). Tumors are known for high interstitial pressures that would reduce or even cease the pressure gradient between the blood and the adjacent tissue. Finally, angiogenic blood flow is likely prone to sudden rises and falls of pressure—almost flood-like properties. Hence, vessels around and in the tumor likely offer different avenues for cancer cells to enter and become swept away.

The properties of the cancer cells themselves likely influence the chances for becoming a CTC. Singlets or small clumps of cells may simply become sheered away from "floods" of blood flow or pulses of pressure. Furthermore, cancer cell densities should be higher closer to blood sources and buildups of cancer cells from bouts of proliferation may literally push cancer cells into the bloodstream. The properties of such cancer cells will simply be those associated with close proximity to blood flow, such as estrogen positivity (Lloyd et al., 2015), high expression of carbonic anhydrase 12 (CAXII) (Lloyd et al., 2016), and cell metabolisms associated with high availability of glucose and oxygen, such as low carbonic anhydrase 9 (CA9/CAIX) and lower glucose transporter (GLUT1). If CTCs are more likely to arise from cells on the margin of the tumor, then such cells will exhibit different properties than those of the interior. In breast cancer, the edge cells exhibit high CAIX, likely high glycolysis, and high expressions of GLUT1 and KI67 (see also Chapter 4).

Besides proximity to the blood vessels, another property of cancer cells that might favor intravasation would be higher motility. Such motility both within and on the edge of the tumor might increase the chances of slipping into the blood system. Motile cells might intentionally move toward vessels as they move along gradients of increased glucose, glutamine and/or oxygen (Amend et al., 2015). Two extreme means for creating CTCs include tumor cells that may intentionally reside within or attached to the walls of blood vessels, or those that use blood vessels as an easy means of short-distance dispersal. Both represent *strategies* for increased resource acquisition, both may carry adaptations for creating gaps in the cells of the blood vessel for movement into and out of the vessels, and both *strategies* could be easily swept away to become CTCs.

The presence of CTCs appears to occur early with the onset of cancer and visible tumors (Fidler, 2003). Those with metastatic cancer seem to have roughly 10^3–10^4 live cancer cells per liter of blood (Miller et al., 2010)—note that the average adult has 4.5–5.5 L of blood total. CTCs seem to exhibit much if not all of the heterogeneity found in the primary tumor as well as any secondary tumors (Marrinucci et al., 2009). Finally, in breast and prostate cancer > 5 CTCs per 7.5 mL and in colon cancer >3 CTCs per 7.5 mL blood are associated with unfavorable prognoses (Miller et al., 2010). What is not known is how long CTCs can remain viable in the blood (one complete round of circulation takes about a minute), and how many exit the tumor per unit time either in total or as a fraction of the live cancer cells of the tumor. These CTCs are the first step of the metastatic process.

Establishment—The Soil of Metastasis

The next filter requires that live cancer cells in the blood exit the blood and successfully establish a colony of cancer cells in a distant location and/or organ. To metastasize, a CTC or aggregation of cancer cells must extravasate from the vasculature into a new secondary site and survive within the secondary site. As described earlier, the probability of successful establishment likely relies on the preadaptations of the particular CTC, the characteristics of the new organ in general, and equally important the specifics of the site within the organ. Hence, metastatic progression relies on a "special" cell leaving the circulatory system into a "special" site of an appropriate organ. This requires a lot of happenstance.

In invasion ecology many establishment events are propagule limited—that means, limited by the number of dispersers that arrive at the distant site. The establishment of primates into South America from Africa some 40 million years ago is an example of likely propagule limitation (e.g. Bond et al., 2015). (Despite the fact that the two continents may have been 1000 km closer with a

chain islands providing stepping-stones along the mid-Atlantic ridge.) Primates accidentally finding themselves on vegetation rafts floating into the Atlantic would have been an occasional event. But, survival for more than a few days would have been unlikely. Even rarer would have been the need for either a pregnant female (with uncanny luck in offspring gender!) or a small group to arrive near simultaneously. But it did happen, and millions of years later this single invasion event resulted in an adaptive radiation of abundant, widespread and diverse primate species that we see in South and Central America to this day.

Unlike the primate example, metastases are not propagule limited. The continuous, well-mixed, and large population of CTCs insures that nearly all corners of the human body likely see the arrival of living tumor cells. Like the New World primates, each metastasis seems to be the result of a single or very small number of founders. Metastasis may become the near certain outcome of many attempts each of which has a near zero chance of success.

For success, the founding cell or clump of cells require safety from the immune system, resources, and a favorable and perhaps malleable community of normal cells. If they exist, cancer cells that live in or move through the local vasculature might have an advantage to emerge from the vasculature into a distant site. Cells that occupy the margins of tumors, sometimes referred to as the invasive phenotype, may have the right mix of traits to metastasize. They may already have strong anti-immune adaptations, such as being cryptic and/or glycolytically producing a low pH moat of protection, these may have upregulated CAIX and exaggerated Warburg effects (discussed in detail in Chapter 4). Coming from the margins of tumors, such cells may have adaptations for living among normal cells without the positive aggregation effects that accrue from neighborhoods of cancer cells promoting angiogenesis and other aspects of tumor architecture. For these reasons, cancer cells strongly adapted for the highly competitive, relatively safe, and highly engineered conditions of tumor interiors may be poor to nil candidates for metastases. We suggest that the "seeds" are a special subset of the diverse subpopulations of cancer cells that populate advanced tumors.

What of the soil? In oncology the metastatic site is quite organ-specific with its own limitations. It is known that organs of the body, such as the brain and bone receive the most invasive cancer cells while other organs including the lung, breast and gastrointestinal and genitourinary organs shed/disperse the most cancer cells (Scott and Kuhn, 2012). In ecology, the same patterns are observed. Certain types of habitats seem to have higher numbers of established nonindigenous species than others. For example, islands are more vulnerable to invasion because they usually have fewer resident species leading to the conjecture that simpler systems have less biotic resistance to invaders (Brown, 1989; Elton, 1958). Larger islands can support more species and have lower extinction rates than small ones because they cover larger areas, with a greater diversity of resources and habitats (Robinson and Dickerson, 1987). Less isolated islands tend to support more species than remote ones, because they have higher rates of immigration. The same ideas can be reflected on the organs of the body. For example, the brain could be considered a large isolated island that is seen to receive many invasive cancer cells, yet seldom releases tumor cells. Tumor size likely contributes to heterogeneity and hence the diversity of tumor cells. Testicular cancer produces small, homogeneous tumors whereas colon and breast cancer can result in large, heterogeneous primary tumors.

There are consistent examples of metastatic patterns from one primary tumor to one or many secondary sites (Budczies et al., 2015; Chen et al., 2009). For example, metastatic breast cancer accounts for 20–30% of the 170,000 cases of brain metastases diagnosed annually, and as improvements in systemic therapy prolong survival, brain metastasis in breast cancer patients is becoming more evident (Vallow, 2007). Estimates suggest that 35% of lung cancers, 10–30% of breast cancers, 30–40% of melanomas, 5–10% of kidney cancers, and 5% of colon cancers metastasize to the brain (Table 11.1).

We predict that the reason such organs are the main sites of metastases is because their normal tissue density is medium to low and the amount of nutrients available is very high (high vasculature to normal cell density ratio) (Guyton and Hall, 2000). This may create favorable conditions for a metastatic tumor cell. Also, because these areas are generally highly vascularized, the cell density per vessel is not very high, thereby allowing vacant niches for the tumor cells with abundant access to space and nutrients. These dynamics may be evident in the metastatic disease of the liver. As noted earlier, the normal liver is a common site for metastases from many disparate organs while primary cancer in normal liver is quite rare. In contrast, when the liver becomes cirrhotic, it contains dense scarring and diminished perfusion due to portal venous hypertension. A liver in this physiological state commonly forms primary tumors but is rarely the site of metastatic disease. Donor organs are considered more of the "desert" type of organs where the normal cell density is very high and very limited nutrients are available, encouraging the tumor cells to evolve preadaptations for dispersal to nutrient rich organs. The high cell density allows more tumor cells to be sloughed off, if near a blood vessel or angiogenic vessel. Blood flow within regions of tumors typically results in temporal and spatial variations in concentrations of growth factors, substrate, and metabolite concentrations (Lloyd et al., 2014). Each of these variations represents selection forces that could drive the adaptive evolution of local tumor populations.

TABLE 11.1 Major Site of Metastasis as Recreated From National Cancer Institute: Metastatic Cancer Fact Sheet

Cancer type	Main sites of metastasis
Bladder	Bone, liver, lung
Breast	Bone, brain, liver, lung
Colorectal	Liver, lung, peritoneum
Kidney	Adrenal gland, bone, brain, liver, lung
Lung	Adrenal gland, bone, brain, liver, other lung
Melanoma	Bone, brain, liver, lung, skin/muscle
Ovary	Liver, lung, peritoneum
Pancreas	Liver, lung, peritoneum
Prostate	Adrenal gland, bone, liver, lung
Stomach	Liver, lung, peritoneum
Thyroid	Bone, liver, lung
Uterus	Bone, liver, lung, peritoneum, vagina

SUBSEQUENT EVOLUTION AND THERAPY

Metastatic tumor sites ultimately are the result of a successful adaptation of the tumor to the unique physiological state of the invaded tissue. As noted earlier, this often requires the tumor cells to deploy new *strategies* that optimally interact with the endothelial cells, fibroblasts, immune cells, and intercellular signaling pathways in the new tissue site. It has been suggested (Cunningham et al., 2015) that, in fact, cells from diverse primary organs may begin to converge on common *strategies* (phenotypic traits) that are optimally suited to the environment. Thus, for example, tumors from lung cancer, breast cancer, and colon cancer that grow at metastatic sites may over time diverge from the tumor cells that remain at the primary site and converge on a common "liver morph" *strategy*. This may have clinical significance because treatment for metastatic disease is currently determined by the site of the primary tumor. If correct, these confirmed evolutionary dynamics suggest that optimally successful treatment strategies may require classification that includes both the primary site and the site(s) of metastatic disease.

CONCLUSIONS

There are many analogies between metastatic spread of cancer and the spread of exotic species in nature. There are, however, important differences which should be clearly recognized. Perhaps this is most evident in the ecological and evolutionary dynamics of the recipient tissues. In colonization events in nature, the introduced species is both the product of evolution and must successfully compete with the foreign ecosystem. However, normal cells and normal tissues are not ordinarily subject to evolution by natural selection. Normal cells can change over time and normal tissues can be disrupted due to external perturbation but the temporal variations do not fulfill the criteria for evolution described earlier. Thus, tumor cells must adapt to the new environment but do not compete with other evolving species as is the case with species introductions.

That said, cancer cells do evolve and compete with each other and, thus, understanding cancer biology must begin with understanding the population dynamics and evolutionary parameters that govern neoplastic processes. Invasion ecology can be used to understand biological invasions by applying niche concepts, comparing the two processes to one another, and evaluating species heterogeneity. These ideas lead to the study of tumor cell heterogeneity and the types of communities (organs) that they invade successfully. Thus, investigating cancer metastasis through the lens of invasion ecology helps us grasp the reasons behind why certain dispersal and establishment processes occur successfully. Such knowledge may help in taming the ultimate killers, the malignant cancer cells that disseminate from the primary tumor, invade distant organs, form metastases, and result in lethal tumor burdens.

References

Amend, S.R., Pienta, K.J., 2015. Ecology meets cancer biology: the cancer swamp promotes the lethal cancer phenotype. Oncotarget 6 (12), 9669.

Amend, S.R., Roy, S., Brown, J.S., Pienta, K.J., 2015. Ecological paradigms to understand the dynamics of metastasis. Cancer Lett. 380 (1), 237–242.

Ashworth, T.R., 1869. A case of cancer in which cells similar to those in the tumours were seen in the blood after death. Aust. Med. J. 14, 146–147.

Barbiero, R.P., Rockwell, D.C., 2008. Changes in the crustacean communities of the central basin of Lake Erie during the first full year of the Bythotrephes longimanus invasion. J. Great Lakes Res. 34 (1), 109–121.

Barney, L.E., Jansen, L.E., Polio, S.R., Galarza, S., Lynch, M.E., Peyton, S.R., 2016. The predictive link between matrix and metastasis. Curr. Opin. Chem. Eng. 11, 85–93.

Barriere, G., Fici, P., Gallerani, G., Fabbri, F., Rigaud, M., 2015. Epithelial mesenchymal transition: a double-edged sword. Clin. Transl. Med. 4 (1), 1.

Bond, M., Tejedor, M.F., Campbell, Jr., K.E., Chornogubsky, L., Novo, N., Goin, F., 2015. Eocene primates of South America and the African origins of New World monkeys. Nature. 520 (7548), 538–541.

Brown, J.H., 1989. Patterns, Modes and Extents of Invasions by VertebratesJohn Wiley & Sons, Chichester.

Budczies, J., Winterfeld, M.V., Klauschen, F., Bockmayr, M., Lennerz, J.K., Denkert, C., Wolf, T., et al., 2015. The landscape of metastatic progression patterns across major human cancers. Oncotarget 6 (1), 570–583.

Carlton, J., 1996. Pattern, process, and prediction in marine invasion ecology. Biol. Conserv. 78 (1–2), 97–106.

Chaffer, C.L., Weinberg, R.A., 2011. A perspective on cancer cell metastasis. Science 331 (6024), 1559–1564.

Chakraborty, R., Nei, M., 1977. Bottleneck effects on average heterozygosity and genetic distance with the stepwise mutation model. Evolution, 347–356.

Chambers, A.F., Groom, A.C., MacDonald, I.C., 2002. Metastasis: dissemination and growth of cancer cells in metastatic sites. Nat. Rev. Cancer 2 (8), 563–572.

Chen, K.W., Pienta, K.J., 2011. Modeling invasion of metastasizing cancer cells to bone marrow utilizing ecological principles. Theor. Biol. Med. Model. 8 (1), 1.

Chen, L.L., Blumm, N., Christakis, N.A., Barabasi, A.L., Deisboeck, T.S., 2009. Cancer metastasis networks and the prediction of progression patterns. Br. J. Cancer 101 (5), 749–758.

Childs, D.Z., Metcalf, C.J.E., Rees, M., 2010. Evolutionary bet-hedging in the real world: empirical evidence and challenges revealed by plants. Proc. R. Soc. Lond. [Biol]. 277 (1697), 3055–3064.

Colautti, R.I., et al., 2004. Is invasion success explained by the enemy release hypothesis? Ecol. Lett. 7 (8), 721–733.

Colautti, R.I., Grigorovich, I.A., MacIsaac, H.J., 2006. Propagule pressure: a null model for biological invasions. Biol. Invasions 8 (5), 1023–1037.

Cunningham, J.J., Brown, J.S., Vincent, T.L., Gatenby, R.A., 2015. Divergent and convergent evolution in metastases suggest treatment strategies based on specific metastatic sites. Evol. Med. Public Health 2015, 76–87.

Eigen, M., 1971. Self organization of matter and the evolution of biological macromolecules. Naturwissenschaften 58 (10), 465–523.

Elton, C.S., 1958. The Ecology of Invasions by Plants and AnimalsMethuen, London.

Fidler, I.J., 2003. The pathogenesis of cancer metastasis: the 'seed and soil' hypothesis revisited. Nat. Rev. Cancer 3 (6), 453–458.

Galhardo, R.S., Hastings, P.J., Rosenberg, S.M., 2007. Mutation as a stress response and the regulation of evolvability. Crit. Rev. Biochem. Mol. Biol. 42 (5), 399–435.

Gatenby, R.A., Brown, J., Vincent, T., 2009. Lessons from applied ecology: cancer control using an evolutionary double bind. Cancer Res. 69 (19), 7499–7502.

Gibson, J.P., Tomlinson, A.D., 2002. Genetic diversity and mating system comparisons between ray and disc achene seed pools of the heterocarpic species Heterotheca subaxillaris (Asteraceae). Int. J. Plant. Sci. 163 (6), 1025–1034.

Greaves, M., Maley, C.C., 2012. Clonal evolution in cancer. Nature 481 (7381), 306–313.

Greenlee, R.T., Hill-Harmon, M.B., Murray, T., Thun, M., 2001. Cancer statistics, 2001. CA Cancer J. Clin. 51 (1), 15–36.

Gupta, G.P., Massagué, J., 2006. Cancer metastasis: building a framework. Cell 127 (4), 679–695.

Guyton, A.C., Hall, J.E., 2000. Local control of blood flow by the tissues; and humoral regulation. Medical Physiology. tenth ed. W.B. Saunders Company, Philadelphia, pp. 175–182.

Hess, K.R., Varadhachary, G.R., Taylor, S.H., Wei, W., Raber, M.N., Lenzi, R., Abbruzzese, J.L., 2006. Metastatic patterns in adenocarcinoma. Cancer 106 (7), 1624–1633.

Jeschke, J.M., et al., 2014. Defining the impact of non-native species. Conserv. Biol. 28 (5), 1188–1194.

Keane, Ryan M., Crawley, Michael J., 2002. Exotic plant invasions and the enemy release hypothesis. Trends Ecol. Evol. 17 (4), 164–170.

Kenny, Paraic A., Bissell, Mina J., 2003. Tumor reversion: correction of malignant behavior by microenvironmental cues. Int. J. Cancer 107 (5), 688–695.

King, J.R., Tschinkel, W.R., Ross, K.G., 2009. A case study of human exacerbation of the invasive species problem: transport and establishment of polygyne fire ants in Tallahassee, Florida, USA. Biol. Invasions 11 (2), 373–377.

Lavergne, S., Molofsky, J., 2007. From the cover: increased genetic variation and evolutionary potential drive the success of an invasive grass. Proc. Natl. Acad. Sci. 104 (10), 3883–3888.

Lee, C.E., 2002. Evolutionary genetics in invasive species. Trends Ecol. Evol. 17, 386–391.

Lloyd, M.C., et al., 2014. Vascular measurements correlate with estrogen receptor status. BMC Cancer 14 (1), 1.

Lloyd, M.C., et al., 2015. Pathology to enhance precision medicine in oncology: lessons from landscape ecology. Adv. Anat. Pathol. 22 (4), 267–272.

Lloyd, M.C., Cunningham, J.J., Bui, M.M., Gillies, R.J., Brown, J.S., Gatenby, R.A., 2016. Darwinian Dynamics of Intratumoral Heterogeneity: Not Solely Random Mutations but Also Variable Environmental Selection Forces. Cancer Res. 76 (11), 3136–3144.

Lockwood, J.L., et al., 2001. How many, and which, plants will invade natural areas? Biol. Invasions 3 (1), 1–8.

Lockwood, J.L., Cassey, P., Blackburn, T., 2005. The role of propagule pressure in explaining species invasions. Trends Ecol. Evol. 20 (5), 223–228.

Lockwood, J.L., Hoopes, M.F., Marchetti, M.P., 2007. Invasion Ecology. Blackwell Publishing, Malden, MA.

Lockwood, J.L., Cassey, P., Blackburn, T.M., 2009. The more you introduce the more you get: the role of colonization pressure and propagule pressure in invasion ecology. Divers. Distrib. 15 (5), 904–910.

Marrinucci, D., Bethel, K., Luttgen, M., Bruce, R.H., Nieva, J., Kuhn, P., 2009. Circulating tumor cells from well-differentiated lung adenocarcinoma retain cytomorphologic features of primary tumor type. Arch. Pathol. Lab. Med. 133 (9), 1468–1471.

Merlo, L.M.F., Pepper, J.W., Reid, B.J., Maley, C.C., 2006. Cancer as an evolutionary and ecological process. Nat. Rev. Cancer 6 (12), 924–935.

Miller, M.C., Doyle, G.V., Terstappen, L.W.M.M., 2010. Significance of circulating tumor cells detected by the cellsearch system in patients with metastatic breast colorectal and prostate cancer. J. Oncol. 2010, 617421.

Moreno-Smith, M., Lutgendorf, S.K., Sood, A.K., 2010. Impact of stress on cancer metastasis. Future Oncol. 6 (12), 1863–1881.

Nei, M., Maruyama, T., Chakraborty, R., 1975. The bottleneck effect and genetic variability in populations. Evolution 29, 1–10.

Norton, L., 2005. Conceptual and practical implications of breast tissue geometry: toward a more effective, less toxic therapy. Oncologist 10 (6), 370–381.

Paget, S., 1889. The distribution of secondary growths in cancer of the breast. Lancet 133 (3421), 571–573.

Paterlini-Brechot, P., Benali, N., 2007. Circulating tumor cells (CTC) detection: clinical impact and future directions. Cancer Lett. 253 (2), 180–204.

Phillips, B.F., Kittaka, J., 2008. Spiny Lobsters: Fisheries and Culture, second ed. John Wiley & Sons, Hoboken, 704 p.

Pospischil, R., 2014. Invading species: a challenge for pest management, Proceedings of the Eighth International Conference on Urban Pests, 20–23 July 2014, Zurich, Switzerland, pp. 303–308.

Poste, G., Fidler, I.J., 1980. The pathogenesis of cancer metastasis. Nature 283 (5743), 139–146.

Rack, B., et al., 2014. Circulating tumor cells predict survival in early average-to-high risk breast cancer patients. J. Natl. Cancer Inst. 106 (5), dju066.

Random House Dictionary, 2016. Random House Webster's Unabridged Dictionary. Random House Reference, New York.

Rius, M., Darling, J.A., 2014. How important is intraspecific genetic admixture to the success of colonising populations? Trends Ecol. Evol. 29 (4), 233–242.

Robinson, J.F., Dickerson, J.E., 1987. Does invasion sequence affect community structure? Ecology 68 (3), 587–595.

Rosenberg, S.M., 2001. Evolving responsively: adaptive mutation. Nat. Rev. Genet. 2 (7), 504–515.

Rosenberg, S.M., Queitsch, C., 2014. Combating evolution to fight disease. Science 343 (6175), 1088–1089.

Rothlisberger, J.D., et al., 2010. Aquatic invasive species transport via trailered boats: what is being moved, who is moving it, and what can be done. Fisheries 35 (3), 121–132.

Sakai, A.K., Allendorf, F.W., Holt, J.S., et al., 2003. The population biology of invasive species. Annu. Rev. Ecol. Syst. 32, 305–332.

Scott, J., Kuhn, P., Anderson, A.R.A., 2012. Unifying metastasis—integrating intravasation, circulation and end-organ colonization. Nat. Rev. Cancer 12 (7), 445–446.

Shiraishi, T., Verdone, J.E., Huang, J., Kahlert, U.D., Hernandez, J.R., Torga, G., Zarif, J.C., et al., 2014. Glycolysis is the primary bioenergetic pathway for cell motility and cytoskeletal remodeling in human prostate and breast cancer cells. Oncotarget 6 (1), 130–143.

Showalter, S., Hager, E., Yeo, C., 2008. Metastatic disease to the pancreas and spleen. Semin. Oncol. 35 (2), 160–171.

Siegel, R.L., Miller, K.D., Jemal, A., 2015. Cancer statistics, 2015. CA Cancer J. Clin. 65 (1), 5–29.

Simberloff, D., 2013. Invasive species: what everyone needs to know. Oxford University Press, Oxford, 329 p.

Starrfelt, J., Kokko, H., 2012. Bet-hedging—a triple trade-off between means, variances and correlations. Biol. Rev. Camb. Philos. Soc. 87, 742–755.

Suarez, Andrew V., Tsutsui, Neil D., 2008. The evolutionary consequences of biological invasions. Mol. Ecol. 17 (1), 351–360.

Thellung, A., 1915. Pflanzenwanderungen unter dem Einfluß des Menschen. Englers Bot. Jahrb. 53 (3/5), 37–66.

Theoharides, K.A., Dukes, J.S., 2007. Plant invasion across space and time: factors affecting nonindigenous species success during four stages of invasion. New Phytol. 176 (2), 256–273.

Vallow, 2007. Survival shortened when ER/PR negative breast cancer spreads to the brain, studies show. Science Daily.

Weiss, L., Grundmann, E., Torhorst, J., Hartveit, F., Moberg, I., Eder, M., Fenoglio-Preiser, C.M., et al., 1986. Haematogenous metastastic patterns in colonic carcinoma: an analysis of 1541 necropsies. J. Pathol. 150 (3), 195–203.

Williamson, M., 1996. Biological InvasionsSpringer Science & Business Media, London, UK.

Woodward, S.L., Quinn, J.A., 2011. Encyclopedia of Invasive Species: From Africanized Honey Bees to Zebra Mussels. Greenwood, Westport, CT, 764 p.

12

Transmissible Cancer: The Evolution of Interindividual Metastasis

Beata Ujvari*, Robert A. Gatenby**, Frédéric Thomas[†,‡]

*Centre for Integrative Ecology, School of Life and Environmental Sciences, Deakin University, Geelong, VIC, Australia
**Department of Radiology, H. Lee Moffitt Cancer Center and Research Institute, Tampa, FL, United States
[†]MIVEGEC (Infectious Diseases and Vectors: Ecology, Genetics, Evolution and Control), UMR IRD/CNRS/UM 5290, Montpellier, France
[‡]CREEC (Centre for Ecological and Evolutionary Research on Cancer), Montpellier, France

INTRODUCTION

Neoplasms are composed of an admixture of clones competing for resources, avoiding immune recognition, and cooperating to disperse and to colonize novel organs. Intratumor genetic, epigenetic, and phenotypic heterogeneity provide individual cancer cells the ability to adapt to selection imposed by the microenvironment, rapidly acquire novel cancer phenotypes, immortalization, and increased invasiveness to other organs (Merlo et al., 2006; Podlaha et al., 2012). However, the ultimate fate of malignant cells is to expire with the death of its host. Evolutionary theory postulates, that cancer cell lineages that are able to become transmissible and hence escape the demise of its host, will acquire higher fitness, and consequently be favored by selection (Ujvari et al., 2016b,e). In spite of this, only eight naturally occurring transmissible contagious cancers (one lineage in dogs, two lineages in Tasmanian devils, and five lineages in bivalves), with no underlying pathogen infections, have been recorded in the wild (Metzger et al., 2016).

Transmissible cancers have, however, been recorded under laboratory conditions and on rare occasions in humans. Here we review the transmissible cancers so far observed in humans, under laboratory settings and in the wild and discuss the evolutionary and ecological impacts of the latter.

CONTAGIOUS CANCERS WITH NO UNDERLYING INFECTIOUS ETIOLOGIES

Human Cases

Pregnancy

Although rare, in 0.1% of pregnancies, mother-to-fetus as well as fetus-to-fetus transmission of melanoma, lymphoma, leukemia, and carcinoma have been documented (Dingli and Nowak, 2006; Tolar and Neglia, 2003; Welsh, 2011). Due to evolutionary constraints the placenta and the fetus are immune privileged, protected from immune rejection, therefore facilitating the transmission of malignant cells at the maternal–fetal intrauterine interface (Dingli and Nowak, 2006; O'Neill, 2010; Warning et al., 2011).

Organ Transplant Recipients

Artificial cancer transmission between individuals may result from specific medical practices, such as organ transplants. By matching the immune profile of the donor and the recipient, and by suppressing the immune system of the recipient, organ transplantation artificially creates an immunocompromised state. Reduced immune function of the recipient may result in an increased risk of (1) de novo occurrence (increased risk for diverse infection-related cancers and oncogenic viral infections), (2) recurrence of malignancy, and (3) donor-related malignancy

Ecology and Evolution of Cancer. http://dx.doi.org/10.1016/B978-0-12-804310-3.00012-0

[transmission of donor tumors, donor-derived malignancy, such as donor-derived lymphoproliferative disorders, and transmission of oncogenic viruses e.g., human papilloma virus (HPV)] (Engels et al., 2011; Gandhi and Strong, 2007). Prior to screening of donor malignancies, organs with disseminated malignancies were frequently transplanted resulting in a 1/3 of recipients developing donor-derived cancers (Gandhi and Strong, 2007; Penn, 1978; Penn and Starzl, 1972). However, since the Council of Europe published the International Consensus Document for Standardization of Organ Donor Screening to Prevent Transmission of Neoplastic Diseases in 1997 (reviewed in Gandhi and Strong, 2007) the prevalence of tumors arising in organ transplant recipients has been reduced to 0.05%.

Accidental and Experimental Cases of Human-to-Human Transmission

A few rare cases of accidentally and experimentally transmitted tumors, with no clear underlying immunological etiologies, have been reported (reviewed in Welsh, 2011), such as: (1) accidental transmission of a malignant fibrous histiocytoma from a patient to a surgeon (Gärtner et al., 1996), (2) an unintentional needle transmission of human colonic adenocarcinoma to a laboratory technician (Gugel and Sanders, 1986), (3) experimental homotransplantation of melanoma from a daughter to her volunteering mother (Scanlon et al., 1965), (4) transfer of allogeneic tumor cells to healthy volunteers and patients with advanced debilitating neoplastic disease (Southam and Moore, 1958), and (5) lastly but not least the deliberate injection of prisoners in the 1950s and 1960s with blood from patients suffering from leukemia and/or liver cancer (Loue, 2000).

Laboratory Animals

Syrian Hamster—Contagious Reticulum Cell Sarcoma

A spontaneous reticulum cell sarcoma of the Syrian hamster (*Mesocricetus auratus*) arose in the laboratory colony of the National Institutes of Health in the 1960s (Brindley and Banfield, 1961). Detailed chromosomal studies confirmed the absence of any subcellular oncogenic agent, determined that direct cell implantation via mosquitoes and/or biting/cannibalism was necessary for an among-individual transmission (Banfield et al., 1965; Copper et al., 1964; Nachtigal, 1965).

Genetically Engineered Laboratory Animal Models

Surgical, orthotopic implantation of human tumors into immunodeficiency (NON/SCID) mice has become a standard model for personalized medicine. The system mimics tumor cell stromal interactions and permits preclinical analysis of therapeutic compounds (Hoffman, 2015; Manzotti et al., 1993). Furthermore, the NON/SCID mouse models have been used to investigate not only tumor development and progression but also transmission potential. For example, Kreiss et al. (2011) was able to reproduce the devil facial tumor disease (DFTD) in a NON/SCID murine xenograft mouse model resulting in a histologically indistinguishable DFTD to that observed in Tasmanian devils (*Sarcophilus harrisii*).

Naturally Occurring Transmissible Cancers

Canine Transmissible Venereal Tumor

Canine transmissible venereal tumor (CTVT) is a globally distributed sexually transmitted naturally occurring tumor in dogs that can be experimentally transmitted to other canines, such as jackals and coyotes (Strakova and Murchison, 2014). This transmissible cancer is considered to be the oldest known somatic cell line and has been proposed to have appeared between 6,000 and 11,000 years ago (exact date is still debated) in a postdomestication canid (Decker et al., 2015; Murchison et al., 2014; Murgia et al., 2006; Ostrander et al., 2016). The origin of CTVT is uncertain; it was first described as histiocytic (Mozos et al., 1996; Mukaratirwa and Gruys, 2003), lymphatic, or reticuloendothelial (Gimeno et al., 1995) while the latest analyses suggest a macrophage or a myeloid origin (Albanese et al., 2002; Sandusky et al., 1987). The neoplasms are localized mainly on the external genitalia and the cancer cells are transmitted during coitus which often results in inflamed tissue hence facilitating the transmission of the malignant cells (Fig. 12.1). Experimental transplantation studies have revealed three distinct growth phases (1) a progressive, (2) a static, and (3) a regressive stage (Das and Das, 2000; Mukaratirwa and Gruys, 2003). Following 2–4 months of progressive

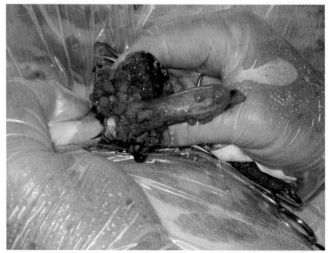

FIGURE 12.1 Canine transmissible venereal tumor (CTVT). *Source: Photo was taken by and being used with the permission of Dr. Elizabeth Murchison.*

growth the tumor regresses (Yang, 1988). Metastasis/fatal outcome have only been recorded in puppies and immunosuppressed adult dogs (Das and Das, 2000; Mukaratirwa and Gruys, 2003; Yang, 1988).

Devil Facial Tumor Disease

This contagious cancer was first observed in Tasmanian devils (*Sarcophilus harrisii*) in north-eastern Tasmania in 1996 (Hawkins et al., 2006). Similar to CTVT, direct contact is required for DFTD transmission. The disease is passed between devils by biting during social interactions (Pearse and Swift, 2006). The name of the disease originates from the large ulcerating tumors primarily caused around the face and jaws of the devils (Fig. 12.2). Since the first description of DFTD as a contagious malignant cell allograft in 2006 (Pearse and Swift, 2006), a second variant of DFTD (now described as DFT2, and the previous lineage renamed as DFT1) was described in 2015 (Pye et al., 2016). Due to its recent discovery, there is limited information available about the etiology of DFT2. DFT1 frequently metastasizes to distant organs, and in most cases results in death, due to starvation and organ failure, within 6 months after the emergence of the first lesions (Pyecroft et al., 2007). Molecular genetic studies suggest that both DFT1 and DFT2 originated from peripheral nervous system cells, either Schwann cells or Schwann cell precursors (Murchison et al., 2010; Pye et al., 2016). Although DFT1 and DFT2 both cause phenotypically similar facial tumors, they differ in histology, karyotype, and genotypes (Pye et al., 2016). The presence of X and the lack of Y chromosomes suggest that DFT1, most likely, have originated from a female, while the existence of Y chromosome in DFT2 shows that the latter cell line has arisen in a male devil (Murchison et al., 2012; Pye et al., 2016). During the last two decades

the disease has reached epidemic proportion decimating the devil population to an extent that presently the Tasmanian devil might be facing extinction in the coming decades (McCallum et al., 2009).

Disseminated Neoplasia of Bivalves

Since the 1960s disseminated, hematopoietic, or hemic neoplasia have been described in numerous bivalves, such as clams, oysters, and cockles (reviewed in Carballal et al., 2015). In disseminated neoplasia of bivalves (DNB) excessively proliferating abnormal cells of the hemolymph disseminate through the circulatory system and infiltrate multiple organs ultimately causing the death of the bivalves. While previous studies suggested a retrovirus or retrotransposon etiology (Oprandy and Chang, 1983), recent studies have shown a horizontal transmission of malignant cells in soft-shell clams (*Mya arenaria*), mussels (*Mytilus trossulus*) (Fig. 12.3), cockles (*Cerastoderma edule*), and golden carpet shell clams (*Politipapes aureus*) (Metzger et al., 2015, 2016). Importantly, the cancer cells observed in golden carpet shell clams were all derived from a different species, the pullet shell clam (*Venerupis corrugata*), showing the presence of cross-species transmission (Metzger et al., 2016).

FIGURE 12.3 *Mytilus trossulus,* **one of the hosts of bivalve disseminated neoplasia.** *Source: Photo was taken by and being used with the permission of Dr. Michael J. Metzger.*

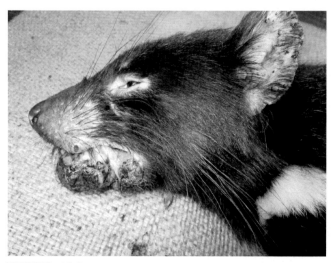

FIGURE 12.2 **Tasmanian devil facial tumor disease (DFTD).** *Source: Photo was taken by and being used with the permission of Dr. Elizabeth Murchison.*

DNB in soft-shell clams was first documented in the 1970s, and has since spread extensively along the North Atlantic coast, where the cancer shows signs of significant genetic divergence resulting in the emergence of region-specific subgroups (Metzger et al., 2015). Transplantation of seed stocks along the coast of Maine, USA during the 1990s has most likely accelerated the spread of the disease in soft-shell clams (Beal and Gayle Kraus, 2002).

Analysis of mussels (*M. trossulus*) from the Pacific Northwest Coast revealed the existence of a *M. trossulus*–derived transmissible cancer lineage in the area, while two distinct lineages of transmissible cancers, with distinct morphologies and genotypes, have been observed in cockles (*C. edule*) on the Galician Coast (Metzger et al., 2016). In the same habitat, golden carpet shell clams (*P. aureus*) contained *V. corrugate*–derived transmissible cancer lineages, while no disseminated neoplasia has so far been found in the cohabiting *V. corrugate* (Metzger et al., 2016).

Bivalves being filter feeders, DNB has been proposed to spread via filtration of seawater contaminated with neoplastic cells. Since hemocytes from leukemic bivalves can survive in seawater for >6 h, currents have been suggested to be the potential environmental vectors facilitating the colonization of novel bivalve beds (Metzger et al., 2015).

The latest findings of widespread bivalve transmissible cancers indicate that transmissible cancers can originate independently and may be more common than thought in marine environments (Metzger et al., 2016; Ujvari et al., 2016c).

Transmission Across the Animal–Human Interface

In 2015 cancer cell transmissions from abnormal malignant tapeworm cells (*Hymenolepis nana*) were recorded to invade the lymph nodes, lungs, liver, and adrenal glands of an HIV patient (Muehlenbachs et al., 2015). Although the patient's immune-compromised state may have facilitated the proliferation of *H. nana* stem cells leading to the malignant transformation and transmission of worm cells, the transmission of cancer cells across the human–animal interface suggests the possibility that transmissible cancers may occur across species in other organisms.

MECHANISMS INVOLVED IN CANCER TRANSMISSION

Transmission of Parasitic Cancer Clones

Although viral agents have been suspected to be involved in the spread of CTVT, DFTD, DNB, and in hamster, CRCS, no evidence for the presence and involvement of viruses has so far been confirmed (Banfield et al., 1965; Murchison et al., 2012, 2014; Ostrander et al., 2016; Pearse and Swift, 2006; Taraska and Anne Böttger, 2013). In all cases, allogenic transfer of living cancer cells is necessary for transmission that is, through biting in DFTD, sexual contact in CTVT, through biting/cannibalism or via mosquito transfer in CRCS, and via filter feeding in DNB (Metzger et al., 2015). The transmission cascade is a multistep biological process that requires high cellular plasticity, such as, survival in transit, breaking down immunological barriers, the initiation and growth of cancer cells at primary tumor sites, followed by dissemination of cancer cells and establishment and adaptation to a novel host environment. Similar to host–parasite interactions, successful transmission of cancer cells hence requires a "perfect storm" and hence a confluence of multiple tumor and host traits (reviewed in Ujvari et al., 2016b).

Dissemination of Cancer Cells to the New Host

The probability of a successful colonization in novel hosts depends on the introduction effort, that is, (1) the number of cells shredded at introduction, (2) the number of introduction events, and (3) the temporal and spatial patterns of propagule arrival (Barfield et al., 2015; Britton-Simmons and Abbott, 2008; Leggett et al., 2012; Simberloff, 2009). Since, both CTVT and DFTD are transmitted via frequent social contacts and DNB via filter feeding, the potential for cancer cell transmission is high. However, to ensure successful transmission the tumor must release a large number of cells with each contact (Barfield et al., 2015; Leggett et al., 2012). This is supported by the fact that the number of DFTD cells inoculated into immunocompromised mice (NOD/SCID) not only determine the success of colonization, but also the speed of tumor establishment [1×10^5 DFTD resulted visible tumors between 7 and 17 weeks after inoculation, whereas 1×10^6 DFTD cells developed between 5 and 15 weeks (Kreiss et al., 2011)]. Interestingly, 1×10^5 tumor cells derived from these inoculations were later passaged into novel NOD/SCID mice resulting in even faster tumor development (between 4 and 12 weeks after inoculation) suggesting that the xenografted DFTD cells had acquired additional characteristics (potentially triploidy) to speed up the invasion process (Kreiss et al., 2011). Unfortunately, DFTD tumor cell viability after inoculation was not recorded, however, in a similar experiment conducted on CTVT cells, approximately 10% of injected CTVT cells remained viable 7 days after inoculation (Holmes, 1981), suggesting that probably only a small population of viable cells would be required to establish a tumor mass (Kreiss et al., 2011).

Recent field data provide support for the importance of propagule quality and quantity for invasion success of contagious cancers. In most of Tasmania DFTD has affected between 40% and 80% of the devil population within a year, whereas at West Pencil Pine (WPP) in north-western Tasmania disease prevalence remained 20% despite the cancer being in the area for over 5 years (Coupland and Anthony, 2007; Hamede et al., 2012). However, following the initial low DFTD infection rates, the prevalence and impact of the diseases have dramatically increased 6 years after the DFTD outbreak (Hamede et al., 2015). The different epidemiology of DFTD at WPP compared to other areas in Tasmania was initially postulated to be either due to genetic differences in the hosts and/or due to genetic differences in the tumors. However, no genetic divergence has so far been observed between diseased and healthy devils (Lane et al., 2012) whereas recent analyses of the WPP tumor types revealed a shift of tumor karyotype variants at the site coincided with a rapid increase in disease prevalence and population decline (Hamede et al., 2015).

The latter results indicate that, either the arrival of novel cancer variants, or increased propagule pressure (the host population being exposed to infection pressure from DFTD for >6 years) has contributed to the colonization of this devil population. The laboratory experiments and field data obtained from CTVT and DFTD clearly demonstrate that the quality and number of individuals (cancer cells) in a transmission event, as well as the number of release events strongly influence the success of colonization (Kreiss et al., 2011; Simberloff, 2009). Furthermore, the field data from WPP construe the possibility that spatial and temporal variation in propagule pressure might be a driving force behind the spread of DFTD.

Establishment in and Adaptation to a Hostile Host Environment

Breaking Down Histocompatibility Barriers

As histocompatibility barriers prevent the transplantation of foreign grafts between individuals, the appearance and success of vertebrate contagious cancers to invade >150,000 hosts have been attributed to the low genetic diversity of the host species (Belov, 2012). Loss of genetic diversity at key immune genes (as seen in wildlife cancers and in serotyping organ donors–recipients), the artificial suppression of immune response (as seen in laboratory animals, or the use of immunosuppressive agents during organ transplantation), and downregulation of histocompatibility molecules on the cell surface (as in DFTD and CTVT) create an immune permissive myopic environment for the allograft to colonize its novel host (Siddle and Kaufman, 2013a,b; Siddle et al., 2010, 2013). In order to overcome the histocompatibility barrier, vertebrate transmissible cancers, similar to primary and metastatic cancers in humans (Fassati and Mitchison, 2010), employ epigenetic immune modulation by downregulating genes involved

in the antigen-processing pathways resulting in the concomitant loss of cell surface expression of *Major Histocompatibility Complex Class I (MHC)*, and active evasion of immune recognition (Belov, 2011; Fassati and Mitchison, 2010; Siddle and Kaufman, 2013a,b, 2015). Intriguingly, although invertebrates do not possess major histocompatibility complex molecules, they may employ other mechanisms (e.g., NK-like cell-mediated processes) to recognize and combat transformed malignant cells. (Metzger et al., 2015; Robert, 2010; Voskoboynik et al., 2013). Whether these histocompatibility systems are genetically and/or epigenetically modulated in bivalve leukemia and in other invertebrate cancers remains to be answered. However, the cross-species transmission of neoplastic cells between two bivalve species suggest that histocompatibility barriers in invertebrates may not be as efficient as that observed in vertebrates. Nevertheless, the genetic similarity of the hosts and the hosts' immune competence are two of the crucial factors that underpin the ability of cancer transmission (Siddle and Kaufman, 2013b).

Phenotypic Plasticity: Adaptive Life-History Traits

Since colonizing cancer cells must be able to cope with a range of environmental conditions, phenotypic plasticity (change in phenotypic expression in response to environmental variations) has been singled out as one of the key life-history traits needed in successful colonization of novel environments (Ujvari et al., 2016b). Epigenetic variation regulating complex gene expression signatures superimposed over the primary tumor genotype allows the invasive cancer genomes to respond to developmental and environmental cues (Jaenisch and Bird, 2003; Rodenhiser, 2009). Epigenetic regulators (as described in Chapter 6), including DNA methylation, histone and chromatin modifications, and miRNA regulation orchestrate dynamic, adaptive, and reversible phenotypic differences (Bird, 2007) and facilitate and ensure transition between fast and slow life-history strategies of colonizing cells (Aktipis et al., 2013). For example, a study by Ingles and Deakin (2015) revealed that a hypermethylated active X chromosome was shattered during DFTD formation. Although immunofluorescent staining of DFTD chromosomes showed stable temporal methylation patterns, methylome fingerprinting showed dynamic changes in the methylome, particularly a temporal increase in hypomethylation (Ujvari et al., 2013). While loss of genome-wide methylation is a typical feature of early cancer development (Ehrlich, 2002; Shao et al., 2009), DNA hypomethylation may also occur late in the metastatic stages of cancer progression and hence in cancer with poor prognosis (Ehrlich, 2002; Yegnasubramanian et al., 2008). In addition the principal enzyme responsible for maintaining CpG methylation (DNMT1)

(Rehen et al., 2005; Singer et al., 2010) has been shown to be upregulated in devil facial tumors compared to peripheral nerves [the sites of DFTD origin (Ujvari et al., 2013)]. The combination of cancer instigation by shattering of a hypermethylated X-chromosome followed by increasing loss of methylation might be a significant feature of DFTD evolution, that is, a punctuated initial stage followed by a step-wise phase, strongly suggesting that DFTD is a dynamically changing entity with a high phenotypic plasticity and evolutionary potential (Ujvari et al., 2013, 2016e).

Permissive Tissue Environment

Since chronic irritation and inflammation are critical components of tumor emergence and progression (Coussens and Werb, 2002), inflamed, infected, and/or injured host tissues may provide a more permissive environment for cancer cell growth and establishment (O'Neill, 2010). Inflammatory cells orchestrate tumor microenvironment, foster neoplastic cell proliferation, survival, and migration; malignant cells hitchhike these signaling molecules and their receptors of the innate immune system to support migration, metastasis, and invasion (Coussens and Werb, 2002). Therefore it is not surprising, that both CTVT and DFTD transmitted via physical contact (sexual and/or by biting) affect areas (genitals and mucous membrane) prone to infections, injury, and/or inflammation. The repair of tissue injury involves enhanced cell proliferation, activation of a multifactorial network of chemical signals, and inflammatory responses (Coussens and Werb, 2002). Malignant cells have shown to proliferate and thrive in microenvironments rich in inflammatory cells (Coussens and Werb, 2002). It has therefore been proposed that "tumours act as wounds that fail to heal" (Coussens and Werb, 2002; Dvorak, 1986), and contagious cancers certainly bear the hallmark of unceasingly fraying and ulcerating lacerations.

TUG-OF-WAR: EVOLUTION OF TRANSMISSIBLE CANCERS AND THEIR HOSTS

The Emergence of Transmissible Cancer Cell Lines

Exposure to novel biotic (e.g., pathogens and overcrowding) or abiotic (e.g., UV exposure) conditions frequently induces stress and concomitantly affects organismal genome stability (Bond and Finnegan, 2007; Molinier et al., 2006). Stress-induced genome modification may generate epigenetic alterations, or retrotransposon activation and reinsertion, and the inheritance of such modification can be transferred through several generations

(Bond and Finnegan, 2007; Kashkush et al., 2003). These transgenerational changes demonstrate the importance of environmental perturbations in combination with genomic instability in influencing phenotypic plasticity and adaptive potential of transmissible cancers (Rando and Verstrepen, 2007). Although the exact origin and the cause of initiation are not known in any of the naturally occurring transmissible cancers, the combination of environmental factors and instable genomes have been implicated in both DNB and CTVT (Ostrander et al., 2016). The origin of soft-shell clam leukemia has been linked to the activation of the retroelement Streamer (Arriagada et al., 2014), which might have been initiated by environmental stressors, such as pollution, temperature, and overcrowding (Arriagada et al., 2014; Barber, 2004). Arriagada et al. (2014) proposed that the activation of Steamer in M. arenaria bears the signatures of a "catastrophic genomic instability," which could have contributed to the initiation and development of DNB. Elevated Steamer mRNA and DNA-copy numbers potentially drive further genomic instability and accelerate disease progression. However, no amplification of Steamer-like elements have been observed in the remaining bivalve transmissible cancers (M. trossulus, C. edule, P. aureus), and hence Metzger et al. (2016) proposed that retrotransposon amplification might not be the driving force in bivalve transmissible clone development.

CTVT also carries the marks of ultraviolet light exposure, high mutation burden, and retrotransposon insertions (Decker et al., 2015; Murchison et al., 2014; Ostrander et al., 2016). All CTVT tumors show the presence of a long interspersed nuclear element (LINE-1) inserted near the c-myc oncogene, which is absent at the corresponding position in the germ line (Katzir et al., 1985; Liao et al., 2003). The structure of LINE-1 bears the signs of mRNA origins, and hints the possibility that transposition of this movable element close to an oncogene could have affected its activity and instigated the emergence of CTVT. While expansion of short transposable elements have been observed in the Tasmanian devil genome, so far no cancer inducing transposons have been identified in DFTD tumors (Murchison et al., 2012; Nilsson et al., 2012).

The Evolution of Transmissible Cancer Cell Lines

In contrast to nontransmissible cancer cells, transmissible cancer cell lines emerge, maintain their malignant genotypes and phenotypes across generations, and evolve on the evolutionary landscape of host genomes (Haig, 2015). Provided the prevalence of transmissible cancer is sufficiently high, the tug-of-war between selfish malignant cell lines and host genomes resemble to that observed in host–parasite interactions (sensu

"Red-Queen" dynamics; Schmid-Hempel, 2011; van Valen, 1973), hence both hosts and contagious cancer cells have to change continuously to keep up with each other's adaptations (Ujvari et al., 2016b). A recent study by Decker et al. (2015) demonstrated that the CTVT genome bears evidence to such an ongoing arms-race. By comparing two CTVT genomes to 186 canid whole-genome sequences, Decker et al. (2015) found somatic mutations not only in the self-antigen presentation pathway (MHC), but also in all aspects of somatic cell participation in immune surveillance, genes involved in maintenance of genome integrity, and the regulation of cell apoptosis. The etiology of CTVT (initial progression followed by regression) in combination with comprehensive combination of mutations observed in its genome carry the signature of thousands of years of selection and coevolution between the host and the parasitic malignant allograft (Decker et al., 2015; Ostrander et al., 2016).

Overcoming Genomic Decay

Similar to asexual reproducing organisms, the absence of recombination of both nuclear and mitochondrial genes leads to irreversible accumulation of deleterious mutations in neoplastic cells (Merlo et al., 2010). Sequencing the mitochondrial genomes of the hosts and CTVT revealed that CTVT periodically captures mtDNA from its host, and hence CTVT has been able to overcome the effect of Muller's ratchet, that is, CTVT is able to rejuvenate its mtDNA to escape genomic decay (Ertel et al., 2012; Murchison et al., 2012; Rebbeck et al., 2011), a process not yet observed in DFTD, or in bivalve leukemia.

Similar to mtDNA hybridization, increasing chromosome numbers may benefit asexually reproducing organisms exposed to loss of heterozygosity and the emergence of recessive mutations (Merlo et al., 2010; Orr and Otto, 1994; Otto, 2007; Otto and Whitton, 2000). The increased genetic polymorphism associated with polyploidy may stabilize the genomic configuration of clonal cells, and hence promote the survival and selection of polyploid cancer cells (Merlo et al., 2010; Orr and Otto, 1994). Indeed, the presence of polyploid karyotypes in malignant cells has been proposed to be the result of adaptation to their asexual lifestyle (Das and Das, 2000; Deakin et al., 2012; Merlo et al., 2010; Metzger et al., 2015). Importantly, polyploid cancer cells have been found to sustain a higher mutation rate, and provide an adaptive advantage by masking deleterious mutations and ameliorating the genomic decay process (Davoli and de Lange, 2011; Orr, 1995; Orr and Otto, 1994; Otto, 2007; Otto and Whitton, 2000). Interestingly recently elevated chromosome numbers have been observed in both DFTD (Pearse et al., 2012; Ujvari et al., 2014) and in DNB (Metzger et al., 2015).

Malignant Transformation: A Form of Speciation?

Carcinogenesis, particularly the evolution of transmissible cancers, has been described as speciation events (Duesberg et al., 2011). Duesberg et al. (2011) postulated that carcinogenesis is initiated by loss or gain of chromosomes (aneuploidy), which results in destabilization of karyotypes followed by strong selection for reproductive autonomy—the primary characteristics of species and cancer cells. Reproductive autonomy will result in stabilization of aneuploid genomes and the development of novel stable karyotypes, "species" (Duesberg et al., 2011). As transmissible cancers have individual clonal karyotypes (highly stable but bearing the signatures of aneuploidy), are immortal and remain stable in countless natural transmissions—manifesting the lifestyle of extrinsic infectious microorganism—they have been proposed to be natural examples of "immortal fully speciated cancers" (Duesberg et al., 2011; Vincent, 2010). Once the neoplastic process has crossed the threshold of autonomy, and cells have acquired germ line properties they have transcended the host and became something novel, that is, a new "species" (Vincent, 2010).

The Evolution of Host in Response to Parasitic Cell Lines

Since transmissible cancers negatively affect host fitness, by either reducing survival (DFTD and DNB) or by impeding sexual intercourse and hence reproductive output (CTVT, DFTD) selection will either evoke host evolutionary responses, or will drive the host species to extinction (see more in Chapter 13). CTVT, the oldest living cell line, provides a clear example of the evolutionary tug-of-war between the host and its malignant invasive allograft. Over the course of >5000 years the canine immune system has evolved to override the initial immune-modulatory signals of invading CTVT cells, supress tumor growth, and provide immune memory and protection to subsequent reinfections (Rebbeck et al., 2009). Although no such clear changes of the immune system have so far been observed in DFTD, two recent studies of Tasmanian devils show signature of potential adaptation by the devils to the cancerous cell. Devils with higher relative natural antibody (IgM) levels compared to specific antibody (IgG) levels appear to be less susceptible to the contagious cancer (Ujvari et al., 2016d). Thus higher IgM levels provide a fitness advantage, and hence potentially being the first step of host's evolutionary response to combat the deleterious effects of DFTD (Ujvari et al., 2016d). Furthermore, another recent study identifies two genomic regions under selection in response to DFTD and suggests that devils' are evolving immune-modulated resistance to the devastating cancer (Epstein et al., 2016). Epstein et al. (2016) propose that the potential genomic responses have evolved within only a few generations [4–6 generations calculated based on devils' reaching sexual maturity at an age of 2 years (McCallum et al., 2009)]. Such a rapid evolutionary response most likely has arisen due to the extreme selection imposed by DFTD (Epstein et al., 2016).

Notably, the strong selection pressure not only invokes genetic and phenotypic changes, but also shifts in the devils' life-history strategies. In response to the disease-induced high mortality the devils' reproductive strategy has transitioned from an iteroparous (multiple reproductive cycles) to a semelparous (single breeding) breeding pattern (Jones et al., 2008) (see more in Chapter 13).

The impact of evolutionary pressure to combat parasitic clonal cell lines is most strikingly shown by the recently discovered cross-species transmission between two mollusks (*P. aureus* and *V. corrugata*) belonging to the same family (Veneridae) and coexisting in shared marine beds (Metzger et al., 2016). While the disease bears the signature of *V. corrugate* genotypes, it has so far only been found in *P. aureus* (but not in *V. corrugate*). Metzger et al. (2016) proposed that the neoplasia has originated in *V. corrugata*, but the high fitness cost of the fatal disease led to the removal of susceptible animals, and to strong selection for resistant individuals in the species of origin. The cancer cell line has potentially overcome/escaped the evolutionary tug-of-war by spreading to a genetically closely related and spatially nearby species.

THE ECOLOGICAL IMPACT OF CANCER CELLS ON THE HOSTS' MICRO- AND MACROENVIRONMENT

Impact on the Microenvironment

Tumor microenvironments are known to be active contributors to cancer development, by supporting carcinogenesis, progression, and metastasis (Polyak et al., 2009). The communication between cancer cells and their microenvironment is bidirectional: tumor cells adapt to circumvent the normalizing cues of the microenvironment, and in turn, the microenvironment evolves to accommodate the neoplastic cells (Polyak et al., 2009). Tumor–stromal cell interactions result in gene expression changes in both malignant and normal epithelial cells (Bhowmick et al., 2004; Hu and Polyak, 2008; Nelson and Bissell, 2006; Orimo and Weinberg, 2006; Tlsty and Coussens, 2006). Similarly to invasive species, cancer cells often cause (micro-)environmental changes (e.g., via ecological facilitation and allelopathy in invasive species and via hypoxia and acidosis in cancer), creating novel niches less favorable to competitors and reducing the risk of predation by the immune system

(Alfarouk et al., 2013; Lowry et al., 2013). For example, variations in the tumor vascular network and associated blood flow, substrate and metabolite availability (such as oxygen and acid, respectively) serve as significant selective agents in the tumor ecosystems (Alfarouk et al., 2013). Furthermore, cancer cells use aerobic glycolysis or fermentation of glucose to lactic acid in the presence of oxygen ("Warburg effect"; Warburg, 1930), to produce energy. The result of this abnormal glycolysis is enhanced lactic acid production and the creation of a hostile hypoxic and acidic microenvironment. The generated acidic environment reduces the viability and function of normal cells, including immune cells, hence concomitantly protecting the cancer cells from the host's immune system. Additionally, hypoxia and/or the acidophilic environment also supports stromal remodeling, growth of blood vessels, and the escape of immune surveillance by recruiting myeloid-derived suppressor cells (Gabrilovich and Nagaraj, 2009; Murdoch et al., 2008). By producing ATP at a higher rate, but at lower yields, results in selective advantage for malignant cells, such as the inducement of oncogenes by lactate, the by-product of anaerobic glycolysis (Hans et al., 2009). The complex allopathic interactions of tumor and microenvironment resulting in reversed intraextracellular pH gradients and immune system evasion aid cancer cells to out-compete normal cells to proliferate and survive in a hostile environment. Since low extracellular pH of tumors also promotes invasiveness and metastatic behavior (Gatenby and Gawlinski, 1996; Gatenby et al., 2006; Martínez-Zaguilán et al., 1996), the initiation of an invasive phenotype may be one of the first steps along the path to the emergence of transmissible cancers (Gatenby and Gillies, 2008).

As mentioned earlier, injured, inflamed areas are characterized by superfluous blood flow, chemical signaling for cell proliferation, and access to an oversupply of biosynthetic building blocks, clearly providing a permissive resource rich-environment for colonizing tumor cells. In addition, since blood vessels, the suppliers of nutrients and cleaners of metabolites, are critically important for cancer cell development, tumors remain limited in size, and will not become clinically significant in the absence of angiogenesis (Alfarouk et al., 2013). Therefore it is not surprising that both DFTD and CTVT primarily affect body parts with dense vasculature which will nurture the establishment of allografts. Similarly, by being directly involved with the hemolymphatic circulatory system, the bivalve leukemia cells have direct access to high levels of crucial resources.

Impact on the Macroenvironment

Unfortunately documentation of the ecological impact of transmissible cancers is rare. However, as discussed in Chapter 13 cancer and oncogenic phenomena in general may have a significant impact on ecosystem functioning. Similar to infectious diseases, transmissible cancers may have a significant ecological impact on trophic cascades, niche allocation, host–parasite dynamics, and disease epidemiology of the ecosystem occupied by the affected host species (Ujvari et al., 2016a,b; Vittecoq et al., 2015).

Following the extinction of the thylacine (*Thylacinus cynocephalus*), the Tasmanian devil, is presently the world's largest extant marsupial carnivore, and hence the apex predator on the island of Tasmania. Apart from the devils, the Tasmanian carnivore guild includes three additional predators: the native marsupial spotted-tailed quoll (*Dasyurus maculatus*), the introduced feral cat (*Felis catus*), and the significantly smaller mesopredator eastern quoll (*Dasyurus viverrinus*) (Hollings et al., 2014; Jones, 1997). Following the emergence of DFTD, devil numbers have decreased dramatically that concomitantly resulted in an increase in feral cat numbers. Interestingly the number of the smallest predators, the eastern quolls has declined in parallel with the decline in devil numbers suggesting that the presence of the devils provided indirect benefits to the eastern quolls (potentially by devils supressing cat and spotted-tailed quolls numbers) (Hollings et al., 2014). The decline of the apex predator ultimately led to increased activity of invasive species, such as cats and black rats (*Rattus rattus*) and to substantial loss of prey species, altering the biodiversity of one of Australia's most pristine ecosystem (Hollings et al., 2016).

In contrast to Tasmanian devils, clams and cockles occupy a different level of the food chain and serve as prey to various predators. A recent study demonstrated that the collapse of prey numbers led to starvation and subsequent massive decline in predator numbers in a terrestrial ecosystem (Ujvari et al., 2016b). Whether the decline of bivalve populations due to DNB will cause a similar impact on predators, such as the blue crab (*Callinectes sapidus*) (Seitz et al., 2001), remains to be answered.

In addition to the trophic cascade, the immunocompromised state of animals affected by contagious cancer may affect parasite communities, parasite transmission, and disease epidemiology. For example, Tasmanian devils are hosts to various external and internal parasites, including trematodes, cestodes, lice, ticks, fleas, Mycobacterial infection, *Salmonella* (Beveridge and Spratt, 2003; Holz, 2008); thus future studies should focus on how DFTD and the dramatic reduction in devil numbers may affect the dynamics and distribution of these parasites. The scenario outlined earlier suggests that the removal of a given species from an ecosystem, being either predator or prey, will have both top-down and bottom-up effects impacting on species diversity, competition for resources (e.g., food, habitat), niche allocation, and

hence will significantly alter the ecosystem (Hollings et al., 2016; Petchey, 2000).

CONCLUSIONS

Successful invasion of novel host ecosystem by malignant cells is a multiplayer and multistage process, which requires the contagious allograft to succeed transmission "steps" in order to become a successful invader. The elusive nature of the invasion process arises from the necessity of the contemporaneous availability of resource-rich permissive environments, release of sufficient number of propagules and the clonal cancer cell lineage to procure all the essential genotypes/phenotypes, and life-history strategies. To become invasive, cancer cells first have to acquire a highly plastic, migratory, and immunomodulatory phenotype. Depending on the availability of permissive environments (immune privileged and/or inflamed sites) malignant allografts must acquire the ability to evade histocompatibility barriers. Once successful, the transmissible cancer and its host are subjected to an evolutionary tug-of-war, where the fitness reducing impact of the cancer serves as potent selective force evoking genotypic and phenotypic adaptations in the host, barriers which have to be overcome by the malignant cell lines to survive and to achieve successful transmission (potentially to a new host species). Thus, in order to evolve and persist, successful transmission of cancer cells requires a "perfect storm" with an optimal confluence of multiple host and tumor cell traits (Ujvari et al., 2016b). This scenario suggests that the likelihood of transmissible cancers emerging at high frequencies is indeed limited. Nevertheless, similar to host–parasite interactions, the impact of transmissible cancers on ecosystem functioning cannot and should not be underestimated. Transmissible cancers have been shown to cause massive population declines (and potential species extinction), and hence have an overarching effect on trophic cascades and ecosystem dynamics. Consequently, despite of their rarity, transmissible cancers provide both fascinating examples of cancer evolution and ecology as well as have a major impact on the ecology and evolution of the affected host species and its ecosystem.

Acknowledgments

This work was supported by the Australian Academy of Science's French–Australian Science Innovation Collaboration Program Early Career Fellowship and an International Associated Laboratory Project France/Australia.

References

Aktipis, C.A., Boddy, A.M., Gatenby, R.A., Brown, J.S., Maley, C.C., 2013. Life history trade-offs in cancer evolution. Nat. Rev. Cancer 13, 883–892.

Albanese, F., Poli, A., Millanta, F., Abramo, F., 2002. Primary cutaneous extragenital canine transmissible venereal tumour with Leishmania-laden neoplastic cells: a further suggestion of histiocytic origin? Vet. Dermatol. 13, 243–246.

Alfarouk, K.O., Ibrahim, M.E., Gatenby, R.A., Brown, J.S., 2013. Riparian ecosystems in human cancers. Evol. Appl. 6, 46–53.

Arriagada, G., Metzger, M.J., Muttray, A.F., Sherry, J., Reinisch, C., Street, C., Lipkin, W.I., Goff, S.P., 2014. Activation of transcription and retrotransposition of a novel retroelement, Steamer, in neoplastic hemocytes of the mollusk Mya arenaria. Proc. Natl. Acad. Sci. USA 111, 14175–14180.

Banfield, W.G., Woke, P.A., Mackay, C.M., Cooper, H.L., 1965. Mosquito transmission of a reticulum cell sarcoma of hamsters. Science 148, 1239–1240.

Barber, B.J., 2004. Neoplastic diseases of commercially important marine bivalves. Aquat. Living Resour. 17, 449–466.

Barfield, M., Orive, M.E., Holt, R.D., 2015. The role of pathogen shedding in linking within- and between-host pathogen dynamics. Math. Biosci. 270 (Part B), 249–262.

Beal, B.F., Gayle Kraus, M., 2002. Interactive effects of initial size, stocking density, and type of predator deterrent netting on survival and growth of cultured juveniles of the soft-shell clam, Mya arenaria L., in eastern Maine. Aquaculture 208, 81–111.

Belov, K., 2011. The role of the major histocompatibility complex in the spread of contagious cancers. Mamm. Genome 22, 83–90.

Belov, K., 2012. Contagious cancer: lessons from the devil and the dog. BioEssays 34, 285–292.

Beveridge, I., Spratt, D.M., 2003. Parasites of carnivorous marsupials. In: Jones, M., Dickman, C., Archer, M. (Eds.), Predators with pouches. CSIRO Publishing, Melbourne.

Bhowmick, N.A., Neilson, E.G., Moses, H.L., 2004. Stromal fibroblasts in cancer initiation and progression. Nature 432, 332–337.

Bird, A., 2007. Perceptions of epigenetics. Nature 447, 396–398.

Bond, D.M., Finnegan, E.J., 2007. Passing the message on: inheritance of epigenetic traits. Trends Plant Sci. 12, 211–216.

Brindley, D.C., Banfield, W.G., 1961. A contagious tumor of the hamster. J. Natl. Cancer Inst. 26, 949–954.

Britton-Simmons, K.H., Abbott, K.C., 2008. Short- and long-term effects of disturbance and propagule pressure on a biological invasion. J. Ecol. 96, 68–77.

Carballal, M.J., Barber, B.J., Iglesias, D., Villalba, A., 2015. Neoplastic diseases of marine bivalves. J. Invertebr. Pathol. 131, 83–106.

Copper, H.L., Mackay, C.M., Banfield, W.G., 1964. Chromosome studies of a contagious reticulum cell sarcoma of the Syrian hamster. J. Natl. Cancer Inst. 33, 691–706.

Coupland, C., Anthony, W., 2007. Devils of the Alpine Project: field monitoring program. Tasmanian Nat. 129, 65–81.

Coussens, L.M., Werb, Z., 2002. Inflammation and cancer. Nature 420, 860–867.

Das, U., Das, A.K., 2000. Review of canine transmissible venereal sarcoma. Vet. Res. Commun. 24, 545–556.

Davoli, T., de Lange, T., 2011. The causes and consequences of polyploiin normal development and cancer. Annu. Rev. Cell Dev. Biol. 27, 585–610.

Deakin, J.E., Bender, H.S., Pearse, A.-M., Rens, W., O'Brien, P.C.M., Ferguson-Smith, M.A., Cheng, Y., Morris, K., Taylor, R., Stuart, A., Belov, K., Amemiya, C.T., Murchison, E.P., Papenfuss, A.T., Marshall Graves, J.A., 2012. Genomic restructuring in the Tasmanian devil facial tumour: chromosome painting and gene mapping provide clues to evolution of a transmissible tumour. PLoS Genet. 8, e1002483.

Decker, B., Davis, B.W., Rimbault, M., Long, A.H., Karlins, E., Jagannathan, V., Reiman, R., Parker, H.G., Drögemüller, C., Corneveaux, J.J., Chapman, E.S., Trent, J.M., Leeb, T., Huentelman, M.J., Wayne, R.K., Karyadi, D.M., Ostrander, E.A., 2015. Comparison against 186 canid whole-genome sequences reveals survival strategies of an ancient clonally transmissible canine tumor. Genome Res. 25, 1646–1655.

Dingli, D., Nowak, M.A., 2006. Cancer biology: infectious tumour cells. Nature 443, 35–36.

Duesberg, P., Mandrioli, D., Mccormack, A., Nicholson, J.M., 2011. Is carcinogenesis a form of speciation? Cell Cycle 10, 2100–2114.

Dvorak, H.F., 1986. Tumors: wounds that do not heal. N. Engl. J. Med. 315, 1650–1659.

Ehrlich, M., 2002. DNA methylation in cancer: too much, but also too little. Oncogene 21, 5400–5413.

Engels, E.A., Pfeiffer, R.M., Fraumeni, J.F., Kasiske, B.L., Israni, A.K., Snyder, J.J., Wolfe, R.A., Goodrich, N.P., Bayakly, A.R., Clarke, C.A., Copeland, G., Finch, J.L., Fleissner, M.L., Goodman, M.T., Kahn, A., Koch, L., Lynch, C.F., Madeleine, M.M., Pawlish, K., Rao, C., Williams, M.A., Castenson, D., Curry, M., Parsons, R., Fant, G., Lin, M., 2011. Spectrum of cancer risk among U.S. solid organ transplant recipients: the transplant cancer match study. JAMA 306, 1891–1901.

Epstein, B., Jones, M., Hamede, R., Hendricks, S., McCallum, H., Murchison, E.P., Schönfeld, B., Wiench, C., Hohenlohe, P., Storfer, A., 2016. Rapid evolutionary response to a transmissible cancer in Tasmanian devils. Nat. Commun. 7, 12684.

Ertel, A., Tsirigos, A., Whitaker-Menezes, D., Birbe, R.C., Pavlides, S., Martinez-Outschoorn, U.E., Pestell, R.G., Howell, A., Sotgia, F., Lisanti, M.P., 2012. Is cancer a metabolic rebellion against host aging? In the quest for immortality, tumor cells try to save themselves by boosting mitochondrial metabolism. Cell Cycle 11, 253–263.

Fassati, A., Mitchison, N.A., 2010. Testing the theory of immune selection in cancers that break the rules of transplantation. Cancer Immunol. Immunother. 59, 643–651.

Gabrilovich, D.I., Nagaraj, S., 2009. Myeloid-derived-suppressor cells as regulators of the immune system. Nat. Rev. Immunol. 9, 162–174.

Gandhi, M., Strong, D.M., 2007. Donor derived malignancy following transplantation: a review. Cell Tissue Bank. 8, 267–286.

Gärtner, H.-V., Seidl, C., Luckenbach, C., Schumm, G., Seifried, E., Ritter, H., Bültmann, B., 1996. Genetic analysis of a sarcoma accidentally transplanted from a patient to a surgeon. N. Engl. J. Med. 335, 1494–1497.

Gatenby, R.A., Gawlinski, E.T., 1996. A reaction–diffusion model of cancer invasion. Cancer Res. 56, 5745–5753.

Gatenby, R.A., Gawlinski, E.T., Gmitro, A.F., Kaylor, B., Gillies, R.J., 2006. Acid-mediated tumor invasion: a multidisciplinary study. Cancer Res. 66, 5216–5223.

Gatenby, R.A., Gillies, R.J., 2008. A microenvironmental model of carcinogenesis. Nat. Rev. Cancer 8, 56–61.

Gimeno, E.J., Massone, A.R., Marino, F.P., Idiart, J.R., 1995. Intermediate filament expression and lectin histochemical features of canine transmissible venereal tumour. APMIS 103, 645–650.

Gugel, E.A., Sanders, M.E., 1986. Needle-stick transmission of human colonic adenocarcinoma. N. Engl. J. Med. 315, 1487–11487.

Haig, D., 2015. Maternal–fetal conflict, genomic imprinting and mammalian vulnerabilities to cancer. Philos. Trans. R. Soc. Lond. B 370, 20140178.

Hamede, R., Lachish, S., Belov, K., Woods, G., Kreiss, A., Pearse, A.-M., Lazenby, B., Jones, M., McCallum, H., 2012. Reduced effect of Tasmanian devil facial tumor disease at the disease front. Conserv. Biol. 26, 124–134.

Hamede, R.K., Pearse, A.-M., Swift, K., Barmuta, L.A., Murchison, E.P., Jones, M.E., 2015. Transmissible cancer in Tasmanian devils: localized lineage replacement and host population response. Proc. R. Soc. Lond. B 282, 20151468.

Hans, H.K., Taeho, K., Euiyong, K., Ji Kyoung, P., Seok-Ju, P., Hyun, J., Han Jip, K., 2009. The mitochondrial Warburg effect: a cancer enigma. IBC 1, 7.

Hawkins, C.E., Baars, C., Hesterman, H., Hocking, G.J., Jones, M.E., Lazenby, B., Mann, D., Mooney, N., Pemberton, D., Pyecroft, S., Restani, M., Wiersma, J., 2006. Emerging disease and population decline of an island endemic, the Tasmanian devil Sarcophilus harrisii. Biol. Conserv. 131, 307–324.

Hoffman, R.M., 2015. Patient-derived orthotopic xenografts: better mimic of metastasis than subcutaneous xenografts. Nat. Rev. Cancer 15, 451–452.

Hollings, T., Jones, M., Mooney, N., McCallum, H., 2014. Trophic cascades following the disease-induced decline of an apex predator, the Tasmanian devil. Conserv. Biol. 28, 63–75.

Hollings, T., Jones, M., Mooney, N., McCallum, H., 2016. Disease-induced decline of an apex predator drives invasive dominated states and threatens biodiversity. Ecology 97, 394–405.

Holmes, J.M., 1981. Measurement of the rate of death of canine transmissible venereal tumour cells transplanted into dogs and nude mice. Res. Vet. Sci. 30, 248.

Holz, P. (Ed.), 2008. Dasyurids. CSIRO Publishing, Collingwood, Victoria.

Hu, M., Polyak, K., 2008. Microenvironmental regulation of cancer development. Curr. Opin. Genet. Dev. 18, 27–34.

Ingles, E.D., Deakin, J.E., 2015. Global DNA methylation patterns on marsupial and devil facial tumour chromosomes. Mol. Cytogenet. 8, 1–11.

Jaenisch, R., Bird, A., 2003. Epigenetic regulation of gene expression: how the genome integrates intrinsic and environmental signals. Nat. Genet. 33 (Suppl.), 245–254.

Jones, M., 1997. Character displacement in Australian dasyurid carnivores: size relationships and prey size patterns. Ecology 78, 2569–2587.

Jones, M.E., Cockburn, A., Hamede, R., Hawkins, C., Hesterman, H., Lachish, S., Mann, D., McCallum, H., Pemberton, D., 2008. Life-history change in disease-ravaged Tasmanian devil populations. Proc. Natl. Acad. Sci. USA 105, 10023–10027.

Kashkush, K., Feldman, M., Levy, A.A., 2003. Transcriptional activation of retrotransposons alters the expression of adjacent genes in wheat. Nat. Genet. 33, 102–106.

Katzir, N., Rechavi, G., Cohen, J.B., Unger, T., Simoni, F., Segal, S., Cohen, D., Givol, D., 1985. "Retroposon" insertion into the cellular oncogene c-myc in canine transmissible venereal tumor. Proc. Natl. Acad. Sci. USA 82, 1054–1058.

Kreiss, A., Tovar, C., Obendorf, D.L., Dun, K., Woods, G.M., 2011. A murine xenograft model for a transmissible cancer in Tasmanian devils. Vet. Pathol. 48, 475–481.

Lane, A., Cheng, Y., Wright, B., Hamede, R., Levan, L., Jones, M., Ujvari, B., Belov, K., 2012. New insights into the role of MHC diversity in devil facial tumour disease. PLoS One 7, e36955.

Leggett, H.C., Cornwallis, C.K., West, S.A., 2012. Mechanisms of pathogenesis, infective dose and virulence in human parasites. PLoS Pathog. 8, e1002512.

Liao, K.-W., Lin, Z.-Y., Pao, H.-N., Kam, S.-Y., Wang, F.-I., Chu, R.-M., 2003. Identification of canine transmissible venereal tumor cells using in situ polymerase chain reaction and the stable sequence of the long interspersed nuclear element. J. Vet. Diagn. Invest. 15, 399–406.

Loue, S., 2000. Textbook of Research Ethics: Theory and Practice. Springer Science & Business Media, New York, NY.

Lowry, E., Rollinson, E.J., Laybourn, A.J., Scott, T.E., Aiello-Lammens, M.E., Gray, S.M., Mickley, J., Gurevitch, J., 2013. Biological invasions: a field synopsis, systematic review, and database of the literature. Ecol. Evol. 3, 182–196.

Manzotti, C., Audisio, R.A., Pratesi, G., 1993. Importance of orthotopic implantation for human tumors as model systems: relevance to metastasis and invasion. Clin. Exp. Metastasis 11, 5–14.

Martínez-Zaguilán, R., Seftor, E.A., Seftor, R.E., Chu, Y.W., Gillies, R.J., Hendrix, M.J., 1996. Acidic pH enhances the invasive behavior of human melanoma cells. Clin. Exp. Metastasis 14, 176–186.

McCallum, H., Jones, M., Hawkins, C., Hamede, R., Lachish, S., Sinn, D.L., Beeton, N., Lazenby, B., 2009. Transmission dynamics of Tasmanian devil facial tumor disease may lead to disease-induced extinction. Ecology 90, 3379–3392.

Merlo, L.M.F., Pepper, J.W., Reid, B.J., Maley, C.C., 2006. Cancer as an evolutionary and ecological process. Nat. Rev. Cancer 6, 924–935.

Merlo, L.M.F., Wang, L.-S., Pepper, J.W., Rabinovitch, P.S., Maley, C.C., 2010. Polyploidy, aneuploidy and the evolution of cancer. In: Poon, R.Y.C. (Ed.), Polyploidization and Cancer. Springer, New York, NY.

Metzger, M.J., Reinisch, C., Sherry, J., Goff, S.P., 2015. Horizontal transmission of clonal cancer cells causes leukemia in soft-shell clams. Cell 161, 255–263.

Metzger, M.J., Villalba, A., Carballal, M.J., Iglesias, D., Sherry, J., Reinisch, C., Muttray, A.F., Baldwin, S.A., Goff, S.P., 2016. Widespread transmission of independent cancer lineages within multiple bivalve species. Nature 534, 705–709.

Molinier, J., Ries, G., Zipfel, C., Hohn, B., 2006. Transgeneration memory of stress in plants. Nature 442, 1046–1049.

Mozos, E., Méndez, A., Gómez-Villamandos, J.C., Mulas, J.M.D.L., Pérez, J., 1996. Immunohistochemical characterization of canine transmissible venereal tumor. Vet. Pathol. 33, 257–263.

Muehlenbachs, A., Bhatnagar, J., Agudelo, C.A., Hidron, A., Eberhard, M.L., Mathison, B.A., Frace, M.A., Ito, A., Metcalfe, M.G., Rollin, D.C., Visvesvara, G.S., Pham, C.D., Jones, T.L., Greer, P.W., Vélez Hoyos, A., Olson, P.D., Diazgranados, L.R., Zaki, S.R., 2015. Malignant transformation of Hymenolepis nana in a human host. N. Engl. J. Med. 373, 1845–1852.

Mukaratirwa, S., Gruys, E., 2003. Canine transmissible venereal tumour: cytogenetic origin, immunophenotype, and immunobiology. A review. Vet. Q. 25, 101–111.

Murchison, E.P., Schulz-Trieglaff, O.B., Ning, Z., Alexandrov, L.B., Bauer, M.J., Fu, B., Hims, M., Ding, Z., Ivakhno, S., Stewart, C., Ng, B.L., Wong, W., Aken, B., White, S., Alsop, A., Becq, J., Bignell, G.R., Cheetham, R.K., Cheng, W., Connor, T.R., Cox, A.J., Feng, Z.-P., Gu, Y., Grocock, R.J., Harris, S.R., Khrebtukova, I., Kingsbury, Z., Kowarsky, M., Kreiss, A., Luo, S., Marshall, J., Mcbride, D.J., Murray, L., Pearse, A.-M., Raine, K., Rasolonjatovo, I., Shaw, R., Tedder, P., Tregidgo, C., Vilella, A.J., Wedge, D.C., Woods, G.M., Gormley, N., Humphray, S., Schroth, G., Smith, G., Hall, K., Searle, S.M.J., Carter, N.P., Papenfuss, A.T., Futreal, P.A., Campbell, P.J., Yang, F., Bentley, D.R., Evers, D.J., Stratton, M.R., 2012. Genome sequencing and analysis of the Tasmanian devil and its transmissible cancer. Cell 148, 780–791.

Murchison, E.P., Tovar, C., Hsu, A., Bender, H.S., Kheradpour, P., Rebbeck, C.A., Obendorf, D., Conlan, C., Bahlo, M., Blizzard, C.A., Pyecroft, S., Kreiss, A., Kellis, M., Stark, A., Harkins, T.T., Graves, J.A.M., Woods, G.M., Hannon, G.J., Papenfuss, A.T., 2010. The Tasmanian devil transcriptome reveals schwann cell origins of a clonally transmissible cancer. Science 327, 84–87.

Murchison, E.P., Wedge, D.C., Alexandrov, L.B., Fu, B., Martincorena, I., Ning, Z., Tubio, J.M.C., Werner, E.I., Allen, J., de Nardi, A.B., Donelan, E.M., Marino, G., Fassati, A., Campbell, P.J., Yang, F., Burt, A., Weiss, R.A., Stratton, M.R., 2014. Transmissible dog cancer genome reveals the origin and history of an ancient cell lineage. Science 343, 437–440.

Murdoch, C., Muthana, M., Coffelt, S.B., Lewis, C.E., 2008. The role of myeloid cells in the promotion of tumour angiogenesis. Nat. Rev. Cancer 8, 618–631.

Murgia, C., Pritchard, J.K., Kim, S.Y., Fassati, A., Weiss, R.A., 2006. Clonal origin and evolution of a transmissible cancer. Cell 126, 477–487.

Nachtigal, M., 1965. Chromosome analysis of a spontaneous, transplantable golden hamster tumor. J. Natl. Cancer Inst. 35, 381–411.

Nelson, C.M., Bissell, M.J., 2006. Of extracellular matrix, scaffolds, and signaling: tissue architecture regulates development, homeostasis, and cancer. Annu. Rev. Cell Dev. Biol. 22, 287–309.

Nilsson, M., Janke, A., Murchison, E., Ning, Z., Hallstrom, B., 2012. Expansion of CORE-SINEs in the genome of the Tasmanian devil. BMC Genomics 13, 172.

O'Neill, I.D., 2010. Tasmanian devil facial tumor disease: insights into reduced tumor surveillance from an unusual malignancy. Int. J. Cancer 127, 1637–1642.

Oprandy, J.J., Chang, P.W., 1983. 5-Bromodeoxyuridine induction of hematopoietic neoplasia and retrovirus activation in the soft-shell clam, *Mya arenaria*. J. Invertebr. Pathol. 42, 196–206.

Orimo, A., Weinberg, R.A., 2006. Stromal fibroblasts in cancer: a novel tumor-promoting cell type. Cell Cycle 5, 1597–1601.

Orr, H.A., 1995. Somatic mutation favors the evolution of diploidy. Genetics 139, 1441–1447.

Orr, H.A., Otto, S.P., 1994. Does diploidy increase the rate of adaptation? Genetics 136, 1475–1480.

Ostrander, E.A., Davis, B.W., Ostrander, G.K., 2016. Transmissible tumors: breaking the cancer paradigm. Trends Genet. 32, 1–15.

Otto, S.P., 2007. The evolutionary consequences of polyploidy. Cell 131, 452–462.

Otto, S.P., Whitton, J., 2000. Polyploid incidence and evolution. Annu. Rev. Genet. 34, 401–437.

Pearse, A.M., Swift, K., 2006. Allograft theory: transmission of devil facial-tumour disease. Nature 439, 549.

Pearse, A.-M., Swift, K., Hodson, P., Hua, B., McCallum, H., Pyecroft, S., Taylor, R., Eldridge, M.D.B., Belov, K., 2012. Evolution in a transmissible cancer: a study of the chromosomal changes in devil facial tumor (DFT) as it spreads through the wild Tasmanian devil population. Cancer Genet. 205, 101–112.

Penn, I., 1978. Tumors arising in organ transplant recipients. Adv. Cancer Res. 28, 31–61.

Penn, I., Starzl, T.E., 1972. Malignant tumors arising de novo in immunosuppressed organ transplant recipients. Transplantation 14, 407–417.

Petchey, O.L., 2000. Species diversity, species extinction, and ecosystem function. Am. Nat. 155, 696–702.

Podlaha, O., Riester, M., De, S., Michor, F., 2012. Evolution of the cancer genome. Trends Genet. 28, 155–163.

Polyak, K., Haviv, I., Campbell, I.G., 2009. Co-evolution of tumor cells and their microenvironment. Trends Genet. 25, 30–38.

Pye, R.J., Pemberton, D., Tovar, C., Tubio, J.M, Dun, K.A., Fox, S., Darby, J., Hayes, D., Knowles, G.W., Kreiss, A., Siddle, H.V.T., Swift, K., Lyons, A.B., Murchison, E.P., Woods, G.M., 2016. A second transmissible cancer in Tasmanian devils. Proc. Natl. Acad. Sci. 113, 374–379.

Pyecroft, S.B., Pearse, A.M., Loh, R., Swift, K., Belov, K., Fox, N., Noonan, E., Hayes, D., Hyatt, A., Wang, L., Boyle, D., Church, J., Middleton, D., Moore, R., 2007. Towards a case definition for devil facial tumour disease: what is it? EcoHealth 4, 346–351.

Rando, O.J., Verstrepen, K.J., 2007. Timescales of genetic and epigenetic inheritance. Cell 128, 655–668.

Rebbeck, C.A., Leroi, A.M., Burt, A., 2011. Mitochondrial capture by a transmissible cancer. Science 331, 303.

Rebbeck, C.A., Thomas, R., Breen, M., Leroi, A.M., Burt, A., 2009. Origins and evolution of a transmissible cancer. Evolution 63, 2340–2349.

Rehen, S.K., Yung, Y.C., Mccreight, M.P., Kaushal, D., Yang, A.H., Almeida, B.S.V., Kingsbury, M.A., Cabral, K.M.S., Mcconnell, M.J., Anliker, B., Fontanoz, M., Chun, J., 2005. Constitutional aneuploidy in the normal human brain. J. Neurosci. 25, 2176–2180.

Robert, J., 2010. Comparative study of tumorigenesis and tumor imminity in invertebrates and nonmammlian vertebrates. Dev. Comp. Immunol. 34, 915–925.

Rodenhiser, D., 2009. Epigenetic contributions to cancer metastasis. Clin. Exp. Metastasis 26, 5–18.

Sandusky, G.E., Carlton, W.W., Wightman, K.A., 1987. Diagnostic immunohistochemistry of canine round cell tumors. Vet. Pathol. 24, 495–499.

Scanlon, E.F., Hawkins, R.A., Fox, W.W., Smith, W.S., 1965. Fatal homotransplanted melanoma. A case report. Cancer 18, 782–789.

Schmid-Hempel, P., 2011. Evolutionary Parasitology: The Integrated Study of Infections, Immunology, Ecology, and Genetics. Oxford University Press, New York, NY.

Seitz, R.D., Lipcius, R.N., Hines, A.H., Eggleston, D.B., 2001. Density-dependent predation, habitat variation, and the persistence of marine bivalve prey. Ecology 82, 2435–2451.

Shao, C., Lacey, M., Dubeau, L., Ehrlich, M., 2009. Hemimethylation footprints of DNA demethylation in cancer. Epigenetics 4, 165–175.

Siddle, H.V., Kaufman, J., 2013a. How the devil facial tumor disease escapes host immune responses. OncoImmunology 2, e25235.

Siddle, H.V., Kaufman, J., 2013b. A tale of two tumours: comparison of the immune escape strategies of contagious cancers. Mol. Immunol. 55, 190–193.

Siddle, H.V., Kaufman, J., 2015. Immunology of naturally transmissible tumours. Immunology 144, 11–20.

Siddle, H.V., Kreiss, A., Tovar, C., Yuen, C.K., Cheng, Y., Belov, K., Swift, K., Pearse, A.-M., Hamede, R., Jones, M.E., Skjødt, K., Woods, G.M., Kaufman, J., 2013. Reversible epigenetic down-regulation of MHC molecules by devil facial tumour disease illustrates immune escape by a contagious cancer. Proc. Natl. Acad. Sci. 110, 5103–5108.

Siddle, H.V., Marzec, J., Cheng, Y., Jones, M., Belov, K., 2010. MHC gene copy number variation in Tasmanian devils: implications for the spread of a contagious cancer. Proc. R. Soc. B 277, 2001–2006.

Simberloff, D., 2009. The role of propagule pressure in biological invasions. Annu. Rev. Ecol. Evol. Syst. 40, 81–102.

Singer, T., Mcconnell, M.J., Marchetto, M.C.N., Coufal, N.G., Gage, F.H., 2010. LINE-1 retrotransposons: mediators of somatic variation in neuronal genomes? Trends Neurosci. 33, 345–354.

Southam, C.M., Moore, A.E., 1958. Induced immunity to cancer cell homografts in man. Ann. NY Acad. Sci. 73, 635–653.

Strakova, A., Murchison, E.P., 2014. The changing global distribution and prevalence of canine transmissible venereal tumour. BMC Vet. Res. 10, 1–11.

Taraska, N.G., Anne Böttger, S., 2013. Selective initiation and transmission of disseminated neoplasia in the soft shell clam *Mya arenaria* dependent on natural disease prevalence and animal size. J. Invertebr. Pathol. 112, 94–101.

Tlsty, T.D., Coussens, L.M., 2006. Tumor stroma and regulation of cancer development. Annu. Rev. Pathol. 1, 119–150.

Tolar, J., Neglia, J.P., 2003. Transplacental and other routes of cancer transmission between individuals. J. Pediatr. Hematol. Oncol. 25, 430–434.

Ujvari, B., Beckmann, C., Biro, P.A., Arnal, A., Tasiemski, A., Massol, F., Salzet, M., Mery, F., Boidin-Wichlacz, C., Misse, D., Renaud, F., Vittecoq, M., Tissot, T., Roche, B., Poulin, R., Thomas, F., 2016a. Cancer and life-history traits: lessons from host–parasite interactions. Parasitology 143, 533–541.

Ujvari, B., Gatenby, R.A., Thomas, F., 2016b. The evolutionary ecology of transmissible cancers. Infect. Genet. Evol. 39, 293–303.

Ujvari, B., Gatenby, R.A., Thomas, F., 2016c. Transmissible cancers, are they more common than thought? Evol. Appl. 9, 633–634.

Ujvari, B., Hamede, R., Peck, S., Pemberton, D., Jones, M., Belov, K., Madsen, T., 2016d. Immunoglubolin dynamics and cancer prevalence in Tasmanian devils (*Sarcophilus harrisii*). Sci. Rep. 6, 25093.

Ujvari, B., Papenfuss, A.T., Belov, K., 2016e. Transmissible cancers in an evolutionary context. BioEssays 38, S14–S23.

Ujvari, B., Pearse, A.-M., Peck, S., Harmsen, C., Taylor, R., Pyecroft, S., Madsen, T., Papenfuss, A.T., Belov, K., 2013. Evolution of a contagious cancer: epigenetic variation in devil facial tumour disease. Proc. R. Soc. B 280, 20121720.

Ujvari, B., Pearse, A.-M., Swift, K., Hodson, P., Hua, B., Pyecroft, S., Taylor, R., Hamede, R., Jones, M., Belov, K., Madsen, T., 2014. Anthropogenic selection enhances cancer evolution in Tasmanian devil tumours. Evol. Appl. 7, 260–265.

van Valen, L., 1973. A new evolutionary law. Evol. Theor. 1, 1–30.

Vincent, M.D., 2010. The animal within: carcinogenesis and the clonal evolution of cancer cells are speciation events sensu stricto. Evolution 64, 1173–1183.

Vittecoq, M., Ducasse, H., Arnal, A., Møller, A.P., Ujvari, B., Jacqueline, C.B., Tissot, T., Missé, D., Bernex, F., Pirot, N., Lemberger, K., Abadie, J., Labrut, S., Bonhomme, F., Renaud, F., Roche, B., Thomas, F., 2015. Animal behaviour and cancer. Anim. Behav. 101, 19–26.

Voskoboynik, A., Newman, A.M., Corey, D.M., Sahoo, D., Pushkarev, D., Neff, N.F., Passarelli, B., Koh, W., Ishizuka, K.J., Palmeri, K.J., Dimov, I.K., Keasar, C., Fan, H.C., Mantalas, G.L., Sinha, R., Penland, L., Quake, S.R., Weissman, I.L., 2013. Identification of a colonial chordate histocompatibility gene. Science 341, 384–387.

Warburg, O., 1930. Ueber den stoffwechsel der tumoren, Constable, London.

Warning, J.C., Mccracken, S.A., Morris, J.M., 2011. A balancing act: mechanisms by which the fetus avoids rejection by the maternal immune system. Reproduction 141, 715–724.

Welsh, J.S., 2011. Contagious cancer. Oncologist 16, 1–4.

Yang, T.J., 1988. Immunobiology of a spontaneously regressive tumor, the canine transmissible venereal sarcoma (review). Anticancer Res. 8, 93–95.

Yegnasubramanian, S., Haffner, M.C., Zhang, Y., Gurel, B., Cornish, T.C., Wu, Z., Irizarry, R.A., Morgan, J., Hicks, J., Deweese, T.L., Isaacs, W.B., Bova, G.S., de Marzo, A.M., Nelson, W.G., 2008. DNA hypomethylation arises later in prostate cancer progression than CpG island hypermethylation and contributes to metastatic tumor heterogeneity. Cancer Res. 68, 8954–8967.

13

Cancer in Animals: Reciprocal Feedbacks Between Evolution of Cancer Resistance and Ecosystem Functioning

Benjamin Roche*,**, *Anders Pape Møller*†, *James DeGregori*‡,
Frédéric Thomas*,§

*CREEC (Centre for Ecological and Evolutionary Research on Cancer), Montpellier, France
**UMMISCO (International Center for Mathematical and Computational
Modeling of Complex Systems), UMI IRD/UPMC UMMISCO, Bondy, France
†Ecology Systematic Evolution, CNRS UMR 8079, University Paris-Sud, Orsay, France
‡Department of Biochemistry and Molecular Genetics, University of Colorado School of Medicine,
Aurora, CO, United States
§MIVEGEC (Infectious Diseases and Vectors: Ecology, Genetics, Evolution and Control),
UMR IRD/CNRS/UM 5290, Montpellier, France

INTRODUCTION

Cancer, is often seen as an exclusively human disease. While this anthropocentrism is natural given the extremely high burden of this disease on human populations (Ferlay et al., 2010), this reveals nevertheless how much cancer in wildlife has been ignored by the scientific community. This ignorance is extremely detrimental for our understanding of this disease because we know that cancer occurs across almost the entire animal kingdom, from bivalves to whales through dogs and felines among others (see Chapter 2).

Cancer is a disease largely of the elderly. "Old age" is now common for humans, and for animals kept in captivity, but is much less common for animals in the wild, where death by other causes typically prevents survival to ages where the risk of cancer is greatest (DeGregori, 2011). If cancer primarily affects the elderly, and old animals are rare in the wild and even those that make it to old age should be less likely to reproduce successfully, then one might assume that cancer has had little impact on the evolution of these animals. However, this would be incorrect. The relative paucity of cancer in youth is an indication of the strong selection against cancer during periods when reproductive success is sufficiently likely.

In addition, as discussed later, changing environmental conditions (often due to human activities) can increase cancer risk even in young animals, with impacts on species survival and overall ecology.

Cancer in animals has received slightly more attention over the last few years due to two different topics. First, a large number of studies have attempted to figure out why the naked-mole rat (*Heterocephalus glaber*) was considered almost completely resistant to cancer (Tian et al., 2013) until recently (Delaney et al., 2016). Second, the absence of a positive relationship between body mass and cancer incidence across species, known as Peto's paradox, has attracted the attention of many evolutionary biologists. Indeed, if larger body mass corresponds to more cell divisions and each have the same probability of generating a cancer, whales should have much more cancer than humans and mice. The absence of a positive relationship suggests that the mechanisms of cancer resistance must have been more strongly selected in larger species (Caulin and Maley, 2011; Roche et al., 2012, see also Chapters 1 and 22). Nevertheless, such resistance has mostly been studied with the aim of identifying resistance mechanisms that could be translatable to humans. Alternatively, Peto's paradox is based on a cell intrinsic view of cancer formation, assuming that cancer is limited by the occurrence of oncogenic mutations. More recent studies have challenged this paradigm, and noncell autonomous factors like cell competition, tissue maintenance, and even immunity, all in correlation with body size, can provide a potential answer to this paradox (Rozhok and DeGregori, 2015).

Since applying evolutionary ecology concepts to cancer has been instrumental for our understanding of cancer and its treatment (Gatenby et al., 2009; Greaves and Maley, 2012), it is also likely that cancer can have implications for the evolutionary ecology of animals, which could in turn affect the dynamics of oncogenic phenomena. In this chapter, we review the potential impact of cancer on ecology, behavior and evolution of animal species. Furthermore, we discuss how this could impact the structure of ecological communities. We finally highlight that, in a similar way as parasites were previously neglected by ecologists, oncogenic processes are also crucial for understanding ecosystem functioning.

CANCER IN WILDLIFE: WHERE ARE WE?

Cancers are quite well monitored in human populations, thanks to the numerous prevention and surveillance programs around the world. Consequently, data are available about the incidence and mortality of most cancers and for different age classes in human populations throughout the world (Ferlay et al., 2010). Such datasets, although not available with the same level of details, also exist for domestic animals (Meuten, 2008). However, the rates of occurrence and the impact of cancer in wildlife is mostly a mystery (see Chapter 2).

The minimal data reported for cancer in animals include only scattered studies on dogs, cats, humans, elephants, and Beluga whales [reviewed in (Ducasse et al., 2015)]. The first reason for the paucity of studies is obviously logistic, since cancer in wildlife is extremely difficult to observe. While cancer can be lethal in wild animal species, quantifying its contribution to mortality is extremely challenging since death in natural conditions can be attributed to numerous other factors (e.g., predation, infectious diseases, somatic diseases, abiotic perturbations), and identification of the presence of one disease cannot be used as evidence that others did not play a role. Finally, predation and parasitism may have a particularly strong impact on individuals with cancer, resulting in a significant interaction between cancer incidence and interspecific ecological interactions.

It is therefore important to extrapolate what could be the impact of cancer in wildlife based on the knowledge accumulated for humans. Indeed, the extensive surveillance of cancer in humans has told us at least two important things. First, a lot of individuals are accumulating malignant lesions through their life, which fortunately for most individuals do not progress to invasive cancer, and these individuals largely remain asymptomatic (Folkman and Kalluri, 2004). Second, cancer and other malignancies in humans can have highly variable consequences for health and vigor, depending on the type of cancer and its stage (Vittecoq et al., 2013). Such impacts, even if manageable and probably underreported in humans, may have dramatic fitness outcomes for wild animals since a decrease in vigor can reduce defense mechanisms against predators or immunocompetence against parasites. Therefore, progression of malignancies, depending on the species and the organ considered, could impact host fitness more or less gradually, and not only during the metastatic and eventually fatal stage. As a consequence, cancer progression should be detrimental for the affected individuals, which would select for sophisticated resistance mechanisms, and which in turn could impact ecological interactions between affected species.

HOW DOES NATURAL SELECTION DEAL WITH CANCER RISK?

Resistance Mechanisms

Studies on Peto's paradox have revealed some of the mechanisms used by animals to achieve protection from cancer, especially cellular mechanisms that can block proliferation of cancerous cells (see also Chapters 1 and 7). One of the most striking examples is the resistance of naked-mole rats to cancer that seems to be in part the consequence of production of a large quantity of hyaluronic acid. The molecular mass of hyaluronic acid in naked mole rats have been observed to be five times larger than in humans, and hence, could act as a "cage" around the extracellular matrix of cells and therefore isolate tumor development (Tian et al., 2013).

Despite this singular example (naked-mole rats also have a very specific ecology and is the only species where the absence of cancer was assumed until very recently), other mechanisms could potentially contribute, most likely in a nonmutually exclusive way. For instance, redundancy of tumor-suppressor genes (TSG) has been suggested as a potential explanation for Peto's paradox (Nunney, 2003). Recently, this explanation has received empirical support with the observation that elephants, which have specifically high resistance against cancer, possess 20 copies of the *p53* anticancer gene (Abegglen et al., 2015) (see also Chapters 1 and 7). However, it remains to be seen whether these extra p53 genes (which all possess mutations leading to truncation of the proteins before the crucial DNA binding domain) all encode for proteins that promote tumor suppression. This crucial gene, involved in DNA repair in all mammals, has been nicknamed the "guardian of genome", but is only present in a single copy in humans. Similar patterns have been suggested with the gene *MAL* in horses and *FBXO31* in microbats to explain partial resistance to cancer, but this mechanism of increased TSG number does not seem to be a common explanation across a large range of species (Caulin et al., 2015).

To become cancer cells, stem cells need to mutate. Therefore, a slower somatic mutation rate per cell-generation in large animals could decrease the occurrence of cancer. Moreover, the somatic mutation rate, in addition to being species-dependent, could also differ among organs when cells have differentiated (Ducasse et al., 2015). Nevertheless, mutation rate seems to be of similar magnitude in mice and humans, which does not support mutation rate as an exclusive solution for Peto's paradox, even if it could be an additional mechanism involved in cancer resistance in very large and long-lived mammals, such as elephants or whales. In fact, many single celled organisms, like yeast, exhibit lower mutation rates than mammals (Lynch, 2010), and thus the evolution of multicellularity, complexity, and long lives has not come hand and hand with a reduction in mutation rates (DeGregori, 2013).

Apoptosis, even for cells becoming cancerous, could be triggered in many ways. One of the "inherited" possibilities to initiate apoptosis is when telomeres, which shorten after each cell division, become too short to protect the end of chromosomes. Therefore, it would be possible that individuals with shorter telomeres belonging to species with large body mass may have profited from a selective advantage, resulting in selection for shorter telomeres (or faster telomere reduction) (Caulin and Maley, 2011; Gorbunova et al., 2014) (see Chapter 1).

Another interesting explanation could rely on the modulation of metabolic rhythm (Dang, 2015). Indeed, reactive oxygen species (ROS), which can cause DNA damage, are more abundant in individuals with high basal metabolic rate (Ku et al., 1993), and in areas with higher background radiation. Therefore, a high metabolic rate could result indirectly in more DNA damage and then in cancerous cells. Since the basal metabolic rate is negatively associated with body mass (Savage et al., 2007), large mammals should on average produce less ROS and therefore decreasing the occurrence of cancer cells than small mammals. However, there are many exceptions to this correlation, such as birds (which have relatively long lives) (Munshi-South and Wilkinson, 2010), and large variance in lifespans among similarly sized rodents like mice and moles (Gorbunova et al., 2014).

Finally, it is worth mentioning that a neutral explanation for Peto's paradox could simply rely on the fact that the number of stem cells does not necessarily correlate with body mass (Abkowitz et al., 2002). This can cast doubt on the fact that Peto's paradox is a relevant null hypothesis to reveal selection for resistance mechanisms.

Indeed, while body mass does not correlate with cancer incidence, there appears to be some association between the cumulative number of stem cell divisions and cancer risk for some organs within humans (Noble et al., 2015; Pepper et al., 2007; Roche et al., 2015; Tomasetti and Vogelstein, 2015). This suggests that stem cell abundance could be a more relevant null hypothesis, although this idea remains poorly explored.

Today, Peto's paradox has stimulated research on the selection of extreme resistance mechanisms through focusing on cellular causes of cancer because of evident expectations for medical treatments. However, it is very likely that, apart from cellular controls, other resistance mechanisms may have been selected above the cell levels, selective pressures that concomitantly also lead to increase in body mass (Ducasse et al., 2015). This is the reason why many studies have attempted to improve our knowledge on the different selective pressures enhancing cancer resistance and to identify potential animal species with particular tumor suppressor capacities (Nagy et al., 2007; Roche et al., 2012). For example, models of cancer that describe tissue landscape maintenance as a key mechanism of preventing somatic evolution of cancer may provide a potential solution to Peto's paradox (Rozhok and DeGregori, 2015). Basically, this model postulates that different animals have evolved different strategies (largely due to different external hazards) for somatic tissue maintenance, and that these same strategies serve to limit somatic evolution largely until postreproductive periods (old age) by a somatic stabilizing selection mechanism.

As described in the "Introduction" (as well as in the Introductory Chapter and Chapter 7), natural selection has worked to limit (but not eliminate) cancer incidence through the period of life when reproduction was most likely, which for most animals would not include ages where physiological decline (i.e., senescence) is evident. As described previously, this selection could reflect different evolutionary strategies, including: (1) evolution of unique strategies for different animals (such as for TSG number in elephants and high molecular weight hyaluronic acid in naked mole rats), or (2) modulation and tuning of common strategies that are used by many animals (e.g., all vertebrates or at least all mammals), such as by limiting the number of stem cells or by the maintenance of tissue landscapes that disfavor somatic evolution.

Behavioral Adjustments to Cancer

There are numerous ways for an individual to decrease exposure to the source that may contribute to cancer initiation and development. We should expect that animal host species should be under selective pressures to first avoid the sources of pathology (prophylactic behavior), then to prevent its progression if avoidance is unattainable (evolved tumor suppressive strategies) and finally alleviate the fitness costs if lethal development is not preventable (postcancer behavior).

Many prophylactic behaviors may have been selected to protect against cancer because the large role of environmental factors involved in carcinogenesis, which could be naturally present (e.g., natural radiation levels, oncogenic pathogens, transmissible cancers, and secondary compounds of plants) or due to anthropogenic alterations including diet and pollution in its many forms. Therefore, any form of behavior that may potentially contribute to reduce or increase an individual's risk of exposure (and concomitantly its cancer consequences) could be selected or counter-selected, which will obviously be modulated by the necessity of the initial behavior for the individual's survival and reproduction.

A striking example of *habitat selection* relies on the reduction of radiation encountered from natural or anthropogenic sources, the first being surprisingly more important than the latter (Aarkrog, 1990). While the health consequences for animal species of such exposure are poorly understood, they are presumed to be substantial given the strong association between radiation exposure and cancer-related mortality observed in humans (e.g., Brenner et al., 2003; Lubin and Boice, 1997; Møller and Mousseau, 2011; Prasad et al., 2004). Therefore, such habitat selection has been suggested to occur in extreme environments, such as the Chernobyl area where birds prefer to breed in sites with the lowest radioactivity (Møller and Mousseau, 2007a). Similarly, in highly polluted habitats where some fish and aquatic invertebrate species are able to detect and to avoid polluted areas (da Luz et al., 2004; De Lange et al., 2006; Giattina and Garton, 1983), including sediments contaminated by Polycyclic aromatic hydrocarbons (PAH) that have been shown to induce cancer in fish and marine mammals. While these examples could represent the first evidence of habitat selection to avoid carcinogenic exposures, it is nevertheless important to consider hormetic effects since low doses of X-ray radiation may protect individuals from the subsequent development of tumors, as suggested for mice (Yu et al., 2013), although this remains controversial (Costantini et al. 2010, Costantini and Macri 2013).

As cancer development can be initiated and accelerated by the accumulation of DNA mutations, cancer avoidance could have in part selected for *antioxidant consumption*. Indeed, DNA damage is frequently caused by free radicals (von Sonntag, 2006) for which antioxidants, especially some carotenoids, can encapsulate and consequently mitigate damage to DNA sequences (Møller et al., 2000). This could explain why many animal species tend to show a strong preference for food containing antioxidants (Senar et al., 2010), although the health benefits of these foods clearly extends beyond cancer prevention.

Avoidance of contagious cancers could also be a selective force for specific behavior, although only two examples of such disease exist today in mammals: Tasmanian devil facial tumor disease (DFTD), which is transmitted among Tasmanian devils (*Sarcophilus harrisii*) through biting during social interactions (Welsh, 2011), and the canine transmissible venereal tumor (CTVT) (Murchison, 2008) that is transmitted during sexual intercourse in dogs (see also Chapter 12). As these cancers decrease individual fitness significantly, by affecting sexual intercourse in dogs for CTVT and through survival reduction of devils to 6 months upon infection for DFTD (Murchison, 2008), it is expected that natural selection should favor susceptible individuals that are capable of recognizing infectious conspecifics in order to avoid risky contact (Boots et al., 2009; O'Donnell, 1997). These benefits could be counter-balanced by costs driven by sexual selection. For instance, some beneficial traits (such as the extended and rough sexual intercourse in dogs and aggressive biting behavior of devils) could improve breeding success (Hamede et al., 2009). To illustrate this potential counter-balancing force, it seems that aggression in dogs is not modulated by CTVT frequency, despite CTVT being the oldest and most widely disseminated cancer in the world (Murchison et al., 2014).

Such avoidance behavior can also modulate exposure to oncogenic pathogens, which are increasingly recognized in cancer epidemiology (Hausen and Zur Hausen, 2009), despite being poorly documented in wild animals. As for any other virulent contagious pathogens, prophylactic behaviors should be selected to reduce the risk of infection by oncogenic pathogens (e.g., Kavaliers et al., 2004). A potential example exists for cancer with visible symptoms, as for green sea turtles (*Chelonia mydas*) that are frequently affected by fibropapillomatosis (caused by a sexually-transmitted herpesvirus), which is characterized by multiple external epithelial tumors reliably indicating the infectious status of the individual [(Aguirre and Lutz, 2004), Fig. 13.1].

Despite its plausibility, it is worth mentioning that examples of prophylactic behaviors against cancer appear to be very rare. As mentioned, this could be due to the cost of modulation behavior. Another possibility is that prophylactic behaviors for avoiding cancer could be rarer than behaviors that prevent tumor progression and/or that compensate for their fitness consequences. A first example could rely on *self-medication*, which is now abundantly described in the animal kingdom as a defense against parasites (de Roode et al., 2013; Huffman, 2001; Lozano, 1998). A classic example in great apes is leaf ingestion that helps decrease intestinal parasite load, especially nematodes (Huffman, 2001). Regarding the large number of anticancer substances present in many ecosystems (ranging from leaves and bark to fungi with potential effective cancer treatment properties

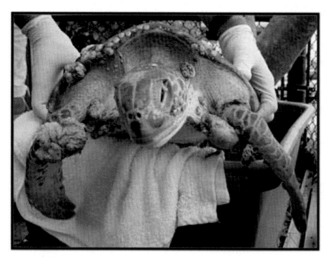

FIGURE 13.1 **A green sea turtle (*Chelonia mydas*) affected by fibropapillomatosis.** *Source: Reproduced with permission from Dr. N. Mettee, WIDECAST.*

(Vittecoq et al., 2015), self-medication could have been selected to slow down tumor progression associated with fitness reductions. For instance, chimpanzees (*Pan troglodytes)* could be suspected to use self-medication by consuming different plants with antitumoral properties (Masi et al., 2012). Nevertheless, it is also important to keep in mind that, as for any treatment targeting symptoms rather than sources, self-medication against cancer would make selection more "myopic" to the genes responsible for the disease, thus increasing persistence of inherited oncogenic mutations in populations and therefore maintaining a higher frequency of individuals affected by cancer.

Another kind of behavior that may have been selected to decrease cancer burden is the *increase in sleep duration.* Indeed, sleep duration seems strongly associated with a stronger immune system (Bryant et al., 2004), corroborated by the observations that mammals sleeping longer exhibit lower levels of parasitic infections (Preston et al., 2009). Regarding the extremely important role on cancer cell destruction (Grivennikov et al., 2010), a more efficient immune system could be a significant selective force increasing sleep duration. Moreover, sleep duration is also linked by the production of different hormones that are considered important antitumor agents (Blask, 2009). It is also interesting that species with short sleep duration, such as herbivorous elephants (*Loxodonta africana*) that only sleep 3 h per day, because of constant foraging, may have necessitated the selection for other kinds of cancer resistance mechanisms (Belyi et al., 2010). Obviously, the respective contribution of sleep duration and other factors, such as body size on cancer risk, is challenging to quantify, knowing that they interact in a complex fashion and have coevolved. Indeed, there are complex strategies to optimize the balance of energy

allocated to fighting disease relative to other essential activities, such as reproduction [see (Aubert, 1999) for a review on host-parasite interactions].

Finally, in addition to selection for processes that prevent cancer initiation, or that slow down its progression, natural selection may potentially also favor individuals displaying parental behavior that reduces cancer risks for offspring. For instance, preference for breeding habitats with low mutagen exposure could have a weak impact on parental survival, but could potentially be important for reducing the subsequent development of cancers in offspring. This suggests that adaptive behavior to cancer avoidance could be a *trans*-generational process. Another possibility is through sexual selection. According to the "good genes" hypothesis (Møller and Alatalo, 1999), healthy individuals will provide their progeny with a superior genetic background, such as efficient genetic defenses against cancer. Therefore, offspring have a direct benefit by avoiding partners with malignancies and/or genetic vulnerabilities to cancer because individuals mating with sick conspecifics will acquire a mate unable to provide high quality parental care, which will lower the survival prospects of their offspring [efficient parent hypothesis (Hoelzer, 1987)]. However, this would depend of the benefit of such genetic susceptibility, as demonstrated in male fish *Xiphophorus* spp., in which the cost of bearing an oncogene associated with a very high risk of melanoma seems to be overcome by its strong positive effect on male reproductive success, especially caused by more aggressive behavior (Fernandez and Bowser, 2010).

Evolution of Life History Traits

Cellular mechanisms and behavioral adjustments are two powerful possibilities for avoiding cancer and/or slowing its progression. Nevertheless, in some cases, the optimal solution may have been evolution of life history (LH) traits that alleviate the fitness costs if further development of the disease is not preventable (Thomas et al., 2009).

Despite that the shape of the relationship between fitness and cancer progression has not been quantified for most cancers and species, individuals with tumors are likely to be, at some point, in worse condition than the average healthy individual. In humans, extreme tiredness (fatigue) or weight loss has been documented in affected individuals, resulting from cancer cells taking much of the body's energy supply or releasing substances altering how the body derives energy from food (Ryan et al., 2007; Wagner and Cella, 2004). Cancer can also reduce immune surveillance and increase susceptibility to infectious diseases (de Visser et al., 2006). This altered allocation of energy acquisition is fundamental for figuring out how competing LH traits might be affected by cancer progression. Therefore, this represents a very promising research focus.

In addition to the direct effects of genetic inheritance on their offspring's phenotype, the influence of parental phenotypes is increasingly recognized (i.e., parental effects, see Mousseau et al., 2009; Wolf and Wade, 2009). It is therefore important to understand what could be the consequences of having "cancerous" parents for their offspring. While it would remain dependent on the relationship between fitness and cancer progression, parents with advanced disease stages are likely to be less able to deliver adequate resources/parental care for their offspring. Despite this question never having been addressed for cancer, the numerous examples documented for host-parasite interactions (Ujvari et al., 2016) strongly suggest that this research avenue is promising.

Finally, alteration of life-history strategies could also be the consequence of avoidance exposure to the disease for the specific case of contagious cancers. A striking example involves the DFTD transmitted between Tasmanian devils, for which the emergence of DFTD has been associated with precocious breeding by 1-year-old females that survive to produce only one litter instead of several before DFTD emergence (Jones et al., 2008). Thus, this suggests that Tasmanian devils have responded to the cancer-induced mortality through a radical change in reproductive strategies, from an iteroparous (multiple reproductive cycles) to a semelparous (single breeding) strategy (Jones et al., 2008).

IMPACT OF CANCER ON ECOLOGICAL COMMUNITIES

As previously described, cancer may impact individual fitness, and consequently could interfere with many ecological interactions. Depending on the magnitude (and the nature) of this fitness decrease, cancer may play a significant role in structuring animal communities and ultimately affecting biodiversity.

Predation is one of the core ecological interactions structuring ecological communities. It has already been documented that individuals in poor health condition, such as those that are heavily parasitized, such as birds infected by *Plasmodium* parasites (Møller and Nielsen, 2007; Murray et al., 1997), suffer from predation more than healthy ones. Thus, predators could also disproportionately catch prey affected by advanced malignancies, which are generally associated with reduced viability. This phenomenon could be exacerbated if predators use cues to detect cancerous prey, as suggested by the difference in odors between tumor-bearing and healthy mice (Alves et al., 2010). It is nevertheless worth mentioning that oncogenic phenomena can also impact predators themselves and may reduce the ability

of predators to catch prey, as shown for captive wild felids that suffer from anorexia when affected by different types of cancers (Dorso et al., 2008; Finotello et al., 2011). The net outcome of cancer on predator/prey relationships would then depend on the balance between the impacts of oncogenic phenomena in predators and in prey.

This predation relationship can also be embedded in other indirect interspecific interactions, which are frequently observed in nature (Walsh, 2013). A striking example is the process of *apparent competition*, that is, when interactions between two different species is mediated by a third species (Paine, 1966) like when two prey are attacked by the same predator (Holt and Lawton, 1994; Price et al., 1986). Since oncogenic processes can differentially affect individuals belonging to different species, cancer progression can, for instance, make a prey species more vulnerable to the predator, which would increase predator abundance, but consequently decrease the abundance of the second prey species. Therefore, species whose fitness is the most impaired by cancer could experience a strong selective disadvantage in competition with less affected species, but also for other species.

Many interactions also rely on available *abiotic resources*. The increase in pollution in a large range of habitats, as documented for instance by the increase of toxic contaminants in freshwater ecosystems (Lebarbenchon et al., 2008), can dramatically scramble interspecific interactions through cancer initiation. Indeed, pollution and radioactive contamination, as observed across tens of thousands of square kilometers of land worldwide because of the Chernobyl disaster and nuclear weapons testing (Møller and Mousseau, 2007b), can induce somatic mutations that are sometimes the source of genetic diseases including cancer (Brown et al., 1973; Yablokov, 2010). Therefore, cancer observations in wildlife are significantly more frequent in wildlife species inhabiting environments that are heavily affected with such contaminants (McAloose and Newton, 2009). It is worth pointing out that all species are not equally affected. Within the Chernobyl area, birds show frequent tumors on their feet, beak or head (Mousseau and Moller, 2011). Nevertheless, bird species that are the most affected by radiation are those that are the most susceptible to high substitution rates in mitochondrial DNA (Møller et al., 2010). For fish communities, a benthic lifestyle increases chronic exposure to contaminated sediment and contaminated invertebrates, and they are therefore more affected by cancer (Black and Baumann, 1991; Martineau et al., 2002). A similar pattern is observed in dead adult belugas (*Delphinapterus leucas*) that are heavily affected by cancer in the St Lawrence River estuary (Martineau et al., 2002), a strongly contaminated area.

As for apparent competition, this differential susceptibility of wildlife species to trigger cancer after exposure to contaminants suggests that oncogenic phenomena could scramble the biotic species interactions relying on abiotic contaminated factors, such as competition for shared resources. Nevertheless, species can also rapidly evolve to escape the pressure of such pollution. Such rapid evolution of contaminant tolerance has been shown to be pivotal in shaping fish communities in polluted areas (Elskus, 2001), where three populations of killifish (*Fundulus heteroclitus*) living in different settings differentially contaminated by pollutants have independently converged on an extreme tolerance to the lethal effects of toxic dioxin-like pollutants (Whitehead et al., 2012). Therefore, such rapid evolution of tolerance mechanisms could increase the competitive ability of mutagen tolerant species in contaminated areas, despite there undoubtedly being a cost for tolerance, such as compromised immune function (Hammouda et al., 2012) or decreased growth rate (Sibly et al., 2012).

Infectious diseases also represent an ecological pressure comparable to predators since they are a major cause of mortality and morbidity in wildlife (Gulland, 1995). This pressure could be exacerbated if individuals are immunosuppressed, increasing the probability of becoming infected. Interestingly, immunosuppression can be triggered by many processes, especially factors produced by tumors or by host responses to the presence of a tumor (Pollock and Roth, 1989). Therefore, individuals bearing tumors generally suffer from a higher vulnerability to infections (Aguirre and Lutz, 2004; Bernstorff et al., 2001; Kim et al., 2006). Since parasite, communities are known to be an important component affecting host community structure and functioning (Hudson et al., 2006; Lafferty et al., 2006), the possible role of cancer on pathogen communities could be a key determinant for host community structure as well.

Oncogenic phenomena and pathogens can also be involved in "vicious circles" where hosts in poor condition are more susceptible to higher pathogen occurrence and infection intensity, further weakening host condition (Beldomenico and Begon, 2010), as well as increasing probability of cancer occurrence. Such interactions could be even worse since several cancers in wildlife are initiated by infections (Møller and Nielsen, 2007; Murray et al., 1997), such as otarine herpesvirus-1 infections that seem to favor carcinoma formation in California sea lions [*Zalophus californianus*; (Newman and Smith, 2006)] (see also Chapter 3).

To conclude, the large number of *reciprocal interactions* between cancer progression and ecological interactions reveals that oncogenic phenomena cannot be considered in isolation of the ecosystem surrounding the affected individual. Reciprocally, it has become clear that understanding ecosystem functioning without considering cancer could provide an extremely incomplete picture because of missed crucial components.

Ecological interactions are typically studied separately (Møller, 2008). This is why cancer should be included within any kind of framework aiming at understanding how community structure results from these "interactions between interactions". To illustrate this importance, one could imagine that predation risk perceived by prey through different cues can modulate prey immune response (Navarro et al., 2004), which could in turn enhance cancer proliferation (Reiche et al., 2004) and parasite loads. Nevertheless, this increased parasite load can also increase the probability of infection in other wildlife species, and therefore decrease predator capability. The outcomes of such complex interactions, while challenging to predict, could nevertheless open new avenues in both community ecology and oncology.

DISCUSSION

Regarding the fact that almost all multicellular organisms can be affected by cancer, the influence of malignancy progressions on the ecology and evolution of wildlife species is a very promising research area. As reviewed in this chapter, wildlife species may have selected many different ways of modulating the potentially devastating impact of cancer, from cellular and tissue mechanisms constraining cancer cell proliferation to prophylactic behavior through evolution of LH traits. The impact of cancer itself, by decreasing the health condition of those affected, or various resistance mechanisms can both scramble ecological interactions between wildlife species, mostly because cancer may have an extremely diverse impact across different species. Such noise in ecological interactions, even if not quantified yet, has the potential to significantly shape the structure of animal communities (Fig. 13.2).

Several deficits in our knowledge are needed to be addressed before we are able to quantify the role of cancer on evolution of species and community structure. First, the shape of the relationship between cancer progression and fitness decrease should be quantified for a range of animal species and cancer types. As stated previously, there is accumulating evidence that cancer progression, even if it apparently remains "silent" for a long time, could have a large impact on health conditions. Such fitness-impairment, within an animal community, could be exacerbated according to the other species present (predators, prey, parasites, etc.). Nevertheless, this could depend on each organ, since some "key" organs could, when affected by cancer degrade health conditions more rapidly than others, such as pancreas versus gallbladder.

Obviously, the main limitation in our current understanding relies on the lack of empirical data (as highlighted in Chapter 2). Few examples exist, but cancers in wild animals are extremely difficult to study because of the lack of apparent symptoms and/or due to the presence

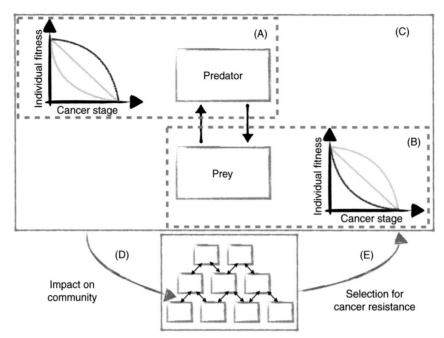

FIGURE 13.2 **Reciprocal interactions between cancer and structure of the ecological community.** (A) Possible relationship between cancer progression and impact on individual fitness with an example of a cancer that has a slow impact on predator (*dark blue line*). (B) Possible relationship between cancer progression and impact on individual fitness with an example of a cancer that has a fast impact on prey (*dark blue line*). *Light blue-lines* represent the other possible impacts of cancer on fitness that could be selected through resistance/tolerance mechanisms. (C) A given cancer having different impact on each animal species could scramble interspecific interactions (e.g., predation). (D) If such an impact significantly changes the outcome of species interactions, it could therefore impact the whole structure of the animal community, and then (E) trigger a selection for cancer resistance.

of scavengers that rapidly remove corpses (among other factors). To this extent, the development of noninvasive diagnostic technologies, as developed for wildlife infectious diseases, could be extremely insightful. Nevertheless, current methodologies, such as circulating tumor cells in blood (Plaks et al., 2013), today require logistic conditions that are extremely hard to satisfy in the field. One way to compensate for this lack could rely on the observation of cancer in captive animals, especially in zoos (see Chapter 2). Even if captivity conditions are extremely benign compared to natural environments, such analysis could possibly inform on susceptibility to cancer for a large spectrum of species and cancer types.

Overall, insufficient consideration of cancer in wild animals needs to be remedied for at least two reasons. First, mutagenic pollutants heavily affect a large number of ecosystems on our planet and this trend is probably not going to be reversed any time soon. The first consequence of this pollution is the degradation of abiotic resources, which is known to contribute to a decrease in biodiversity. Nevertheless, the concomitant abrupt increase in cancer frequency observed in wildlife has the potential to trigger a "cascade effect" on biodiversity decrease at all scales. During our current "anthropocenosis" (Monastersky, 2015), defined as this new ecological era where Earth is mostly impacted by human activities,

it seems crucial to anticipate the role of cancer in community structure. Reciprocally, understanding how wildlife species may have selected mechanisms against cancer can provide important insights into different ways to avoid cancer. This win–win interaction between oncology and evolutionary ecology is ripe for further study.

References

Aarkrog, A., 1990. Environmental radiation and radioactive releases. Int. J. Radiat. Biol. 57, 619–631.
Abegglen, L.M., Caulin, A.F., Chan, A., Lee, K., Robinson, R., Campbell, M.S., Kiso, W.K., Schmitt, S.L., Waddell, P.J., Bhaskara, S., Jensen, S.T., Maley, C.C., Schiffman, J.D., 2015. Potential mechanisms for cancer resistance in elephants and comparative cellular response to DNA damage in humans. JAMA 314, 1850–1860.
Abkowitz, J.L., Catlin, S.N., McCallie, M.T., Guttorp, P., 2002. Evidence that the number of hematopoietic stem cells per animal is conserved in mammals. Blood 100, 2665–2667.
Aguirre, A.A.A., Lutz, P.L., 2004. Marine turtles as sentinels of ecosystem health: is fibropapillomatosis an indicator? Ecohealth 1, 275–283.
Alves, G.J., Vismari, L., Lazzarini, R., Merusse, J.L.B., Palermo-Neto, J., 2010. Odor cues from tumor-bearing mice induces neuroimmune changes. Behav. Brain Res. 214, 357–367.
Aubert, A., 1999. Sickness and behaviour in animals: a motivational perspective. Neurosci. Biobehav. Rev. 23, 1029–1036.
Beldomenico, P.M., Begon, M., 2010. Disease spread, susceptibility and infection intensity: vicious circles? Trends Ecol. Evol. 25, 21–27.

Belyi, V., Ak, P., Markert, E., Wang, H., Hu, W., Puzio-Kuter, A., Levine, A.J., 2010. The origins and evolution of the p53 family of genes. Cold Spring Harb. Perspect. Biol. 2, a001198.

Bernstorff, W.von, Voss, M., Freichel, S., Schmid, A., Vogel, I., Jöhnk, C., Henne-Bruns, D., Kremer, B., Kalthoff, H., 2001. Systemic and local immunosuppression in pancreatic cancer patients. Clin. Cancer Res. 7, 925s–932s.

Black, J.J., Baumann, P.C., 1991. Carcinogens and cancers in freshwater fishes. Environ. Health Perspect. 90, 27.

Blask, D.E., 2009. Melatonin, sleep disturbance and cancer risk. Sleep Med. Rev. 13, 257–264.

Boots, M., Best, A., Miller, M.R., White, A., 2009. The role of ecological feedbacks in the evolution of host defence: what does theory tell us? Philos. Trans. R. Soc. Lond. B. Biol. Sci. 364, 27–36.

Brenner, D.J., Doll, R., Goodhead, D.T., Hall, E.J., Land, C.E., Little, J.B., Lubin, J.H., Preston, D.L., Preston, R.J., Puskin, J.S., 2003. Cancer risks attributable to low doses of ionizing radiation: assessing what we really know. Proc. Natl. Acad. Sci. 100, 13761–13766.

Brown, E.R., Hazdra, J.J., Keith, L., Greenspan, I., Kwapinski, J.B.G., Beamer, P., 1973. Frequency of fish tumors found in a polluted watershed as compared to nonpolluted Canadian waters. Cancer Res. 33, 189–198.

Bryant, P.A., Trinder, J., Curtis, N., 2004. Sick and tired: does sleep have a vital role in the immune system? Nat. Rev. Immunol. 4, 457–467.

Caulin, A.F., Graham, T.A., Wang, L.-S., Maley, C.C., 2015. Solutions to Peto's paradox revealed by mathematical modelling and cross-species cancer gene analysis. Philos. Trans. R. Soc. Lond. B. Biol. Sci. 370, 20140222.

Caulin, A.F., Maley, C.C., 2011. Peto's Paradox: evolution's prescription for cancer prevention. Trends Ecol. Evol. 26, 175–182.

Costantini, D., Macri, S. (Eds.), 2013. Adaptive and Maladaptive Aspects of Developmental Stress. Springer, Berlin.

Costantini, D., Metcalfe, N.B., Monaghan, P., 2010. Ecological processes in a hormetic framework. Ecol. Lett. 13, 1435–1447.

da Luz, T.N., Ribeiro, R., Sousa, J.P., 2004. Avoidance tests with Collembola and earthworms as early screening tools for site-specific assessment of polluted soils. Environ. Toxicol. Chem. 23, 2188–2193.

Dang, C.V., 2015. A metabolic perspective of Peto's paradox and cancer. Philos. Trans. R. Soc. B Biol. Sci. 370, 20140223.

De Lange, H.J., Sperber, V., Peeters, E.T.H.M., 2006. Avoidance of polycyclic aromatic hydrocarbon–contaminated sediments by the freshwater invertebrates *Gammarus pulex* and *Asellus aquaticus*. Environ. Toxicol. Chem. 25, 452–457.

de Roode, J.C., Lefèvre, T., Hunter, M.D., 2013. Self-medication in animals. Science 340, 150–151.

de Visser, K.E., Eichten, A., Coussens, L.M., 2006. Paradoxical roles of the immune system during cancer development. Nat. Rev. Cancer 6, 24–37.

DeGregori, J., 2011. Evolved tumor suppression: why are we so good at not getting cancer? Cancer Res. 71, 3739–3744.

DeGregori, J., 2013. Challenging the axiom: does the occurrence of oncogenic mutations truly limit cancer development with age? Oncogene 32, 1869–1875.

Delaney, M.A., Ward, J.M., Walsh, T.F., Chinnadurai, S.K., Kerns, K., Kinsel, M.J., Treuting, P.M., 2016. Initial case reports of cancer in naked mole-rats (*Heterocephalus glaber*). Vet. Pathol. 53 (3), 691–696.

Dorso, L., Risi, E., Triau, S., Labrut, S., Nguyen, F., Guigand, L., Wyers, M., Abadie, J., 2008. High-grade mucoepidermoid carcinoma of the mandibular salivary gland in a lion (*Panthera leo*). Vet. Pathol. 45, 104–108.

Ducasse, H., Ujvari, B., Solary, E., Vittecoq, M., Arnal, A., Bernex, F., Pirot, N., Misse, D., Bonhomme, F., Renaud, F., Thomas, F., Roche, B., 2015. Can Peto's paradox be used as the null hypothesis to identify the role of evolution in natural resistance to cancer? A critical review. BMC Cancer 15, 792.

Elskus, A.A., 2001. Toxicant resistance in wildlife: fish populations. In: Susan Schantz et al., (Ed.), PCBs: Recent Advances in Environmental Toxicology and Health Effects. University Press of Kentucky, pp. 273–276.

Ferlay, J., Shin, H.R., Bray, F., Forman, D., Mathers, C., Parkin, D.M., 2010. Estimates of worldwide burden of cancer in 2008: GLOBOCAN 2008. Int. J. Cancer 127, 2893–2917.

Fernandez, A.A., Bowser, P.R., 2010. Selection for a dominant oncogene and large male size as a risk factor for melanoma in the *Xiphophorus* animal model. Mol. Ecol. 19, 3114–3123.

Finotello, R., Ressel, L., Verin, R., Di Lollo, S., Baroni, G., Piccinini, R., Poli, A., 2011. Mammary carcinoma in a tiger (*Panthera tigris*): morphological and immunohistochemical study. J. Zoo Wildl. Med. 42, 134–138.

Folkman, J., Kalluri, R., 2004. Cancer without disease. Nature 427, 787.

Gatenby, R.A., Silva, A.S., Gillies, R.J., Frieden, B.R., 2009. Adaptive therapy. Cancer Res. 69, 4894–4903.

Giattina, J., Garton, R., 1983. A review of the preference-avoidance response of fishes to aquatic contaminants. Residue Rev. 87, 43–90.

Gorbunova, V., Seluanov, A., Zhang, Z., Gladyshev, V.N., Vijg, J., 2014. Comparative genetics of longevity and cancer: insights from long-lived rodents. Nat. Rev. Genet. 15, 531–540.

Greaves, M., Maley, C.C., 2012. Clonal evolution in cancer. Nature 481, 306–313.

Grivennikov, S.I., Greten, F.R., Karin, M., 2010. Immunity, inflammation, and cancer. Cell 140, 883–899.

Gulland, F.M., 1995. Impact of infectious diseases on wild animal populations: a review. In: Ecology of Infectious Diseases in Natural Populations. Cambridge University Press, Cambridge, UK pp. 20–51.

Hamede, R.K., Bashford, J., McCallum, H., Jones, M., 2009. Contact networks in a wild Tasmanian devil (*Sarcophilus harrisii*) population: using social network analysis to reveal seasonal variability in social behaviour and its implications for transmission of devil facial tumour disease. Ecol. Lett. 12, 1147–1157.

Hammouda, A., Selmi, S., Pearce-Duvet, J., Chokri, M.A., Arnal, A., Gauthier-Clerc, M., Boulinier, T., 2012. Maternal antibody transmission in relation to mother fluctuating asymmetry in a long-lived colonial seabird: the yellow-legged gull Larus michahellis. PLoS One 7, e34966.

Hausen, H., Zur Hausen, H., 2009. The search for infectious causes of human cancers: where and why. Virology 392, 1–10.

Hoelzer, G., 1987. The good parent model of sexual selection. Anim. Behav. 38 (6), 1067–1078.

Holt, R.D., Lawton, J.H., 1994. The Ecological Consequences of Shared Natural Enemies. Annu. Rev. Ecol. Syst. 25, 495–520.

Hudson, P.J., Dobson, A.P., Lafferty, K.D., 2006. Is a healthy ecosystem one that is rich in parasites? Trends Ecol. Evol. 21, 381–385.

Huffman, M.A., 2001. Self-medicative behavior in the African great apes: an evolutionary perspective into the origins of human traditional medicine. Bioscience 51, 651.

Jones, M.E., Cockburn, A., Hamede, R., Hawkins, C., Hesterman, H., Lachish, S., Mann, D., McCallum, H., Pemberton, D., 2008. Life-history change in disease-ravaged Tasmanian devil populations. Proc. Natl. Acad. Sci. USA 105, 10023–10027.

Kavaliers, M., Choleris, E., Ågmo, A., Pfaff, D.W., 2004. Olfactory-mediated parasite recognition and avoidance: linking genes to behavior. Horm. Behav., 272–283.

Kim, R., Emi, M., Tanabe, K., 2006. Cancer immunosuppression and autoimmune disease: beyond immunosuppressive networks for tumour immunity. Immunology 119, 254–264.

Ku, H.H., Brunk, U.T., Sohal, R.S., 1993. Relationship between mitochondrial superoxide and hydrogen peroxide production and longevity of mammalian species. Free Radic. Biol. Med. 15, 621–627.

Lafferty, K.D., Dobson, A.P., Kuris, A.M., 2006. Parasites dominate food web links. Proc. Natl. Acad. Sci. USA 103, 11211–11216.

Lebarbenchon, C., Brown, S.A.M.P.A.M., Poulin, R., Gauthier-Clerc, M., Thomas, F., 2008. Evolution of pathogens in a man-made world. Mol. Ecol. 17, 475–484.

Lozano, G.A., 1998. Parasitic Stress and Self-Medication in Wild Animals. In: Møller, A.P. (Ed.), Advances in the Study of Behavior. Academic Press, London, pp. 291–317.

Lubin, J.H., Boice, J.D., 1997. Lung cancer risk from residential radon: meta-analysis of eight epidemiologic studies. J. Natl. Cancer Inst. 89, 49–57.

Lynch, M., 2010. Evolution of the mutation rate. Trends Genet. 26, 345–352.

Martineau, D., Lemberger, K., Dallaire, A., Labelle, P., Lipscomb, T.P., Michel, P., Mikaelian, I., 2002. Cancer in wildlife, a case study: Beluga from the St. Lawrence Estuary, Québec, Canada. Environ. Health Perspect. 110, 285–292.

Masi, S., Gustafsson, E., Saint Jalme, M., Narat, V., Todd, A., Bomsel, M.-C., Krief, S., 2012. Unusual feeding behavior in wild great apes, a window to understand origins of self-medication in humans: Role of sociality and physiology on learning process. Physiol. Behav. 105, 337–349.

McAloose, D., Newton, A.L., 2009. Wildlife cancer: a conservation perspective. Nat. Rev. Cancer 9, 517–526.

Meuten, D.J. (Ed.), 2008. Tumors in Domestic Animals. fourth ed. Iowa State University Press.

Møller, A.P., 2008. Interactions between interactions. Ann. NY Acad. Sci. 1133, 180–186.

Møller, A.P., Alatalo, R.V., 1999. Good-genes effects in sexual selection. Proc. R. Soc. Lond. Ser. B Biol. Sci. 266, 85–91.

Møller, A.P., Biard, C., Blount, J.D., Houston, D.C., Ninni, P., Saino, N., Surai, P.F., 2000. Carotenoid-dependent signals: indicators of foraging efficiency, immunocompetence or detoxification ability? Avian Poult. Biol. Rev. 11, 137–159.

Møller, A.P., Erritzøe, J., Karadas, F., Mousseau, T.A., 2010. Historical mutation rates predict susceptibility to radiation in Chernobyl birds. J. Evol. Biol. 23, 2132–2142.

Møller, A.P., Mousseau, T.A., 2007a. Birds prefer to breed in sites with low radioactivity in Chernobyl. Proc. R. Soc. B Biol. Sci. 274, 1443–1448.

Møller, A.P., Mousseau, T.A., 2007b. Determinants of interspecific variation in population declines of birds after exposure to radiation at Chernobyl. J. Appl. Ecol. 44, 909–919.

Møller, A.P., Mousseau, T.A., 2011. Efficiency of bio-indicators for low-level radiation under field conditions. Ecol. Indic. 11, 424–430.

Møller, A.P., Nielsen, J.T., 2007. Malaria and risk of predation: a comparative study of birds. Ecology 88, 871–881.

Monastersky, R., 2015. Anthropocene: the human age. Nature 519, 144–147.

Mousseau, T.A, Uller, T., Wapstra, E., Badyaev, A.V., 2009. Evolution of maternal effects: past and present. Philos. Trans. R. Soc. Lond. B. Biol. Sci. 364, 1035–1038.

Mousseau, T.A., Moller, A.P., 2011. Landscape portrait: a look at the impacts of radioactive contaminants on Chernobyl's wildlife. Bull. At. Sci. 67, 38–46.

Munshi-South, J., Wilkinson, G.S., 2010. Bats and birds: exceptional longevity despite high metabolic rates. Ageing Res. Rev. 9, 12–19.

Murchison, E.P., 2008. Clonally transmissible cancers in dogs and Tasmanian devils. Oncogene 27, S19–S30.

Murchison, E.P., Wedge, D.C., Alexandrov, L.B., Fu, B., Martincorena, I., Ning, Z., Tubio, J.M., Werner, E.I., Allen, J., De Nardi, A.B., 2014. Transmissable dog cancer genome reveals the origin and history of an ancient cell lineage. Science 343, 437–440.

Murray, D.L., Cary, J.R., Keith, L.B., 1997. Interactive effects of sublethal nematodes and nutritional status on snowshoe hare vulnerability to predation. J. Anim. Ecol. 66, 250–264.

Nagy, J.D., Victor, E.M., Cropper, J.H., 2007. Why don't all whales have cancer? A novel hypothesis resolving Peto's paradox. Integr. Comp. Biol. 47, 317–328.

Navarro, C., Lope, F.de, Marzal, A., Møller, A.P., 2004. Predation risk, host immune response, and parasitism. Behav. Ecol. 15, 629–635.

Newman, S.J., Smith, S.A., 2006. Marine mammal neoplasia: a review. Vet. Pathol. Online 43, 865–880.

Noble, R., Kaltz, O., Hochberg, M.E., 2015. Peto's paradox and human cancers. Philos. Trans. R. Soc. B Biol. Sci. 370, 20150104.

Nunney, L., 2003. The population genetics of multistage carcinogenesis. Proc. Biol. Sci. 270, 1183–1191.

O'Donnell, S., 1997. How parasites can promote the expression of social behaviour in their hosts. Proc. R. Soc. London. Ser. B Biol. Sci. 264, 689–694.

Paine, R.T., 1966. Food web complexity and species diversity. Am. Nat. 100, 65–75.

Pepper, J.W., Sprouffske, K., Maley, C.C., 2007. Animal cell differentiation patterns suppress somatic evolution. PLoS Comput. Biol. 3, e250.

Plaks, V., Koopman, C.D., Werb, Z., 2013. Cancer. Circulating tumor cells. Science 341, 1186–1188.

Pollock, R.E., Roth, J.A., 1989. Cancer-induced immunosuppression: Implications for therapy? Semin. Surg. Oncol. 5 (6), 414–419.

Prasad, K.N., Cole, W.C., Hasse, G.M., 2004. Health risks of low dose ionizing radiation in humans: a review. Exp. Biol. Med. 229, 378–382.

Preston, B.T., Capellini, I., McNamara, P., Barton, R.A., Nunn, C.L., 2009. Parasite resistance and the adaptive significance of sleep. BMC Evol. Biol. 9, 7.

Price, P.W., Westoby, M., Rice, B., Atsatt, P.R., Fritz, R.S., Thompson, J.N., Mobley, K., 1986. Parasite mediation in ecological interactions. Annu. Rev. Ecol. Syst. 17, 487–505.

Reiche, E.M.V., Nunes, S.O.V., Morimoto, H.K., 2004. Stress, depression, the immune system, and cancer. Lancet Oncol. 5, 617–625.

Roche, B., Hochberg, M.E.M.E., Caulin, A.F.A.F., Maley, C.C.C.C., Gatenby, R.A.R., Misse, D., Thomas, F., Missé, D., Bondy Cedex, F., Thomas, F., Bondy Cedex, F., 2012. Natural resistance to cancers: a Darwinian hypothesis to explain Peto's paradox. BMC Cancer 12, 387.

Roche, B., Ujvari, B., Thomas, F., 2015. Bad luck and cancer: does evolution spin the wheel of fortune? BioEssays 37 (6), 586–587.

Rozhok, A.I., DeGregori, J., 2015. Toward an evolutionary model of cancer: considering the mechanisms that govern the fate of somatic mutations. Proc. Natl. Acad. Sci. 112, 8914–8921.

Ryan, J.L., Carroll, J.K., Ryan, E.P., Mustian, K.M., Fiscella, K., Morrow, G.R., 2007. Mechanisms of cancer-related fatigue. Oncologist 12 (Suppl. 1), 22–34.

Savage, V.M., Allen, A.P., Brown, J.H., Gillooly, J.F., Herman, A.B., Woodruff, W.H., West, G.B., 2007. Scaling of number, size, and metabolic rate of cells with body size in mammals. Proc. Natl. Acad. Sci. USA 104, 4718–4723.

Senar, J.C., Møller, A.P., Ruiz, I., Negro, J.J., Broggi, J., Hohtola, E., 2010. Specific appetite for carotenoids in a colorful bird. PLoS One 5, e10716.

Sibly, R.M., Walker, C.H., Hopkin, S.P., Peakall, D.B., 2012. Principles of Ecotoxicology. CRC Press, New York.

Thomas, F., Guégan, J.F., Renaud, F., 2009. Ecology and evolution of parasitism. Oxford University Press, Oxford, UK.

Tian, X., Azpurua, J., Hine, C., Vaidya, A., Myakishev-Rempel, M., Ablaeva, J., Mao, Z., Nevo, E., Gorbunova, V., Seluanov, A., 2013. Cancer resistance of the naked mole rat. Nature, 1–6.

Tomasetti, C., Vogelstein, B., 2015. Variation in cancer risk among tissues can be explained by the number of stem cell divisions. Science 347, 78–81.

Ujvari, B., Beckmann, C., Biro, P.A., Arnal, A., Tasiemki, A., Massol, F., Salzet, M., Mery, F., Boidin-Wichlasez, C., Missé, D., Renaud, F., Vittecoq, M., Tissot, T., Roche, B., Poulin, R., Thomas, F., 2016. Cancer and life-history traits: lessons from host-parasite interactions. Parasitol. 143 (5), 533–541.

Vittecoq, M., Ducasse, H., Arnal, A., Møller, A.P., Ujvari, B., Jacqueline, C.B., Tissot, T., Missé, D., Bernex, F., Pirot, N., Lemberger, K., Abadie, J., Labrut, S., Bonhomme, F., Renaud, F., Roche, B., Thomas, F., 2015. Animal behaviour and cancer. Anim. Behav. 101, 19–26.

Vittecoq, M., Roche, B., Daoust, S.P., Ducasse, H., Missé, D., Abadie, J., Labrut, S., Renaud, F., Gauthier-Clerc, M., Thomas, F., 2013. Cancer: a missing link in ecosystem functioning? Trends Ecol. Evol. 28, 628–635.

von Sonntag, C., 2006. Free-Radical-Induced DNA Damage and Its Repair. Springer Verlag, Berlin.

Wagner, L.I., Cella, D., 2004. Fatigue and cancer: causes, prevalence and treatment approaches. Br. J. Cancer 91, 822–828.

Walsh, M.R., 2013. The evolutionary consequences of indirect effects. Trends Ecol. Evol. 28, 23–29.

Welsh, J.S., 2011. Contagious cancer. Oncologist 16, 1–4.

Whitehead, A., Pilcher, W., Champlin, D., Nacci, D., 2012. Common mechanism underlies repeated evolution of extreme pollution tolerance. Proc. R. Soc. B Biol. Sci. 279, 427–433.

Wolf, J.B., Wade, M.J., 2009. What are maternal effects (and what are they not)? Philos. Trans. R. Soc. L. B. Biol. Sci. 364, 1107–1115.

Yablokov, A.V., 2010. Chernobyl's radioactive impact on fauna. Ann. NY Acad. Sci. 1181, 255–280.

Yu, H.-S., Liu, Z.-M., Yu, X.-Y., Song, A.-Q., Liu, N., Wang, H., 2013. Low-dose radiation induces antitumor effects and erythrocyte system hormesis. Asian Pacific J. Cancer Prev. 14, 4121–4126.

14

Applying Tools From Evolutionary Biology to Cancer Research

Pedro M. Enriquez-Navas, Robert A. Gatenby

H. Lee Moffitt Cancer Center and Research Institute, Tampa, FL, United States

It is not the strongest of the species that survives, nor the most intelligent that survives. It is the one that is the most adaptable to change—**Charles Darwin**

INTRODUCTION

Cancer is often described as a "collection of diseases"(NCI, 2015) reflecting the multiscalar spatial and temporal complexity of typical malignant tumors. At a molecular scale, a number of studies have demonstrated marked variations in the genotypes of cancer cells between different tumor sites (e.g., the primary and metastatic tumors) and even within regions of the same tumor (Bundschuh et al., 2014; Gerlinger et al., 2012). On a larger scale, areas of varying blood flow, including necrosis, are commonly observed in cancer imaging. Furthermore, intratumoral blood flow can exhibit temporal variation as the often-chaotic organization of tumor blood vessels results in periods of blood flow cessation or even reversal (Nagy et al., 2009) (see also Chapter 19). In turn, interruption of blood flow can produce transient episodes of hypoxia and acidosis leading to tumor death or evolution of adaptive strategies leading to molecular heterogeneity (see also Chapter 4). Finally, tumor therapy imposes a new selection pressure as cytotoxicity selects for resistant phenotypes and local cell death can produce secondary effects in the tumor environment (Bundschuh et al., 2014) (see also Chapter 10).

Thus, most cancers represent a complex and often patient-specific tumor-host ecology as interactions among tumor and host cells results in formation of probably several tissue "niches" with distinctive microenvironmental characteristics reflecting formation of blood vessels, synthesis of extracellular matrix to support tumor growth (Martinez-Zaguilan et al., 1996), and deployment of local predators in the form of immune system effectors. In turn, as in most of existing ecologies, it is likely that each tissue niche selects for distinct heritable tumor properties that correspond to a distinctive local "species." Thus, while individual malignant cells are the evolutionary unit of selection in cancer populations, in situ they commonly use ecological strategies that coopt normal stromal tissue to generate a loosely organized multicellular tissue (Tarin, 2012) (see also Introductory Chapter and Chapter 11).

Importantly, spatial heterogeneity in cancers can also significantly alter the efficacy of tumor therapies. For example, hypoxic tumor regions are typically resistant to radiation therapy (Horsman et al., 2012) and some chemotherapies (Cosse and Michiels, 2008). Local acidosis can both promote efficacy of some treatments (hyperthermia, e.g., Freeman et al., 1980) and reduce efficacy of other treatments [e.g., acid pH reduces the uptake of anthracyclines, anthraquinones, and vinca alkaloids in solid tumors (Wojtkowiak et al., 2011)]. Furthermore, spatial and temporal variations in blood flow can profoundly affect the local delivery of systemic drugs.

Temporal variations in the tumor ecosystem are readily evident during cancer treatment as therapy is often highly effective initially but becomes progressively less so as the tumor cells evolve resistant strategies (Housman et al., 2014). In fact, the evolutionary capacity of tumor cells to use the extant molecular machinery in the human genome and to develop novel strategies through mutations is arguably the proximate cause of death in most cancer patients.

Interestingly, while evolutionary dynamics allow tumor cells to evade host response and iatrogenic perturbations, they can also be exploited to successfully treat cancers with sufficient understanding of the underlying Darwinian interactions that govern tumor growth. This reflects a commonly unrecognized weakness inherent in the evolutionary capacity of cancer cells: *Evolving*

Ecology and Evolution of Cancer. http://dx.doi.org/10.1016/B978-0-12-804310-3.00014-4

populations can only adapt to current and local conditions, they can never anticipate future or distant environmental factors. In contrast, with sufficient understanding of the underlying Darwinian dynamics, physicians treating cancer can plan therapies that strategically use initial therapies to induce adaptive strategies that can be exploited. That is, oncologists can change the tumor environment (by using different therapies and the timing of the same ones) in such a way that proliferation of resistant clones can be suppressed for prolonged periods of time.

In devising evolution-based cancer treatments, guidance is available in extensive prior experience in controlling crop pests and exotic species (Renton et al., 2014), as well as bacterial resistance to antibiotics (Nichol et al., 2015), and so on (Hendry et al., 2011) (see also Chapter 21). A critical lesson from these experiences is that, while evolution of resistant phenotypes following some perturbation is virtually inevitable, proliferation of the resistant clones is not and can, in fact, be delayed or even suppressed through thoughtful manipulation of the underlying Darwinian dynamics (Gatenby, 2009). Of particular importance for controlling unwanted populations is detailed understanding of the mechanism(s) of resistance to any given therapy and the phenotypic cost of resistance. For example, cancer cells often become resistant to chemotherapy by expressing large numbers of membrane extrusion pumps, such as p-glycoprotein, that move cytotoxic drugs from the cytoplasm into the tumor extracellular space. While this confers substantial resistance to treatment, it also requires significant resources to synthesize, localize, maintain, and operate these pumps. Clearly, the benefits of the membrane pumps exceed the costs in the presence of cytotoxic drugs. However, in the absence of drugs, the cost of maintaining the pumps when they provide no benefit reduces fitness so that, in general, treatment-sensitive cells are fitter than treatment-resistant cells. We propose that this fitness trade-off presents an opportunity to use evolutionary strategies to suppress or eliminate the proliferation of resistant populations.

To understand the complex dynamics that are present in the tumor microenvironment, mathematical models are of increasing importance to cancer biologists and oncologists (Anderson and Quaranta, 2008). This highly multidisciplinary approach requires continuous interactions between biologists and mathematicians. Although they are often not consciously aware, experimentalists and oncologists routinely apply conceptual models to understand the underlying dynamics of cancer and its therapy. The role of the mathematician is to quantitatively frame the "verbal" models of interactions that occur between tumor cell, stromal tissue, and treatments. Such models are often motivated by a need to understand existing data but with the added advantage that they allow computer simulations that can explore the often vast parameter space uncovering novel and unexpected dynamics. Finally, it is important that the mathematical models form predictions that can then be tested empirically (Scott, 2012). Thus, multidisciplinary combination of disparate subjects (biology, mathematics, medicinal chemistry, physics, and medicine, of course) is increasingly important in devising evolution-based treatment strategies in cancer patients.

While it seems clear that evolutionary dynamics generally govern the development of resistance to cancer therapy, it is also clear that the ecoevolutionary details, such as the specific mechanism for resistance, proliferation time of tumor subpopulations, tumor vascularity, and so on, are highly patient-specific. Thus, the need to identify the "right treatment for the right patient at the right time," often termed "precision medicine," both applies to and is informed by evolutionary principles. While the focus of precision medicine is often the design of a specific drug for a specific molecular target, an evolutionary approach is primarily focused on the population dynamics following initiation of treatment in which the tumor inevitably evolves resistant mechanisms and, ultimately, progresses even when the treatment was highly successful. Here, we demonstrate that precision medicine must also integrate the obvious but often ignored reality that complex ecoevolutionary changes are elicited by any treatment, and successful long-term control of cancer requires dynamic treatment strategies that accommodate, anticipate, and exploit the intratumoral Darwinian dynamics that govern tumor response and resistance to treatment.

In summary, tumors are spatially and temporally heterogeneous complex dynamic systems that respond to iatrogenic perturbations in complex and, often, nonlinear ways. Consequently, therapies must integrate these dynamics to achieve the principle goal: maximizing patient survival. While preventing evolution of tumor adaptive responses to therapies is virtually impossible, we propose that using evolutionary principles will delay, guide, and exploit those adaptive dynamics and can substantially improve cancer treatment outcomes.

Herein, we introduce methods by which the principles of evolutionary biology can be used to improve cancer treatment and, consequently, increase the progression free survival of the patients.

ADAPT OR DIE (ADAPTIVE THERAPY)

As stated previously, tumors are governed by Darwinian principles that allow them to evolve, often rapidly, to perturbations that result in cell death and, therefore, impose strong evolutionary selection forces. In contrast, current cancer therapies typically apply the same drugs and doses through multiple cycles until the tumor progresses. This strategy is based on the principle that a tumor must be eradicated as quickly as possible to prevent evolution of resistance and dissemination to other organs. If, in fact, the tumor is potentially curable, then maximally aggressive therapy seems entirely reasonable. However, in treating disseminated epithelial cancers for which cure is no longer possible, we propose that maximum dose treatment is evolutionarily unwise. The Darwinian dynamics behind this statement can be summarized by a term from pest management: "competitive release."

Briefly, we accept that, in metastatic cancers, the underlying heterogeneity insures that resistant clones are present prior to therapy. This has been documented in several empirical studies and is, in any case, obvious by the simple fact that nearly 5 decades of clinical experiences with maximum dose density therapy has clearly demonstrated that it does not cure metastatic epithelial cancers. In fact, maximum dose density therapy actually promotes the growth of resistant population because it both strongly selects for adapted phenotypes and eliminates all potentially competing populations. These dynamics by which attempts to eradicate a population lead, ironically, to rapid proliferation of resistant phenotypes is commonly observed in nature and termed "competitive release." Similarly, we propose that, by this process of competitive release, traditional cancer high dose treatment strategies actually accelerates the progression of resistant populations.

The design of evolution-informed therapies focuses on applying evolutionary principles to prolong the time to progression. This approach is designed only for incurable disseminated cancers and, thus, explicitly abandons any pretext of "aggressive" treatment with the expectation of cure when such an outcome is not possible. It also abandons the traditional goal of treating with the intent of obtaining the maximum regression of tumor volume. Such treatment strategies and goals, we propose, are largely irrelevant to our primary goal of maximizing the time of tumor *control* by using the tumor cells that are sensitive to treatment as agents that can suppress the proliferation of the resistant cells. In effect, our goal is to maintain significant residual populations of tumor cells that we can control (i.e., can be killed by therapy) to inhibit the growth of cells that we otherwise cannot control (i.e., are resistant to therapy).

Perhaps an obvious, but still often ignored, fundamental component of the evolutionary dynamics within in vivo cancers is that resources are limited. That is, chaotic blood flow and reduced vascular density result in necrotic regions in which resources are so limited that only a few cancer cells survive. Furthermore, even in well-vascularized regions of tumors, substrate concentrations can never exceed physiological levels unlike, for example, culture media for in vitro studies in which

substrate, such as glucose are maintained in concentrations that are many fold larger than those in vivo. Thus, regions of tumor appear to experience stochastic variations in blood flow resulting in often rapid temporal changes in, for example, oxygen concentration (Zhang et al., 2014).

Consequently, in vivo cancer cells must evolve strategies to optimally use limited and variably available resources to maximally increase their fitness. This *plasticity* becomes a particularly important component of adaptations to therapy because any such strategy typically requires expenditure of resources. For example, a common mechanism of resistance requires synthesis, maintenance, localization, and operation of membrane extrusion pumps, such as the multidrug resistance (MDR) proteins. However, the operation of these proteins requires significant commitment of resources (up to 50% of total cell's energy budget) (Kam et al., 2015).

Therefore, any new expenditure of energy and substrate requires a reconfiguration of the cell's energy budget to either acquire additional resources or reallocate resources from some other functions. In an environment of limited resources, such as an in vivo tumor, the latter strategy is usually required. Thus, a tumor cell can expend energy on a resistant mechanism only if it can divert resources from other "optional" (i.e., not necessary for survival) expenses, such as proliferation and invasion (Silva et al., 2015). We have proposed that this evolutionary trade-off provides an opportunity to exploit Darwinian dynamics to optimize cancer treatment (Gatenby et al., 2009b).

Adaptive Therapy was designed to exploit the evolutionary cost of resistance. It is based on the admittedly oversimplified premise that, for the purposes of treatment design, each tumor is composed of two competing cancer cell types: therapy resistant and therapy sensitive. As of the cost of resistance outlined previously, we assume that sensitive cells are fitter than resistant ones in the absence of treatment. This assumption is supported if initial tumor response to a treatment is significant since this model predicts that, because of the associated fitness cost, resistant cells will be rare, but present, prior to therapy. Of course, the resistant cells are fitter than the sensitive ones in the presence of the same treatment and come to dominate the tumor population during prolonged therapy.

The goal of adaptive therapy is quite simple: (1) therapy is applied in small doses to reduce the tumor population only sufficiently to improve symptoms. Drugs, thus, should be administered at a dose that is not the maximum possible but the minimum necessary. (2) Treatment is then withdrawn. In the absence of chemotherapy, the treatment-sensitive cells will proliferate at the expense of resistant ones. Thus, while tumor will increase in size between treatments, the extant tumor cells will continue to be sensitive to therapy (Fig. 14.1).

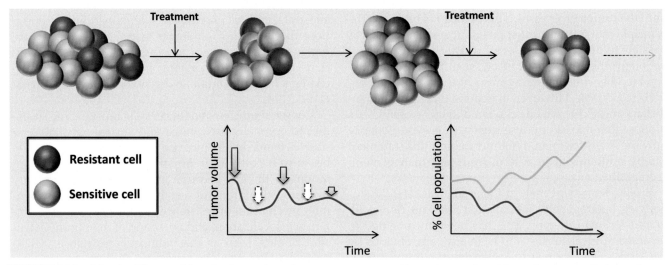

FIGURE 14.1 **Adaptive Therapy.** Adapt the therapy intensity to a tumor composed by a mixture of sensitive and resistant cells is mandatory in order to give sensitive cells the opportunity to outcompete the resistant ones. In the presence of treatment, resistant cells will be fitter than sensitive ones, which will die by the effect of drug. However, the opposite occurs when the therapy is withdrawn and the sensitive cells are fitter than the resistant ones, competing then for space and resources with the resistant ones. Thus, oncologists, based on evolutionary principles, may have to reduce (*smaller arrows* in central graph) or even withdrawn *(white dashed arrows)* the treatment. Consequently, the tumor cell population will be driven to a situation in which resistant cells will disappear with the pass of the time (right graph).

This treatment strategy was originally framed mathematically and demonstrated to be feasible using computational simulations (Gatenby et al., 2009b). It was first tested in vivo using an ovarian cancer (OVCAR-3 cells) xenograft preclinical model treated with carboplatin. The treatment algorithm was based solely and directly on tumor volume, which was measured by caliper. Thus, if the tumor increases its size during two consecutives measurements, the dose of carboplatin was increased. Consequently, the drug dose was reduced if the tumor volume was decreased in the same period of time. This treatment was compared with a standard high-dose treatment of carboplatin and the conclusion was that the Adaptive Therapy was able to keep the tumor under control for an extended prolonged time than the standard therapy (Gatenby et al., 2009b). Hence, producing a long-term overall survival in those mice under the Adaptive Therapy.

More recently, Adaptive Therapy has been applied to two orthotopic preclinical models: (1) triple-negative and (2) estrogen receptor positive (ER+) breast cancers (Enriquez-Navas et al., 2016). In both cases, the mice were treated with paclitaxel and, again, the treatment algorithms were based on the tumor volumes, which, in this case, were measured by magnetic resonance imaging (MRI). Two different treatment algorithms were tested: (1) one closer to a possible clinical translation in which a fixed drug dose was only applied if the tumor grew more than 25% of the previous measurements, and (2) another one in which after starting with the maximum tolerable dose, it allows to increase by 50% the dose if the tumor grew more than 20% and decrease it by a 50% if the tumor reduces its volume by 20%, always

respect to the previous measurement and never giving more dose than the maximum tolerable dose. Also, in this case, a lower tumor volume threshold was used and below it the therapy was withdrew. Both therapies were compared to a standard high-dose therapy of paclitaxel.

Interestingly, the data demonstrated that the therapy with a modulated dose schedule (the second one explained previously) was most successful in maintaining tumor control. In addition, the experimental results, similar to those in the OVCAR model, found that adaptive therapy could be divided into two phases; first, initial treatment when tumor growth was exponential required aggressive therapy with frequent application of treatment although the dose could be decreased with time. Second, in a "maintenance" period after the tumor growth was controlled a progressively lower and less frequent doses (even including several withdrawing of the treatment) were sufficient to keep the tumor under control (Enriquez-Navas et al., 2016).

ALL AT ONCE IT IS NOT ALWAYS THE BEST THING (THE DOUBLE BIND THEORY)

A second evolution-based treatment strategy relies on strategically planning which coordinates first and second line therapies to exploit the adaptive tumor cell strategies. Consider a hypothetical case in which a tumor can be treated by two drugs with different mechanisms of action and for which different cellular adaptive strategies are required. A conventional treatment strategy would

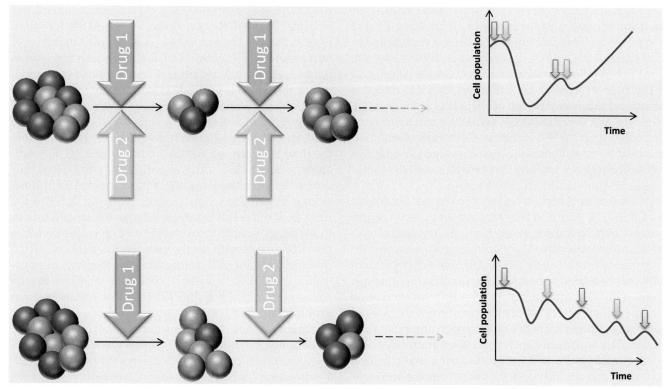

FIGURE 14.2 **The "double-bind" therapy.** When a tumor is sensitive to two or more drugs, evolutionary principles have demonstrated that the application of these drugs at the same time will end in the apparition of individuals resistant to both therapies. However, if these drugs are applied in an alternative way, it will be always a population sensitive to one or the other drug and, delaying the apparition of the double-resistant cell clone. The graphics represent how the treatment with both drugs (upper graph) gives a fast response with also a fast and treatment-resistant tumor relapse, or how the sequential (bottom graph) application of both drugs ends in a more treatment-sensitive and under control tumor.

apply both drugs at the same time to kill as many tumor cells as possible. This is the rationale for multidrug combination therapies commonly employed in clinical cancer treatments. In general, combination therapies result in greater responses compared to monotherapies as measured by change in tumor volume. However, this does not typically result in a cure as the tumor generally recurs through proliferation of cell clones that evolve resistant mechanisms to both drugs (Fig. 14.2, upper graph).

Rather than applying multiple drugs simultaneously, here we ask: can sequential application of multiple drugs extend progression free or overall survival more than the simultaneous application of the same drugs? In our analysis of the evolutionary dynamics, we propose that, under some circumstances, sequential application of treatments can be far more effective than when they are combined (Fig. 14.2, bottom graph).

An interesting and useful analogy in nature is termed "predator facilitation." Similar to chemotherapies, predators can be introduced into an ecosystem to eliminate a pest population. However, like cancer cells, pests typically evolve resistant strategies allowing their population to return to the initial size. For example, the introduction of an owl to control a population of field mice can be limited when the prey population adapts by hiding

under bushes. However, this adaptive response can be exploited by introducing a predator that typically hunts under bushes—a snake for example. This dynamic exploits the adaptive strategy to an initial treatment strategy to increase the efficacy of the second treatment (Burt and Kotler, 1992). Thus, we propose that "evolutionarily coupled" treatments in which the adaptation to one therapy renders the resistant phenotype more vulnerable to the second therapy and vice versa is an effective strategy for long-term population control that could be applied to multidrug therapy or combinations of treatment strategies, such as immunotherapy and chemotherapy. Interestingly, a study examining *Escherichia coli* treatment in an in vitro setting with two antibiotics, concluded that the sequential treatment is more efficient than the synergistic one (Fuentes-Hernandez et al., 2015).

We propose that tumor populations can be forced into a similar "double-bind" evolutionary dynamic (Gatenby et al., 2009a) through thoughtful sequential application of treatments with sufficient understanding of the underlying evolutionary dynamics (Fig. 14.2, bottom graph).

The potential clinical benefit of this "double bind" strategy is suggested in a study performed at the Moffitt Cancer Center using a combination of immunotherapy and chemotherapy. This study (Antonia et al., 2006) investigated a tumor protein P53 (P53) cancer vaccine for treatment of small-cell lung cancer. In a treatment cohort of 29 patients, a significant immune response was elicited in most patients but only one partial response was observed. However, in follow-on therapy with cytotoxic drugs, 62% of the patients exhibited a partial or complete response (compared to a response rate of 5% in historical controls). While the precise evolutionary dynamics remain unclear, we propose that the P53 vaccine and cytotoxic chemotherapy may be evolutionarily coupled so that the adaptive response necessary to defeat vaccine rendered the tumor cells more vulnerable to chemotherapy.

A similar approach is demonstrated by Schweizer et al. (2015) in a clinical trial focused on prostate cancer patients that have progressed through first line chemical castration therapy. They administered testosterone to promote proliferation of the testosterone-dependent cells that had been suppressed by the first line therapy. They then readministered the androgen-deprivation therapy. This "bipolar" study then continued to cycle patients through the sequence of androgen administration followed by androgen deprivation. They found that 50% of the patients who had progressed on initial chemical castration therapy responded to this "bipolar strategy" and that, in this group, tumor control was maintained for a mean of 11 months (Schweizer et al., 2015). However, further studies are needed to understand if this "bipolar therapy" will improve patient survival respect to the traditional androgen deprivation therapy.

NO DOUBLE PAIN, NO DOUBLE GAIN (THE COST OF BEING RESISTANT)

As described earlier, a common mechanism of tumor resistance to chemotherapy is upregulation of the expression of the membrane extrusion pump Multidrug Resistant Protein 1 (MDR-1). MDR-1, also known as P-glycoprotein, P-gp, is a cell membrane protein that belongs to the superfamily of ATP-binding cassette (ABC) transporter family, that can extrude a variety of compounds, including chemotherapy drugs (Fletcher et al., 2010), from the cytosol into the interstitial space through an ATP-dependent pumping mechanism. Expression and operation of MDR-1 is a common molecular resistance strategy used by cancer cells. However, this incurs a significant resources cost, as MDR-1 functions can consume up to 50% of a cell's energy budget (Broxterman et al., 1988).

We have explored ancillary treatment strategies that increase the cost of this resistance strategy thus further decreasing the evolutionary fitness of the resistant population in the absence of therapy. This work builds on prior studies that have identified substrates of the MDR-1 protein to act as competitive inhibitors. This approach

simultaneously administers a chemotherapy agent and an additional MDR-1 substrate. As the MDR-1 protein must extrude both compounds, it is assumed that greater intracellular concentration of the chemotherapy agents is achieved. We have examined an alternative strategy in which the MDR-1 substrate is administered *between* applications of chemotherapy. Specifically, we have tested this strategy in both, in vitro and in vivo, breast cancer models, and the preliminary results have shown us that the time between the administration of the MDR-1 substrate and cytotoxic drug is critical (*in prep*). Here, our goal is to force the resistant cells to expend additional energy to extrude a noncytotoxic compound. Since the drug does not result in any damage to the sensitive cells, this strategy simply increases the cost of resistance while providing no benefit. In the competition between sensitive and resistant cells during adaptive therapy, administration of a MDR-1 substrate, because it increases energy consumption and requires diversion of resources from proliferation and invasion, confers a further evolutionary disadvantage on the resistant phenotype (Fig. 14.3).

Recently, Kam et al. (2015) have shown some promising results in in vitro experiments in wild type and MDR-1-expressing human breast adenocarcinoma (MCF7) cell lines. In in vitro media with physiological concentration of substrate administration of a nontoxic MDR-1 substrate dramatically increased the glucose uptake and decreased the proliferation rate of the MDR-1-expressing, but not wild-type, MCF7 cells (Kam et al., 2015).

CONCLUDING REMARKS, DRAWBACKS, AND FUTURE DIRECTIONS

Most clinical cancers are heterogeneous ecosystems composed of normal host cells and different tumor cell clones. The latter continuously evolve to adapt to the changing (and challenging) environment. This population and environmental heterogeneity, as well as the evolutionary capacity of tumor cells significantly limit the long-term efficacy of systemic therapies. Remarkably, however, despite the obvious role of evolution in treatment failure, Aktipis et al. (2011) has found that less than 1% of current clinical trials consider evolutionary principles—a number that is identical to clinical trials from 30 years ago (Aktipis et al., 2011). In contrast, we propose that cancer therapists, with sufficient understanding of the underlying Darwinian dynamics, can harness and exploit evolutionary forces to optimize therapy.

As summarized in this chapter, when curative therapy is not available, application of evolutionary principles can significantly prolong tumor control. However, this approach requires explicit rejection of common practices and paradigms in cancer therapy and increasingly on a new multidisciplinary paradigm that incorporates

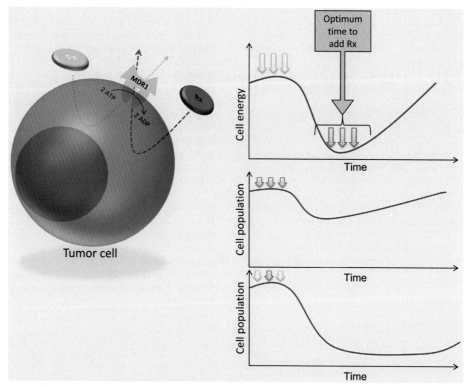

FIGURE 14.3 **The cost of resistance can be exploited to control tumor growth.** Here, based on MDR-1 resistance mechanism, two concepts are represented to use these resistant mechanisms for the therapy advantage. It is possible to use a drug that keeps the resistant mechanisms working (Fx, *green*) with the only purpose of (1) exhaust the cell energy (upper graphic) or (2) increase the pharmacokinetics of the cytotoxic drug (Rx, *red*) (middle and bottom graphics) and, thus, increase the percentage of tumor cell killing.

Darwinian first principles and application of mathematical models to define the complex, often nonlinear, dynamics that govern tumor response and resistance to treatment.

The challenge is opened.

References

Aktipis, C.A., Kwan, V.S., Johnson, K.A., Neuberg, S.L., Maley, C.C., 2011. Overlooking evolution: a systematic analysis of cancer relapse and therapeutic resistance research. PLoS One 6, e26100.

Anderson, A.R., Quaranta, V., 2008. Integrative mathematical oncology. Nat. Rev. Cancer 8, 227–234.

Antonia, S.J., Mirza, N., Fricke, I., Chiappori, A., Thompson, P., Williams, N., Bepler, G., Simon, G., Janssen, W., Lee, J.H., Menander, K., Chada, S., Gabrilovich, D.I., 2006. Combination of p53 cancer vaccine with chemotherapy in patients with extensive stage small cell lung cancer. Clin. Cancer Res. 12, 878–887.

Broxterman, H.J., et al., 1988. Induction by verapamil of a rapid increase in ATP consumption in multidrug-resistant tumor cells. FASEB J. 2 (7), 2278–2282.

Bundschuh, R.A., Dinges, J., Neumann, L., Seyfried, M., Zsoter, N., Papp, L., Rosenberg, R., Becker, K., Astner, S.T., Henninger, M., Herrmann, K., Ziegler, S.I., Schwaiger, M., Essler, M., 2014. Textural parameters of tumor heterogeneity in (1)(8)F-FDG PET/CT for therapy response assessment and prognosis in patients with locally advanced rectal cancer. J. Nucl. Med. 55, 891–897.

Burt, P., Kotler, L.B.J.S.B., 1992. Predator facilitation: the combined effect of snakes and owls on the foraging behaviour of gerbils. Ann. Zool. Fennnici 29, 199–206.

Cosse, J.P., Michiels, C., 2008. Tumour hypoxia affects the responsiveness of cancer cells to chemotherapy and promotes cancer progression. Anti-Cancer Agents Med. Chem. 8, 790–797.

Enriquez-Navas, P.M., Kam, Y., Das, T., Hassan, S., Silva, A., Foroutan, P., Ruiz, E., Martinez, G., Minton, S., Gillies, R.J., Gatenby, R.A., 2016. Exploiting evolutionary principles to prolong tumor control in preclinical models of breast cancer. Sci. Transl. Med. 8, 327ra24.

Fletcher, J.I., et al., 2010. ABC transporters in cancer: more than just drug efflux pumps. Nat. Rev. Cancer 10 (2), 147–156.

Freeman, M.L., Raaphorst, G.P., Hopwood, L.E., Dewey, W.C., 1980. The effect of pH on cell lethality induced by hyperthermic treatment. Cancer 45, 2291–2300.

Fuentes-Hernandez, A., Plucain, J., Gori, F., Pena-Miller, R., Reding, C., Jansen, G., Schulenburg, H., Gudelj, I., Beardmore, R., 2015. Using a sequential regimen to eliminate bacteria at sublethal antibiotic dosages. PLoS Biol. 13, e1002104.

Gatenby, R.A., 2009. A change of strategy in the war on cancer. Nature 459, 508–509.

Gatenby, R.A., Brown, J., Vincent, T., 2009a. Lessons from applied ecology: cancer control using an evolutionary double bind. Cancer Res. 69, 7499–7502.

Gatenby, R.A., Silva, A.S., Gillies, R.J., Frieden, B.R., 2009b. Adaptive therapy. Cancer Res. 69, 4894–4903.

Gerlinger, M., Rowan, A.J., Horswell, S., Larkin, J., Endesfelder, D., Gronroos, E., Martinez, P., Matthews, N., Stewart, A., Tarpey, P., Varela, I., Phillimore, B., Begum, S., McDonald, N.Q., Butler, A., Jones, D., Raine, K., Latimer, C., Santos, C.R., Nohadani, M., Eklund, A.C., Spencer-Dene, B., Clark, G., Pickering, L., Stamp, G., Gore, M., Szallasi, Z., Downward, J., Futreal, P.A., Swanton, C., 2012. Intratumor heterogeneity and branched evolution revealed by multiregion sequencing. N. Engl. J. Med. 366, 883–892.

Hendry, A.P., Kinnison, M.T., Heino, M., Day, T., Smith, T.B., Fitt, G., Bergstrom, C.T., Oakeshott, J., Jørgensen, P.S., Zalucki, M.P., Gilchrist, G., Southerton, S., Sih, A., Strauss, S., Denison, R.F., Carroll, S.P., 2011. Evolutionary principles and their practical application. Evol. Appl. 4, 159–183.

Horsman, M.R., Mortensen, L.S., Petersen, J.B., Busk, M., Overgaard, J., 2012. Imaging hypoxia to improve radiotherapy outcome. Nat. Rev. Clin. Oncol. 9, 674–687.

Housman, G., Byler, S., Heerboth, S., Lapinska, K., Longacre, M., Snyder, N., Sarkar, S., 2014. Drug resistance in cancer: an overview. Cancers 6, 1769.

Kam, Y., Das, T., Tian, H., Foroutan, P., Ruiz, E., Martinez, G., Minton, S., Gillies, R.J., Gatenby, R.A., 2015. Sweat but no gain: inhibiting proliferation of multidrug resistant cancer cells with "ersatzdroges". Int. J. Cancer 136, E188–E196.

Martinez-Zaguilan, R., et al., 1996. Acidic pH enhances the invasive behavior of human melanoma cells. Clin. Exp. Metastasis 14 (2), 176–186.

Nagy, J.A., Chang, S.H., Dvorak, A.M., Dvorak, H.F., 2009. Why are tumour blood vessels abnormal and why is it important to know? Br. J. Cancer 100, 865–869.

NCI, 2015. What is cancer? Available from: http://www.cancer.gov/about-cancer/what-is-cancer

Nichol, D., Jeavons, P., Fletcher, A.G., Bonomo, R.A., Maini, P.K., Paul, J.L., Gatenby, R.A., Anderson, A.R., Scott, J.G., 2015. Steering evolution with sequential therapy to prevent the emergence of bacterial antibiotic resistance. PLoS Comput. Biol. 11, e1004493.

Renton, M., Busi, R., Neve, P., Thornby, D., Vila-Aiub, M., 2014. Herbicide resistance modelling: past, present and future. Pest Manag. Sci. 70, 1394–1404.

Schweizer, M.T., et al., 2015. Effect of bipolar androgen therapy for asymptomatic men with castration-resistant prostate cancer: results from a pilot clinical study. Sci. Transl. Med. 7 (269), 269ra262.

Scott, J., 2012. Phase I trialist. Lancet Oncol. 13, 236.

Silva, R., et al., 2015. Modulation of P-glycoprotein efflux pump: induction and activation as a therapeutic strategy. Pharmacol. Ther. 149, 1–123.

Tarin, D., 2012. Clinical and biological implications of the tumor microenvironment. Cancer Microenviron. 5 (2), 95–112.

Wojtkowiak, J.W., Verduzco, D., Schramm, K.J., Gillies, R.J., 2011. Drug resistance and cellular adaptation to tumor acidic pH microenvironment. Mol. Pharm. 8, 2032–2038.

Zhang, Z., Hallac, R.R., Peschke, P., Mason, R.P., 2014. A noninvasive tumor oxygenation imaging strategy using magnetic resonance imaging of endogenous blood and tissue water. Magn. Reson. Med. 71, 561–569.

PERSPECTIVES

15

Understanding Ancient Legacies to Expose and Exploit Cancer's Evolutionary Vulnerabilities

Aurora M. Nedelcu

Department of Biology, University of New Brunswick, Fredericton, NB, Canada

"Rather than being an interpreter, the scientist who embraces a new paradigm is like the man wearing inverting lenses."
Thomas Kuhn

With so much interest in understanding cancer from both a mechanistic and an evolutionary perspective, it is to be expected that a number of paradigms in the cancer field have emerged in the last few decades. As with all paradigms, while they serve an important role in defining common ways of thinking, assumptions, methodologies, they can also hinder progress by distracting from potential new views with possible positive outcomes. Here, I discuss three current paradigms, highlight several implications and how they can impact/bias our understanding of cancer as an evolutionary process, and provide alternative views that might open new research and potential therapeutic directions (note that the repetitive structure of the discussion is meant to stress the logical framework and the commonalities in addressing the issues). Then, I briefly describe several examples from the unicellular world that can serve as case-studies and argue for the benefits of understanding ancient legacies that still live on in multicellular organisms. Last, I suggest that evolutionary vulnerabilities inherited from our unicellular ancestors could be exposed and exploited to decrease the fitness of cancer cells.

A NEW LIGHT ON OLD PARADIGMS

Paradigm 1. Cancer is a Disease of Multicellularity

Cancer is generally seen as intrinsically connected to the evolution of multicellularity (e.g., Aktipis et al., 2015; Chen et al., 2015; Davies and Lineweaver, 2011; also see Chapter 1). That is, it is often stated that cancer only makes sense in a multicellular organism; it is a by-product of the evolution of multicellularity; it is a problem of multicellularity, and so on. This is the most common view of cancer, and is built on several well-accepted premises. (1) The evolution of multicellularity was contingent on the emergence and enforcement of cooperative interactions among previously independent single-celled individuals, and cancer reflects the breakdown of these cooperative behaviors. (2) The evolution of multicellularity was dependent on the ability to regulate the cell cycle (especially the control of cell proliferation and the loss of immortality). (3) The emergence of complex multicellularity required the evolution of cell differentiation. (4) To ensure cooperation among cells and the stability of multicellular individuals, a "cell suicide" mechanism to remove damaged/mutated/"selfish" cells had to evolve. (5) The functionality of multicellular organisms required the evolution of cell–cell communication pathways (see also Chapter 7).

While these statements are all valid, they contain several implicit assumptions with implications that are not fully appreciated. Specifically, all these traits are assumed to have emerged during the evolution of multicellularity—that is, they are specific to multicellular organisms. Consequently, their loss is presumed to result in a reversal to an ancestral-like unicellular state (e.g., Chen et al., 2015; see also Chapter 16); that is, cancer cells are acting like ancestral single-celled organisms, enjoying their solitary existence and immortality as free-living unicellular individuals increasing their individual fitness.

Ecology and Evolution of Cancer. http://dx.doi.org/10.1016/B978-0-12-804310-3.00015-6

However, cooperative behaviors/interactions are also known in many unicellular species, from bacteria to algae (for a review, see, for instance, West et al., 2006, 2007; Xavier, 2011). In fact, many unicellular organisms exist in multicellular communities (e.g., bacterial biofilms, yeast colonies) or form multicellular assemblages in specific life-stages (e.g., slime molds, myxobacteria) (e.g., Webb et al., 2003; West et al., 2007). Regulation of cell cycle is also not specific to multicellular organisms; it is a vital mechanism for unicellular individuals especially when dealing with stress or limited resources. Likewise, cell differentiation—in a temporal context (i.e., as a function of time, not space as in multicellular individuals) is part of the life history of many unicellular lineages, especially in response to environmental stress (e.g., spore/cyst formation and sexual differentiation in most unicellular lineages). Furthermore, the "cell suicidal" program that was initially suggested to have evolved with the evolution of multicellularity is now known to be also present in many unicellular lineages, where it is thought to serve both similar (e.g., safe elimination of damaged, old or mutant individuals) and distinct roles (e.g., releasing nutrients) (e.g., Deponte, 2008; Gordeeva et al., 2004; Nedelcu et al., 2011). Last, cell–cell communication is very important in unicellular populations, as best exemplified by quorum sensing mechanisms identified in many bacterial and eukaryotic species (e.g., Keller and Surette, 2006; West et al., 2007).

In multicellular organisms, mutations in the genes associated with any of these traits are known to result in loss of cooperative behavior, uncontrolled proliferation, de-differentiation, evasion of death, and unresponsiveness to signals—which are among the recognized hallmarks of cancer (Hanahan and Weinberg, 2000, 2011). Nevertheless, mutations in these genes also occur in unicellular species and result in "cancer-like mutants"; such mutants are known as cheaters or opportunistic mutants (discussed later). In fact, it has been suggested that "it is quite possible that the main danger that unicellular organisms face is not competitors, pathogens, or lack of nutrients, but their own kin turning into "unhopeful monsters" and causing the death of the population" (Lewis, 2000).

What does this mean and why is this distinction important? First, cancer cells in multicellular organisms should not be seen as analogous to regular single-celled individuals in a unicellular species; that is, they are not a reversal to an ancestral-like single-celled life style. Rather, cancer cells are analogous to single-celled opportunistic (cheater) mutants. Consequently, evolutionary dynamics in cancer cell populations are likely different from those of "wild-type" single-celled populations. Furthermore, understanding the mechanistic basis for the emergence of such mutants in unicellular lineages and being able to predict or affect their evolutionary fate could provide new ways to think about cancer and therapy.

Paradigm 2. The Stability of Multicellular Individuals was Dependent on the Evolution of Tumor Suppression Mechanisms

The evolutionary stability of multicellular individuals is thought to require a combination of mechanisms that prevent the occurrence of selfish cell mutants (i.e., mechanisms that decrease genetic variation within the group), as well as strategies that lower the negative effects of such mutant cells (i.e., lower their selective advantage) (e.g., Nunney, 1999). As most of cancer's manifestations are due to mutations in somatic cells, many of the described "anticancer mechanisms" are involved in preventing somatic mutations from occurring or being propagated; in other words, they suppress genetic variation in somatic cell lineages. These mechanisms are usually known as tumor suppressor mechanisms and include both "caretakers" (e.g., DNA damage sensing and DNA repair) and "gatekeepers" (premature senescence and apoptosis) (e.g., Campisi, 2003; Kinzler and Vogelstein, 1997) (see also Chapter 8).

However, many genes and pathways associated with tumor suppression in multicellular lineages are also present in unicellular species, and some of them have been shown to play roles similar to those of their homologs in multicellular lineages (e.g., Laun et al., 2008; Madeo et al., 2002; Nedelcu and Tan, 2007; Umen and Goodenough, 2001). Indeed, as caretaker genes function in indispensable cellular processes, many evolved a long time before the origin of multicellularity (Domazet-Loso and Tautz, 2010). Also, although the evolution of gatekeeper genes was thought to overlap with the emergence of metazoans (and has been interpreted to reflect the need for both increased cooperation and cheating prevention), senescence and apoptosis/programmed cell death (PCD)-like processes have been described in many unicellular lineages (e.g., Bidle and Falkowski, 2004; Deponte, 2008; Gordeeva et al., 2004; Nedelcu et al., 2011).

For instance, homologs of *p53*—the most studied tumor suppressor gene (and most frequently mutated gene in cancer; Hollstein et al., 1991) are present in the closest unicellular relatives of animals, the choanoflagellates (Nedelcu and Tan, 2007). Likewise, a retinoblastoma (*Rb*) homolog—another important tumor suppressor gene that acts as a negative regulator of cell division (Weinberg, 1995), has been shown to be involved in controlling cell proliferation in the green alga *Chlamydomonas reinhardtii* (Umen and Goodenough, 2001). Lastly, homologs of many PCD genes have been found in various unicellular lineages, and some are known to also perform additional PCD-unrelated roles (see Nedelcu, 2009a;

Nedelcu et al., 2011 for references). Notably, mutants in several of these genes have also been described in single-celled species (discussed later).

What does this mean and why is this important? These findings argue that unicellular organisms—either as independent individuals or as members of a group/population, are subjected to similar pressures as cells in a multicellular organism (e.g., sensing and responding adaptively to signals from environment or neighbors, maintaining genome integrity). Understanding the specific roles these pathways play in unicellular lineages, how their loss affects individual fitness, and how they have been coopted into new roles during the evolution of multicellularity could be relevant to understanding how their loss affects the fitness of cancer cells as individual entities.

Paradigm 3. Cancer Cells are Infallible

This statement is based on the fact that cancer cells are very successful in the competition with normal cells by employing a variety of "unorthodox" strategies. For instance, cancer cells are thought to "disable" the cell suicidal program, "evade" the immune system, "trick" other cells into "working" for them, and so on. Overall, these statements imply that cancer cells "evolve" new traits.

However, in many cases, these "acquired" traits are not new traits in the sense of gain-of-function mutations/alleles; most are simply effects of loss-of-function mutations (Chen et al., 2015). What does this mean and why is this important? The affected genes are likely part of complex gene/adaptive networks, and they likely coevolved with other genes. Also, most genes affect more than one trait, and in many cases, the traits affected have opposite effects on fitness (known as antagonistic pleiotropy). Consequently, the same mutated gene that confers a selective advantage in one context might be deleterious in other contexts. Addressing this possibility can open up different approaches to therapy based on identifying and exploiting such pleiotropic effects.

MOVING FORWARD BY LOOKING BACK

As with other major evolutionary transitions, the emergence of multicellularity required that the reproduction potential (and thus fitness) of each constituent cell be adjusted to increase the fitness of the newly emerged individual (Maynard Smith and Szathmáry, 1997). During this process, a change in levels of selection occurred, with selection at the higher level (i.e., the multicellular individual) overriding selection at the lower level (the individual cells). Such a change was dependent on

cooperation among lower levels (Maynard Smith and Szathmáry, 1997), and required the reorganization of fitness (Michod and Nedelcu, 2003). However, in addition to its benefits, cooperation also sets the stage for defection, since cheaters that increase their own fitness at a cost to group fitness can occur (Maynard Smith, 1964; Michod and Roze, 2001).

As cooperative interactions are frequent in unicellular populations (e.g., West et al., 2007), cheater mutants are also known in many unicellular lineages that engage in social interactions or exhibit social behaviors in some phase of their life cycle (e.g., West et al., 2006). Many opportunities for cheating (associated with most aspects of social behavior from resource sharing, cell–cell communication, cell adhesion, to division of labor and altruism) occur in these lineages and many cheater mutants have been described. However, many of the social defect mutations are known to generate cheaters that increase transiently within a group during one life cycle phase but decrease to an even greater degree during a different phase. Furthermore, in lab settings, such mutants often drove themselves to extinction. Below, I am presenting several examples from the unicellular world that are relevant to two of the hallmarks of cancer and the paradigms discussed previously.

Uncontrolled Proliferation

The ability to control proliferation is thought to be paramount to the evolution of multicellularity. The basis for this seemingly reasonable view is the assumption that the regulation of the cell cycle is only important for multicellular organisms, and thus uncontrolled cell proliferation can be detrimental only to a multicellular organism. Moreover, the evolution of mechanisms to prevent the uncontrolled proliferation of cells is thought to reflect strong selective pressures to avoid cancer in multicellular lineages. An evolutionary approach to understanding the role of these mechanisms in multicellular lineages—involving both proximate and ultimate questions—might help reveal aspects that are being currently overlooked and that might have the potential to change our view on cancer and even on anticancer therapeutic strategies.

Although single-celled species are thought to be released from the danger of cancer, this is not entirely true. Variants that manifest uncontrolled cell replication can also occur in unicellular populations, and at least in theory, they can take over the wild-type strain. For instance, in *Escherichia coli*, mutant cells that proliferate faster during stationary phase increase in number and can dominate in the population (see, for instance, Finkel, 2006). Additional mutations accumulate in these growth advantage in stationary phase (GASP) variants, which further increase their competitive advantage during stationary phase. These proliferating mutants arising during stationary phase can become dead-end variants that could lead to a hostile takeover resulting in the demise of the population. The GASP mutants are thus analogous to cancer cells (for a more detailed discussion and references see Lewis, 2000; Finkel, 2006), and are even used experimentally to model cancer (Lambert et al., 2011). However, once conditions change, these mutants can lose out to a wild-type as they are unable—among other things—to upregulate catalase (see discussion in Lewis, 2000).

Likewise, in yeast, high nutrient conditions favor high levels of cell division and cellular cyclic adenosine monophosphate (cAMP), while nutrient-deprived environments trigger a drop in the concentration of cAMP and cells cease to divide (Eraso & Gancedo, 1985). Interestingly, yeast strains that cannot properly regulate their cAMP levels have been described; they continue to divide during nutrient deprivation (uncontrolled cell proliferation), but ultimately die due to starvation (Wilson & Tatchell, 1988). While such mutants can have a short-term fitness advantage in terms of reproductive output, their inability to adaptively regulate cell division in limiting conditions detracts from their long-term survival. Opportunistic mutants that cannot control their proliferation are also known in *Chlamydomonas reinhardtii*. However, the inability to down-regulate reproduction during nutrient deprivation results in a redox imbalance that culminates in oxidative stress and the demise of these mutants when grown in light (Chang et al., 2005; Davies et al., 1996; Moseley et al., 2006).

These examples argue that, in contrast to what is generally held, the regulation of cell proliferation, and thus mechanisms that regulate the cell cycle, are also very important in single-celled species. The control of cell division is usually mediated by the environment (e.g., nutrient availability, environment-induced metabolic imbalances, or DNA damage), but signals from other individuals (e.g., quorum sensing) can also affect the progression through the cell cycle (Keller and Surette, 2006; West et al., 2007). Thus, mutations that affect the regulation of these processes (or the ability to sense or respond to signals from the environment or their kin) can also occur and have the potential to be detrimental at either the individual level or the group/population level. These aspects of unicellular life suggest that the view of uncontrolled proliferation being relevant only to multicellular life and the implication that mechanisms to control cell proliferation have evolved specifically in multicellular lineages to suppress cancer needs to be reconsidered. Most importantly, understanding how unicellular populations avoid extinction in the face of opportunistic mutants exhibiting uncontrolled proliferation could provide insights into new therapeutic strategies.

Death Evasion

PCD has been initially associated with the evolution of multicellular organisms, where it can serve both developmental and cancer suppression roles. However, PCD-like processes have been later described in many unicellular lineages, both prokaryotic and eukaryotic (see Nedelcu et al., 2011 for a review). Although the adaptive significance of such processes in single-celled species is not always obvious, many experimental studies investigating potential benefits have been reported, including studies that use mutants capable to avoid PCD. Interestingly, while searching for the benefits of PCD such studies have revealed potential trade-offs/antagonistic pleiotropic effects (see Nedelcu et al., 2011 for discussion and references) that can be relevant to cancer cells.

For instance, yeast mutants that avoid death under oxidative stress conditions or during chronological aging have been described. However, these mutants also lost the ability to grow when transferred to fresh medium, accumulated more mutations than the wild-type, and although they had a short-term advantage, ultimately lost in competition with the wild-type (Vachova and Palkova, 2005). Likewise, in the slime mold, *Dictyostelium discoideum*, mutants that ignore signals from kin to escape death (and thus somatic cell-fate) during the formation of the stalk form fewer or no spores (e.g., Strassmann and Queller, 2011). Again, their short-term advantage comes with a cost in the long-term, and in some cases this is thought to be the result of antagonistic pleiotropy (Foster et al., 2004).

Overall, the few examples presented above argue that although cancer-like mutants do occur in unicellular populations, their fate appears to be determined by a series of evolutionary vulnerabilities (caused by antagonistic pleiotropy or life history trade-offs), such that their evolutionary success is restricted or short-lived. Could the same evolutionary vulnerabilities be exposed and be exploited to restrict the evolutionary success of cancer cells?

EXPOSING AND EXPLOITING EVOLUTIONARY VULNERABILITIES

A full understanding of the basic cellular mechanisms (e.g., cell replication, responses to stress) and life traits (survival and reproduction) that evolved early in the evolution of life might allow us to uncover the ancient evolutionary legacies that still "live on" in cells that are now part of a multicellular organism, and that cancer cells are apparently evading. Exposing these legacies could be exploited to decrease the fitness of cancer cells.

Life History Trade-Offs

Reproduction and survival represent the two main components of fitness in any individual. While in unicellular organisms the same cell is associated with both fitness components, in multicellular individuals the two components of fitness are generally reorganized between two distinct cell lineages—the germ and soma (Michod and Nedelcu, 2003). Life history theory states that investment in one fitness component detracts from the other (i.e., life history trade-offs) (Stearns, 1992). Reduced nutrient availability is known to substantially magnify, while increased nutrient availability can diminish (or obviate), an apparent trade-off; such plastic responses are thought to be determined by priority rules that govern the relative allocation of resources to organismal processes as a function of nutrient input—known as resource allocation (Zera & Harshman, 2001). In other instances, life history trade-offs appear to be the result of one activity negatively influencing another activity (Monaghan et al., 2009) or of genes with antagonistic pleiotropic effects on fitness (Leroi, 2001). Taking into account the reproductive success of cancer, such trade-offs between reproduction and survival might also exist and they could be exposed to reduce the fitness of cancer cells. Furthermore, during cancer progression distinct life-history strategies might evolve (i.e., fast/proliferative vs. slow/resistant) and understanding the selective pressures and life history trade-offs underlying these adaptive strategies might help designing more effective therapies (Aktipis et al., 2013).

Although cancer cells are not a reversal to an ancestral-like single-cell existence, the emergence of cancer is associated with a reversal to cell-level selection. When behaving as single-celled individuals (either solitary or as part of a group), cancer cells are likely to be subjected to selective pressures and constraints generally associated with single-celled life style. That is, as mutant somatic cells reclaim their individual-level fitness cancer cells have to carry out functions related to both components of fitness, and thus they are likely to encounter a series of new life history trade-offs. Thus, it is conceivable that the apparent reproductive success of cancer cells would incur a survival cost in specific settings; or some aspects of increased survival are likely to come with a cost in either the long-term or in a different setting. Moreover, it is possible that such life history trade-offs can be magnified under certain environmental conditions. This is even more likely the case as cancer cells are not simply reversals to an ancestral-like single-celled life-style and their fitness is not optimized. Identifying such conditions might allow us to manipulate tumor microenvironment to expose and/or magnify these trade-offs. In fact, there is experimental evidence that this strategy can be effective as it has been shown that, in contrast to healthy cells, cancer cells fail or only partial respond to limiting conditions, and continue to proliferate instead of promoting mechanism to prevent oxidative stress (Raffaghello et al., 2008).

Antagonistic Pleiotropy

Although the functional basis of a trade-off is often understood in terms of competition for limited resources among competing traits (such as reproduction, somatic growth, and maintenance) within an organism (Zera & Harshman, 2001), trade-offs could also be the result of signaling genes or pathways that simultaneously regulate two life-history traits in opposite directions, independent of resource allocation (Leroi, 2001). Also, as many cancer traits are consequences of gene loss—and since many genes are known to have pleiotropic effects (Leroi et al., 2005), it is possible that, such losses would also affect cancer-unrelated traits. Moreover, as antagonistic pleiotropy is thought to have played a role in the early stabilization of cooperative behaviors (Foster et al., 2004) it is likely that the evolution of cooperative interactions during the emergence of multicellularity has involved many genes with such effects. Thus, the loss of cooperative genes could have, besides the implicit benefits to the "selfish" mutants, additional negative effects on other aspects of fitness at the cell level—either in the long run or in different settings. As a proof of principle, in the multicellular green alga *Volvox carteri*, the loss of *regA*—the "cooperation" gene responsible for somatic cell differentiation in this alga (Kirk et al., 1999), results in uncontrolled proliferation of somatic cells, but the "cheater" cells are more sensitive to stress (König, 2015).

Gene Cooption

As discussed earlier, many tumor suppressor genes have been coopted from genes already present in our unicellular ancestors, where they function in non-multicellularity related activities. It is possible that some of those functions are still expressed in multicellular organisms and the loss of such genes can affect more than "multicellular"/cooperative traits. For instance, the closest homolog of *regA* in the unicellular *Chlamydomonas reinhardtii* is expressed under environmental stress as an adaptive response to enhance long-term survival at a cost to immediate reproduction (i.e., it acts as a life history trade-off gene; Nedelcu and Michod, 2006; Nedelcu, 2009b). Interestingly, in addition to its developmental regulation, *regA* in the multicellular *Volvox carteri* can still be induced under stress, which is consistent with the increased sensitivity to stress observed in *regA* mutants (König, 2015; König and Nedelcu, 2016). Other multicellular development pathways are thought to have evolved from pathways present in unicellular lineages. Notably, the developmental cyclic AMP signaling in the cellular slime mold *D. discoideum* also evolved from a stress response in single-cell amoebae (Schaap, 2011). Thus, it is conceivable that some of the pathways that are affected in cancer have their roots in responses to environmental stimuli. Identifying the ancestral roles of "cancer" genes and how they have been coopted during the evolution of multicellularity can open new therapeutic avenues by imposing changes in the tissue microenvironment that could differentially affect cancer cells.

CONCLUDING REMARKS

In the last few decades, evolutionary and ecological approaches and questions have been increasingly associated with cancer research. However, potential limitations of the current paradigms should be acknowledged. New evolutionary perspectives should be developed that can make use of our knowledge of homologous mechanisms/processes in unicellular lineages to both further our understanding of cancer and design new therapies. Specifically, understanding how ancestral pathways that regulate basic life history traits have been coopted into new roles during the evolution of multicellularity will help us realize how their loss affects negatively not only the fitness of the multicellular organism but also that of individual cancer cells. In particular, understanding the biology of cancer cells not only in terms of their success/ strengths but also in terms of their weaknesses/limitations (resulting from life history trade-offs and/or antagonistic pleiotropy) might help develop strategies that could take advantage of potential evolutionary vulnerabilities inherited from our unicellular ancestors and be used towards decreasing cancer cells' fitness.

References

Aktipis, C.A., Boddy, A.M., Gatenby, R.A., Brown, J.S., Maley, C.C., 2013. Life history trade-offs in cancer evolution. Nat. Rev. Cancer 13, 883–892.

Aktipis, C.A., Boddy, A.M., Jansen, G., et al., 2015. Cancer across the tree of life: cooperation and cheating in multicellularity. Phil. Trans. Roy. Soc. B. Biol. Sci. 1673, 20140219.

Bidle, K.D., Falkowski, P.G., 2004. Cell death in planktonic, photosynthetic microorganisms. Nat. Rev. Microbiol. 2, 643–655.

Campisi, J., 2003. Cancer and ageing: rival demons? Nat. Rev. Cancer 3, 339–349.

Chang, C.W., Moseley, J.L., Wykoff, D., Grossman, A.R., 2005. The LPB1 gene is important for acclimation of *Chlamydomonas reinhardtii* to phosphorus and sulfur deprivation. Plant Physiol. 138, 319–329.

Chen, H., Lin, F., Xing, K., et al., 2015. The reverse evolution from multicellularity to unicellularity during carcinogenesis. Nat. Commun. 6, 637.

Davies, P.C.W., Lineweaver, C.H., 2011. Cancer tumors as Metazoa 1.0: tapping genes of ancient ancestors. Phys. Biol. 8, 015001.

Davies, J.P., Yildiz, F.H., Grossman, A.R., 1996. Sac1, a putative regulator that is critical for survival of *Chlamydomonas reinhardtii* during sulfur deprivation. EMBO J. 15, 2150–2159.

Deponte, M., 2008. Programmed cell death in protists. Biochim. Biophys. Acta 1783, 1396–1405.

Domazet-Loso, T., Tautz, D., 2010. Phylostratigraphic tracking of cancer genes suggests a link to the emergence of multicellularity in metazoa. BMC Biol. 8, 66.

Eraso, P., Gancedo, J.M., 1985. Use of glucose analogs to study the mechanism of glucose-mediated cAMP increase in yeast. FEBS Lett. 191, 51–54.

Finkel, S.E., 2006. Long-term survival during stationary phase: evolution and the GASP phenotype. Nat. Rev. Microbiol. 4, 113–120.

Foster, K.R., Shaulsky, G., Strassmann, J.E., Queller, D.C., Thompson, C.R.L., 2004. Pleiotropy as a mechanism to stabilize cooperation. Nature 431, 693–696.

Gordeeva, A.V., Labas, Y.A., Zvyagilskaya, R.A., 2004. Apoptosis in unicellular organisms: mechanisms and evolution. Biochemistry 69, 1055–1066.

Hanahan, D., Weinberg, R.A., 2000. Hallmarks of cancer. Cell 100, 57–70.

Hanahan, D., Weinberg, R.A., 2011. Hallmarks of cancer: the next generation. Cell 144, 646–674.

Hollstein, M., Sidransky, D., Vogelstein, B., Harris, C.C., 1991. p53 mutations in human cancers. Science 253, 49–53.

Keller, L., Surette, M.G., 2006. Communication in bacteria: an ecological and evolutionary perspective. Nat. Rev. Microbiol. 4, 249–258.

Kinzler, K.W., Vogelstein, B., 1997. Cancer-susceptibility genes. Gatekeepers and caretakers. Nature 386, 761–763.

Kirk, M., Stark, K., Miller, S., Muller, W., Taillon, B., Gruber, H., Schmitt, R., Kirk, D.L., 1999. regA, a *Volvox* gene that plays a central role in germ soma differentiation, encodes a novel regulatory protein. Development 126, 639–647.

König, S.G., 2015. The genetic and evolutionary basis for somatic cell differentiation in the multicellular alga *Volvox carteri*: investigations into the regulation of *regA* expression. PhD Thesis. University of New Brunswick, Fredericton, Canada.

König, S.G., Nedelcu, A.M., 2016. The mechanistic basis for the evolution of soma during the transition to multicellularity in the volvocine algae. In: Niklas, K., Newman, S. (Eds.), Multicellularity: Origins and Evolution. MIT Press, Cambridge, pp. 43–70.

Lambert, G., Estevez-Salmeron, L., Oh, S., Liao, D., Emerson, B.M., Tlsty, T.D., Austin, R.H., 2011. An analogy between the evolution of drug resistance in bacterial communities and malignant tissues. Nat. Rev. Cancer 11, 375–382.

Laun, P., Heeren, G., Rinnerthaler, M., Rid, R., Kossler, S., Koller, L., Breitenbach, M., 2008. Senescence and apoptosis in yeast mother cell-specific aging and in higher cells: a short review. Biochim. Biophys. Acta 1783, 1328–1334.

Leroi, A.M., 2001. Molecular signals versus the Loi de Balancement. Trends Ecol. Evol. 16, 24–29.

Leroi, A.M., Bartke, A., De Benedictis, G., et al., 2005. What evidence is there for the existence of individual genes with antagonistic pleiotropic effects? Mech. Ageing Dev. 126, 421–429.

Lewis, K., 2000. Programmed death in bacteria. Microbiol. Mol. Biol. Rev. 64, 503–514.

Madeo, F., Herker, E., Maldener, C., Wissing, S., Lachelt, S., 2002. A caspase-related protease regulates apoptosis in yeast. Mol. Cell 9, 911–917.

Maynard Smith, J., 1964. Group selection and kin selection. Nature 201, 1145–1147.

Maynard Smith, J., Szathmáry, E., 1997. The Major Transitions in Evolution. Oxford University Press, Oxford.

Michod, R.E., Nedelcu, A.M., 2003. On the reorganization of fitness during evolutionary transitions in individuality. Integr. Comp. Biol. 43, 64–73.

Michod, R.E., Roze, D., 2001. Cooperation and conflict in the evolution of multicellularity. Heredity 86, 1–7.

Monaghan, P., Metcalfe, N.B., Torres, R., 2009. Oxidative stress as a mediator of life history trade-offs: mechanisms, measurements and interpretation. Ecol. Lett. 12, 75–92.

Moseley, J.L., Chang, C.W., Grossman, A.R., 2006. Genome-based approaches to understanding phosphorus deprivation responses and PSR1 control in *Chlamydomonas reinhardtii*. Eukaryotic Cell 5, 26–44.

Nedelcu, A.M., 2009a. Comparative genomics of phylogenetically diverse unicellular eukaryotes provide new insights into the genetic basis for the evolution of the programmed cell death machinery. J. Mol. Evol. 68, 256–268.

Nedelcu, A.M., 2009b. Environmentally-induced responses co-opted for reproductive altruism. Biol. Lett. 5, 805–808.

Nedelcu, A.M., Michod, R.E., 2006. The evolutionary origin of an altruistic gene. Mol. Biol. Evol. 23, 1460–1464.

Nedelcu, A.M., Tan, C., 2007. Early diversification and complex evolutionary history of the p53 tumor suppressor gene family. Dev. Genes Evol. 217, 801–806.

Nedelcu, A.M., Driscoll, W.W., Durand, P.M., Herron, M.D., Rashidi, A., 2011. On the paradigm of altruistic suicide in the unicellular world. Evolution 65, 3–20.

Nunney, L., 1999. Lineage selection and the evolution of multistage carcinogenesis. Proc. R. Soc. Lond. B 266, 493–498.

Raffaghello, L., Changhan, L., Fernando, M.S., Min, W., Federica, M., Giovanna, B., Longo, V.D., 2008. Starvation-dependent differential stress resistance protects normal but not cancer cells against high-dose chemotherapy. Proc. Natl. Acad. Sci. USA 105, 8215–8220.

Schaap, P., 2011. Evolution of developmental cyclic AMP signalling in the *Dictyostelia* from an amoebozoan stress response. Dev. Growth Differ. 53, 452–462.

Stearns, S.C., 1992. The evolution of life histories. Oxford University Press, Oxford.

Strassmann, J., Queller, D.C., 2011. Evolution of cooperation and control of cheating in a social microbe. Proc. Natl. Acad. Sci. USA 108, 10855–10862.

Umen, J.G., Goodenough, U.W., 2001. Control of cell division by a retinoblastoma protein homolog in *Chlamydomonas*. Genes Dev. 15, 1652–1661.

Vachova, L., Palkova, Z., 2005. Physiological regulation of yeast cell death in multicellular colonies is triggered by ammonia. J. Cell Biol. 169, 711–717.

Webb, J.S., Givskov, M., Kjelleberg, S., 2003. Bacterial biofilms: prokaryotic adventures in multicellularity. Curr. Opin. Microbiol. 6, 578–585.

Weinberg, R.A., 1995. The retinoblastoma protein and cell cycle control. Cell 81, 323–330.

West, S.A., Griffin, A.S., Gardner, A., Diggle, S.P., 2006. Social evolution theory for microorganisms. Nat. Rev. Microbiol. 4, 597–607.

West, S.A., Diggle, S., Buckling, A., Gardner, A., Griffins, A.S., 2007. The social lives of microbes. Ann. Rev. Ecol. Evol. Syst. 38, 53–77.

Wilson, R.B., Tatchell, K., 1988. Sra5 encodes the low-km cyclic-Amp phosphodiesterase of *Saccharomyces cerevisiae*. Mol. Cell. Biol. 8, 505–510.

Xavier, J.B., 2011. Social interaction in synthetic and natural microbial communities. Mol. Syst. Biol. 7, 483.

Zera, A.J., Harshman, L.G., 2001. The physiology of life history trade-offs in animals. Ann. Rev. Ecol. Syst. 32, 95–126.

16

Atavism Theory—An Introductory Discourse

Mark Vincent

Department of Oncology, University of Western Ontario, London, ON, Canada

Nothing in life is to be feared, it is only to be understood. Now is the time to understand more, so that we may fear less. **Marie Curie**

INTRODUCTION

The evolutionary nature of cancer holds the promise of providing insight into the vexed questions of intratumoral heterogeneity, target selection, and drug resistance, and is appropriately the subject of incremental attention. Evolutionary dynamics is a good, perhaps, even the only way to describe complex lineage relationships in clinical tumors, with important implications for therapeutic choice and, as is still all too often the case, a deeper understanding of therapeutic failure. Moreover, classical Darwinian concepts appear adequate for these tasks, particularly the seminal ideas of inherited variation, competition and selection, and gel nicely with the currently regnant somatic mutation theory (M-Theory), which holds that cancer is causally initiated and maintained by acquired "driver" mutations, involving inappropriate activation (oncogenes) and/or inactivation (tumor suppressor genes) of key growth control genes, with additional mutations providing the substrate for further evolution toward increasing aggressivity over time (Nowell, 1976).

What more is there to add to this perspective? In my opinion, there is much to add. For while there can be no denying the high value of a comprehensive descriptive approach to clonal evolution (including, critically, the elucidation of driver vs. passenger derangements), there are four additional questions, which deal more fundamentally with the nature of cancer. These questions are subterranean, and at a more conceptual level than the predominantly descriptive information associated with genomic sequencing. They have received scant attention either in the literature or in academic meetings, to the extent one suspects they may border on being scientifically off-limits, for a variety of reasons: they are mathematically intractable, seem to have little obvious therapeutic or commercial value, require an unusual type of speculative curiosity, may not be amenable to ready proof, and deal with subject matter that is arcane, tangential and remote. What are these questions (Table 16.1)?

What Form of Life?

The first question is what form of life is represented by a cancer? In other words, where on the Tree of Life should cancer cells be placed, and what does this imply? This question implies that cancers may not be the same form of life as their hosts of origin, which is indeed reflected in the word "transformed." In turn this implies that cancers have to have speciated in some sense, and perhaps even more than speciated, given what appears to be a vast taxonomic gap between the unicellular/colonial cancer cells and the metazoan host of origin. A subsidiary question is whether, given the intratumoral and even intercellular heterogeneity, the cancer cells within any given tumor are indeed speciated with respect to each other.

I have previously discussed the speciation question in relation to cancer, concluding that carcinogenesis fulfills the various definitional criteria, including those that apply to sexually reproducing organisms (Vincent, 2010). To put this on a more secure footing, what is needed is a formalized mathematical comparison of genomic differences between cancers and the host of origin, such as represented by Birky's θ, for instance, to document whether these increasingly well-documented genomic alterations cross a mathematical threshold for speciation: "The ratio of the mean pairwise sequence difference between a pair of clades (K) to the mean pairwise sequence difference within a clade (θ) can be used to determine whether

Ecology and Evolution of Cancer. **http://dx.doi.org/10.1016/B978-0-12-804310-3.00016-8**

TABLE 16.1 Four Fundamental Questions

- What form of life is represented by cancer cells?
- Why, despite the myriad of genomic derangements and the long list of oncogenes and tumor suppressed genes which have each been published as causally contributory, is the MP so paradoxically stereotyped in all cancers?
- Are the characteristics of the MP primitive, and/or adapted for a primeval environment, in respect of the earth's geochemical history?
- Is there, or was there ever, a conceivable biological purpose to the MP, in the interests of survival, or is the MP merely a collection of errors?

MP, Malignant phenotype.

the clades are samples from different species (K/θ ≥ 4) or the same species (K/θ < 4) with probability ≥0.95" (Birky, 2013). My prediction is that, for the comparison of a cancer cell and its host of origin, they would, but this effort would face a practical obstacle in the intratumoral heterogeneity, since it might be difficult to decide which clone in the tumor to select. Furthermore, the interclonal heterogeneity might itself even cross the speciation threshold; if this were so, there is an interesting implication in that logically it would imply that at least some of the cancer cells, if speciated from each other, also have to be speciated from the host of origin.

As suggested previously, however, the speciation threshold is only a minimalist taxonomic perspective. This is because cancer cells appear most like a type of protozoan, facultatively colonial, but no longer metazoan, and hence at least a phylum away from the originating metazoan hosts, and perhaps even occupying a niche within a separate kingdom of life. Consequently, speciation it may be, but carcinogenesis is categorically not the usual type of speciation, which leads gradually to two similar types of organism but still within the same genus.

Furthermore, the relative rapidity of the carcinogenesis violates one of Darwin's dearly held beliefs, which is that taxonomic change is always gradualistic; carcinogenesis, by contrast, is actually saltationist, which is another clue that classical Darwinian principles may not apply to the act of carcinogenesis itself, even if they do apply to what happens in an extant cancer afterward. Without Darwin, evolutionary discourse becomes strained and tentative; yet it cannot be denied that carcinogenesis, out of a previously normal cell, is a radical act of evolution, notwithstanding the result might be distasteful to progressivists.

Are we then, any closer to answering this first question: what form of life is a cancer (cell)? Complicating this issue is the now well-established fact of entropic genome instability in cancers, and which, in turn, has to imply an unstable identity. This is a problem for the tree of life concept, depending as it does on identitarian stability. Indeed, I am unaware of any other protozoan with a similarly

unstable genome, although it is possible that this question has not been fully explored, and/or that that some degree of genomic instability might have been normal in early eukaryotes. The best that I can offer is that phenotypically, cancer cells should be positioned as a type of dyskaryote holozoan opisthokont protist, close to the basal trunk of Animalia. A recent and excellent review on the ubiquity of cancer and like phenomena across the Tree of Life (Aktipis et al., 2015), although advocating a "cheater" paradigm for cancer, emphatically represents all forms of neoplasia as an assault on multicellularity per se. Given the categorical gap between unicellular and multicellular animals, this review provides support for the pervasive availability of cancer as a taxonomic alternative deeply entrenched in the genome of every metazoan. Since the well-known and ubiquitous dedifferentiation characteristic of cancer is convincingly and strongly advocated by these authors as a *prima facie* manifestation of metazoan deconstruction, there is, therefore an unambiguous directionality evident, which points away from organizational complexity and sophistication, and toward simplicity, consistent with the more primitive unicellular and colonial morphology and behavior of all cancers.

One is struck particularly by the transmissible cancers (see Chapter 12 in this book), which, despite their unusual communicability, are unlikely to be inherently different from any other cancers, notwithstanding the accidental features that favor transmission (such as location, mode of spread, and relatedness of their host population). It is very difficult to doubt that these transmissible cancers do indeed represent anything other than a grossly different form of life from their originating hosts; if they do, and since they are in all likelihood inherently similar to any other cancer (most of which can be artificially transmitted in the laboratory in nude mice, for instance), then so should every other types of cancer also represent a different form of life, from the originating metazoan.

If this set of arguments is credible, and the most reasonable locus on the tree of life for cancer been near the base of the eukaryotes, the implication of reprimitivisation seems inescapable. If these arguments are not credible, then we are thrust back on the "Error Hypothesis" implicit in M-Theory, that is, the malignant phenotype is merely a stochastic phenomenon, the result of selection for survival and aggressivity, fueled by random genetic errors (mutations) and has no other significance; or worse, on delegitimizing this question altogether.

The Commonality of the Malignant Phenotype

The second question concerns the strange commonality of the malignant phenotype (MP), quite paradoxical in the face of so much causal heterogeneity. The point here is not so much the diversity of the remote, original causes (smoking, environmental radiation, chemical carcinogens, etc.) but the multiplicity of genomic lesions (drivers) which mediate the stereotypical MP, on an ongoing basis. This is widely acknowledged to be true, whether by the majority who support conventional "M-Theory," or those who believe that aneuploidy per se is the key driver (Nicholson and Duesberg, 2009). Either way the purported causal mediators are understood to be highly variable between different cancers, and even, quite possibly, between different cancer cells within the same tumor (Ling et al., 2015); this observation relates to the causal mediating genetic lesions, such as they are understood, but is also true of the noncontributing "passenger" mutations as well. Passenger mutation heterogeneity is a different issue, to which I shall return, but for now the key point is this: how can such variability in proximate genomic causation be reduced, in all cases of cancer, to such a remarkably similar phenotype, in all cancers? The commonality of the core MP, despite some phenotypic heterogeneity in respect of noncore characteristics, is, of course, what enables all these entities to be definitionally grouped under one banner (cancer), a good example of what is known in philosophy as a "natural kind" (Boyd, 1991). That this is true by definition does not excuse us from having to answer as to how this happens. However, first, we should be clear about what it is we are addressing: what really is the MP?

The common, essential attributes of cancer have been outlined by Hanahan and Weinberg (H–W) in several of the most widely quoted articles in the literature (Hanahan and Weinberg, 2000, 2011). However, their achievement was to describe the core element of cancer as a disease process; I have elsewhere delivered a different set of core attributes (adaptive resilience, reprimitivisation, and phylogenation) (Vincent, 2012), and also outlined a core agenda of every cancer as a combination of biomass accumulation and genomic rescrambling (together amounting to low-fidelity iteration), along with anatomic escape (Vincent, 2011). This perspective, which is more ecological than medical, priorizes the cancer cell as an organism in its own right, rather than as a bit player in a disease process.

However, whichever perspective is adopted, it is the case that a myriad of genomic rearrangements always converge on a common set of (relatively few) phenotypic attributes. Since this cannot be a coincidence, there has to be an explanation. Under "M-Theory," the explanation is said to be "selection" among a plethora of randomly generated mutations, which in one way or another, all converge on a small number of key pathways which disrupt control of the cell cycle; that is, a type of convergent evolution. While there is doubtless a substantial element of truth in this explanation, it should be noted that there is an alternative (or complementary) explanation, which is more parsimonious: that there exists a common, underlying program in every eukaryotic cell that can be released, when enough of the evolved superstructure of

control has been damaged, or in response to a deliberate switch, and which results, by a release mechanism, in the core manifestations of the MP.

As noted (Aktipis et al., 2015), cancer, or at least neoplasia, occurs throughout the animal kingdom, as well as in plants, and even in simple hydra (Domazet-Lošo et al., 2014) (see Chapter 2); the implication, therefore, is that the mediating program is itself ancient. This is one way it can be in-common, and the most parsimonious explanation for the constant "reinvention of the wheel" in every cancer, in different species, across vast expanses of geography and time. However, an alternative explanation exists, which is that the genomic diversity characterizing intratumoral heterogeneity is sufficient to generate evolutionary convergence; this topic is therefore debatable (see for instance Chapter 17). In the case of the former explanation, that there is an ancient, in-common program underpinning the malignant phenotype, what the multiple diverse genomic drivers do, singly, or (more likely) in combination, is to either damage the (later-evolved) managerial superstructure of control, or activate a switch; in either case, the MP is derepressed. That the genomic drivers are so heterogeneous is no surprise, since there are many ways to damage a sophisticated control system; yet there is relative homogeneity in the core program which is released, resulting in the survival of the MP. This could be the solution to the paradox.

The Malignant Phenotype: Primitive and/or Adapted to Primeval Geochemistry?

The third question is whether the characteristics of the MP are inherently primitive and/or represent adaptations to the geochemistry of the ancient earth. This question was stimulated by an observation of Otto Warburg, who believed that cancer cell metabolism (i.e., his "Warburg Effect") was a primitive feature characteristic of early life forms extant prior to the oxygenation of the earth's atmosphere (Warburg, 1969); however, previous authors had also alluded to the resemblance of cancer cells to primitive, unicellular organisms like amoebae (Roberts, 1926; Snow, 1893). It is certainly possible to list multiple features of the MP which might suggest reprimitivisation, and/or adaptation to the Proterozoic eon in which eukaryotic cells first evolved (~1.6 BYA) (Table 16.2).

The Proterozoic (early life) eon [2.5 billions of years ago (BYA)–0.542 millions of years ago (MYA)] was characterized by hypoxia (~1% of the present atmospheric level), but not anoxia; volcanic eruptions spewing vast amounts of sulfur into the atmosphere and oceans, leading to pyritic precipitation, and removal from solution of trace metals like iron and molybdenum (required for nitrogen fixation at the prokaryote base of the food chain); extraterrestrial radiation unimpeded by any ozone layer; and consequently, a tenuous and resource-depleted environment, in which

TABLE 16.2   Features of the Malignant Phenotype That Might Be Primitive, and/or Adaptations to Proterozoic Oceanic Geochemistry; Actual or Possible Therapeutic Implications

Feature	Therapeutic implications
Taxonomic shift	Implies the inevitable existence of significant genomic and phenotypic differences, which may be targetable; opposes "Type I Nihilism" that cancer cells and normal cells are too similar
Genomic instability	May mediate much of therapeutic resistance to common cancer drugs; but also an opportunity to push the cancer beyond the edge of survivability (probably how chemotherapy works); enables strategies that further disable DNA repair (induced synthetic lethality); generation of neoantigens permits immune recognition; opposes "Type II Nihilism" that resistance will always necessarily develop
Tendency to grow as colonies (tumors)	Permits local ablative therapies, such as surgery, radiotherapy, radio-frequency ablation, and so on
Dedifferentiation and loss of specialized functions	Permits removal or killing of tumor cells without further loss of normal function, that is, the tumor cells are nonfunctional anyway so their removal per se does not lead to further functional decline (although therapeutic toxicity may lead to this, this is conceptually different)
Bloom-like growth	Inability to enter G_0 of the cell cycle, with impaired capacity to repair DNA damage, leading to heightened vulnerability to chemotherapy (no off-switch)
Hyperphagia (ceaseless ingestion of nutrients)	"Poison its food' strategies—preferential drug delivery (e.g., melphalan, albumin-bound paclitaxel, various antimetabolites e.g., pemetrexed, possibly deoxyglucose); as well as radioactive glucose for PET scanning or radioactive phenylalanine for scanning and therapy (exp.)
Warburg-type metabolism	Tumor-specific isoforms of enzymes, for example, GLUT 12, M2-PK, are potential targets; ketogenic diets (exp.)
Hypoxia-adaptation	Inhibitors of HIF (exp.); hypoxia activated drugs; hyperbaric oxygen (exp.). But hypoxia is responsible for resistance to radiotherapy
ROS susceptibility	ROS mediate the efficacy of radiotherapy; agents to deplete ROS defenses (exp.); ROS generator agents "mitocans" (exp.); certain chemotherapies, for example, anthracycline-generated ROS
Milieu acidification	May be partly responsible for localized immunodepression; milieu alkalinization (exp.); selective drug delivery (e.g., 3-bromopyruvate)
Immortalization	Telomerase inhibitors (exp.); reversal of apoptotic refractoriness associated with the Warburg Effect (exp.)
Asexual reproduction	Probably enables some cancer cells to accumulate deleterious mutations over time (Muller's Ratchet), compromising fitness in the face of stressors, for example, therapy
Motility (amoeboid)	Possible antimetastatic strategies involving targeting the apparatus for motility (exp.)
Prespeciation and self-recognition	Do cancer cells recognize each other as "self" and host cells as "foreign?"Given genomic instability, could cancer cells be induced to suppress/attack other clones of cancer cells? (speculative)

even bacteria might have struggled (Canfield, 2014). This is the "euxinic" environment in which unicellular eukaryotes evolved; multicellular animals (i.e., metazoa) only evolved late in the Proterozoic, contemporaneous with a second uptick in the atmospheric oxygen concentration to present levels, and which soon led to the "Cambrian explosion" of species diversity, heralding the opening of our current eon, the Phanerozoic, some 542 MYA (Mills and Canfield, 2014).

As noted, cancer cells are obviously unicellular, or at most, quasicolonial, and do not sustain tissue differentiation; their morphology resembles that of simple organisms. Furthermore, their metabolism is hypoxia-adapted, although not in the simplistic way hypothesized by Warburg (1969); they reproduce asexually, which may have predated sexual reproduction. They vigorously and excessively pump protons into their environment (Parks et al., 2013), so that, despite the proton flux from enhanced glycolysis, the status quo ante for the pH gradient is not

only restored but markedly exaggerated; the extracellular milieu becomes quite acid and the intracellular pH quite alkaline. This peculiar feature may be originally derived from an early eukaryotic strategy to solubilize the iron and molybdenum required for nitrogen fixation by their bacterial prey (Anbar and Knoll, 2002), or detoxify the sulfide ion, but is otherwise unexplained in modern cancer cells. The growth of cancer cells resembles that of an algal bloom, being only constrained by resource availability. Furthermore, they import huge quantities of carbon skeleton in the form of glucose and glutamine (Gatenby and Gillies, 2004; Wise et al., 2008), as if they were perpetually resource-depleted. More tenuously, the genomic instability might resemble the situation before sophisticated DNA repair systems evolved. Furthermore, genetic instability could be represented as "prespeciation," reflecting a time before genomic stability was achieved, or a time when some degree of genetic instability was retained as advantageous in the harsher and more unpredictable environment of the mid-Proterozoic (Goodman, 2016). It is also possible to consider cancer cells as type of microcarnivore, given the successful assault on the host body, and which likely involves at least some direct tissue destruction and not merely diversion of ambient nutrients. Furthermore, the known sensitivity toward, and vigorous defense mechanisms against ROS in cancer cells (Marengo et al., 2016) may hark back to a time when oxygen was more of a toxin than a resource. Finally, cancer cells are generally considered to be immortalized, another characteristic of unicellular organisms (Chen et al., 2015). These resemblances between cancer cells and extant primitive organisms, or to what we would predict to be early eukaryotes evolving in the Proterozoic, are at least provocative.

Interestingly, the extracellular tumor milieu in cancer patients may in fact represent a type of recreation of the Proterozoic. Intratumoral oxygen levels are known to be low, often profoundly so (Semenza, 2013), given the bloom-like tendency of cancer cells to outgrow their blood supply. Despite being some 500 MY from the end of the Proterozoic, because the diffusion capacity of oxygen in tissue is very limited, we are always only less than 1 mm away from it. Other features, such as the extracellular acidosis in tumors, may also resemble zones within the Proterozoic oceans (Perez-Jiminez et al., 2011).

These analogies between the malignant phenotype and its milieu on the one hand, and Proterozoic organisms and their geochemistry on the other, are to some extent tenuous and speculative. However, they span morphology, metabolism and life history, and are, taken together, sufficiently striking to warrant serious consideration. Much more empirical research is necessary, so that they may be either refuted, or placed on a more secure scientific footing. Our understanding of the Proterozoic oceans and their ecology is expanding (Johnston et al., 2009), to the point at which it might be experimentally tractable to consider whether cancer cells could have survived as free-living plankton in such an environment under low oxygen, in an acidotic, sulfidic, trace-metal depleted, and nutrient-poor aquatic medium, and feeding off scarce bacterial and other microplankton with the aid of the proton pump. The discovery of transmissible clam leukemia off the eastern seaboard of North America (Metzger et al., 2015), and which inescapably must involve a planktonic phase in its life-cycle, suggests this is less far-fetched than it sounds. However, to the extent any of this speculation can be confirmed, the notion of the MP as an ancient, in-common program would be further supported.

Biological Purpose to the Malignant Phenotype?

Since cancer normally occurs in the elderly, it is tempting to suggest that its "biological purpose" (perhaps "function" is a better word) is to turn over the population so that excessive investment in frail, nonreproductive individuals is prevented. However, this is doubtful since individuals within ancestral populations rarely lived that long and most wild animals are caught in food webs or other competitive situations in which even minor reductions in fitness are promptly rewarded with elimination and recycling.

This leaves the conventional and nihilistic explanation under "M-Theory," that cancer is merely the result of a series of mistakes: The "Error Hypothesis." Organisms are said to have evolved a range of defenses against rogue cellular behavior, both before and after cellular transformation, that are "good enough" to prevent cancer from being a major problem during their reproductive lives (see the Introductory Chapter and Chapter 7); further investment to eradicate it entirely may be unnecessary and is probably counter-productive (since some degree of genomic instability may be necessary to optimizes evolvability, for instance) (Goodman, 2016; Lehman and Miikkulainen, 2015).

However, it may be possible to sustain another hypothesis entirely (Vincent, 2012). Let us start by assuming the perspective of the individual cell, rather than that of the metazoan of which it is a part. The story of the emergence of metazoan life must have involved the transfer of sovereignty from the individual cell, to the colony, to the whole organism (and even beyond that, to the species); this process may not always have been congenial to the individual cell. It may have been, as with slime molds, that there are decisions to be made as to whether to individuate or to aggregate, or whether to self-sacrifice or not. It is, furthermore, highly likely that individual cells might be able to sense whether "their" metazoan was doing well or not, and if not, to activate a survival program which logically would involve certain tactical expedients: biomass accumulation (via conversion of the host's biomass); slaughter of potential competition (killing the host); low-fidelity iteration (to maximize survival prospects in an environment that may be changing unpredictably), and anatomic escape and dispersal of the rogue cell progeny (into the water). Those functions happen to be identical to the core universal attributes the MP (except that in humans, this drama unfolds on the dry land, where the strategy cannot ultimately succeed). This hypothesis in some ways contains elements of the "cheater" hypothesis published by Aktipis et al. (2015), but sees the rogue cancer cell more as a survivalist than as a cheat.

Although I will immediately acknowledge that this scenario is conjectural, it should again be noted that the documented planktonic spread of clam leukemia (Metzger et al., 2015) is predicted by, and is hence supportive of, this "lifeboat" hypothesis, as is the recent discovery of neoplasia in hydra (Domazet-Lošo et al., 2014). The corollary of this hypothesis is that all of extant metazoa derive from cells which depended on this ability at some critical but remote junction in deep time; also, it implies the ability to revert back later to the domesticated metazoan phenotype, as very occasionally documented in some forms of clinical and experimental cancer (Bolande, 1991; Parikh et al., 2014).

Again, this is a speculative hypothesis, not a set of facts. The notion that the MP might have served a survival purpose at one time is an off-shoot of the umbrella atavism hypothesis, and is not a requirement for it. It is, however, as with this basic atavism hypothesis, also more parsimonious than the "Error Hypothesis" (Bolande, 1991) of M-Theory, with which is not necessarily mutually exclusive. Furthermore, it makes sense, and may be the only hypothesis that makes complete sense, at least to this author.

THE NATURE OF CANCER

In aggregate, the answers I have provided to these four questions, albeit tentative and conjectural, could suggest an explanation for the nature of cancer which involves the derepression of an ancient modus operandi laid down at the dawn of eukaryotic life. This perspective envisions the cancer cell as a prespeciated, unicellular/quasicolonial protozoan, adapted for the harsh life of mid-Proterozoic, euxinic ocean geochemistry. This hypothesis may also explain the refractoriness of cancer cells to modern therapies, some of which work (or attempt to work) by recreating stresses (radiation, ROS, nutrient deprivation, chemical warfare) similar or identical to those that primitive eukaryotes had to be able to survive as part of their natural environment: the "eukaryotic extremophile" theory of therapeutic resistance. Along with this is the analogy of some successful cancer treatments which utilize combined modality treatment, such as chemotherapy plus radiotherapy, with the "press-pulse" extinction mechanism, of acute-on-chronic stress, which is hypothesized to have been responsible for some of the major extinction events in prehistory (Arens and West, 2008; Vincent, 2011).

In summary, this "ancient derepression" model is Atavism Theory, or "A-Theory." It is more complementary to, than contradictory to M-Theory, providing "mutation" is interpreted broadly. Like any other hypothesis, it should be rigorously examined and, to the extent possible, investigated empirically, despite the challenges inherent in inferring the molecular biology of Proterozoic eukaryotes and their environs. Although there is broad agreement on major aspects of Proterozoic geochemistry (as noted previously, featuring hypoxic, sulfidic, trace-metal depleted, nutrient-scarce, acidotic oceans), there are certainly ongoing controversies about details (Lyons and Reinhard, 2009) which might warrant reevaluation of some aspects of A-Theory.

THERAPEUTIC IMPLICATIONS

M-Theory, despite the mass of genetic sequencing data documenting the strong association between the MP and various genomic derangements, has had surprisingly little impact on the clinical management, or natural history of cancer as a disease. There are only a handful of cancer types which have been unambiguously associated with a driver mutant oncogene amenable to therapeutic inhibition (e.g., chronic myeloid leukemia and the BCR–ABL fusion oncogene; certain relatively uncommon forms of nonsmoking associated lung cancer with either EGFR mutations or ALK rearrangements; about half of melanomas with BRAF mutations). Although excellent drugs have been designed and approved in these indications, they generally work well only for about a year or so before resistance emerges, and have led to long-term control and possible cure in just one form of cancer, chronic myeloid leukemia. While several more of these drugs are in development (e.g., FLT3 inhibitors in acute myeloid leukemia), most common cancers remain without approved drugs directed against putative oncogenes like K-RAS. Furthermore, loss-of-function mutations (like P53 e.g.), are very difficult to drug. So-called "targeted drugs" are usually not oncogene inhibitors, but directed against pathways which are overactive in cancer and/or on which cancer cells rely more than normal cells, such as the VEGF pathway. Further complicating the application of M-theory to drug design is the rarity of incommon mutations which are also causal drivers.

On the other hand, one of the chief theoretical attractions of M-theory for drug design relates to the potential for efficacy via causality reversal, and the simultaneous potential for selectivity via the fact that the target mutation happens to be restricted to the cancer. Paradoxically however, it now appears that the major therapeutic victory for M-Theory may be substantially related to the vast number of passenger mutations which do nothing to contribute causally to the MP, but merely flag the cancer a "foreign" to the immune system. Of course, causally implicated driver mutations and other contributory genomic derangements can also generate neo-antigenic signals; the point is, neither the nature of these mutations, nor their function, if any, are of any concern, providing the immune system can be pharmacologically triggered to act (i.e., kill) the malignant cells exhibiting neo-antigens encoded by these multiple genetic alterations, and do so without cross-reactivity to critical normal tissues. The DNA changes involved may or may not be related to the causal mediation of the cancer but that is not what is necessary for the immune system to recognize cancer cells as foreign (Ilias and Yang, 2015). This T cell–mediated anticancer efficacy can be greatly increased in the clinic by the use of immune checkpoint inhibitory drugs, some of which are now approved in common cancers, such as advanced nonsmall cell lung cancer, bladder cancer, melanoma, kidney cancers, and Hodgkin's disease (Callahan et al., 2015). A massive effort is underway to extend this slate of approvals, by increasingly complex and sophisticated ways to activate the immune system to recognize, engage, and destroy a wide variety of lethal cancers (Sharma and Alison, 2015). This endeavor is the most important recent development, and probably the single best hope in the war on cancer.

What has the immunological treatment of cancer got to do with A-Theory? The basis for the therapeutic manipulation of immune recognition and killing is thought to have evolved in the earliest metazoa to eradicate not only the phenomena of infection, infestation, and germline parasitism/microchimerism (Poudyal et al., 2007) but also self-originated rogue cells identifiable by altered protein expression (neo-antigens) coded for by genomic alterations, that is, transformed and potentially cancerous cells (Jack, 2015).

The success of this effort so far, and in the future, will depend on the marker principle rather than the driver-inhibitory principle, the idea being that as long as the cancer cell can be identified by some sort of signature difference from normal tissues, and some form of destructive energy brought in selective proximity to it, then the cancer can be effectively treated (and maybe, even eradicated).

What then, has A-theory to say about therapy? Unsurprisingly, this has not often been discussed in the literature, except as noted previously in relation to resistance, and extinction, and also by collaborators of mine who have proposed a strategy of exploiting vulnerabilities based on atavism (Lineweaver et al., 2014), that is, on the assumption that certain later-evolved attributes (e.g., efflux pumps) might be preferentially lost in the cancer cell, allowing drug design to be oriented toward cancer's weaknesses rather than its strengths. To some extent, this theory assumes that carcinogenesis involves a linear reversion of evolution, a thesis that still requires formal confirmation but might be experimentally approached by a kind of genomic stratigraphy; however, some indirect experimental support for this general approach has recently emerged correlating the appearance of drug resistance in human myeloma with a suite of changes involving subsets of genes older, and in some cases considerably older, than the average age of the human genome (Wu et al., 2015).

If, on the other hand, the atavistic features of the MP are simply listed (Table 16.2), it can be seen that significant therapeutic implications exist, some of which have already been empirically exploited and have been for decades, without overt recognition that the therapeutic index involved might rely on an archaic origin of the cancer cell. Most of these applications listed are based on "marker" distinctions between cancer and normal cells,

and fewer on "causality-inhibition." Either way, it can be seen that cancer cells are inextricably associated with signature vulnerabilities, many of which can arguably be best rationalized in an atavistic framework; the various vulnerabilities of the cancer cell vis-à-vis the normal cells crystalize out of the temporal differences between the relatively modernity of the host's tissues, and the archaic framework within which the cancer may operate.

In our laboratory, we have pursued the opportunity provided by the well-documented genomic disarray and instability in cancer cells, by seeking, not to remedy it, but rather to increase it further beyond the bounds of viability; therapeutic inhibition of certain DNA repair enzymes has resulted in single-agent cytotoxicity, as well as exploitable potentiation of standard DNA-damaging therapy and, in addition, induced synthetic lethality with PARP inhibitors (Rytelewski et al., 2016).

CONCLUSIONS

Most of the recent evolutionary work in cancer relates to the dynamics of clonal evolution after the original carcinogenesis event; this is potentially valuable for clinical practice, since intratumoral heterogeneity is a significant impediment to the success of cancer treatments. However, I have here presented a different perspective, which is not only to see the cancer cell as an organism, but to suggest it might also be understood as an organism which is fundamentally driven by the derepression of an ancient, simplified genomic program, inherently primitive and/or adapted to the exigencies of the Proterozoic and its harsh geochemistry. A further extension of this line of thinking allows speculative consideration of a possible survival value to the core agenda of the MP in the remote past.

Although other interpretations are possible, this conclusion follows, I believe, most logically from the posing of four key questions: What form of life does cancer embody? Why is the core MP always the stereotypically the same, despite causal molecular heterogeneity? Do the traits of the MP resemble adaptations to the geochemistry of the Precambrian? And finally, could there be a survival purpose to the MP?

My bias is that the answers to these questions, although substantially speculative, do support an atavistic (A-Theoretical) interpretation of cancer as the most parsimonious explanation of many otherwise mysterious features, as well as point to a variety of therapeutic opportunities, some of which have already been empirically (and successfully) exploited for decades. The task now is to obtain experimental support for A-Theory, which will not be easy, since the complexity of the Proterozoic cannot readily be recapitulated in a wet lab. However, elements of it may be experimentally tractable, especially presumptive aquatic geochemistry and food webs, and it is perhaps here that we should start by testing whether modern cancer cells are able to survive in synthetic versions of this environment, better than nontumorigenic or other near-normal cell lines. Beyond that, what is needed is a systematic analysis of the apparent parallelisms between cancer cells and putative eukaryotic ancestral organisms, the better to establish whether these resemblances are superficial or substantive, and if the latter, to help identify the inherited core molecular apparatus of the MP, and its mode of detachment from the subsequently evolved metazoan control system. This would then constitute a true explanation for the nature of cancer (Hempel, 1963): why it exists at all, and why it always is the way it is.

References

Aktipis, C.A., Boddy, A.M., Jansen, G., et al., 2015. Cancer across the tree of life: cooperation and cheating in multicellularity. Philos. Trans. R. Soc. Lond. B Biol. Sci. 370 (1673), 20140219.

Anbar, A.D., Knoll, A.H., 2002. Proterozoic ocean chemistry and evolution: a bioinorganic bridge? Science 297 (5584), 1137–1142.

Arens, N.C., West, I.D., 2008. Press-pulse: a general theory of mass extinction. Paleobiology 34, 456–471.

Birky, C.W., 2013. Species detection and identification in sexual organisms using population genetic theory and DNA sequences. PLoS One 8 (1), e52544.

Bolande, R.P., 1991. The spontaneous regression of neuroblastoma. Experimental evidence for a natural host immunity. Pathol. Annu. 26 (Pt. 2), 187–199.

Boyd, R., 1991. Realism, anti-foundationalism and the enthusiasm for natural kinds. Philos. Stud. 61 (1/2), 127–148.

Callahan, M.K., Postow, M.A., Wolchok, J.D., 2015. Targeting T cell co-receptors for cancer therapy. Immunity 44 (5), 1069–1078.

Canfield, D.E., 2014. Oxygen. A Four Billion Year History. Princeton University Press, Princeton and Oxford, pp. 123–158.

Chen, H., Lin, F., Xing, K., et al., 2015. The reverse evolution from multicellularity to unicellularity during carcinogenesis. Nat. Commun. 6, 6367.

Domazet-Lošo, T., Klimovich, A., Anokhin, B., et al., 2014. Naturally occurring tumours in the basal metazoan Hydra. Nat. Commun. 5, 4222.

Gatenby, R.A., Gillies, R.J., 2004. Why do cancers have high aerobic glycolysis? Nat. Rev. Cancer 4 (11), 891–899.

Goodman, M.F., 2016. Better living with hyper-mutation. Environ. Mol. Mutagen. 57 (6), 421–434.

Hanahan, D., Weinberg, R.A., 2000. The hallmarks of cancer. Cell 100 (1), 57–70.

Hanahan, D., Weinberg, R.A., 2011. Hallmarks of cancer: the next generation. Cell 144 (5), 646–674.

Hempel, C., 1963. Aspects of Scientific Explanation and Other Essays in the Philosophy of Science. Free Press, New York, pp. 335–338, 380–384, and 394–403.

Ilias, S., Yang, J.C., 2015. Landscape of tumor antigens in T cell immunotherapy. J. Immunol. 195 (11), 5117–5122.

Jack, R.S., 2015. Evolution of immunity and pathogens. Results Probl. Cell Differ. 57, 1–20.

Johnston, D.T., Wolfe-Simon, F., Pearson, A., et al., 2009. Anoxygenic photosynthesis modulated Proterozoic oxygen and sustained Earth's middle age. Proc. Natl. Acad. Sci. USA 106 (40), 16925–16929.

Lehman, J., Miikkulainen, R., 2015. Extinction events can accelerate evolution. PLoS One 10 (8), e0132886.

Lineweaver, C.H., Davies, C.W., Vincent, M.D., 2014. Targeting cancer's weaknesses (not its strengths): therapeutic strategies suggested by the atavistic model. Bioessays 36 (9), 827–836.

Ling, S., Hu, Z., Yang, Z., et al., 2015. Extremely high genetic diversity in a single tumor points to prevalence of non-Darwinian cell evolution. Proc. Natl. Acad. Sci. USA 112 (47), E6496–E6505.

Lyons, T.W., Reinhard, C.T., 2009. An early productive ocean unfit for aerobics. Proc. Natl. Acad. Sci. USA 106 (43), 18045–18046.

Marengo, B., Nitti, M., Furfaro, F.L., et al., 2016. Redox homeostasis and cellular antioxidant systems: crucial players in cancer growth and therapy. Oxid. Med. Cell Longev. 2016, 6235641.

Metzger, M.J., Reinisch, C., Sherry, J., et al., 2015. Horizontal transmission of clonal cancer cells causes leukemia in soft-shell clams. Cell 161 (2), 255–263.

Mills, D.B., Canfield, D.E., 2014. Oxygen and animal evolution: did a rise of atmospheric oxygen "trigger" the origin of animals? Bioessays 36 (12), 1145–1155.

Nicholson, J.M., Duesberg, P., 2009. On the karyotypic origin and evolution of cancer cells. Cancer Genet. Cytogenet. 194 (2), 96–110.

Nowell, P.C., 1976. The clonal evolution of tumor cell populations. Science 194, 23–28.

Parikh, A.P., Curtis, R.E., Kuhn, I., et al., 2014. Network analysis of breast cancer progression and reversal using a tree-evolving network algorithm. PLoS Comput. Biol. 10 (7), e1003713.

Parks, S.K., Chiche, J., Pouysségur, J., 2013. Disrupting proton dynamics and energy metabolism for cancer therapy. Nat. Rev. Cancer 13 (9), 611–623.

Perez-Jiminez, R., Inglés-Prieto, A., Zhao, Z.M., et al., 2011. Single-molecule paleoenzymology probes the chemistry of resurrected enzymes. Nat. Struct. Mol. Biol. 18 (5), 592–596.

Poudyal, M., Rosa, S., Powell, A.E., et al., 2007. Embryonic chimerism does not induce tolerance in an invertebrate model organism. Proc. Natl. Acad. Sci. USA 104 (11), 4559–4564.

Roberts, M., 1926. Malignancy and Evolution. Grayson and Grayson Publishers, London, pp. 308–310.

Rytelewski, M., Maleki Vareki, S., Mangala, L.S., et al., 2016. Reciprocal positive selection for weakness—preventing olaparib resistance by inhibiting BRCA2. Oncotarget 7 (15), 20825–20839.

Semenza, G.L., 2013. HIF-1 mediates metabolic responses to intratumoral hypoxia and oncogenic mutations. J. Clin. Invest. 123 (9), 3664–3671.

Sharma, P., Alison, J.P., 2015. The future of immune checkpoint therapy. Science 348 (6230), 56–61.

Snow, H., 1893. Cancers and the Cancer Process. J. & A. Churchill Publishers, London.

Vincent, M.D., 2010. The animal within: carcinogenesis and the clonal evolution of cancer cells are speciation events sensu stricto. Evolution 64 (4), 1173–1186.

Vincent, M., 2011. Cancer: beyond speciation. Adv. Cancer Res. 112, 283–350.

Vincent, M., 2012. Cancer: a de-repression of a default survival program common to all cells? A life-history perspective on the nature of cancer. Bioessays 34 (1), 72–82.

Warburg, O., 1969. The prime cause and prevention of cancer (Lindau Lecture), Revised ed. Konrad Triltsch, Würzburg, Germany.

Wise, D.R., DeBardinis, R.J., Mancuso, A., et al., 2008. Myc regulates a transcriptional program that stimulates mitochondrial glutaminolysis and leads to glutamine addiction. Proc. Natl. Acad. Sci. USA 105 (48), 18782–18787.

Wu, A., Zhang, Q., Lambert, G., et al., 2015. Ancient hot and cold genes and chemotherapy resistance emergence. Proc. Natl. Acad. Sci. USA 112 (33), 10467–10472.

17

Toward an Ultimate Explanation of Intratumor Heterogeneity

Frédéric Thomas*, Beata Ujvari**, Cindy Gidoin*,
Aurélie Tasiemski†, Paul W. Ewald‡, Benjamin Roche*,§

*MIVEGEC (Infectious Diseases and Vectors: Ecology, Genetics, Evolution and Control),
UMR IRD/CNRS/UM 5290, Montpellier, France
**Centre for Integrative Ecology, School of Life and Environmental Sciences,
Deakin University, Geelong, VIC, Australia
†Lille1 University, UMR CNRS 8198, Evolution, Ecology and Paleontology Unit,
Villeneuve d'Ascq, France
‡Department of Biology and the Program on Disease Evolution,
University of Louisville, Louisville, KY, United States
§UMMISCO (International Center for Mathematical and Computational Modeling of
Complex Systems), UMI IRD/UPMC UMMISCO, Bondy, France

No biological problem is solved until both the proximate and the evolutionary causation has been elucidated. Furthermore, the study of evolutionary causes is as legitimate a part of biology as is the study of the usually physico-chemical proximate causes. **Mayr (1982)**

It is well established that extensive variation that can be inherited across cell cycles (including genetic, cytogenetic, and epigenetic) and phenotypic diversity exist within tumors (i.e., intratumor heterogeneity, referred to as ITH hereafter) (Fidler, 1978; Heppner, 1984; Marusyk and Polyak, 2010; Marusyk et al., 2012; Yates and Campbell, 2012). The implications and clinical importance of ITH are considerable since it affects key oncogenic pathways, drives phenotypic changes over time, and also poses a significant challenge to cancer medicine, being the primary underlying cause of resistance to systemic therapies (McGranahan and Swanton, 2015; Merlo et al., 2006).

The origins of ITH have been the subject of much discussion by investigators from diverse fields but no consensus has emerged. From a mechanistic perspective, it has been suggested that genomic instability is *the* major process generating ITH (Burrell et al., 2013). Apart from this proximate explanation, several nonmutually exclusive models have been proposed to explain the

establishment and the maintenance of ITH (Michor and Polyak, 2010). For instance, Waclaw et al. (2015) described a model for tumor evolution that shows how short-range migration and cell turnover can account for rapid cell mixing inside the tumor. The cancer stem cell hypothesis postulates that the differentiation of few cells with stem cell properties, notably unrestricted self-renewal abilities, generate various cell types in the tumor, leading to ITH (Campbell and Polyak, 2007). In parallel, the linear clonal evolution hypothesis suggests that the accumulation of various hereditary changes over time confer different selective advantages to premalignant and malignant cells, and hence gives rise to ITH (Gerlinger and Swanton, 2010). The plasticity cell hypothesis states that the majority of tumor cells have varying degrees of stem cell–like characteristics, depending on cell intrinsic stochasticity and/or microenvironmental conditions (Michor and Polyak, 2010). For instance, Lloyd et al. (2016) recently proposed that at least some intratumoral heterogeneity in the molecular properties of cancer cells is governed by predictable regional variations in environmental selection forces. Broadly, these hypotheses argue that due to ITH's key role in neoplasia, cancer progression, and therapeutic resistance, its persistence, once initiated, is supported by various selective benefits. It is apparent that ITH may contribute to the persistence and continued evolution of a cancer (e.g., through drug resistance, immune evasion, and metastasis); however, because environments change unpredictably and evolution cannot anticipate the future, it is challenging to explain the occurrence of ITH at the very first steps of tumorigenesis. Michor and Polyak (2010) proposed the importance of stem cells when discussing ITH in early stage of in situ cancers because only stem cells with unlimited self-renewal capability are able to persist over time and accumulate the genetic and epigenetic changes that will potentially be retained by selection. Conversely, Ling et al. (2015) recently argued that ITH is so extreme, even in tiny tumors, that it implies evolution under a "non-Darwinian mode" because genetic diversity observed would be orders of magnitude lower than predicted by simple classic Darwinian selection, suggesting that the observed diversity is much less structured than expected theoretically.

Here, we propose a novel parsimonious Darwinian scenario, nonmutually exclusive with the others, to explain elevated ITH not only in late but also in early stage tumors. We argue that such heterogeneity could be the result of a selective process, bet-hedging, that starts from the very first steps of oncogenesis for specific reasons. In evolutionary ecology, bet-hedging is defined as a strategy that reduces the temporal variance in an organism's fitness at the expense of lowered arithmetic mean fitness. We adapt this definition to oncogenesis by considering it to be a process that reduces the temporal variance in the success of a cancerous or precancerous cell relative to that of other cells in the organism. Success, being the analog of evolutionary fitness, takes into account both the reproduction and survival of the abnormal cells within the organism. In oncogenesis, selection for bet-hedging favors cells that are more prone to generate variation in their progeny. The most obvious source of this variation would be a proneness to mutation (e.g., a mutator phenotype), but other sources of variation, such as increased tendency to undergo epigenetic changes could also play a role.

FROM COOPERATION TO SELFISHNESS AND BACK TO (MALIGNANT) COOPERATION

Neoplasia originates from normal cells that lose their typical cooperative behavior, become malignant, and hence proliferate to greater numbers than would normal cells. Diversification of cellular activities and tissue development in solid tumors (e.g., angiogenesis) suggest that cancer cells increase in abundance in part by engaging in cooperative activities. At many levels of life organization, from plants to human societies, emergence of cooperation in communities of selfish individuals occurs under adverse environmental conditions

(Andras et al., 2007; Nowak and Highfield, 2011). Similarly, genetically distinct tumor cells are able to cooperate via the exchange of different diffusible products to overcome host defenses and fluctuating microenvironmental challenges (Axelrod et al., 2006), providing malignant cells, that have both selfish and cooperative characteristics, with a selective advantage.

As solid tumors develop only in nonliquid tissues, we can hypothesize that competition with healthy cells and space constraints (which are greater in nonliquid tissues than in liquid environments) are the main microenvironmental factors that favor the selection of cooperative behaviors in malignant cells and lead to tumor formation.

DE NOVO TUMOR DEVELOPMENT— FROM AN INDIVIDUAL CELL TO A COOPERATING ORGANIZED SYSTEM

Solid cancers are not simply clones of cancerous cells; they are complex and well-organized systems that have been compared to functional, though abnormal, organs (Egeblad et al., 2010). Oncogenesis not only generates sophisticated levels of convergent organization within a few months or years (Chen and He, 2016), but also often does so de novo in most cancers; each cancer must reinvent the wheel because its evolutionary products will die with the host (Arnal et al., 2015) (see Chapter 12). In this context, bet-hedging appears to be a widespread mechanism to produce de novo such malignant, elaborate, cooperative cell populations. From the viewpoint of evolutionary biology, bet-hedging is traditionally viewed as an adaptation to environmental uncertainty (Simons, 2011), and phenotypic diversification enables species to survive environmental fluctuations. Here we argue that bet-hedging arises when complex cancers are generated through oncogenesis, despite the fact that this process is inefficient and wasteful. Although bet-hedging does not maximize expected fitness within a generation, it reduces fitness variance and hence maximizes fitness across generations under environmental unpredictability (Simons, 2011).

BET-HEDGING AS EVOLUTIONARY RESPONSE TO UNPREDICTABLE TUMOR ENVIRONMENT

In ecological settings, long-term unpredictable selection is expected to result in the evolution of bet-hedging strategies and development of either or both conservative (i.e., insurance policy) and diversifying (i.e., risk spreading) bet-hedging traits (Simons, 2011). Progenitor cancer cells in de novo tumors could employ either, but most likely both, adaptive strategies: (1) short-range migration/dissemination of early metastatic cancer cells (that potentially lie dormant for decades) to provide a hedge against seasonal, but unpredictable onset of disastrous "predation" by the immune system (Eyles et al., 2010; Röcken, 2010), or (2) generation of an array of phenotypes (subclones of primary tumor cells) to facilitate cooperation and to reduce the risk of extinction (Simons, 2011). The evolution and appearance of late disseminating metastatic phenotypes may potentially be an adaptive bet-hedging response (both conservative and diversifying) to declining resources once primary tumors reach their growth limit and the carrying capacity of their microenvironment. Other evolutionary responses to environmental variance, such as adaptive tracking and adaptive phenotypic plasticity, are constrained by various factors: adaptive tracking is impeded under extreme environmental changes (e.g., switching between the hypoxic and oxygen-rich environments of primary tumors and the bloodstream), while adaptive phenotypic plasticity is only an ideal solution in a conceivable/predictable environment. Thus, development and survival under the broad array of circumstances that cancer cells experience in the lifetime of their host may be enhanced by bet-hedging strategies generating high ITH. It is important to mention that this perspective does not exclude the possibility that adaptive tracking and phenotypic plasticity also applies to cancer cells (but at later stages of tumor progression), but rather we propose that bet-hedging (and hence ITH) is the initial attribute of de novo cancer cells as adaptation to the organisms' unpredictable environment.

PREDICTIONS AND IMPLICATIONS

Even though bet-hedging could facilitate de novo development of tumor complexity, it is expected to be a slow process given that only a few useful components are produced per time unit. The formation of solid tumors by bet-hedging is thus expected to be a slow phenomenon. This is in accordance with solid tumors generally occurring late during the life, even if several other explanations are also fully valid (Merlo et al., 2006). In addition, in contrast with organisms that have practiced bet-hedging for millions of years, bet-hedging in our hypothesis is a de novo trait favored by oncogenic selection (i.e., because each cancer reinvents the wheel). This short evolutionary time may explain why bet-hedging here yields an abundance of nonfunctional/aberrant cells, instead of producing a variety of different but functional entities, as is usually observed when bet-hedging arises through natural selection on organisms. This explanation explains the variety of genomic aberrations among cancers of the

same histological type, to the extent that no two tumors are thought to show an identical somatic genetic aberration profile (Lipinski et al., 2016). This model also predicts that in liquid environments like blood, where cell competition for resources and space limitations are relaxed, oncogenic selection should rather favor highly proliferative clones that do not cooperate. The futility of building cooperative tumor systems in this case may explain why cancer cells from liquid tumors do not aggregate and are on average less heterogeneous than in solid tumors (Alexandrov et al., 2013) (i.e., low selection for bet-hedging). Also because the dynamics of liquid tumors would not rely on a slow bet-hedging process, it could explain why they can reach, all things being equal, stages that are detrimental for health earlier in life (e.g., leukemia) compared to solid tumor cancers.

CONCLUDING REMARKS

Although the applications of an evolutionary perspective in human health research vary depending on the disease under study (Williams and Nesse, 1991), it is increasingly accepted that cancer is a process that follows Darwinian evolution (Aktipis and Nesse, 2013; Thomas et al., 2013). ITH is central in this reasoning since it provides the substrate from which somatic cellular selection and evolution can occur, leading to malignancy, with its many manifestations: neoangiogenesis, evasion of the immune system, metastasis, and resistance to therapies, and sometimes contagion (Ujvari et al., 2016). Recently, Ling et al. (2015) cast doubt on the selected nature of ITH, mainly because the great variation that can occur even in small tumors seems incompatible with predictions made by classical Darwinian reasoning. Here, we provide a conceptual framework proposing that ITH itself could also directly result from a selective process: bet-hedging. We cannot exclude in our hypothesis that there is an underlying ancestral bet-hedging program normally repressed that cancer cells are able to reactivate following mutations (Soto and Sonnenschein, 2011; Vincent, 2012) (see Chapter 16). We cannot exclude either that ultimate reasons other than the one proposed here (e.g., immune escape) also explain why tumorigenesis implies the concomitant selection of bet-hedging from the first malignant steps. Further work is necessary to determine whether or not classical evolutionary/ecological scenario (e.g., tumor heterogeneity would be the consequence of a heterogeneous landscape selecting for different phenotypes that are adaptive to these different microenvironments) can successfully explain the striking level of ITH observed in most tumors, or whether it is necessary to invoke bet-hedging processes. In the latter scenario, the true ultimate causation will remain to be determined.

References

Aktipis, C.A., Nesse, R.M., 2013. Evolutionary foundations for cancer biology. Evol. Appl. 6, 144–159.

Alexandrov, L.B., Nik-Zainal, S., Wedge, D.C., Aparicio, S.A., Behjati, S., Biankin, A.V., et al., 2013. Signatures of mutational processes in human cancer. Nature 500, 415–421.

Andras, P., Lazarus, J., Roberts, G., 2007. Environmental adversity and uncertainty favour cooperation. BMC Evol. Biol. 7, 240.

Arnal, A., Ujvari, B., Crespi, B., Gatenby, R.A., Tissot, T., Vittecoq, M., et al., 2015. Evolutionary perspective of cancer: myth, metaphors, and reality. Evol. Appl. 8, 541–544.

Axelrod, R., Axelrod, D.E., Pienta, K.J., 2006. Evolution of cooperation among tumor cells. Proc. Natl. Acad. Sci. 103, 13474–13479.

Burrell, R.A., McGranahan, N., Bartek, J., Swanton, C., 2013. The causes and consequences of genetic heterogeneity in cancer evolution. Nature 501, 338–345.

Campbell, L.L., Polyak, K., 2007. Breast tumor heterogeneity: cancer stem cells or clonal evolution? Cell Cycle 6, 2332–2338.

Chen, H., He, X., 2016. The convergent cancer evolution toward a single cellular destination. Mol. Biol. Evol. 33, 4–12.

Egeblad, M., Nakasone, E.S., Werb, Z., 2010. Tumors as organs: complex tissues that interface with the entire organism. Dev. Cell 18, 884–901.

Eyles, J., Puaux, A.L., Wang, X., Toh, B., Prakash, C., Hong, M., et al., 2010. Tumor cells disseminate early, but immunosurveillance limits metastatic outgrowth, in a mouse model of melanoma. J. Clin. Invest. 120, 2030–2039.

Fidler, I.J., 1978. Tumor heterogeneity and the biology of cancer invasion and metastasis. Cancer Res. 38, 2651–2660.

Gerlinger, M., Swanton, C., 2010. How Darwinian models inform therapeutic failure initiated by clonal heterogeneity in cancer medicine. Br. J. Cancer 103, 1139–1143.

Heppner, G.H., 1984. Tumor heterogeneity. Cancer Res. 44, 2259–2265.

Ling, S., Hu, Z., Yang, Z., Yang, F., Li, Y., Lin, P., et al., 2015. Extremely high genetic diversity in a single tumor points to prevalence of non-Darwinian cell evolution. PNAS 112, E6496–E6505.

Lipinski, K.A., Barber, L.J., Davies, M.N., Ashenden, M., Sottoriva, A., Gerlinger, M., 2016. Cancer evolution and the limits of predictability in precision cancer medicine. Trends Cancer 2, 49–63.

Lloyd, M.C., Cunningham, J.J., Bui, M.M., Gillies, R.J., Brown, J.S., Gatenby, R.A., 2016. Darwinian dynamics of intratumoral heterogeneity: not solely random mutations but also variable environmental selection forces. Cancer Res. 76, 3136–3144.

Marusyk, A., Almendro, V., Polyak, K., 2012. Intra-tumour heterogeneity: a looking glass for cancer? Nat. Rev. Cancer 12, 323–334.

Marusyk, A., Polyak, K., 2010. Tumor heterogeneity: causes and consequences. Biochim. Biophys. Acta 1805, 105–117.

Mayr, E., 1982. The Growth of Biological Thought: Diversity, Evolution, and Inheritance. American Anthropologist. The Belknap Press of Harvard University Press, Cambridge.

McGranahan, N., Swanton, C., 2015. Biological and therapeutic impact of intratumor heterogeneity in cancer evolution. Cancer Cell 27, 15–26.

Merlo, L.M.F., Pepper, J.W., Reid, B.J., Maley, C.C., 2006. Cancer as an evolutionary and ecological process. Nat. Rev. Cancer 6, 924–935.

Michor, F., Polyak, K., 2010. The origins and implications of intratumor heterogeneity. Cancer Prev. Res. 3, 1361–1364.

Nowak, B.M.A., Highfield, R., 2011. SuperCooperators: Altruism, Evolution, and Why We Need Each Other to Succeedvol. 46Free Press, New York, NY, pp. 1003-1007.

Röcken, M., 2010. Early tumor dissemination, but late metastasis: insights into tumor dormancy. J. Clin. Invest. 120, 1800–1803.

Simons, A.M., 2011. Modes of response to environmental change and the elusive empirical evidence for bet hedging. Proc. R. Soc. Lond. B Biol. Sci. 278, 1601–1609.

Soto, A.M., Sonnenschein, C., 2011. The tissue organization field theory of cancer: a testable replacement for the somatic mutation theory. Bioessays 33, 332–340.

Thomas, F., Fisher, D., Fort, P., Marie, J.-P., Daoust, S., Roche, B., et al., 2013. Applying ecological and evolutionary theory to cancer: a long and winding road. Evol. Appl. 6, 1–10.

Ujvari, B., Gatenby, R.A., Thomas, F., 2016. The evolutionary ecology of transmissible cancers. Infect. Genet. Evol. 39, 293–303.

Vincent, M., 2012. Cancer: a de-repression of a default survival program common to all cells?: A life-history perspective on the nature of cancer. Bioessays 34, 72–82.

Waclaw, B., Bozic, I., Pittman, M.E., Hruban, R.H., Vogelstein, B., Nowak, M.A., 2015. Spatial model predicts dispersal and cell turnover cause reduced intra-tumor heterogeneity. Nature 525, 261–267.

Williams, G.C., Nesse, R.M., 1991. The dawn of Darwinian medicine. Q. Rev. Biol. 66, 1–22.

Yates, L.R., Campbell, P.J., 2012. Evolution of the cancer genome. Nat. Rev. Genet. 13, 795–806.

18

Obstacles to the Darwinian Framework of Somatic Cancer Evolution

Andriy Marusyk

Department of Cancer Imaging and Metabolism, H Lee Moffitt Cancer Center and Research Institute,
Tampa, FL, United States

INTRODUCTION

The widely cited phrase of Theodsius Dobzhansky "Nothing makes sense in biology except in the light of evolution" captures the key importance of the framework of Darwinian evolution in understanding the forces that shape the staggering diversity and complexity of species in nature. The remarkable explanatory power of Darwinian evolution is based on a relatively simple concept: heritable phenotypic diversity within populations that stems from stochastic mutational processes enables selection for phenotypes that are most fit to a given context. In stable environments, selective pressures stabilize optimal phenotypes, whereas changes in the context create novel selective pressures that lead to changes in the phenotypic composition of populations. Obviously, the exact evolutionary processes are complex and nuanced. Moreover, changes in the genetic make-up of populations reflect not only deterministic selection pressures, but also stochastic genetic drift. However, the central idea of the interplay between stochastic diversification and directional context-specified selection remains a guiding light in understanding evolution.

Following the seminal 1976 paper by the codiscoverer of Philadelphia chromosome Peter Nowell (1976), the applicability of the Darwinian framework has been extended toward the initiation and progression of cancers (Greaves and Maley, 2012). Research over the following decades has resulted in the discovery of a large number of recurrent genetic alterations, leading to the general acceptance of the evolutionary underpinning of cancer progression. Somatic evolution is explicitly called a Darwinian process in perhaps the most thorough and influential cancer biology textbook (Weinberg, 2013). However, despite the lack of an overt disagreement with the Darwinian underpinning of cancer initiation and progression, application of the conceptual framework of evolutionary theory is largely missing from mainstream cancer research. Even within the research focused on therapeutic failure that is unambiguously attributable to Darwinian selection of therapy-resistant phenotypes, only about 1% of published research considers the evolutionary perspective (Aktipis et al., 2011). This omission is especially disturbing considering that the evolutionary perspective could substantially impact our approaches to cancer prevention and treatment.

The lack of evolutionary thinking can be attributed to three major factors. First, the dominance of gene- and mutation-centric views on cancer evolution leads to neglecting or underappreciating natural selection. If cancer evolution can be reduced to the acquisition of genetic mutations that drive the process, then cancer evolution can be understood through genetics, leaving no need to consider selection forces. Second, tumor cells retain many of the characteristics of their counterparts from normal tissues. Therefore, phenotypes of tumor cells integrate the impact of genetic and epigenetic alterations associated with somatic evolution, with partially retained mechanisms responsible for shaping phenotypes of normal cells. Third, the genomes of somatic cells encode for a much larger range of phenotypic manifestations than expressed in a given tissue. This excess of information enables phenotypic diversification through chromosomal instability and epigenetic dysregulations, sources of variability that are not readily available to natural populations. Considerations of these factors and their incorporation into the framework of somatic cancer evolution will be required for a wider acceptance of evolutionary thinking within the cancer research community (see also the Introductory chapter).

CHALLENGE #1: THE DOMINANCE OF GENE AND MUTATION-CENTRIC PARADIGMS

The prevailing way to conceptualize somatic evolution in the cancer biology field is through the framework of "driver" mutations (Vogelstein and Kinzler, 2004) (see also Chapter 5). This framework implies that fitness of precancer cells can always be improved at least until complete malignancy is reached as long as an appropriate type of driver mutation occurs. The driver mutations framework has strong foundations. The discovery of recurrent genetic mutations and characterization of their impact on malignancy-associated phenotypes has been the main focus of cancer research and a major driver of clinical advances in diagnostics/therapy. The concepts of oncogenes and tumor suppressor genes have become the keystones of the conceptual framework of cancer biology. Given the profound phenotypic consequences of highly recurrent genetic aberrations and their ability to induce malignant transformation of normal cells by introducing defined genetic alterations, the importance of oncogenes and tumor suppressor genes for the initiation and progression of cancers appears to be self-evident. Moreover, the discoveries of cancer-associated genetic alterations and characterization of their physiological impact not only preceded the idea of cancers being a result of somatic clonal evolution, but also provided a bulk of evidence on which the concept of somatic clonal evolution is based.

Yet, the idea that evolution can be driven by mutations conflicts with both Darwin's original ideas and modern evolutionary biology. Instead, it repeats the ideas that were widespread among many prominent geneticists before they were refuted by evolutionary synthesis (Mayr, 1982). Unless somatic evolution is fundamentally different from the evolutionary processes in the rest of biology, the view of mutations driving evolution must be incorrect, and selection, rather than mutations, must be driving the process, with mutational diversification providing a substrate for natural selection to work on. A large body of evidence points out that somatic evolution is not fundamentally different from evolution in natural populations. The effects of oncogenic mutations, including highly recurrent ones, are strongly dependent on tissue, cell type, and developmental stage (Sieber et al., 2005), suggesting that context-specific selection forces operate. Moreover, a growing body of evidence associated with the rediscovery of intratumor heterogeneity points to convergent evolution within tumors, whereby functionally similar phenotypic adaptations are acquired through recurrent mutations that independently occur in different tumor subclones (Anderson et al., 2011; Martins et al., 2012). Convergent evolution within tumors dilutes the meaning of "driver mutations"

and points to the importance of selective pressures that promote convergent evolution with tumors. A shift toward applying evolutionary frameworks is contingent on abandoning the driver mutation view of somatic evolution. Considerations of proximal genetic mechanisms that underlie phenotypic adaptations need to be combined with considerations of context-dependent forces of natural selection.

CHALLENGE #2: COMPLEX INPUTS SHAPING THE PHENOTYPES OF CANCER CELLS

Considering selection pressures provides unprecedented explanatory power toward understanding complex and diverse phenotypes of species in natural populations. If cancer evolution is a Darwinian process, they should also be applicable to the phenotypic traits of cancer cells. For example, the development of aerobic glycolysis (the Warburg effect) likely reflects selection pressures imposed by hypoxic and acid microenvironments in premalignant lesions (Gatenby and Gillies, 2004). However, human cancers start and perish during ontogeny and the time window available for selection forces in somatic evolution is dramatically smaller compared to evolutionary forces that shape phenotypes in natural populations. To a substantial degree, the phenotypes of cancer cells mimic the phenotypic properties of stem/progenitor cells from normal tissues, reflecting an incomplete inactivation of physiological responses to environmental cues. For example, the expression profiles of stem cell-like cells from breast cancers defined by the expression of CD44 and by the lack of expression of CD24, more closely resemble the expression profiles of normal CD44+/CD24− breast tissue stem cells, rather than more differentiated cells from the same tumors (Shipitsin et al., 2007). Similarly, the expression profiles of more differentiated CD24+/CD44+ tumor cells are more similar to the expression profiles of their normal counterparts than to those of stem-like tumor cells. As this example attests, despite the multiple genetic and epigenetic alterations involved in malignant progression, residual differentiation programs have a stronger influence on cellular phenotypes than genetic alterations. The discovery of the retention of some of the normal differentiation responses by cancer cells has paved a way to the development of cancer stem cell theory that considers the range of phenotypes of malignant cells within tumors through the framework of differentiation hierarchies (Kreso and Dick, 2014).

Notably, the hierarchy of partial retention of differentiation programs by cancer cells does not contradict the Darwinian view of somatic clonal evolution. However, it substantially complicates the interpretation of cancer cell phenotypes, as they integrate the influences of genetic and epigenetic alterations acquired in the course of somatic evolution with the effects of residual normal differentiation programs (Marusyk et al., 2012). Furthermore, the existence of differentiation hierarchies raises the question of what constitutes the units of selection in cancers. Within rigid differentiation hierarchies, only stem cells can act as units of selection. However, differentiation hierarchies in cancers are more relaxed, as at least some of the "differentiated" cancer cells retain the ability to dedifferentiate (Chaffer et al., 2011). This plasticity, together with the widespread noncell autonomous interactions between tumor cells (Tabassum and Polyak, 2015), suggest that the question of what constitutes a unit of selection might not have a single simple answer. Further complicating the picture, nongenetic influences on phenotypes cannot be mapped entirely to differentiation hierarchies, and incorporate stochastic gene expression changes that operate within different time scales (Timp and Feinberg, 2013).

CHALLENGE #3: UNIQUE AND POWERFUL SOURCES OF PHENOTYPIC VARIABILITY

Genetic mutations and sexual recombination are the two major sources of heritable phenotypic variability in natural populations. Somatic clonal evolution most closely resembles evolutionary processes in single cell asexual organisms, such as bacteria (Sprouffske et al., 2012), taking sex out of the equation and bringing the spotlight to genetic mutations as the main source of diversity. It should be noted, however, that whereas sexual recombination is not an obligatory part of the life cycle of bacteria, a number of parasexual mechanisms, many of which are activated during conditions of stress, could still act as a substitute. Consequently, horizontal gene transfer resulting from these paraxesual processes is a major factor in bacterial evolution. It is possible that cell fusions between tumor cells or between tumors and nontumor cells can serve as analogues of parasexual processes (Duelli and Lazebnik, 2003), but this possibility remains mostly unexplored.

In addition to point mutations and small-scale genetic rearrangements, tumor cells have another powerful source of genetic diversity—chromosomal instability (CIN). In fact, the majority of sporadic cancers are aneuploid, and aneuploidy is strongly linked with CIN (Thompson and Compton, 2008) (see also Chapter 5). Whereas chromosomal amplifications, deletions and translocations can be observed in organisms in natural populations, they are much less common, reflecting a high fitness penalty. In contrast to the genomes of single-cell organisms, the genotypes of somatic cells have been optimized by selection that operates at the organismal (germline) level. They have to encode for a multitude of phenotypes defining cells of different tissue lineages. Therefore, the genomes of somatic cells carry excessive information which is not essential for the functionality and survival of "egoistic" malignant clones. Moreover, genome doubling, a common step in the natural history of cancers, further increases the flexibility of their genomes and reduces the fitness penalty for significant chromosomal aberrations (Dewhurst et al., 2014; Kuznetsova et al., 2015).

CIN is often viewed through the same driver mutation-centric lenses as genetic point mutations (Michor et al., 2005). Namely, CIN is viewed as a condition that increases the probability of oncogene activation (by fusions, such as Bcr–Abl, or by increased expression due to chromosomal duplications) and tumor suppressor gene inactivation (due to chromosomal deletions). However, the impact of CIN is much more complex, as changes in copy numbers of whole chromosomes and large chromosomal regions impact the expression of multiple genes, leading to substantial changes in gene regulatory networks. Furthermore, cytogenetic evidence suggests that cancer cells might have very high rates of chromosomal aberrations, enabling much higher levels of genetic and phenotypic diversification compared to cells without CIN and thus elevating their evolvability potential (Heng et al., 2013; Li et al., 2009). Notably, CIN comes at a price of fitness penalty in stem cells (Pfau et al., 2016). Therefore, there appears to be a "sweet spot" of CIN levels in cancer cell populations, as intermediate levels of CIN provide for the highest tumorigenic potential (Godek et al., 2016).

The excess of information content carried by the genomes of somatic cells is likely responsible for enabling another unique source of phenotypic heterogeneity in cancer cells – epigenetic variability. The phenotypes of somatic cells are tightly defined through sophisticated epigenetic control mechanisms, so that genes are expressed at levels and times that are optimal for organismal fitness. However, these mechanisms are commonly altered in cancers, often by genetic mutations (Plass et al., 2013) (see also chapter 6). As a result, cancer cells display highly elevated levels of cell-to-cell variability in the expression of many genes, thereby increasing population-wide phenotypic heterogeneity (Hansen et al., 2011; Landau et al., 2014). Some of the epigenetic changes, such as inactivation of the p16 gene, are practically irreversible and can be considered analogous to genetic inactivation. However, many of the changes are reversible, complicating the applicability of a strict Darwinian framework. On one hand, selection pressures can enrich for phenotypes defined by stochastic epigenetic aberrations. On the other, due to the lack of strict heritability, these phenotypes can be lost once selective pressures change (Sharma et al., 2010). The high recurrence of mutations that alter epigenetic programs suggests the importance of epigenetic regulation in cancer evolution. As an extreme example, rhabdoid pediatric cancers maintain diploid genome and display very few genetic changes. The only recurrent genetic aberration is a biallelic loss of a subunit of the SWI/SNF chromatin remodeling complex (Lee et al., 2012), suggesting that epigenetic alterations might be the main source of phenotypic diversification in these malignancies.

CONCLUSIONS

As in the rest of biology, the framework of Darwinian evolution provides an explanatory power that can illuminate the causation of cancers. Unfortunately, applications of this framework in cancer research have been rather limited. The preferential focus on the proximal mechanisms of cancer bears a large part of the responsibility for this omission. However, it is not the only reason, as unique features of somatic evolution present strong challenges toward the application of the lessons learned from studying evolution in natural populations. For the Darwinian paradigm to provide inferences that can be used to improve cancer treatment and prevention, these unique features need to be considered and incorporated into a better developed and more nuanced conceptual framework of somatic evolution. Hopefully, the recent rediscovery of genetic heterogeneity in tumors and the lessons from the predictable failures of targeted therapies will provide sufficient motivation to focus experimental and clinical attention on efforts required to implement these changes.

Acknowledgments

This work was supported by funding received from Shula Breast Cancer Award. I thank Andrii Rozhok for his critical comments and suggestions.

References

Aktipis, C.A., Kwan, V.S., Johnson, K.A., Neuberg, S.L., Maley, C.C., 2011. Overlooking evolution: a systematic analysis of cancer relapse and therapeutic resistance research. PLoS One 6, e26100.

Anderson, K., Lutz, C., van Delft, F.W., Bateman, C.M., Guo, Y., Colman, S.M., Kempski, H., Moorman, A.V., Titley, I., Swansbury, J., Kearney, L., Enver, T., Greaves, M., 2011. Genetic variegation of clonal architecture and propagating cells in leukaemia. Nature 469, 356–361.

Chaffer, C.L., Brueckmann, I., Scheel, C., Kaestli, A.J., Wiggins, P.A., Rodrigues, L.O., Brooks, M., Reinhardt, F., Su, Y., Polyak, K., Arendt, L.M., Kuperwasser, C., Bierie, B., Weinberg, R.A., 2011. Normal and neoplastic nonstem cells can spontaneously convert to a stem-like state. Proc. Natl. Acad. Sci. 108, 7950–7955.

Dewhurst, S.M., McGranahan, N., Burrell, R.A., Rowan, A.J., Gronroos, E., Endesfelder, D., Joshi, T., Mouradov, D., Gibbs, P., Ward, R.L.,

Hawkins, N.J., Szallasi, Z., Sieber, O.M., Swanton, C., 2014. Tolerance of whole-genome doubling propagates chromosomal instability and accelerates cancer genome evolution. Cancer Discov. 4, 175–185.

Duelli, D., Lazebnik, Y., 2003. Cell fusion: a hidden enemy? Cancer Cell 3, 445–448.

Gatenby, R.A., Gillies, R.J., 2004. Why do cancers have high aerobic glycolysis? Nat. Rev. Cancer 4, 891–899.

Godek, K.M., Venere, M., Wu, Q., Mills, K.D., Hickey, W.F., Rich, J.N., Compton, D.A., 2016. Chromosomal instability affects the tumorigenicity of glioblastoma tumor-initiating cells. Cancer Discov. 6, 532–545.

Greaves, M., Maley, C.C., 2012. Clonal evolution in cancer. Nature 481, 306–313.

Hansen, K.D., Timp, W., Bravo, H.C., Sabunciyan, S., Langmead, B., McDonald, O.G., Wen, B., Wu, H., Liu, Y., Diep, D., Briem, E., Zhang, K., Irizarry, R.A., Feinberg, A.P., 2011. Increased methylation variation in epigenetic domains across cancer types. Nat. Genet. 43, 768–775.

Heng, H.H., Bremer, S.W., Stevens, J.B., Horne, S.D., Liu, G., Abdallah, B.Y., Ye, K.J., Ye, C.J., 2013. Chromosomal instability (CIN): what it is and why it is crucial to cancer evolution. Cancer Metastasis. Rev. 32, 325–340.

Kreso, A., Dick, J.E., 2014. Evolution of the cancer stem cell model. Cell Stem Cell 14, 275–291.

Kuznetsova, A.Y., Seget, K., Moeller, G.K., de Pagter, M.S., de Roos, J.A., Durrbaum, M., Kuffer, C., Muller, S., Zaman, G.J., Kloosterman, W.P., Storchova, Z., 2015. Chromosomal instability, tolerance of mitotic errors and multidrug resistance are promoted by tetraploidization in human cells. Cell Cycle 14, 2810–2820.

Landau, D.A., Clement, K., Ziller, M.J., Boyle, P., Fan, J., Gu, H., Stevenson, K., Sougnez, C., Wang, L., Li, S., Kotliar, D., Zhang, W., Ghandi, M., Garraway, L., Fernandes, S.M., Livak, K.J., Gabriel, S., Gnirke, A., Lander, E.S., Brown, J.R., Neuberg, D., Kharchenko, P.V., Hacohen, N., Getz, G., Meissner, A., Wu, C.J., 2014. Locally disordered methylation forms the basis of intratumor methylome variation in chronic lymphocytic leukemia. Cancer Cell 26, 813–825.

Lee, R.S., Stewart, C., Carter, S.L., Ambrogio, L., Cibulskis, K., Sougnez, C., Lawrence, M.S., Auclair, D., Mora, J., Golub, T.R., Biegel, J.A., Getz, G., Roberts, C.W., 2012. A remarkably simple genome underlies highly malignant pediatric rhabdoid cancers. J. Clin. Invest. 122, 2983–2988.

Li, L., McCormack, A.A., Nicholson, J.M., Fabarius, A., Hehlmann, R., Sachs, R.K., Duesberg, P.H., 2009. Cancer-causing karyotypes: chromosomal equilibria between destabilizing aneuploidy and stabilizing selection for oncogenic function. Cancer Genet. Cytogenet. 188, 1–25.

Martins, F.C., De, S., Almendro, V., Gonen, M., Park, S.Y., Blum, J.L., Herlihy, W., Ethington, G., Schnitt, S.J., Tung, N., Garber, J.E., Fetten, K., Michor, F., Polyak, K., 2012. Evolutionary pathways in BRCA1-associated breast tumors. Cancer Discov. 2, 503–511.

Marusyk, A., Almendro, V., Polyak, K., 2012. Intra-tumour heterogeneity: a looking glass for cancer? Nat. Rev. Cancer 12, 323–334.

Mayr, E., 1982. The Growth of Biological Thought: Diversity, Evolution, and Inheritance. Belknap Press, Cambridge, MA.

Michor, F., Iwasa, Y., Vogelstein, B., Lengauer, C., Nowak, M.A., 2005. Can chromosomal instability initiate tumorigenesis? Semin. Cancer Biol. 15, 43–49.

Nowell, P.C., 1976. The clonal evolution of tumor cell populations. Science 194, 23–28.

Pfau, S.J., Silberman, R.E., Knouse, K.A., Amon, A., 2016. Aneuploidy impairs hematopoietic stem cell fitness and is selected against in regenerating tissues in vivo. Genes Dev. 30, 1395–1408.

Plass, C., Pfister, S.M., Lindroth, A.M., Bogatyrova, O., Claus, R., Lichter, P., 2013. Mutations in regulators of the epigenome and their connections to global chromatin patterns in cancer. Nat. Rev. Genet. 14, 765–780.

Sharma, S.V., Lee, D.Y., Li, B., Quinlan, M.P., Takahashi, F., Maheswaran, S., McDermott, U., Azizian, N., Zou, L., Fischbach, M.A., Wong, K.K., Brandstetter, K., Wittner, B., Ramaswamy, S., Classon, M., Settleman, J., 2010. A chromatin-mediated reversible drug-tolerant state in cancer cell subpopulations. Cell 141, 69–80.

Shipitsin, M., Campbell, L.L., Argani, P., Weremowicz, S., Bloushtain-Qimron, N., Yao, J., Nikolskaya, T., Serebryiskaya, T., Beroukhim, R., Hu, M., Halushka, M.K., Sukumar, S., Parker, L.M., Anderson, K.S., Harris, L.N., Garber, J.E., Richardson, A.L., Schnitt, S.J., Nikolsky, Y., Gelman, R.S., Polyak, K., 2007. Molecular definition of breast tumor heterogeneity. Cancer Cell 11, 259–273.

Sieber, O.M., Tomlinson, S.R., Tomlinson, I.P., 2005. Tissue, cell and stage specificity of (epi)mutations in cancers. Nat. Rev. Cancer 5, 649–655.

Sprouffske, K., Merlo, L.M., Gerrish, P.J., Maley, C.C., Sniegowski, P.D., 2012. Cancer in light of experimental evolution. Curr. Biol. 22, R762–R771.

Tabassum, D.P., Polyak, K., 2015. Tumorigenesis: it takes a village. Nat. Rev. Cancer 15, 473–483.

Thompson, S.L., Compton, D.A., 2008. Examining the link between chromosomal instability and aneuploidy in human cells. J. Cell Biol. 180, 665–672.

Timp, W., Feinberg, A.P., 2013. Cancer as a dysregulated epigenome allowing cellular growth advantage at the expense of the host. Nat. Rev. Cancer 13, 497–510.

Vogelstein, B., Kinzler, K.W., 2004. Cancer genes and the pathways they control. Nat. Med. 10, 789–799.

Weinberg, R., 2013. The Biology of Cancer, second ed. Taylor & Francis Group, New York, NY.

19

Cancer as a Disease of Homeostasis: An Angiogenesis Perspective

Irina Kareva

Simon A. Levin Mathematical, Computational and Modeling Sciences Center (SAL MCMSC),
Arizona State University, Tempe, AZ, United States

INTRODUCTION TO NEOVASCULARIZATION

Tumors cannot grow beyond a size of 1–2 mm^3 without securing their own blood supply (Naumov et al., 2006a, 2009). Angiogenesis observed in tumors stems from a dysregulated process of normal wound healing (Dvorak, 1986), and thus understanding pathological neovascularization first requires understanding the process of normal blood vessel formation.

Both the normal and pathological variants of angiogenesis are regulated by carefully orchestrated, temporally, and spatially controlled signals from surrounding tissues, which produce a cascade of angiogenesis regulators that are collected and then released by platelets (Italiano et al., 2008; Klement et al., 2009). Platelets are microparticles formed by megakaryocytes in the bone marrow (Italiano and Shivdasani, 2003). They circulate in the body, collecting pro- and angiogenic proteins, which are stored in separate compartments in the platelet, called alpha-granules (Italiano and Battinelli, 2009). In the event of a wound, platelets get summoned to the tissue, where they become activated by undergoing a series of morphological changes, such as extension of lamellipodia and filipodia, which allow attachment to tissues (Italiano and Shivdasani, 2003). Platelets assist in wound healing in two ways: through mechanistically clotting the wound, and through initiating and promoting signals that facilitate formation of new blood vessels.

At the wound, first the proangiogenic regulators are released from the platelet's alpha-granules, containing angiogenesis stimulators, such as vascular endothelial growth factor (VEGF). These angiogenesis stimulators initiate formation of an early vascular sprout (tip cell) (Hellström et al., 2007). This is followed by release of proteins, such as fibroblast growth factor-basic (bFGF), that facilitate sprout growth and lumenization (stalk cells) (Blanco and Gerhardt, 2013; Gerhardt, 2008). I refer to these proteins as stabilizers. Proliferation and stabilization of endothelial cells and tube formation is further facilitated by proteins, such as platelet-derived growth factor subunit b (PDGF-B). Finally, platelets release angiogenesis inhibitors, such as platelet factor 4 (PF-4), endostatin and thrombospondin 1 (TSP-1), which inhibit angiogenesis through promoting vessel pruning (Xie et al., 2011).

In order to signal, these angiogenesis regulators need to attach to tissue glycosaminoglycans (GAGs), such as heparan sulfate (HS) on the cells' surface (Vlodavsky et al.,1988, 1995, 2011). They can be removed by sufficient concentrations of the enzyme heparinase, which is produced by fibroblasts and other stromal elements. All of the angiogenesis regulators discussed here are characterized by varying degrees of affinity for HS: angiogenesis stimulators, such as VEGF, have the lowest affinity for HS and thus require the least amount of heparinase to destroy tissue anchors and release the growth factor. Blood vessel stabilizers, such as bFGF, are characterized by medium affinity for HS. Angiogenesis inhibitors have the highest affinity for HS, requiring the largest amounts of heparinase to release it from the tissue anchor and terminate signaling (Eldor et al., 1987; Vlodavsky et al., 1988, 2011). Normal physiological state of postnatal angiogenesis is "off", as is supported by predominance of angiogenesis inhibitors in platelets of healthy individuals, which is magnitudes higher than that of stimulators (Peterson et al., 2010). These mechanisms are reviewed in greater detail in (Kareva et al., In press).

PATHOLOGICAL ANGIOGENESIS AND MUSICAL CHAIRS

The process of normal wound healing is well orchestrated and regulated, relying on a sequence of events that maintain a homeostatic balance of angiogenesis regulators. However, in the event of active tumor growth, such balance can become upset, resulting in unregulated blood-vessel formation. Noticeably, while all the aforementioned angiogenic regulators have different receptors, which, upon ligand binding, promote intracellular signaling leading to proliferation, migration, and pericyte activation and coverage (Logsdon et al., 2014), binding to HS is almost a prerequisite for effective signaling (Dejana et al., 2009).

Now let us consider the following. Many tumors are characterized by increased production of angiogenesis stimulators, such as VEGF (Bremnes et al., 2006; Carmeliet, 2005; Peterson et al., 2012). Since all the angiogenesis regulators need to bind to HS for effective signaling, increased production of stimulators may lead to angiogenesis inhibitors being outcompeted for binding cites, much like in a game of musical chairs. Inability of angiogenesis inhibitors to bind and signal would preclude vessel pruning, resulting in formation of immature, leaky vessels, which have been observed in tumors (Carmeliet and Jain, 2011). Therefore, it is likely that pathological angiogenesis commences in the same manner as normal wound healing but due to the activity of the tumor cells, the balance of angiogenesis regulators to inhibitors becomes upset to such an extent that the inhibitors would become outcompeted and thus incapable of signaling. The matter is further augmented by the fact that the advancing edge of the growing tumor does not allow time for tissue proteases to create proinhibitory microenvironment.

TUMOR-INDUCED STROMAL STIMULATION AND TIME TO ESCAPE FROM TUMOR DORMANCY

As was aforementioned, tumors cannot grow beyond a size of 1–2 mm^3 without securing blood supply. In fact, many tumors remain at a microscopic size indefinitely, a phenomenon that is known as tumor dormancy (Aguirre-Ghiso, 2007; Brackstone et al., 2007; Folkman and Kalluri, 2004; Klein, 2011) (see also Chapter 20). Tumors that remain at a microscopic size preclinically are sometimes referred to as primary dormant tumors, while persistence of dormant state of disseminated tumors after removal of a primary tumor is referred to as metastatic tumor dormancy (Klein, 2011). Both will be discussed in this perspective.

Let us first focus on primary dormant tumors. Their existence was revealed by a number of autopsy studies of men and women who died of trauma (Black and Welch, 1993; Feldman et al., 1986), which has prompted intensive investigations into the nature of mechanisms that preclude exit from dormancy. Achilles et al. (2001) have shown that different tumors contain subpopulations of cancer cells that differ in their potential to induce angiogenesis. Isolation of subpopulations within a heterogeneous tumor has allowed developing animal models demonstrating spontaneous neovascularization, or "angiogenic switch" (Almog et al., 2006; Naumov et al., 2006a,b). Moreover, dormancy period varied between cell lines by months, a characteristic that has become termed "dormancy clock" (Almog et al., 2006; Naumov et al., 2006b). Since all the cell lines grew at the same rate in vitro but vascularized at different times, if at all, in vivo, a natural conclusion was that something in the tumor microenvironment was affecting time to angiogenic switch. The authors have confirmed a genetic component to this mechanism (Almog et al., 2009). However, it is possible that the aforementioned mechanism of competitive inhibition, or musical chairs, may also contribute to variations in time to angiogenic switch. These considerations are summarized in Fig. 19.1, which was adapted from (Kareva, 2016).

In order to evaluate this hypothesis, we created a mathematical model, which investigated the effects of varying degrees of tumor-induced stromal stimulation on build-up of neovasculature and consequent escape from tumor dormancy (Kareva, 2016). We were able to show that indeed, variations in the degree of tumor-induced stromal stimulation, and specifically, of the amount of angiogenesis stimulators and blood vessel stabilizers, were sufficient to reproduce experimentally observed tumor growth curves (Fig. 19.2).

Interestingly, in Rogers et al. (2014), the authors investigated the question of whether differences in time to angiogenic switch result from either a small number of discrete events, such as mutations, or are more clock-like and governed by accumulation of specific factors until a threshold is reached, thereby triggering the phenotypic change. The authors then investigated the distribution of switch times and came to the conclusion that mutation model is more likely, and on average, approximately two changes were necessary to convert a nonangiogenic tumor to an angiogenic tumor.

The model proposed by Kareva (2016) showed that indeed, exactly two parameters were necessary and sufficient to reproduce variations in onset of "angiogenic switch", namely, parameter 1, which represents tumor-induced production of angiogenesis stimulators (e.g., VEGF), causing formation of tips, and parameter 2, which represents tumor-induced production of blood vessel stabilizers (e.g., PDGF), necessary for formation of stalks. In other words, this model confirmed that a cumulative

Tumor that remains dormant due to insufficient stromal stimulation

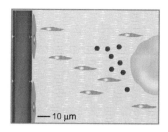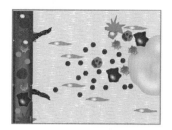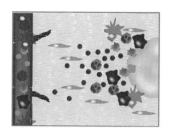

Tumor that escapes dormancy due to sufficient stromal stimulation

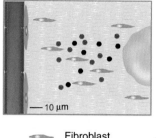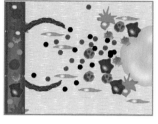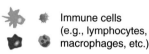

FIGURE 19.1 **The effect of the activity of angiogenesis regulators on possible escape from tumor dormancy.** (A) Release from the platelets and signaling of angiogenesis stimulators, such as VEGF, allows for formation of vascular sprouts (tip cells). However, without release and signaling of stabilizing angiogenesis regulators, such as bFGF, sprout growth and lumenization cannot occur, thus precluding tumor vascularization and subsequent escape from dormancy. (B) In the event of release and active signaling of both angiogenesis stimulators and vessel stabilizing angiogenesis regulators (which, when released in excess, can out-compete angiogenesis inhibitors), neovascularization can occur, allowing angiogenesis-induced escape from tumor dormancy. *Source: Adapted from Kareva, I., 2016. Escape from tumor dormancy and time to angiogenic switch as mitigated by tumor-induced stimulation of stroma. J. Theor. Biol. 395, 11–22.*

effect of the two types of angiogenesis regulators is necessary to promote the onset of angiogenic switch.

Primary and Secondary Tumors: Why do Secondary Tumors Often Start Growing Upon Removal of Primary Tumors?

Metastases are a primary cause of death of cancer. It is now increasingly believed that secondary tumors have been seeded early in tumor development (Hüsemann et al., 2008), suggesting that they remain at a dormant state. There also exist numerous observations that surgical removal of primary tumors can lead to increased growth of secondary tumors (Bedenne et al., 2007; Peeters et al., 2006, 2008; Retsky et al., 2010), suggesting that primary tumors may have some inhibitory effect on metastases. While no suggestion of postponing surgery would of course be made here (Baum et al., 2005; Retsky et al., 2010), it would be beneficial to build a theoretical base to explain this phenomenon to hopefully be able to influence it.

As was discussed previously, the normal physiological state of postnatal angiogenesis is "off", as indicated by orders of magnitude higher levels of angiogenesis inhibitors as compared to angiogenesis stimulators (Peterson et al., 2010). It is a reasonable assumption that the body would try to maintain a homeostatic ratio of angiogenesis stimulators to inhibitors. In fact, some preliminary data suggest that injection of VEGF into mice resulted in increased levels of angiogenesis inhibitors (Giannoula Klement, personal communication), suggesting that increase in level of stimulators would be followed by increase in systemic levels of angiogenesis inhibitors, as the body would attempt to suppress unnecessary angiogenesis.

If the body indeed responds with increasing inhibitor production to compensate for increased stimulator production to restore the state of homeostasis, one can surmise of the following scenario:

Assume the presence in the body of two tumors, a larger primary tumor and a much smaller secondary

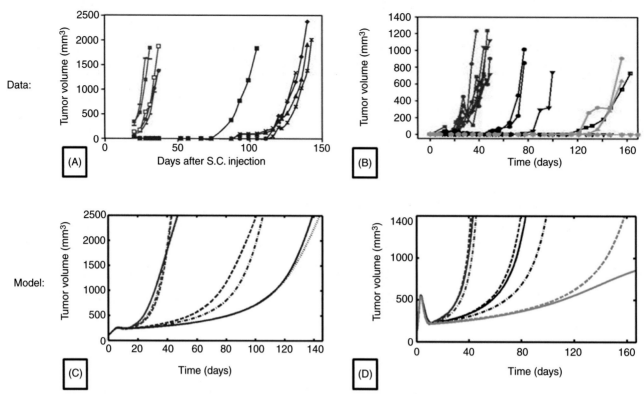

FIGURE 19.2   Comparison of model predictions with previously reported experimental data. The data in (A) was reported in Almog et al. (2006). The authors isolated and reported growth curves for nonangiogenic dormant and angiogenic actively growing tumor clones that were implanted subcutaneously (SC) in severe combined immunodeficiency (SCID) mice. Qualitatively similar growth curves were reproduced with the proposed mathematical model using variations of parameters, which represent tumor-induced stimulation of stroma to produce angiogenesis stimulators (e.g., VEGF), blood vessel stabilizers (e.g., nbFGF), and angiogenesis inhibitors, such as PF-4 (B). The data in (B) was reported by Naumov et al. (2006b). Qualitatively similar growth curves were once again reproduced with the proposed mathematical model using just the variations in the same parameters. *Source: Adapted from Kareva, I., 2016. Escape from tumor dormancy and time to angiogenic switch as mitigated by tumor-induced stimulation of stroma. J. Theor. Biol. 395, 11–22.*

tumor. Both of them are promoting to varying degrees the secretion of some angiogenesis stimulators. At some point, this may become sufficient to trigger the body to commence production of additional angiogenesis inhibitors to restore homeostasis. It is possible that this increase in the systemic level of angiogenesis inhibitors may be sufficient to suppress the smaller tumor.

Surgical removal of primary tumor would remove the source of a large amount of angiogenesis stimulators, effectively reducing or even eliminating the need to further produce additional inhibitors in order to maintain homeostasis. Therefore, the amount of inhibitors would decrease, giving the secondary tumor an opportunity to build up vasculature and grow, perhaps to the point when newly activated systemic production of angiogenesis inhibitors can no longer keep up, and the forward feedback loop of tumor-induced production of self-supporting vasculature has progressed too far. This mechanism, which is based on normal physiological responses and the body's attempts to restore homeostasis, could provide a possible explanation for why removal of primary tumors could result in uprising of secondary tumors. This hypothesis is summarized in Fig. 19.3.

This hypothesis is supported by several studies, where the authors looked at levels of angiogenesis stimulators and inhibitors before and after surgery (e.g., reviewed in Demicheli et al., 2008). Dhar et al. (2002) reported that significant decreases in serum endostatin and bFGF were observed in postoperative samples of patients with hepatocellular carcinoma, compared to the preoperative values. Similar results were observed by Feldman et al. (2001) for patients with colorectal cancer. The authors reported that plasma endostatin levels were significantly higher in the 30 patients compared to controls before surgery. However, none of the patients who remained progression-free had elevated endostatin levels at follow up. Several studies also report that elevated postoperative endostatin levels were associated with poor prognosis for patients with advanced stage nasopharyngeal carcinoma (Mo et al., 2013) and for human malignant gliomas (Morimoto et al., 2002), potentially suggesting presence of metastatic tumors. This area requires further investigation.

However, ingif the proposed hypothesis is true, there may be a need for therapeutic interventions that would allow a more gradual decrease of angiogenesis inhibitors following resection of primary tumor, to avoid triggering

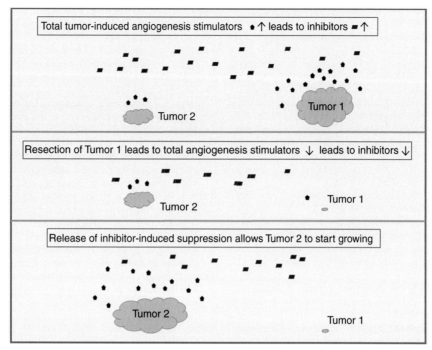

FIGURE 19.3 **Proposed scenario of mechanisms underlying accelerated growth of secondary tumors following resection of a primary tumor.** Suppose the host is harboring two tumors. Cumulative amount of angiogenesis stimulators could trigger a natural response, where the host's body would attempt to restore homeostasis of angiogenesis regulators by producing additional inhibitors. Resection of one of the tumors would reduce the total number of angiogenesis stimulators, causing responsive decrease in angiogenesis inhibitors, thus allowing the secondary tumor to progress.

this response. Metronomic chemotherapy, which aims to gradually reduce tumor size and targets the stroma (and thus the source of angiogenesis regulators) rather than the tumor directly (Kareva et al., 2015), may be an effective approach to reduce chances of accelerated growth of secondary tumors following surgical removal of the primary tumor.

References

Achilles, E.-G., Fernandez, A., Allred, E.N., Kisker, O., Udagawa, T., Beecken, W.-D., et al., 2001. Heterogeneity of angiogenic activity in a human liposarcoma: a proposed mechanism for "no take" of human tumors in mice. J. Natl. Cancer Inst. 93 (14), 1075–1081.

Aguirre-Ghiso, J.A., 2007. Models, mechanisms and clinical evidence for cancer dormancy. Nat. Rev. Cancer 7 (11), 834–846.

Almog, N., Henke, V., Flores, L., Hlatky, L., Kung, A.L., Wright, R.D., et al., 2006. Prolonged dormancy of human liposarcoma is associated with impaired tumor angiogenesis. FASEB J. 20 (7), 947–949.

Almog, N., Ma, L., Raychowdhury, R., Schwager, C., Erber, R., Short, S., et al., 2009. Transcriptional switch of dormant tumors to fast-growing angiogenic phenotype. Cancer Res. AACR 69 (3), 836–844.

Baum, M., et al., 2005. Does surgery unfavourably perturb the "natural history" of early breast cancer by accelerating the appearance of distant metastases? Eur. J. Cancer 41 (4), 508–515.

Bedenne, L., Michel, P., Bouché, O., Milan, C., Mariette, C., Conroy, T., et al., 2007. Chemoradiation followed by surgery compared with chemoradiation alone in squamous cancer of the esophagus: FFCD 9102. J. Clin. Oncol. 25 (10), 1160–1168.

Black, W.C., Welch, H.G., 1993. Advances in diagnostic imaging and overestimations of disease prevalence and the benefits of therapy. New Eng. J. Med. 328 (17), 1237–1243.

Blanco, R., Gerhardt, H., 2013. VEGF and Notch in tip and stalk cell selection. Cold Spring Harb. Perspect. Med. 3 (1), a006569.

Brackstone, M., Townson, J.L., Chambers, A.F., et al., 2007. Tumour dormancy in breast cancer: an update. Breast Cancer Res. 9(3), 208.

Bremnes, R.M., Camps, C., Sirera, R., 2006. Angiogenesis in non-small cell lung cancer: the prognostic impact of neoangiogenesis and the cytokines VEGF and bFGF in tumours and blood. Lung Cancer 51 (2), 143–158.

Carmeliet, P., 2005. VEGF as a key mediator of angiogenesis in cancer. Oncology 69 (Suppl. 3), 4–10.

Carmeliet, P., Jain, R.K., 2011. Principles and mechanisms of vessel normalization for cancer and other angiogenic diseases. Nat. Rev. Drug Discov. 10 (6), 417–427.

Dejana, E., Orsenigo, F., Molendini, C., Baluk, P., McDonald, D.M., 2009. Organization and signaling of endothelial cell-to-cell junctions in various regions of the blood and lymphatic vascular trees. Cell Tissue Res. 335 (1), 17–25.

Demicheli, R., Retsky, M.W., Hrushesky, W.J., Baum, M., Gukas, I.D., 2008. The effects of surgery on tumor growth: a century of investigations. Ann. Oncol. 19 (11), 1821–1828.

Dhar, D.K., Ono, T., Yamanoi, A., Soda, Y., Yamaguchi, E., Rahman, M.A., et al., 2002. Serum endostatin predicts tumor vascularity in hepatocellular carcinoma. Cancer 95 (10), 2188–2195.

Dvorak, H.F., 1986. Tumors: wounds that do not heal. New Eng. J. Med. 315 (26), 1650–1659.

Eldor, A., Bar-Ner, M., Yahalom, J., Fuks, Z., Vlodavsky, I., 1987. Role of heparanase in platelet and tumor cell interactions with the subendothelial extracellular matrix. Semin. Thromb. Hemost., 475–488.

Feldman, A.R., Kessler, L., Myers, M.H., Naughton, M.D., 1986. The prevalence of cancer. New Eng. J. Med. 315 (22), 1394–1397.

Feldman, A.L., Alexander, Jr., H.R., Bartlett, D.L., Kranda, K.C., Miller, M.S., Costouros, N.G., et al., 2001. A prospective analysis of plasma endostatin levels in colorectal cancer patients with liver metastases. Ann. Surg. Oncol. 8 (9), 741–745.

Folkman, J., Kalluri, R., 2004. Cancer without disease. Nature 427 (6977), 787–1787.

Gerhardt, H., 2008. VEGF and endothelial guidance in angiogenic sprouting. Organogenesis 4 (4), 68–78.

Hellström, M., Phng, L.-K., Hofmann, J.J., Wallgard, E., Coultas, L., Lindblom, P., et al., 2007. Dll4 signalling through Notch1 regulates formation of tip cells during angiogenesis. Nature 445 (7129), 776–780.

Hüsemann, Y., Geigl, J.B., Schubert, F., Musiani, P., Meyer, M., Burghart, E., et al., 2008. Systemic spread is an early step in breast cancer. Cancer Cell 13 (1), 58–68.

Italiano, Jr., J., Battinelli, E., 2009. Selective sorting of alpha-granule proteins. J. Thromb. Haemost. 7 (s1), 173–176.

Italiano, J., Shivdasani, R., 2003. Megakaryocytes and beyond: the birth of platelets. J. Thromb. Haemost. 1 (6), 1174–1182.

Italiano, J.E., Richardson, J.L., Patel-Hett, S., Battinelli, E., Zaslavsky, A., Short, S., et al., 2008. Angiogenesis is regulated by a novel mechanism: pro-and antiangiogenic proteins are organized into separate platelet α granules and differentially released. Blood 111 (3), 1227–1233.

Kareva, I., 2016. Escape from tumor dormancy and time to angiogenic switch as mitigated by tumor-induced stimulation of stroma. J. Theor. Biol. 395, 11–22.

Kareva, I., Waxman, D.J., Klement, G.L., 2015. Metronomic chemotherapy: an attractive alternative to maximum tolerated dose therapy that can activate anti-tumor immunity and minimize therapeutic resistance. Cancer Lett. 358 (2), 100–106.

Kareva, I., Abou-Slaybi, A., Dodd, O., Dashevsky, O., Klement, G., In press. Normal wound healing and tumor angiogenesis as a game of competitive inhibition. PLoS One

Klein, C.A., 2011. Framework models of tumor dormancy from patient-derived observations. Curr. Opin. Genet. Dev. 21 (1), 42–49.

Klement, G.L., Yip, T.-T., Cassiola, F., Kikuchi, L., Cervi, D., Podust, V., et al., 2009. Platelets actively sequester angiogenesis regulators. Blood 113 (12), 2835–2842.

Logsdon, E.A., Finley, S.D., Popel, A.S., Gabhann, F.M., 2014. A systems biology view of blood vessel growth and remodelling. J. Cell. Mol. Med. 18 (8), 1491–1508.

Mo, H.-Y., Luo, D.-H., Qiu, H.-Z., Liu, H., Chen, Q.-Y., Tang, L.-Q., et al., 2013. Elevated serum endostatin levels are associated with poor survival in patients with advanced-stage nasopharyngeal carcinoma. Clin. Oncol. 25 (5), 308–317.

Morimoto, T., Aoyagi, M., Tamaki, M., Yoshino, Y., Hori, H., Duan, L., et al., 2002. Increased levels of tissue endostatin in human malignant gliomas. Clin. Cancer Res. 8 (9), 2933–2938.

Naumov, G.N., Akslen, L.A., Folkman, J., 2006a. Role of angiogenesis in human tumor dormancy: animal models of the angiogenic switch. Cell Cycle 5 (16), 1779–1787.

Naumov, G.N., Bender, E., Zurakowski, D., Kang, S.-Y., Sampson, D., Flynn, E., et al., 2006b. A model of human tumor dormancy: an angiogenic switch from the nonangiogenic phenotype. J. Natl. Cancer Inst. 98 (5), 316–325.

Naumov, G.N., Folkman, J., Straume, O., 2009. Tumor dormancy due to failure of angiogenesis: role of the microenvironment. Clin. Exp. Metas. 26 (1), 51–60.

Peeters, C.F., De Waal, R.M., Wobbes, T., Westphal, J.R., Ruers, T.J., 2006. Outgrowth of human liver metastases after resection of the primary colorectal tumor: a shift in the balance between apoptosis and proliferation. Int. J. Cancer 119 (6), 1249–1253.

Peeters, C.F., Waal, R.M., de, Wobbes, T., Ruers, T.J., 2008. Metastatic dormancy imposed by the primary tumor: does it exist in humans? Ann. Surg. Oncol. 15 (11), 3308–3315.

Peterson, J.E., Zurakowski, D., Italiano, J.E., Michel, L.V., Fox, L., Klement, G.L., et al., 2010. Normal ranges of angiogenesis regulatory proteins in human platelets. Am. J. Hematol. 85 (7), 487–493.

Peterson, J.E., Zurakowski, D., Italiano, Jr., J.E., Michel, L.V., Connors, S., Oenick, M., et al., 2012. VEGF, PF4 and PDGF are elevated in platelets of colorectal cancer patients. Angiogenesis 15 (2), 265–273.

Retsky, M., Demicheli, R., Hrushesky, W., Baum, M., Gukas, I., 2010. Surgery triggers outgrowth of latent distant disease in breast cancer: an inconvenient truth? Cancers 2 (2), 305–337.

Rogers, M.S., Novak, K., Zurakowski, D., Cryan, L.M., Blois, A., Lifshits, E., et al., 2014. Spontaneous reversion of the angiogenic phenotype to a nonangiogenic and dormant state in human tumors. Mol. Cancer Res. 12 (5), 754–764.

Vlodavsky, I., Eldor, A., Bar-Ner, M., Fridman, R., Cohen, I.R., Klagsbrun, M., 1988. Heparan sulfate degradation in tumor cell invasion and angiogenesis. Adv. Exp. Med. Biol. 233, 201.

Vlodavsky, I., Miao, H., Atzmon, R., Levi, E., Zimmermann, J., Bar-Shavit, R., et al., 1995. Control of cell proliferation by heparan sulfate and heparin-binding growth factors. Thromb. Haemost. 74 (1), 534.

Vlodavsky, I., Elkin, M., Ilan, N., 2011. Impact of heparanase and the tumor microenvironment on cancer metastasis and angiogenesis: basic aspects and clinical applications. Rambam Maimonides Med. J. 2, e0019.

Xie, L., Duncan, M.B., Pahler, J., Sugimoto, H., Martino, M., Lively, J., et al., 2011. Counterbalancing angiogenic regulatory factors control the rate of cancer progression and survival in a stage-specific manner. Proceedings of the National Academy of Sciences. National Academy Sciences 108(24), 9939–9944.

Dormancy: An Evolutionary Key Phenomenon in Cancer Development[a]

Ole Ammerpohl*, Kirsten Hattermann**, Janka Held-Feindt[†],
Christoph Röcken[‡], Heiner Schäfer[§], Christian Schem[¶], Denis
Schewe[††], Hinrich Schulenburg[‡‡], Susanne Sebens[§], Michael
Synowitz[†], Sanjay Tiwari[§§], Arne Traulsen***, Anna
Trauzold[§], Thomas Valerius[†††], Daniela Wesch[‡‡‡]

*Institute of Human Genetics, Christian-Albrechts-University Kiel and University Hospital Schleswig-Holstein, Campus Kiel, Kiel, Germany

**Institute of Anatomy, Christian-Albrechts-University Kiel, Kiel, Germany

[†]Department of Neurosurgery, University Hospital Schleswig-Holstein Campus Kiel, Kiel, Germany

[‡]Department of Pathology, Christian-Albrechts-University Kiel and University Hospital Schleswig-Holstein, Campus Kiel, Kiel, Germany

[§]Institute for Experimental Cancer Research, Christian-Albrechts-University Kiel and University Hospital Schleswig-Holstein, Campus Kiel, Kiel, Germany

[¶]Department of Obstetrics and Gynecology, University Hospital Schleswig-Holstein, Campus Kiel, Kiel, Germany

[††]Department of Pediatrics, University Hospital Schleswig-Holstein, Campus Kiel, Kiel, Germany

[‡‡]Department of Evolutionary Ecology Genetics, Zoological Institute, Christian-Albrechts-University Kiel, Kiel, Germany

[§§]Department of Diagnostic Radiology and Neuroradiology, University Hospital Schleswig-Holstein Campus Kiel, Kiel, Germany

***Department of Evolutionary Theory, Max Planck Institute for Evolutionary Biology, Plön, Germany

[†††]Department of Internal Medicine II, Section for Stem cell transplantation and Immunotherapy, University Hospital Schleswig-Holstein, Campus Kiel, Kiel, Germany

[‡‡‡]Institute of Immunology, Christian-Albrechts-University Kiel and University Hospital Schleswig-Holstein, Campus Kiel, Kiel, Germany

[a]Authors are listed in alphabetical order and are members of the Kiel Oncology Network

Ecology and Evolution of Cancer. http://dx.doi.org/10.1016/B978-0-12-804310-3.00020-X
Copyright © 2017 Elsevier Inc. All rights reserved.

INTRODUCTION

Considerable progress has been made in the treatment of cancer in the last years. Progress in prevention, diagnosis, and therapy of cancer has improved the survival of many cancer patients. However, some tumors still have a dismal prognosis, and can relapse years or decades after initial diagnosis, even though they have been presumably cured (Bragado et al., 2012; Wikman et al., 2008; Yeh and Ramaswamy, 2015). Accordingly, the majority of cancer patients still die from the disease sooner or later. For solid tumors, metastatic disease represents the most challenging clinical problem in the treatment of cancer patients and is the major cause of cancer-related deaths (Yeh and Ramaswamy, 2015). In hematological malignancies, the eradication of minimal residual disease is the key to prevention of relapses and the amelioration of patient prognosis (Sosa et al., 2014).

Dormancy describes a reversible quiescent cellular stage, which plays a fundamental role in tumor evolution. Exit of this dormant stage can lead to proliferation and tumor outgrowth in both primary tumor cells, as well as in metastases (Bragado et al., 2012; Celià-Terrassa and Kang, 2016; Wikman et al., 2008; Yeh and Ramaswamy, 2015). Importantly, dormancy is not restricted to tumors and has been described in a variety of organisms (Roff, 1992; Stearns, 1992). Species which are able to enter a reversible inactive state, such as dormancy when facing unfavorable environmental conditions may have a tremendous evolutionary advantage. Thus, dormancy can be regarded as a key ingredient to evolution, which enables the persistence of populations in adverse conditions.

Given the fact that dormancy and exit from dormancy are crucial steps in tumor evolution, one reason for the failure of current treatment strategies might be the insufficiently understood mechanisms underlying these phenomena. As the fundamental concept of dormancy in oncology is similarly present in evolutionary biology, novel research concepts based on the interdisciplinary work of oncological clinicians, applied researchers, evolutionary biologists, and mathematicians might pave the way to generate a novel understanding of this phenomenon ultimately leading to improved cancer treatments.

DORMANCY IN TUMOR EVOLUTION

For details on the ecology and evolution of cancer, this book contains a number of excellent articles. Briefly, from an evolutionary perspective, a neoplasm is a large, genetically and epigenetically heterogeneous cell population (different tumor cell clones) with a high functional heterogeneity (Merlo et al., 2006). The fitness of each cell to survive and reproduce is highly dependent on its interaction with other cells (within the tumor and the stroma), as well as with the local and systemic microenvironment (its ecology). Important to note, not only the tumor population but also the stromal compartment is characterized by a high dynamic heterogeneity. Besides the noncellular compartment comprised of extracellular matrix and soluble factors, such as growth factors, cytokines, proteases, and chemokines, various nonneoplastic cells (e.g., fibroblasts, myofibroblasts, endothelial cells, immune cells) enrich in the tumor during tumor evolution. Hence, evolution of tumor cell clones occurs within a complex ecosystem and leads to modulation and adaptation of the microenvironment so that progression of tumors is the result of a coevolution of tumor cells and their niche (Barcellos-Hoff et al., 2013; Greaves and Maley, 2012). Most tumors can arise at different sites so that metastatic evolution of different tumor clones is a successful cellular strategy to adapt to different environmental forces, which leads to divergent phenotypes (Gillies et al., 2012). However, the evolution of divergent clonal populations is also possible in the same microenvironment, for example under the pressure of chemotherapy, giving rise to intratumoral heterogeneity. The ability to enter and exit proliferation reflects the phenotypic plasticity of tumor cells and is crucial for initiation, maintenance, and progression of tumors. This highly dynamic state was first described by the Australian pathologist Rupert Willis in 1934 and termed dormancy (Willis, 1934). Although a sustained proliferative signaling leading to uncontrolled cell growth is defined as a hallmark of cancer (Hanahan and Weinberg, 2011), tumors can only progress and successfully evolve when at least some tumor cells of the population retain the capacity to become temporarily quiescent in order to survive hostile conditions. Therefore, dormant tumor cells are fundamental for the fitness of the entire tumor cell population assuring and promoting tumor evolution so that dormancy has been postulated as a novel hallmark of cancer (Yeh and Ramaswamy, 2015).

As a reversible quiescent cellular stage dormancy is pivotal for the entire tumor evolution allowing adaptation and survival during critical phases of tumorigenesis (Crea et al., 2015). Accordingly, three "dormancy phases" have been termed: (1) primary cancer dormancy (gain of genetic and epigenetic alterations "optimizing" the tumor phenotype; Galvao et al., 2014); (2) metastatic dormancy (adaptation to different microenvironments; Giancotti, 2013), and (3) therapy induced dormancy (escape of and developing resistance to (chemo)therapy; Uhr and Pantel, 2011).

As previously mentioned, dormancy is regarded as fundamental for the colonization of secondary organs by tumor cells and thereby the formation of metastases. Meanwhile, it is widely accepted that tumor cell dissemination into distant organs is an early event in tumor evolution not necessarily leading to an immediate outgrowth of overt metastases (Friberg and Nyström, 2015; Rhim et al., 2012). Thus, the capability of early disseminated tumor cells (DTCs) to survive in foreign (unfavorable) microenvironments by acquiring a reversible dormant state is a prerequisite for tumor progression at later time points. Similarly, dormant tumor cells that have not been eradicated by the therapy can survive for prolonged periods of time at the primary tumor location suggesting that dormancy also plays an important role for local relapse. Since many cancer therapies preferentially eliminate proliferating cells, such treatments may select for dormant tumor cells causing a state of minimal residual disease and promoting disease recurrence (Aguirre-Ghiso, 2007).

Like dormant cells, cancer stem cells (CSC) are supposed to be essential for the maintenance, progression, and relapses of tumors. CSC are cells of an undifferentiated phenotype with the unique property to self-renew and to give rise to differentiated tumor cell clones. Besides the fact that dormant cells and CSCs represent minor cell subsets of the tumor cell population, both cell subsets have further characteristics in common: (1) high dependency on a particular niche, (2) exhibition of effective survival strategies, (3) pronounced drug resistance, and (4) ability to undergo a temporary/reversible growth arrest (Ghajar, 2015; Magee et al., 2012). Moreover, recent data indicate that CSC-properties are not fixed to particular cells but can be gained and lost in dependence on the microenvironment (O'Connor et al., 2014). Therefore, it is still a matter of debate whether dormant tumor cells are CSCs or whether these two tumor cell subsets are different but highly dependent on each other.

Importantly, different forms of dormancy have been described in which cellular dormancy can be differentiated from tumor mass dormancy. While cellular dormancy implies the growth arrest of solitary cells, tumor mass dormancy describes a balance of proliferating and dying cells within tumor cell clusters/micrometastases maintained either by a balance of pro- and antiangiogenic factors (Aguirre-Ghiso, 2007; Almog et al., 2009; Hofstetter et al., 2012) or immune surveillance (Aguirre-Ghiso, 2007 and references therein). The latter describes the influence of immune cells on maintaining cancer cells in dormancy or promoting their exit thereof (Romero et al., 2014). In this respect, interferon gamma (IFN-γ)- or tumor necrosis factor (TNF)-mediated signaling is postulated to have a pivotal role in inducing the dormant state. Thus, a balance of cytotoxic, as well as immunosuppressive cells in the tumor microenvironment substantially impact on the growth behavior and survival of tumor cells (Braumüller et al., 2013; Müller-Hermelink et al., 2008; Wieder et al., 2013).

Overall, the microenvironment including resident and infiltrated stromal cells, extracellular matrix proteins, availability of oxygen, nutrients, growth factors, cytokines, and chemokines are thought to essentially determine the acquisition of a dormant stage and with this the equilibrium between proliferation and survival pathways (Aguirre-Ghiso, 2007; Barkan et al., 2010; Hurst et al., 2013; Marsden et al., 2012; Wang et al., 2015). Likewise the reversal of the dormant stage leading to a resumption of proliferation and outgrowth of metastases is postulated to be highly dependent on alterations of the microenvironment, for example, due to aging or inflammation (Aguirre-Ghiso, 2007; Bragado et al., 2012; Yeh and Ramaswamy, 2015). The mechanisms leading to the entry in and exit from dormancy are still poorly understood (Bragado et al., 2013; Sosa et al., 2011; Retsky et al., 2008). At the signaling level, one important mechanism for this switch seems to be the increase in the ratio of the mitogen activated protein kinases (MAPK) p38 to Erk (p38^{high}/Erk^{low} signaling ratio) in solid tumors (Aguirre-Ghiso, 2007; Bragado et al., 2013; Sosa et al., 2011). Furthermore, changes in the epigenome have been shown to essentially impact the growth and survival behavior of tumor cells (Marjanovic et al., 2013). In glioma, the status of the coagulome (expression of coagulation associated genes) was identified as important modulator of the epigenome in dormant tumor cells (Magnus et al., 2013, 2014a). More specifically, the clothing protein tissue factor (TF) triggers the awakening of dormant glioma cells, resulting in lasting changes in cancer cell genome and epigenome. Interestingly, connectivity mapping of molecular changes imparted by TF expression during tumor latency indicates similarities with inflammatory and cell movement pathways, mostly converging on the respective cytokine network (Magnus et al., 2014b). Consistent with TF influence on tumor heterogeneity, coagulation factor gene expression profile reveals unique expression patterns between the four molecular glioblastoma subtypes (Magnus et al., 2013).

Another level which appears to be important for the switch from entry in and exit from dormancy is cellular metabolism. The link between alterations of the metabolism and dormancy is still insufficiently investigated. Nevertheless several principles seem to exist in evolutionary biology pertaining to dormancy which we will outline in more detail. Metabolic reprogramming of tumor cells is a hallmark of cancer and not a simple consequence of their malignant transformation. It represents an essential prerequisite for oncogene driven transformation, as well as manifestation of malignant properties (Galluzzi et al., 2013), such as therapy resistance and CSC-characteristics (Folmes and Terzic, 2016), as well as a reversible state of a quiescent cell phenotype (dormancy). Notably, such phenotypes are not only linked to metabolic reprogramming but may also depend on metabolic symbiosis between tumor cells and their microenvironment (Romero et al., 2015; Xu et al., 2015). Therefore, the principles governing the metabolism of tumor cells are probably very close or even identical to those defined in evolutionary biology.

Thus, it can be envisioned that each tumor cell, residing in a complex ecosystem composed of heterogeneous populations of neighboring tumor cells and stromal cells (e.g., fibroblasts, immune cells), compete for nutrients like glucose and oxygen on the one hand, and mutually provide each other with certain metabolites on the other hand (Moir et al., 2015; Xu et al., 2015). Depending on distinct metabolic conditions, such as a high rate of oxidative phosphorylation or aerobic glycolysis (Warburg effect) (see also Chapters 4 and 8), tumor cells may either need to get rid of certain metabolites, such as lactate or in turn may use them as substrate for their energy metabolism (reverse Warburg effect). Here, the flux of metabolites and thereby, of energy released by one cell type (e.g., stromal fibroblast) and consumed by another one (e.g., cancer cell), greatly adds to the tumoral ecosystem and ensures individual cellular fitness. Together with sporadic genetic alterations these metabolic restraints at the primary tumor site strongly shape the phenotypic heterogeneity (Viale et al., 2015) and may favor the emergence of single cancer cells exhibiting a dormant state. Likewise, these metabolic restraints may exist at distant sites or niches, thereby maintaining dormancy of cancer cells at potential sites for metastasis. Due to environmental changes (e.g., due to inflammatory/oxidative stress) and metabolic alterations, cancer cells may recover from the dormant state (reawakening) giving rise to tumor relapse or metastatic outgrowth.

As a prerequisite for biomass production, aerobic glycolysis is a hallmark of proliferating tumor cells within the tumor. This condition is evolutionarily reminiscent of unicellular organisms growing under optimal nutrient supply (Vander Heiden et al., 2009). Conversely, cells engage in oxidative phosphorylation when nutrient supply is restricted and a maximum of energy yield is required for cell homeostasis and maintenance of a differentiated state. This is evolutionarily reminiscent of unicellular organisms adapting to nutrient restriction (Vander Heiden et al., 2009) or facing otherwise inappropriate growth conditions, for example, inflammatory, xenobiotic (anticancer drugs), or oxidative stress (as further outlined below). A particular manifestation of this adaptation culminates in a state of hypometabolism associated with cellular dormancy. Thereby, cells become resilient to life-threatening hazards until favorable conditions arise. By having the ability to regularly modulate biosynthetic, proliferative, and electrogenic processes that would require high amounts of energy when its gain is greatly restricted, these cells and organisms undergo an ultimate adaptation to exogenous stressors or to seasonal endogenous cues. This metabolic condition in cellular dormancy is further characterized by consolidated mitochondrial functions (Testa et al., 2016; Viale et al., 2015),

particularly including a low mitochondrial membrane potential. The capability of avoiding detrimental/lethal depletion of adenosine triphosphate (ATP) confers an evolutionary advantage to these resilient cells and organisms when compared with those organisms which are incapable (see later).

Another and more complex role in such an adaptation through dormancy is concerned with metabolites that substantially affect the epigenome through interference with epigenetic editors like histone demethylases (HDMs) (Li and Li, 2015). Under normal conditions HDMs remove repressive histone methylation marks and thereby protect gene promoters from aberrant DNA methylation. By inhibiting HDMs and activating DNA methyltransferases (DNMTs) or histone methyltransferases (HMTs), certain metabolites lead to chromatin modifications, as well as to DNA hypermethylation. Along with global downregulation of genes involved in differentiation this condition particularly accounts for a bias in developmental gene-expression patterns (Menendez et al., 2016). By lowering the "reprogramming barrier" of the epigenetic landscape such "oncometabolites" allow tumors cells to switch more easily from differentiated cell states to CSC-like attractors and concomitantly stabilizing the self-maintaining character of CSC-states (Menendez et al., 2016). It can be assumed that the dormant state of cancer cells, which is supposed to unify quiescence and stemness features may similarly relate to metabolite dependent changes in the reprogramming barrier. Therefore, improving our knowledge of the metabolic state and the "mitochondrial fitness" associated with cellular dormancy, as well as the understanding of the evolutionary principles behind it, will provide an important measure to discriminate and target this particular subset of tumor cells.

The fact that breast cancer patients frequently relapse after years, even decades has essentially driven the investigation of dormancy in order to improve long term survival in this tumor entity (Marsden et al., 2012; Wang et al., 2015). Besides breast cancer and hematological cancer diseases, in which dormancy is well known as minimal residual disease (Trumpp et al., 2010), the phenomenon of reversible quiescence has been also described in other tumor entities, such as pancreatic ductal adenocarcinoma (PDAC) and glioma (Lefter et al., 2003; Lin et al., 2013). In gliomas for example, a distinct subset of nonproliferative cells has been identified in murine models (Endaya et al., 2016) and rare cases of metastases have been described in human patients, for example, in the bone marrow (Tanaka et al., 2014). Glioma dormancy is characterized by a nonangiogenic phenotype, a growth arrest of the tumor after a short proliferation period and a prolonged "dormant" period in which the tumor causes no harm to the host (Naumov et al., 2006). These features are accompanied by a specific gene expression signature

(Almog et al., 2009; Endaya et al., 2016; Satchi-Fainaro et al., 2012), including CSC-markers, proteases and hypoxia-regulated genes. Similar dormancy characteristics have been identified in other tumors like osteosarcoma, breast adenocarcinoma, and liposarcoma (Almog et al., 2009) indicating these mechanisms as comprehensive processes applicable to several tumor entities.

Thus, clinical proofs for tumor cell dormancy exist for a wide range of solid and hematological malignancies (Aguirre-Ghiso, 2007) and are mainly corroborated by the fact that many tumor entities can relapse years and even decades after termination of therapy. However, we have to be aware of certain limitations. For example, evidence is still lacking on whether DTC in the bone marrow of breast cancer patients are truly dormant or in an irreversible quiescent stage. Nevertheless, experimental findings from in vitro and preclinical studies along with data obtained from autopsies and reports on inadvertently transmitted tumor cells by organ transplantation strongly support the concept of dormancy (Aguirre-Ghiso, 2007; Friberg and Nyström, 2015).

DORMANCY IN SPECIES EVOLUTION

Dormancy has been defined as "one of two prerequisites for life" (Sneath, 1975). As in tumors, all natural systems underlie spatial and temporal environmental variations which determine survival and reproduction of the species in their ecosystem. Accordingly, many species can enter a dormant phase (e.g., hibernation, diapause) and/or express a specific, long-lasting life-history stage (e.g., dauerstage, spores) during which metabolic activity is reduced. Such a life-history stage is usually highly resistant against various stressors and enables organisms to survive under adverse conditions (e.g., lack of nutrients, oxygen, unfavorable pH, or temperature) (Roff, 1992; Stearns, 1992). Dormancy has been described in a variety of organisms, including representatives of viruses, bacteria, fungi, plants, insects, fish, amphibians, reptiles, birds, mammals (Friberg and Nyström, 2015). Its evolutionary advantages and disadvantages and the underlying genetics have been repeatedly studied in bacteria. For example, *Escherichia coli* and *Salmonella typhimurium* are able to survive over many years in dormancy by entering a stationary phase. A key determinant of survival during such dormancy is protection from oxidative stress (Eisenstark et al., 1992).

Other examples of dormant-like stages include persisters and spores, both often but not exclusively linked to biofilm formation. Biofilms are an exopolymer matrix being produced by assembling bacteria and having several properties in common with the tumoral microenvironment (e.g., heterogeneity, impact on cellular metabolism, drug resistance) (Lambert et al., 2011).

Bacterial spore formation is found in the Firmicutes, usually induced upon unfavorable conditions, such as nutrient limitation, resulting in cells with a thick coat highly resistant against a variety of stressors, for example, including heat, gamma radiation, or toxic substances, such as hydrogen peroxide (Higgins and Dworkin, 2012; Lewis, 2010). Sporulation is a tightly regulated process, based on the master regulator Spo0A and its associated signaling network (Higgins and Dworkin, 2012). Persisters are cells with reduced metabolism and division rates, which can survive antibiotic stress. Their formation can be induced directly by environmental cues or result from stochastic processes (Checinska et al., 2015). The control of these processes is not yet well understood. It can involve the bacterial SOS response (Checinska et al., 2015) or be controlled by stochastic or bistable switches, such as those based on toxin–antitoxin modules (Maisonneuve and Gerdes, 2014). Formation of both spores and especially persisters can be part of a bet-hedging strategy (see later), whereby they are continuously produced at low frequency, allowing the bacteria to survive in fluctuating environments (de Jong et al., 2011). In bacteria, it has been shown that bet-hedging strategies can evolve within a few generations, given that the environmental conditions change sufficiently fast (Beaumont et al., 2009) (see also Chapter 17).

Overall, dormant life-history phases or stages are widespread among living organisms. They represent a highly effective strategy to survive unfavorable environmental conditions, especially if unpredictable, and thus ensure persistence during evolution. At the same time, dormancy comes at the cost of reduced reproduction and growth. Therefore, the decision to enter or exit dormancy is likely subject to strong selection (Walsh, 2013).

LINKING EVOLUTIONARY CONCEPTS TO THE INVESTIGATION OF TUMOR DORMANCY

Even though the concept of tumor dormancy has been well appreciated over the past years, the entire process is still poorly understood. One major limitation is the lack of a defined phenotype or marker profile that clearly identifies dormant tumor cells (and discriminates them from their "normal" counterparts that are physiologically dormant). This is of great importance for therapeutic approaches aimed at targeting dormant tumor cells while sparing noncancerous quiescent cells. A decent amount of knowledge already exists on the mechanism of dormancy in normal hematopoietic stem cells (Trumpp et al., 2010), while less information is available on dormant solid tumor cells and hematopoietic malignancies. Another drawback might be our limited understanding of the dynamics of dormant cells, which might be explained by technological hurdles regarding the availability of patient material and appropriate in vitro and in vivo models on the one hand and neglecting life-history trade-offs on the other hand.

Life-history theory postulates that several trade-offs determine the evolution of phenotypes. This applies to all organisms underlying natural selection (Stearns, 1992). These traits can be mainly seen as investments in growth, reproduction, and survival. Considering this evolutionary framework, tumor cells constantly underlie trade-offs between maximum cell growth and maximum cell survival (Aktipis et al., 2013) (see also the Introductory Chapter). Highly proliferative tumor cells have the advantage to reproduce rapidly, but this happens at the cost of a higher mortality when exposed to unfavorable conditions. In contrast, quiescent/dormant tumor cells exhibit a low reproduction rate and thereby benefit from increased survival under adverse conditions, for example, cancer therapies. The underlying mechanisms have to be fully elucidated, but one can assume that tumor cells are constantly subject to trade-offs among different cellular stages, for example, differentiated versus undifferentiated, stem cell versus nonstem cell, oxidative versus glycolytic (see aforementioned), sessile versus motile phenotypes. In summary, all these phenotypes can be associated with fast and slow life-history strategies in response to extrinsic factors, for example, given by the microenvironment or therapies (Aktipis et al., 2013). Aktipis et al. (2013) postulated that tumor dormancy can be seen as life-history trade-off between immediate and delayed reproduction.

Living organisms can only exist in changing environments when being enabled to adapt to these fluctuating conditions. One adaptive strategy is the acquisition of different phenotypes in response to variable environmental conditions (see aforementioned). However, when reliable indicators for prediction of environmental changes (also between several generations of a population) are not available, individual organisms are constrained to develop a life-history strategy without fully optimizing the fitness advantage (Childs et al., 2010). Such strategy might be bet-hedging which is a stochastic switching between different phenotypes/stages. The bet-hedging theory postulates that altering selection forces given by unpredictable environments favors the evolution of bet-hedging (Beaumont et al., 2009; de Jong et al., 2011). Bet-hedging has been described in a variety of organisms, including bacteria and humans (Beaumont et al., 2009; de Jong et al., 2011; Venable, 2007). Given the fact that cancer cells are only able to adapt to direct selection forces and cannot anticipate future conditions and dynamics of the ecosystem, bet-hedging might be one strategy underlying the entry in and exit from dormancy.

Overall, life-history theory may help us to understand the plasticity of the proliferative phenotype of tumor cells and how this promotes tumor evolution. Incorporating evolutionary methods into cancer research and asking new evolutionary questions may help to explain the phenomenon of dormancy in tumor evolution leading to improved cancer therapies.

Addressing the following evolutionary questions may contribute to a widened understanding of dormancy from an evolutionary perspective:

1. Why does dormancy exist? Is it induced by extrinsic factors, is it a stochastic event or a consequence of both? Is dormancy a bet-hedging strategy for tumors?
2. Do dormant cells need other (nondormant) tumor cells to maintain the tumor cell population?
3. Can every tumor cell become dormant or is dormancy restricted to specific tumor cells? If so, what are the requirements to enter and exit dormancy? How exactly is entry in and exit from dormancy regulated and controlled?
4. What is the best strategy to control tumor evolution? Maintenance or reversal of dormancy?

Considering that inflammatory processes can promote escape from dormancy (fast life-history), therapeutic strategies using antiinflammatory drugs could promote tumor dormancy by maintaining tissue homeostasis and providing growth suppressive cues. Favoring this concept, nonsteroidal antiinflammatory drugs have been shown to prevent the outgrowth of various tumor entities (Rothwell et al., 2012). Moreover, adjuvant application of antiinflammatory drugs in preclinical tumor models has also been shown to be efficient against metastatic outgrowth (Egberts et al., 2008a,b). Due to the high tumor heterogeneity and pronounced resistance to therapy (e.g., in tumors such PDAC or glioma) further investigations are also needed focusing on the impact of therapeutics on entry in and exit from dormancy to improve cancer treatments. Important to note, such strategies may not totally eliminate the cancer rather aiming at tumor control leading to disease chronification (Jansen et al., 2015).

CONCLUSIONS

Integrating the evolutionary perspective in the investigation of tumor dormancy will essentially widen our understanding, of this fundamental concept. We are convinced that this novel avenue of research will provide the basis to develop more efficient therapeutic strategies, which will also be effective in advanced cancer stages. New therapeutic approaches have to consider the complex and dynamic nature of tumors including a more advanced understanding of dormancy. Considering evolutionary life-history approaches, the challenge will be to elaborate whether "keeping dormant cells dormant" or "awakening dormant tumor cells" or "eradicating dormant tumor cells while they are still dormant" will be the best strategy to survive cancer.

References

Aguirre-Ghiso, J.A., 2007. Models, mechanisms and clinical evidence for cancer dormancy. Nat. Rev. Cancer 7 (11), 834–846.

Aktipis, C.A., Boddy, A.M., Gatenby, R.A., Brown, J.S., Maley, C.C., 2013. Life history trade-offs in cancer evolution. Nat. Rev. Cancer 13 (12), 883–892.

Almog, N., Ma, L., Raychowdhury, R., Schwager, C., Erber, R., Short, S., Hlatky, L., Vajkoczy, P., Huber, P.E., Folkman, J., Abdollahi, A., 2009. Transcriptional switch of dormant tumors to fast-growing angiogenic phenotype. Cancer Res. 69 (3), 836–844.

Barcellos-Hoff, M.H., Lyden, D., Wang, T.C., 2013. The evolution of the cancer niche during multistage carcinogenesis. Nat. Rev. Cancer 13 (7), 511–518.

Barkan, D., El Touny, L.H., Michalowski, A.M., Smith, J.A., Chu, I., Davis, A.S., Webster, J.D., Hoover, S., Simpson, R.M., Gauldie, J., Green, J.E., 2010. Metastatic growth from dormant cells induced by a col-I-enriched fibrotic environment. Cancer Res. 70 (14), 5706–5716.

Beaumont, H.J., Gallie, J., Kost, C., Ferguson, G.C., Rainey, P.B., 2009. Experimental evolution of bet hedging. Nature 462 (7269), 90–93.

Bragado, P., Sosa, M.S., Keely, P., Condeelis, J., Aguirre-Ghiso, J.A., 2012. Microenvironments dictating tumor cell dormancy. Recent Results Cancer Res. 195, 25–39.

Bragado, P., Estrada, Y., Parikh, F., Krause, S., Capobianco, C., Farina, H.G., Schewe, D.M., Aguirre-Ghiso, J.A., 2013. TGF-β2 dictates disseminated tumour cell fate in target organs through TGF-β-RIII and p38α/β signalling. Nat. Cell Biol. 15 (11), 1351–1361.

Braumüller, H., Wieder, T., Brenner, E., Aßmann, S., Hahn, M., Alkhaled, M., Schilbach, K., Essmann, F., Kneilling, M., Griessinger, C., Ranta, F., Ullrich, S., Mocikat, R., Braungart, K., Mehra, T., Fehrenbacher, B., Berdel, J., Niessner, H., Meier, F., van den Broek, M., Häring, H.U., Handgretinger, R., Quintanilla-Martinez, L., Fend, F., Pesic, M., Bauer, J., Zender, L., Schaller, M., Schulze-Osthoff, K., Röcken, M., 2013. T-helper-1-cell cytokines drive cancer into senescence. Nature 494, 361–365.

Celià-Terrassa, T., Kang, Y., 2016. Distinctive properties of metastasis-initiating cells. Genes Dev. 30 (8), 892–908.

Checinska, A., Paszczynski, A., Burbank, M., 2015. Bacillus and other spore-forming genera: variations in responses and mechanisms for survival. Annu. Rev. Food Sci. Technol. 6, 351–369.

Childs, D.Z., Metcalf, C.J., Rees, M., 2010. Evolutionary bet-hedging in the real world: empirical evidence and challenges revealed by plants. Proc. Biol. Sci. 277 (1697), 3055–3064.

Crea, F., Nur Saidy, N.R., Collins, C.C., Wang, Y., 2015. The epigenetic/noncoding origin of tumor dormancy. Trends Mol. Med. 21 (4), 206–211.

de Jong, I.G., Haccou, P., Kuipers, O.P., 2011. Bet hedging or not? A guide to proper classification of microbial survival strategies. Bioessays 33 (3), 215–223.

Egberts, J.H., Cloosters, V., Noack, A., Schniewind, B., Thon, L., Klose, S., Kettler, B., von Forstner, C., Kneitz, C., Tepel, J., Adam, D., Wajant, H., Kalthoff, H., Trauzold, A., 2008a. Anti-tumor necrosis factor therapy inhibits pancreatic tumor growth and metastasis. Cancer Res. 68 (5), 1443–1450.

Egberts, J.H., Schniewind, B., Pätzold, M., Kettler, B., Tepel, J., Kalthoff, H., Trauzold, A., 2008b. Dexamethasone reduces tumor recurrence and metastasis after pancreatic tumor resection in SCID mice. Cancer Biol. Ther. 7 (7), 1044–1050.

Eisenstark, A., Miller, C., Jones, J., Levén, S., 1992. Escherichia coli genes involved in cell survival during dormancy: role of oxidative stress. Biochem. Biophys. Res. Commun. 188 (3), 1054–1059.

Endaya, B.B., Lam, P.Y., Meedeniya, A.C., Neuzil, J., 2016. Transcriptional profiling of dividing tumor cells detects intratumor heterogeneity linked to cell proliferation in a brain tumor model. Mol. Oncol. 10 (1), 126–137.

Folmes, C.D., Terzic, A., 2016. Energy metabolism in the acquisition and maintenance of stemness. Semin. Cell Dev. Biol. 52, 68–75, Review.

Friberg, S., Nyström, A., 2015. Cancer metastases: early dissemination and late recurrences. Cancer Growth Metastasis. 8, 43–49.

Galluzzi, L., Kepp, O., Vander Heiden, M.G., Kroemer G, 2013. Metabolic targets for cancer therapy. Nat. Rev. Drug Discov. 12 (11), 829–846.

Galvao, R.P., Kasina, A., McNeill, R.S., Harbin, J.E., Foreman, O., Verhaak, R.G., Nishiyama, A., Miller, C.R., Zong, H., 2014. Transformation of quiescent adult oligodendrocyte precursor cells into malignant glioma through a multistep reactivation process. Proc. Natl. Acad. Sci. USA 111 (40), E4214–4223.

Ghajar, C.M., 2015. Metastasis prevention by targeting the dormant niche. Nat. Rev. Cancer. 15 (4), 238–247.

Giancotti, F.G., 2013. Mechanisms governing metastatic dormancy and reactivation. Cell 155 (4), 750–764.

Gillies, R.J., Verduzco, D., Gatenby, R.A., 2012. Evolutionary dynamics of carcinogenesis and why targeted therapy does not work. Nat. Rev. Cancer 12 (7), 487–493.

Greaves, M., Maley, C.C., 2012. Clonal evolution in cancer. Nature 481 (7381), 306–313.

Hanahan, D., Weinberg, R.A., 2011. Hallmarks of cancer: the next generation. Cell 144 (5), 646–674.

Higgins, D., Dworkin, J., 2012. Recent progress in Bacillus subtilis sporulation. FEMS Microbiol. Rev. 36 (1), 131–148.

Hofstetter, C.P., Burkhardt, J.K., Shin, B.J., Gürsel, D.B., Mubita, L., Gorrepati, R., Brennan, C., Holland, E.C., Boockvar, J.A., 2012. Protein phosphatase 2A mediates dormancy of glioblastoma multiforme-derived tumor stem-like cells during hypoxia. PLoS One 7 (1), e30059.

Hurst, R.E., Hauser, P.J., Kyker, K.D., Heinlen, J.E., Hodde, J.P., Hiles, M.C., Kosanke, S.D., Dozmorov, M., Ihnat, M.A., 2013. Suppression and activation of the malignant phenotype by extracellular matrix in xenograft models of bladder cancer: a model for tumor cell "dormancy". PLoS One 8 (5), e64181.

Jansen, G., Gatenby, R., Aktipis, C.A., Opinion:, 2015. Control vs. eradication: applying infectious disease treatment strategies to cancer. Proc. Natl. Acad. Sci. USA 112 (4), 937–938.

Lambert, G., Estévez-Salmeron, L., Oh, S., Liao, D., Emerson, B.M., Tlsty, T.D., Austin, R.H., 2011 May. An analogy between the evolution of drug resistance in bacterial communities and malignant tissues. Nat. Rev. Cancer 11 (5), 375–382.

Lefter, L.P., Sunamura, M., Furukawa, T., Takeda, K., Kotobuki, N., Oshimura, M., Matsuno, S., Horii, A., 2003. Inserting chromosome 18 into pancreatic cancer cells switches them to a dormant metastatic phenotype. Clin. Cancer Res. 9 (13), 5044–5552.

Lewis, K., 2010. Persister cells. Annu. Rev. Microbiol. 64, 357–372.

Li, L., Li, W., 2015. Epithelial-mesenchymal transition in human cancer: comprehensive reprogramming of metabolism, epigenetics, and differentiation. Pharmacol. Ther. 150, 33–46, Review.

Lin, W.C., Rajbhandari, N., Liu, C., Sakamoto, K., Zhang, Q., Triplett, A.A., Batra, S.K., Opavsky, R., Felsher, D.W., DiMaio, D.J., Hollingsworth, M.A., Morris, 4th, J.P., Hebrok, M., Witkiewicz, A.K., Brody, J.R., Rui, H., Wagner, K.U., 2013. Dormant cancer cells contribute to residual disease in a model of reversible pancreatic cancer. Cancer Res. 73 (6), 1821–1830.

Magee, J.A., Piskounova, E., Morrison, S.J., 2012. Cancer stem cells: impact, heterogeneity, and uncertainty. Cancer Cell 21 (3), 283–296.

Magnus, N., Gerges, N., Jabado, N., Rak, J., 2013. Coagulation-related gene expression profile in glioblastoma is defined by molecular disease subtype. J. Thromb. Haemost. 11 (6), 1197–1200.

Magnus, N., D'Asti, E., Meehan, B., Garnier, D., Rak, J., 2014a. Oncogenes and the coagulation system—forces that modulate dormant and aggressive states in cancer. Thromb. Res. 133 (Suppl 2), S1–S9.

Magnus, N., Garnier, D., Meehan, B., McGraw, S., Lee, T.H., Caron, M., Bourque, G., Milsom, C., Jabado, N., Trasler, J., Pawlinski, R., Mackman, N., Rak, J., 2014 Mar 4b. Tissue factor expression provokes escape from tumor dormancy and leads to genomic alterations. Proc. Natl. Acad. Sci. USA 111 (9), 3544–3549.

Maisonneuve, E., Gerdes, K., 2014. Molecular mechanisms underlying bacterial persisters. Cell 157 (3), 539–548.

Marjanovic, N.D., Weinberg, R.A., Chaffer, C.L., 2013. Cell plasticity and heterogeneity in cancer. Clin Chem. 59 (1), 168–179.

Marsden, C.G., Wright, M.J., Carrier, L., Moroz, K., Rowan, B.G., 2012. Disseminated breast cancer cells acquire a highly malignant and aggressive metastatic phenotype during metastatic latency in the bone. PLoS One 7 (11), e47587.

Menendez, J.A., Corominas-Faja, B., Cuyàs, E., García, M.G., Fernández-Arroyo, S., Fernández, A.F., Joven, J., Fraga, M.F., Alarcón, T., 2016. Oncometabolic Nuclear Reprogramming of Cancer Stemness. Stem Cell Rep. 6 (3), 273–283.

Merlo, L.M., Pepper, J.W., Reid, B.J., Maley, C.C., 2006 Dec. Cancer as an evolutionary and ecological process. Nat Rev Cancer. 6 (12), 924–935.

Moir, J.A., Mann, J., White, S.A., 2015. The role of pancreatic stellate cells in pancreatic cancer. Surg. Oncol. 24 (3), 232–238.

Müller-Hermelink, N., Braumüller, H., Pichler, B., Wieder, T., Mailhammer, R., Schaak, K., Ghoreschi, K., Yazdi, A., Haubner, R., Sander, C.A., Mocikat, R., Schwaiger, M., Förster, I., Huss, R., Weber, W.A., Kneilling, M., Röcken, M., 2008. TNFR1 signaling and IFN-gamma signaling determine whether T cells induce tumor dormancy or promote multistage carcinogenesis. Cancer Cell 13, 507–518.

Naumov, G.N., Bender, E., Zurakowski, D., Kang, S.Y., Sampson, D., Flynn, E., Watnick, R.S., Straume, O., Akslen, L.A., Folkman, J., Almog, N., 2006. A model of human tumor dormancy: an angiogenic switch from the nonangiogenic phenotype. J. Natl. Cancer Inst. 98 (5), 316–325.

O'Connor, M.L., Xiang, D., Shigdar, S., Macdonald, J., Li, Y., Wang, T., Pu, C., Wang, Z., Qiao, L., Duan, W., 2014. Cancer stem cells: a contentious hypothesis now moving forward. Cancer Lett. 344 (2), 180–187.

Retsky, M.W., Demicheli, R., Hrushesky, W.J., Baum, M., Gukas, I.D., 2008. Dormancy and surgery-driven escape from dormancy help explain some clinical features of breast cancer. APMIS 116 (7–8), 730–741.

Rhim, A.D., Mirek, E.T., Aiello, N.M., Maitra, A., Bailey, J.M., McAllister, F., Reichert, M., Beatty, G.L., Rustgi, A.K., Vonderheide, R.H., Leach, S.D., Stanger, B.Z., 2012. EMT and dissemination precede pancreatic tumor formation. Cell 148 (1–2), 349–361.

Roff, D.A., 1992. The evolution of life histories. Chapman & Hall, New York (NY).

Romero, I., Garrido, F., Garcia-Lora, A.M., 2014. Metastases in immune-mediated dormancy: a new opportunity for targeting cancer. Cancer Res. 74, 6750–6757.

Romero, I.L., Mukherjee, A., Kenny, H.A., Litchfield, L.M., Lengyel, E., 2015. Molecular pathways: trafficking of metabolic resources in the tumor microenvironment. Clin. Cancer Res. 21 (4), 680–686, Review.

Rothwell, P.M., Wilson, M., Price, J.F., Belch, J.F., Meade, T.W., Mehta, Z., 2012. Effect of daily aspirin on risk of cancer metastasis: a study of incident cancers during randomised controlled trials. Lancet 379 (9826), 1591–1601.

Satchi-Fainaro, R., Ferber, S., Segal, E., Ma, L., Dixit, N., Ijaz, A., Hlatky, L., Abdollahi, A., Almog, N., 2012. Prospective identification of glioblastoma cells generating dormant tumors. PLoS One 7 (9), e44395.

Sneath, P.H., 1975. Some considerations of the theoretical limits for living organisms. Life Sci. Space Res. 13, 75–82.

Sosa, M.S., Avivar-Valderas, A., Bragado, P., Wen, H.C., Aguirre-Ghiso, J.A., 2011. ERK1/2 and p38α/β signaling in tumor cell quiescence: opportunities to control dormant residual disease. Clin. Cancer Res. 17 (18), 5850–5857.

Sosa, M.S., Bragado, P., Aguirre-Ghiso, J.A., 2014. Mechanisms of disseminated cancer cell dormancy: an awakening field. Nat. Rev. Cancer 14 (9), 611–622.

Stearns, S.C., 1992. The Evolution of Life Histories. Oxford University Press, Oxford, UK.

Tanaka, Y., Nobusawa, S., Ikota, H., Yokoo, H., Hirato, J., Ito, H., Saito, T., Ogura, H., Nakazato, Y., 2014. Leukemia-like onset of bone marrow metastasis from anaplastic oligodendroglioma after 17 years of dormancy: an autopsy case report. Brain Tumor Pathol. 31 (2), 131–136.

Testa, U., Labbaye, C., Castelli, G., Pelosi, E., 2016. Oxidative stress and hypoxia in normal and leukemic stem cells. Exp. Hematol. 44 (7), 540–560, Review.

Trumpp, A., Essers, M., Wilson, A., 2010. Awakening dormant haematopoietic stem cells. Nat. Rev. Immunol. 10 (3), 201–209.

Uhr, J.W., Pantel, K., 2011. Controversies in clinical cancer dormancy. Proc. Natl. Acad. Sci. USA 108 (30), 12396–12400.

Vander Heiden, M.G., Cantley, L.C., Thompson, C.B., 2009. Understanding the Warburg effect: the metabolic requirements of cell proliferation. Science 324 (5930), 1029–1033.

Venable, D.L., 2007. Bet hedging in a guild of desert annuals. Ecology 88 (5), 1086–1090.

Viale, A., Corti, D., Draetta, G.F., 2015. Tumors and mitochondrial respiration: a neglected connection. Cancer Res. 75 (18), 3685–3686, Review.

Walsh, M.R., 2013. The link between environmental variation and evolutionary shifts in dormancy in zooplankton. Integr. Comp. Biol. 53 (4), 713–722.

Wang, H., Yu, C., Gao, X., Welte, T., Muscarella, A.M., Tian, L., Zhao, H., Zhao, Z., Du, S., Tao, J., Lee, B., Westbrook, T.F., Wong, S.T., Jin, X., Rosen, J.M., Osborne, C.K., Zhang, X.H., 2015. The osteogenic niche promotes early-stage bone colonization of disseminated breast cancer cells. Cancer Cell 27 (2), 193–210.

Wieder, T., Braumüller, H., Brenner, E., Zender, L., Röcken, M., 2013. Changing T-cell enigma: cancer killing or cancer control? Cell Cycle 12, 3146–3153.

Wikman, H., Vessella, R., Pantel, K., 2008. Cancer micrometastasis and tumour dormancy. APMIS 116 (7–8), 754–770.

Willis, R.A., 1934. The spread of tumours in the human body. J.&A. Churchill, London.

Xu, X.D., Shao, S.X., Jiang, H.P., Cao, Y.W., Wang, Y.H., Yang, X.C., Wang, Y.L., Wang, X.S., Niu, H.T., 2015. Warburg effect or reverse Warburg effect? A review of cancer metabolism. Oncol. Res. Treat. 38 (3), 117–122.

Yeh, A.C., Ramaswamy, S., 2015. Mechanisms of cancer cell dormancy—another hallmark of cancer? Cancer Res. 75 (23), 5014–5022.

21

Controlling Rogue Cells in Cancer and Bacterial Infections

Wollein Waldetoft*, John F. McDonald*,**,†, Sam P. Brown*

*School of Biological Sciences, Georgia Institute of Technology, Atlanta, GA, United States
**Integrated Cancer Research Center, Georgia Institute of Technology, Atlanta, GA, United States
†Parker H. Petit Institute for Bioengineering and Bioscience, Georgia Institute of Technology, Atlanta, GA, United States

Human health is dependent on an incredible coordination and cooperation among trillions of cells that form our bodies. When human cells break these cooperative rules, we see the emergence of expanding and dangerous cancers. Similarly, when bacterial cells proliferate inappropriately (whether from our microbiome or an external source), we see disease. In this chapter, we explore the broad parallels between cancer and infectious disease control and resistance evolution—and point to potentially productive avenues for the exchange of research and management principles (see also the Introductory Chapter).

Humans are a marvel of multicellularity—around 25–30 trillion eukaryotic cells work together to collectively accomplish the tasks of each and every adult human (Bianconi et al., 2013)—significantly aided by an approximately equal number of bacterial cells that comprise a typical human microbiome (Sender et al., 2016). While there are bacteria that need to cause disease in order to transmit (such as *Mycobacterium tuberculosis*), many infections—from uncomplicated urinary tract infections to sepsis—are due to bacteria that normally reside in or on our bodies as part of the normal commensal microbiota (Lood et al., 2015). These include species typically perceived of as virulent, such as *Streptococcus pneumoniae* and *Staphylococcus aureus*, but also less aggressive species like *Escherichia coli* and *Staphylococcus saprofyticus*.

Human beings generally function for decades as conglomerates of human and microbial cells, but in many cases we meet our end due to an uprising of "rogue" cells that undo the collective interests in the pursuit of their selfish cellular reproduction. If these rogue cells are derived from our own body, we call the consequences cancer; if they are microbial, we call the consequences infectious disease.

There is thus a common theme in cancer and many infections, that cells that were part of a healthy cooperative system revert to short sighted exploitation to the detriment of the patient. Evolution plays an important role in such mutiny. The formation of cancer is an evolutionary process (Merlo et al., 2006), but also in infectious diseases rogue cells may arise by evolution within the host, as can be the case in, for example, invasive group A streptococcal infections (Walker et al., 2007). Evolution is also key to understanding the failure of treatment, as in both cancer and infection culprit cells evolve drug resistance (see also the Introductory Chapter).

Faced with these rogues and their propensity to evolve resistance, we need to develop novel therapeutics and treatment strategies, and in so doing we must address two questions that are shared between infectious diseases and cancer:

1. How can we target pathogenic cells while sparing the rest of the body or microbiome?
2. How can we prevent or delay the evolution of resistance to drugs?

KILLING THE ROGUES—SPARING THE REST

A central problem in cancer chemotherapy is this: how can we kill the cancer without killing the remaining cells and thus the patient? Since cancer cells form a tumor by outgrowing surrounding healthy cells, the

Ecology and Evolution of Cancer. http://dx.doi.org/10.1016/B978-0-12-804310-3.00021-1

traditional solution has been cytotoxics that preferentially damage fast growing cells. This way cancer cells receive a harder blow than most healthy cells, but since the difference between pathologic and healthy cells is only a matter of degree, not kind, the side effects can be very harsh.

By contrast, the treatment of infections took a different route. Here the strategy was to develop drugs against molecular targets that are present in microbes but absent in human cells. This resulted in drugs that spare host cells, but may have indirect side effects by killing or inhibiting members of the commensal and mutualistic microbiota. However, since the growth state of bacteria affects their sensitivity to some antibiotics (Tuomanen, 1986), it is at least conceivable that the effects of antibiotic treatment are biased towardpathogenic bacteria that are outgrowing commensal bystanders.

From these traditional positions both fields have developed in such a way that their tracks have crossed, and there is potential for increased collaboration. In the cancer field there has recently been a strong interest in targeted therapies. One idea has been to target specific genes or molecular mechanisms identified as drivers of cancer (Stegmeier et al., 2010). The archetypical example is Imatinib that inhibits a class of receptor tyrosine kinases that are overexpressed in a subset of chronic myeloid leukemias and some other cancers, and is now in clinical use.

Another approach, which is currently being explored for both cancer and infection, is to exploit structures characteristic of target cells to deliver the drug specifically to these. Cancer cells often have an altered surface proteome as compared to normal cells (Larkin and Aukim-Hastie, 2011), and this can be exploited for the development of vectors that deliver effector molecules more or less specifically to cancer cells. Work along these lines is in progress, and includes nanoparticles coated with peptides that specifically bind to target cells (Hutchinson et al., 2008).

In the field of infection, phage lysins are promising candidates. These proteins often consist of two domains. The binding domain confers much of the specificity for the target bacterium, while the catalytic domain hydrolyses the peptidoglycan, thus killing the cell. Through genetic engineering, domains can be combined in different ways to develop highly specific drugs against a range of pathogens (Pastagia et al., 2013).

PREVENTING RESISTANCE TO TREATMENT

While the principles of evolution by natural selection are general, and apply to both microbes and cancers, there are important evolutionary differences between these two contexts. Infectious microbes can transmit from host to host, while cancer cells generally do not (but see Chapter 12). As a consequence, resistant microbes can spread from one host to another, and accumulate in the host population. This means that antibiotics given to patients today affect the prevalence of resistance, and thus the proportion of patients that will be infected by resistant microbes in the future. With cancer, in contrast, every patient starts anew with a resistance profile that is unaffected by the treatment of other patients. This may seem trivial, but it has important implications for policy. In infectious diseases the evolution of resistance creates a conflict between the interests of current and future patients in that aggressive treatment of current patients worsens the prospects for effective treatment of future patients (Foster and Grundmann, 2006), while in cancer management this is not the case.

Antiresistance strategies generally fall into one of the two classes. Either the strategy aims to prevent resistance from arising or it attempts to limit the expansion and spread of the resistant clone after it has arisen. A classic example of the former is to "hit hard" with aggressive chemotherapeutic strategies in order to rapidly reduce the size of the population of the pathogen or cancer. The small population size means that there are few cells in which resistance mutations can arise. This strategy can be enhanced and complemented with the use of drug combinations. Combination therapy can be very effective in limiting population size, especially if the drugs are synergistic, and can thus reduce the rate of resistance emergence. The simultaneous use of several drugs also lowers the per cell probability of resistance to the drug combination, at least if there are no single mutations that confer cross resistance (Bozic et al., 2013). Combination therapy is used extensively, and successfully, in the treatment of infections, such as HIV and tuberculosis, as well as in cancer care. However, the "hit hard" approach has been questioned for both infection (Read et al., 2011) and cancer (Gatenby et al., 2009) on the grounds that, while limiting the supply of novel mutations, it induces strong competitive release of any resistant mutants that do exist—as drugs act to clear the field of any susceptible cells that would otherwise competitively suppress resistant lineages (Day and Read, 2016) (see also Chapters 10 and 14).

This discussion of competitive release brings us to the second resistance management strategy, which is to limit selection for resistance once it has arisen, an approach that has received more attention among evolutionary biologists. The relevance for infection is obvious, since resistant, including multiresistant, microbes circulate in both hospitals and the community, and are thus often present at the onset of treatment. For cancer the relevance may seem less clear, since resistance arises de novo in each individual patient. However, many cancers are genetically heterogenous at the time of diagnosis, and resistant mutants are present already at the onset of treatment (Bozic et al., 2013). If resistant variants are present within a patient, how can their spread be slowed during effective treatment? The value of combination therapies now flips from the resistance emergence scenario, where synergistic combinations are favored due to their greater efficacy. In a resistance management scenario, the use of synergistically interacting combination therapies can *increase* the spread of a strain that is resistant to one or both drugs—as for instance a strain resistant to drug A will escape the effects of drug A and the synergistically enhanced effect of drug B. In contrast, combination therapies with drug interference can slow the spread of a resistant strain as now resistance to drug A leads to *enhancement* of the effects of drug B (Chait et al., 2007; Yeh et al., 2009). For a recent review of combination therapy strategies to reduce resistance evolution, (Baym et al., 2015).

A relatively new approach to reduce selection for resistance in bacteria is to target the mechanisms of pathology—referred to as virulence factors—using so called antivirulence drugs. Virulence factors present a new class of discriminatory targets that differ from the essential functions targeted by classic antibiotics in that they need not be as strongly connected to the pathogen's fitness (Allen et al., 2014; Clatworthy et al., 2007). A weak, or even negative, association with pathogen fitness can result if the virulence factor is maintained by selection in a setting distinct from that where it causes harm, as in environmental bacteria that only incidentally infect human beings. A strong connection, on the other hand, is expected if, for example, the factor is maintained because it protects the bacterium from immune killing. How strong the connection is, therefore has to be determined for each individual factor. In addition to potentially decreasing the rate of resistance evolution, antivirulence drugs should suffer less from problems with specificity, since the drug targets should be absent in avirulent commensal bacteria.

Some cancers produce hormones that negatively affect the patient, but are unlikely to be strongly connected to cancer cell fitness. An example would be the production of noradrenalin and adrenalin by phaeochromocytomas (endocrine tumors originating from the cells that form the adrenal medulla). Apart from this, however, we are unaware of any suitable candidates for antivirulence drugs in cancer treatment.

Many virulence factors are collective or social traits, in that they or their effects are shared among bacterial cells, making them candidates for the exploitation of social dynamics (Allen et al., 2014). An archetypical example is the production of iron scavanging siderophore molecules by several bacterial pathogens. The cost of siderophore production is felt by individual producer cells, while the benefits are felt by any cell expressing the appropriate siderophore receptor, whether it paid for the siderophore or not. If a drug inhibits the production or activity of such a factor, the benefit of resistance (continued access to iron) would thus be shared, while the cost (production of the siderophores) would be paid by the individual cell. This means that a resistant mutant that arises in a population of sensitive cells will be selected against. There have therefore been attempts to take advantage of collective virulence traits to prevent the evolution of resistance (Allen et al., 2014), and experimental evolution tests have demonstrated significantly greater evolutionary robustness than conventional antibiotics at comparable levels of pathogen control (Ross-Gillespie et al., 2014). There are, however, several problems with this strategy. One is that the inhibition of social bacterial traits may be less efficient in treating bacterial infections than are traditional antibiotics. Another is that the strategy only attenuates resistance evolution in mixed infections when the cooperative resistant mutant is surrounded by "cheating" sensitive cells. This is the case when the mutation first appears, but once a clone of resistant cells has managed to transmit to new individuals, infections can consist purely of resistant cells. However, in cancer treatment the latter problem may be less severe, because, although resistant cells may metastasize and form daughter tumors at other anatomical sites, resistance has to arise de novo in each individual patient.

The growth of a tumor is dependent on a conducive microenvironment, and cancers often recondition their environment by signaling to surrounding cells (see also Chapter 8). The benefit (for the cancer) of this remodeling is shared by cancer cells in close vicinity, but the cost of producing the signals is paid by the individual cells, creating a social dilemma like that described previously for siderophores. Whether this can be exploited for preventing resistance is, however, an empirical question, and depends on, for example, the cost of signaling and the heterogeneity of the cancer. The fact that cancers do evolve signaling suggests that they can in time overcome this social dilemma, but it may nevertheless provide an opportunity for slowing the rate of resistance evolution.

DISCUSSION

Here we have discussed two problems that are common to the treatment of infectious diseases and cancer, namely targeting the correct cells, and avoiding the evolution of resistance. We have also outlined some themes in current research where we think there is potential for these fields to learn from or be inspired by each other.

It seems that, while resistance to treatment has been studied in both medicine and evolutionary biology, the (bio)medical community has focused on the emergence of resistant clones and how this may be prevented, while evolutionary biologists have been more interested in how selection for resistance can be ameliorated once the mechanism has emerged. Ideally, of course, these two strategies should be combined. This will, however, probably require developments in diagnostics. Cancers would need to be detected early enough that resistance mutations have not yet arisen, and in both cancer and infection the emergence of resistance should be rapidly detected, so that clinicians can switch from a strategy optimized for the prevention of emergence to one optimized for the minimization of selection.

We hope that there will be increasing interactions between the fields of infection and oncology, as the problems they have in common suggest there is scope for fruitful exchange. As academic researchers, we need to read each other's work, and attend each other's conferences, but it is our impression that such exchange is relatively sparse. In the clinic interactions may be better developed. These lines are written looking out over a hospital, where infection and oncology wards share a building, and infection consultants work in the cancer clinic on a regular basis. Such cooperation may be beneficial in the academy as well.

Acknowledgments

We thank Rolf Lood for helpful discussion and comments on an earlier draft.

References

Allen, R.C., Popat, R., Diggle, S.P., Brown, S.P., 2014. Targeting virulence: can we make evolution-proof drugs? Nat. Rev. Microbiol. 12, 300–308.

Baym, M., Stone, L.K., Kishony, R., 2015. Multidrug evolutionary strategies to reverse antibiotic resistance. Science 351, aad3292.

Bianconi, E., Piovesan, A., Facchin, F., Beraudi, A., Casadei, R., Frabetti, F., Vitale, L., Pelleri, M.C., Tassani, S., Piva, F., et al., 2013. An estimation of the number of cells in the human body. Ann. Hum. Biol. 40, 463–471.

Bozic, I., Reiter, J.G., Allen, B., Antal, T., Chatterjee, K., Shah, P., Moon, Y.S., Yaqubie, A., Kelly, N., Le, D.T., et al., 2013. Evolutionary dynamics of cancer in response to targeted combination therapy. Elife 2013, 1–15.

Chait, R., Craney, A., Kishony, R., 2007. Antibiotic interactions that select against resistance. Nature 446, 668–671.

Clatworthy, A.E., Pierson, E., Hung, D.T., 2007. Targeting virulence: a new paradigm for antimicrobial therapy. Nat. Chem. Biol. 3, 541–548.

Day, T., Read, A.F., 2016. Does high-dose antimicrobial chemotherapy prevent the evolution of resistance? PLoS Comput. Biol. 12, 1–20.

Foster, K.R., Grundmann, H., 2006. Do we need to put society first? The potential for tragedy in antimicrobial resistance. PLoS Med. 3, 0177–0180.

Gatenby, R.A., Silva, A.S., Gillies, R.J., Frieden, B.R., 2009. Adaptive therapy. Cancer Res. 69, 4894–4903.

Hutchinson, D., Ho, V., Dodd, M., Dawson, H.N., Zumwalt, A.C., Colton, C.A., 2008. NIH Public Access 148, 825–832.

Larkin, S., Aukim-Hastie, C., 2011. In: Cree, I.A. (Ed.), Proteomic Evaluation of Cancer Cells: Identification of Cell Surface Proteins. In Cancer Cell Culture: Methods and Protocols. Humana Press, New York, pp. 395–405.

Lood, R., Wollein Waldetoft, K., Nordenfelt, P., 2015. Localization-triggered bacterial pathogenesis. Future Microbiol. 10, 1659–1668.

Merlo, L.M.F., Pepper, J.W., Reid, B.J., Maley, C.C., 2006. Cancer as an evolutionary and ecological process. Nat. Rev. Cancer 6, 924–935.

Pastagia, M., Schuch, R., Fischetti, V.A., Huang, D.B., 2013. Lysins: the arrival of pathogen-directed anti-infectives. J. Med. Microbiol. 62, 1506–1516.

Read, A.F., Day, T., Huijben, S., 2011. The evolution of drug resistance and the curious orthodoxy of aggressive chemotherapy. Proc. Natl. Acad. Sci. 108, 10871–10877.

Ross-Gillespie, A., Weigert, M., Brown, S.P., Kümmerli, R., 2014. Gallium-mediated siderophore quenching as an evolutionarily robust antibacterial treatment. Evol. Med. Public Heal. 2014, 18–29.

Sender, R., Fuchs, S., Milo, R., 2016. Revised estimates for the number of human and bacteria cells in the body. bioRxiv, 1–21.

Stegmeier, F., Warmuth, M., Sellers, W.R., Dorsch, M., 2010. Targeted cancer therapies in the twenty-first century: lessons from imatinib. Clin. Pharmacol. Ther. 87, 543–552.

Tuomanen, E., 1986. Phenotypic tolerance: the search for β-lactam antibiotics that kill nongrowing bacteria. Rev. Infect. Dis. 8, S279–S291.

Walker, M.J., Hollands, A., Sanderson-Smith, M.L., Cole, J.N., Kirk, J.K., Henningham, A., McArthur, J.D., Dinkla, K., Aziz, R.K., Kansal, R.G., et al., 2007. DNase Sda1 provides selection pressure for a switch to invasive group A streptococcal infection. Nat. Med. 13, 981–985.

Yeh, P.J., Hegreness, M.J., Presser Aiden, A., Kishony, R., 2009. Drug interactions and the evolution of antibiotic resistance. Nat. Rev. Microbiol. 7, 460–466.

22

Searching for a Cancer-Proof Organism: It's the Journey That Teaches You About the Destination

Hanna Kokko, Susanne Schindler, Kathleen Sprouffske
Department of Evolutionary Biology and Environmental Studies, University of Zurich, Winterthurerstrasse, Zurich, Switzerland

To our knowledge, Richard Dawkins never worked on cancer. In his autobiography (Dawkins, 2013), however, he reflects how his field of animal behavior substantially advanced when researchers begun to think about the differences between species. In Dawkins' early career, two different approaches coexisted: psychologists studied rat or pigeon behavior as simple models for human behavior, while what would later be called behavioral ecology (Birkhead and Monaghan, 2009) studied the differences in ecology to explain interspecific variation. Behavioral ecology used the insight that we can better validate our ideas about causalities behind a behavior if we can explain interspecific differences, than if we forever focus on just one particular species' habits, no matter how much detail we gather about its intricacies.

Cancer research might very well be at a stage where it could advance significantly by conceptually reframing it, as was done in behavioral ecology. In this way, we could gain a deeper understanding of cancer by using a comparative approach and looking across species. We can use evolutionary principles to understand cancer at the level of cells or at the level of organisms (see also the Introductory Chapter). Here, we focus on the evolutionary principles at the organismal level to understand cancer both before and after an individual has developed it. The "before" question involves cancer defenses and their evolution; the "after" question explores the effect that an individual's cancer has on its fitness prospects. For instance, what would be the optimal allocation of resources for the remaining lifetime of the organism? We also discuss another aspect on the macroevolutionary timescale, namely, why and how long has cancer been plaguing life?

INSIGHTS FROM LIFE HISTORY THEORY

One useful approach to studying cancer is through life history theory, keeping in mind the different "ecologies" inherent to different species (see also Chapters 7 and 13). Life history theory seeks to understand the scheduling of events in individuals' lives, such as the number of births, the size at which offspring are born (or laid, if they are eggs), the schedule of death, and, in a neighboring field, the patterns of ontogenetic growth. This approach is useful to address questions like: should an organism increase its reproduction or growth rate in an attempt to outcompete other individuals in the population—even if these very increases lead to increased cell division, and thus an elevated risk of cancer?

The answer from life history theory is that "it depends." Life is fundamentally not about striving to perfection when it comes to building an organism adapted to its environment. It is about finding good ways to create gene copies when there are trade-offs, and this can include "accepting" that the death rates of individuals increase.

Countless studies have placed more eggs in birds' nests and documented what happens to the parents who are experiencing a "simulation" of higher reproductive effort in 1 year. Even if the parent did not actually produce the egg, the nestling has to be fed, and the parent's own survival can drop as a consequence (Daan et al., 1996). However, what if the parent also had to produce the egg (that in an experimental setting was just given to it)? It is quite difficult to reproduce without cells dividing, and more generally, achieving reproductive

Ecology and Evolution of Cancer. http://dx.doi.org/10.1016/B978-0-12-804310-3.00022-3

success requires metabolic activity (in itself potentially a cancer risk, Dang, 2015), as well as actual divisions of cells. This is true for divisions both in the germline (egg production for females, or for males, building and maintaining large testes in an effort to outcompete rival males in sperm production; Lewis et al., 2008) and in the soma. Somatic forms of cellular activity are also ultimately selected to improve reproductive success. A cancer perspective on this problem is rarely taken (but see Chapters 7 and 13), but could be exciting, as plenty of testable hypotheses could be generated if there were sufficient data. To mention just one example: are males of species that respond to the risk of sperm competition by evolving larger testes more cancer-prone?

BODY SIZE AND CANCER RISK

Life history theory can also be useful when asking questions like: should an organism grow fast in an attempt to outcompete conspecifics? Even if speeding up growth means that, if we allow ourselves a vaguely expressed moment, something about cell division is done in a "sloppy" way, perhaps leading to mutation accumulation in the soma (and the concomitant increase of risk of cancer)? Would natural selection penalize such sloppiness? In other words, should an organism grow fast, or could it be selected to reduce its growth rate to decrease its cancer risk?

Without accounting for cancer, it is easy to visualize, and relatively easy to collect data on, why being bigger and faster can be better: a male mammal might have little success unless he grows large and maintains a high body size, as these traits help to overcome the mating resistance of females and/or to oust rival males from a harem holding position. Two clear cancer prevention strategies in organisms are either to stay small, or to live at a slower pace with lower metabolism. Most organisms go for one but not the other: metabolic rates tend to scale with body size with an exponent of between 2/3 and 3/4 (Kleiber, 1932; Kolokotrones et al., 2010), meaning that large animals have a lower metabolic rate than small animals.

Until recently, it has not often been appreciated that cancer may itself be a selective force in life history evolution. However, if California sea lions suffer from urogenital carcinomas (Browning et al., 2015), sharks have neoplastic lesions (Huveneers et al., 2016) and even the famously "cancer-free" naked mole rats (Chapters 2 and 7) can have tumors after all (Delaney et al., 2016), when can we state that these are common enough causes of death (or, more generally, a reduction of reproductive success) that natural selection works on reducing them? And can we find theoretical support for a stronger claim: that the outcome is expected to reach an equilibrium

so that the incidence of cancer no longer increases with body size (Nunney, 2013), despite the a priori expectation that larger organisms, based on an immensely larger number of cells, should display more cancer, as stated in the famous Peto's paradox (Caulin and Maley, 2011; Nunney, 1999; Peto, 2015; Chapters 1, 7 and 13).

There are benefits and costs to an organism growing larger, payable at different times during its life. An organism that is actively improving its reproductive effort may reap benefits that are either immediate (reproduction) or delayed (e.g., growth to be bigger and more fecund in the future), and similar temporal complications apply for costs, such as reduced survival. Some activities (e.g., lethal fights) are immediately dangerous; others create costs that only accumulate over time. The costs of cancer to the organism of course depend on the details (for tissue-specific studies or calls for more of them, see e.g., Ducasse et al., 2015; Noble et al., 2015; Nunney and Muir, 2015), but one can still state that *all else being equal*, doubling the number of cell divisions doubles the risk that something goes wrong.

The benefits of growing larger pose an intellectual challenge, because benefits of investing into growth play out at different timescales than the costs of cancer risk, given that time alive increases the probability of death having occurred—by other means than cancer as well. Cancer as a penalty is typically not an immediate one as extant organisms have long evolutionary experience with it: this has led to cancer defenses already being in place (see Chapters 7 and 13). If the same cell lineage takes multiple hits related to tumorigenesis, organismic integrity is eventually predicted to suffer, but for natural selection, competing sources of mortality may remain more important than any sluggishly increasing risk of cancer ending one's reproductively active life. If the organism is small and extrinsic mortality is high, cancer is predicted to only rarely be a relevant source of selection (i.e., rarely leads to selective deaths) (Kokko and Hochberg, 2015). We should also keep in mind that natural selection does not work efficiently to reduce the incidence of rare problems, which predicts that evolved cancer defenses make it easier for cancer-inducing alleles to remain in a population (Arnal et al., 2016). We can also predict that small organisms are only likely to be "interesting" in how they deal with cancer if they are unusually long-lived for their body size because they otherwise die too early for cancer to be a serious threat (hence the interest in bats, Caulin et al., 2015).

There have been critiques of Peto's paradox that, in our view, basically amount to a reminder that "details matter and therefore it is naïve to expect that lifetime cancer risk is exactly equal across species" (e.g., Ducasse et al., 2015). While definitely true (indeed, one of the authors of this chapter has published theoretical expectations where cancer risk can vary from very low to high at evolutionary equilibrium, Kokko and Hochberg, 2015), the lack of an association between body size and cancer nevertheless remains surprising because the predicted increase in cancer risk with size and lifespan is not mild—it is astonishingly steep, all else being equal (Peto, 2015). Therefore, a useful way to express Peto's paradox is that it should be nigh impossible to observe a large, well-differentiated metazoan that also lives long, unless its evolution features a concomitant increase in cancer suppression mechanisms.

In light of our restatement of Peto's paradox, there are two possible paths that an organism, selected to become larger for whatever reason, could take. One is that cancer risk indeed prevents evolution toward larger size; we do not know of studies that would evaluate whether this is an important factor in explaining "what keeps organisms small" (sensu Blanckenhorn, 2000). Another route an organism can take, of course, is to evolve stronger cancer defenses that accompany the evolution of larger size. But how easy is this?

EVOLVED CANCER DEFENSES

Actual protection mechanisms seem to be diverse. Firstly, if large size has the corollary of a slower pace of life, the risk of cancer is already reduced (Kokko and Hochberg, 2015; Maciak and Michalak, 2015; for clock-like mutation processes in human soma see Alexandrov et al., 2015). Interspecific studies show that lower metabolic rate correlates to less DNA-damage causing radicals (see Chapter 7), and tests of whether "slowing down the clock" indeed leads to less cancer would be welcome. The severity of the effects described by Peto, however, requires more adaptations than just slowing the pace of life. For instance, increased number of tumor suppressors may be needed for larger or long-living organisms. Indeed, elephants have gained recent fame for having 20 copies of TP53 (humans have only one), horses seem to have gone for amplification of MAL, and microbats—which are not especially large, but live long for their body size—have amplified FBXO31 (Caulin et al., 2015).

The cellular level of somatic divisions, in particular cellular senescence (arrest of cell division) and programmed cell death, is linked to senescence at the organismal level (increase of mortality rates with age, Smith and Daniel, 2012; see also Chapters 1 and 7). This forms an exciting bridge between evolutionary theories for senescence and those for cancer. One could in principle envisage two solutions to error-prone cell divisions: efficient DNA proofreading and repair might drive the error rate down; or errors might be allowed to happen, but a series of mechanisms prevent further problems: faulty cells may undergo programmed cell death (Vogelstein et al., 2000), faulty cells may be killed (Kang et al., 2012), or mutated proteins within faulty cells may be repaired and regain function by, for example, increasing chaperone expression (Murphy, 2013; Whitesell and Lindquist, 2005) (Chapters 1 and 9 further discuss the role of immune system controlling tumors). There is too little interspecific data to comment on the relative emphasis of each type of solution across taxa that differ in body size and lifespan, though the error rate is thought to run into a lower limit (Lynch, 2011) beyond which handling errors becomes necessary.

There is clear potential for dialogue across the fields of somatic and organismal evolution, as in the case of telomeres and telomerase. The tight connection between telomere length and replicative senescence, which can make a precancerous cell lineage grind to a halt, are attracting attention in the context of cancer (Rodriguez-Brenes et al., 2015). Meanwhile, ecologists are measuring age-dependent changes in telomere length and linking this to life history in a growing number of species (Monaghan, 2014), which could also be used to study the interaction of telomere length and cancer rates across species. Telomerase seems to be upregulated in long-lived bird species relative to short-lived ones (Gomes et al., 2010), and telomeres show a slower age-dependent decline in long-lived birds and mammals than in short-lived ones (Haussmann et al., 2003; see also Chapter 1 for a description of work on rodents). Disentangling the role of telomerase in life histories requires understanding its potential to be a double-edged sword in long-lived species: the inherent upregulation of telomerase in these species may also allow cancers to escape the controls of replicative senescence (Rodriguez-Brenes et al., 2015). If cellular senescence or apoptosis would selectively act on precancerous cells expressing telomerase, this would circumvent the problems inherent with upregulated telomerase. In this way, dangerous cellular developments could be weeded out and prolong the life of the individual.

Tantalizingly, hints that increased apoptosis may be correlated with lower cancer incidence can be found in species of different sizes. Apoptosis occurs at a much higher rate for experimentally irradiated elephant cells than for human cells (Abegglen et al., 2015), and it would be fascinating to know more about the coevolutionary dynamics between cancer resistance and size in elephant lineages. On the other end of the size spectrum, mice live significantly longer with delayed tumorigenesis rates when apoptosis rates are artificially elevated using regular transgene inductions (Baker et al., 2016; see also Chapter 7 discussing an extreme level of induced cell death in naked molerats). The elimination of cells expressing a biomarker for senescence in Baker et al.'s study also delayed dysfunction at the level of organs. Thus, as we've outlined in this section, it is possible to understand the evolution of organismal senescence as a consequence of a cellular cancer defense.

A MACROEVOLUTIONARY PERSPECTIVE

Evolutionary theory excels at giving an ultimate "why" reason for diverse phenomena involving living beings. Diseases are a perfect example of something unpleasant with excellent biological reasons to exist, for life is a tale of conflict in which one organism is another's resource. Transitioning to a parasitic lifestyle is an immensely popular move in the tree of life, having evolved more than 200 times in animals alone (Weinstein and Kuris, 2016); similarly, fungal, bacterial, and viral life forms often (though not nearly always) cause diseases in their hosts. While it is tempting to think of cancer as a parasite, the fact that cancer typically dies with its host makes it rather resemble an evolutionary dead end—which has been argued to strongly limit the applicability of evolutionary thinking to this problem (Gardner, 2015; see also Chapter 1 discussing cellular vs. lineage selection). Before commenting on this, it is useful to remember that there is a terminological gray zone between cancers that occur very early in development and developmental disorders; whichever way we classify them, they can be selected against. A mutation in the germ line or an early somatic mutation can cause developmental disorders (Martincorena and Campbell, 2015), while the phenotypic effects of some later somatic mutations are labeled as cancers if they lead to neoplastic growth.

If we were to brush aside cancer as a mishap with an evolutionary dead end, in our view, we would forego the insights we can gain by considering that evolution can act on the cancer itself (Cunningham et al., 2015). The dead-end nature of competition between cell lineages does not prevent evolutionary processes from occurring. Cell lineages within cancers, like individual parasites and their progeny, can compete for resources and replace each other according to their demographic fitness (Altrock et al., 2015; Cunningham et al., 2015). However, while the short timescale and frequent dead ends experienced by cancer do not prevent evolution, they do—fortunately for us—limit the kinds of adaptations that can be expected, as cancers cannot accumulate innovations across successive generations in different hosts in the same way that host lineages can gain evolutionary experience dealing with harmful agents—including cancer (Haig, 2015). Thus, while we can use evolutionary theory to understand why cancer is plaguing us, evolutionary ideas should not be applied indiscriminately (Arnal et al., 2015). The probably most exciting evolutionary insight is that selection at the cell level really can be in conflict with selection at the organismal level. What we call cancer is the result of a cell lineage that no longer

cooperates to support the fitness of the whole organism, and instead promotes its own (short-term) fitness at the expense of the whole organism.

This conflict escalates to a new scale in cases where a cancerous lineage can escape the death of its host: now the analogy to traditional parasites becomes restored to a much greater extent than in "dead-end" cases. Contagious cancers range from benign (in dogs) to lethal (in Tasmanian devils and clams), and occur in organisms that have vastly different body sizes, lifestyles, and effective population sizes (review Ujvari et al., 2016). Their rarity makes it difficult to state anything general about them; they may simply be idiosyncratic exceptions to the rule that metazoan bodies have evolved to be very "alert" with respect to deviating cell lineages that arise within an individual's own soma. This then extrapolates to an even better ability to detect cancer cells that are "even more different" because they originate from another somatic cell lineage (and, in yet another step of extrapolation, to fierce rejection of tissue transplants). The exciting follow-up question is whether there is a common theme among the known examples in dogs, clams, and Tasmanian devils (Decker et al., 2015; Metzger et al., 2015; Ujvari et al., 2016; Weiss and Fassati, 2015) in terms of how somatic cancer cells transition to a parasitic lifestyle.

SINCE WHEN HAS CANCER BEEN PLAGUING US?

Cancers are usually a metazoan phenomenon, even if cancerous growth can impact other multicellular life. For example, plants also develop cancers, but the impact is much more localized because plants have cell walls, lack circulating cells, and grow according to a modular bauplan. Extant metazoans are highly differentiated, and out-of-control somatic growth is detrimental. Near the origins of multicellularity, on the other hand, we cannot assume a preexisting distinction between soma and germline. This role division itself had to evolve (Barfield et al., 2016).

Going back to our aquatic origins, the question of organismal identity becomes fuzzy. We suspect that there was a gray zone between production of an offspring (which increases fitness) and cancer (detrimental to fitness). In an illustration of how the interests of offspring and cancer are aligned, a recent simple experimental procedure for the initial steps of the evolution of multicellularity evolved yeast that forms snowflake-like clumps. Growth can be followed by branches of the flake breaking off, and the large parent now exists next to a small offspring (Ratcliff et al., 2015). Increased growth in this case would hardly be seen as cancer: in early multicellular organisms with totipotent cells, there is barely a distinction between propagation (offspring) and cancer. The strong conflict between noncancerous tissue (including offspring) and cancerous tissue only arose once there was a clear division of the germ line and soma.

In early multicellular organisms, it might be detrimental for the fitness prospects of all the entities if the emergent "soma" lost its ability to regenerate indefinitely. A clear challenge to evolutionary theory is to explain how a clear division of labor between germline and soma can be selectively advantageous despite this initial cost. Somatic cells not only obey signals that make them lose totipotency, the soma has also been argued to age faster because of the substantial costs of maintaining a functional germline (Maklakov and Immler, 2016).

Dividing into germ and soma cells brings about obvious benefits of distinct tasks performed by different organs, but it also has intriguing consequences in the form of chances of modulating the risks of somatic cancer, as well as inherited cancers (cancer risks and/or developmental failures passed on directly in the germ line). Given the inherent conflict between differentiation and totipotency, it should not come as a surprise that the basal metazoan hydra, whose germ cells and soma originate from the very same stem cells, also develops cancer (Domazet-Lošo et al., 2014). It is intriguing that hydra are apparently not plagued with tremendously high cancer rates, when one might predict it to be particularly cancer-prone because of the combination of morphological differentiation with a less than clear distinction between soma and germline: if germline stem cells are lost during budding, somatic stem cells normally destined to be, for example, nerve or gland cells can replace them (Nishimiya-Fujisawa and Kobayashi, 2012).

Indeed, there seems to be at least one cancer suppressing mechanism in hydra. Stem cells in hydra periodically stop proliferating for several cell cycles (i.e., become quiescent), which can reduce the genotoxic/metabolic stress and DNA-damage accompanying proliferation (Govindasamy et al., 2014). In this way, quiescent hydra stem cells are like quiescent adult mammalian stem cells, which are only recruited for regeneration and wound healing but not for normal physiological maintenance. However, quiescent hydra stem cells restart proliferating to maintain tissue homeostasis, while adult quiescent stem cells do not. As quiescence does not eliminate all kinds of DNA damage and the normal cell cycle offers regular DNA checks, restarting proliferation can repair some of the DNA damage that accumulates during long-term quiescence (Govindasamy et al., 2014). In more complex organisms, adult stem cells exhaust and lose their proliferation ability with time (sometimes through telomere attrition), which may be both an accepted cost of mitigating cancer risk through the soma-germline distinction, and a secondary cancer-prevention mechanism. Hydra, in contrast, does not show stem cell exhaustion and might have developed different secondary cancer suppressors.

SEX AND CANCER

Hydra can readily alternate between asexual and sexual reproduction. Once hydra exhibits cancerous cells, reproduction through both alleys is greatly reduced, but cancerous cells can still spread to vegetative offspring (Domazet-Lošo et al., 2014). It would thus be interesting to know whether a tumorous hydra shifts its resource allocation away from asexual to sexual reproduction, and thus paves a dead end for cancer. Imagine the wave of excitement rolling over the boulevard if we could place sex into the category of a cancer defense. Hydra of course did not invent sex (which is instead an ancient feature of eukaryotic life, Speijer et al., 2015), but a typical sexual life cycle—including that of hydra—involves reverting to a unicellular stage (the zygote, Grosberg and Strathmann, 1998).

Let us add another thought. Hydra, or at least an entire lineage of hydra, might be able to escape cancer but aspects of sexual reproduction can also contribute to cancer risk. Known cancerous cells in hydra seem to originate from female stem cells that have stopped differentiating (Domazet-Lošo et al., 2014), thus cancer might make use of the pathway to suppress female stem cell differentiation when male stem cells are present (Bosch and David, 1986) and of their ability to arrest apoptosis during egg production (Boettger and Alexandrova, 2007). So, what does hydra tell us? Its differentiated soma does senesce and is regularly replaced, but its stem cells do not exhaust like those of more complex organisms. The ability to maintain nonsenescing stem cells may allow hydra to escape ageing (Dańko et al., 2015; Schaible et al., 2015), but on the other hand these perpetually dividing cells can give rise to cancer, which then could be eliminated in progeny arising from sexual reproduction. The process of getting rid of faulty, mutated products (for a model see Dańko et al., 2015) draws a link between Hydra's way of managing stem cell lineages and more general theories of sex as a means to avoid Muller's ratchet—once again highlighting the ontogenetic gray zone between cancer and developmental failures.

Ageing sexually reproducing organisms pose another question: cancer risks might make reducing reproductive output at later ages beneficial. While reproductive senescence can simply be an unavoidable consequence of cellular senescence, selection on life-history scheduling could also be impacted by the chances that an organism develops cancer, either in its own soma (model: Kokko and Hochberg, 2015) or in its offspring, as de novo cancer-inducing mutations may be transmitted more often by older parents. Such risks can make continuing reproduction less valuable relative to early reproduction (or relative to prolonging parental care of already existing offspring, in those species with substantial generational overlap and parental care; see Brown and Aktipis, 2015 for the further idea that shifting resource allocation away from producing new offspring to caring for existing kin could include improved cancer defenses, for only a parent that is alive can provide care).

CAN A CANCER-PROOF ORGANISM REALLY EXIST?

The origins of some cancer-related genes, in particular human oncogenes and tumor suppressor genes, have been placed at our multicellular ancestor (Domazet-Lošo and Tautz, 2010). Theory predicts that under many commonly observed life histories most organisms die from other causes than cancer (Kokko and Hochberg, 2015) yet we only expect them to be cancer-proof in a probabilistic and lifespan-dependent rather than absolute sense. So if we look for a truly cancer-proof organism, we have to search for it among the unicellular organisms where the interests of a cell and its cancerous counterpart are indistinguishable. There, the gray zone between offspring production (i.e., cellular propagation) and cancer propagation does not exist. Indeed, there are close parallels between microbial evolution and somatic evolution (Sprouffske et al., 2012). If we, instead, broaden the search from an absolute absence of cancer to an investigation of interspecific differences in relative cancer risk, we enter a promising new field, fuelled by a resurrected interest in Peto's paradox.

Cope's rule describes the tendency for organisms to grow larger over macroevolutionary scales. It is called Cope's rule sensu stricto when it is a result of selection (for evidence see Kingsolver and Pfennig, 2007) rather than the mere mathematical byproduct of a process that combines increasing variance in body sizes over time with the obvious fact that no organism can diminish its size below zero. The generality and applicability across all taxa is a matter of debate (Butler and Goswami, 2008; Hone and Benton, 2007), but neither side of the debate certainly claims that evolution toward small size cannot occur.

Birds, for example, have dinosaurian ancestry, but the typical bird is much lighter than the average dinosaur was. In principle this could mean two things: (1) despite a difference in body size the evolution of birds did not involve decreasing body size—the particular lineage that led to birds might be one of very small dinosaurs; or (2) miniaturization really did occur. Based on modern phylogenetic methods we now know option (2) to be true (Lee et al., 2014). A cancer-related

thought arises immediately. To survive, lineages leading to birds must have been in the possession of evolutionary innovations that gave them cancer robustness appropriate for a large-bodied organism. Are these still present in extant birds? The likely answer depends on whether they are costly enough to be selected against when needed less than before, but we suspect being built "too well" for the current body size might have been retained in bird lineages, especially given the consistently longer lifespan of birds compared with similar-sized mammals (Lindstedt and Calder, 1981). Thus, there is another place to look for relatively cancer-proof organisms, namely miniaturized species. Looking for genomic and life history data on cancer traits of birds and other miniaturized species would be an exciting endeavor.

CONCLUSIONS

Studying the ecology and evolution of cancer across the tree of life not only complements our direct studies of cancer in humans—it could also inform us of major evolutionary principles. Unfortunately, we are plagued by a lack of quantitative data (Chapters 2 and 13). For example, it would be wonderful to be able to plot the age (or current size) of an organism on an x axis and the cumulative risk of cancer, ideally tissue-specifically (Noble et al., 2015 and also Chapters 1 and 7), on the y axis. In reality, we only have very blunt tools at hand for such assessments: lump measures, such as "percentages of deaths caused by cancer" are rarely available for nonhumans, let alone time graphs truly needed to evaluate the strength of cancer defenses, such as being able to graph cancer incidence conditional on the individual being alive at age t. Real data are of more anecdotal nature, and one would certainly welcome projects, such as the Golden Retriever Lifetime Study (Guy et al., 2015) but conducted for wild organisms—even if logistic issues pose some obvious limits on the quality of data outside the realm of domestication.

Finally, most cancer treatments focus on the somatic level—killing a particular type of cell, or disrupting a particular pathway. However, by identifying the organizing principles of cancer across taxa, in addition to learning something general about biology and cancer evolution, we may also identify new treatment and prevention approaches. The growing body of research across taxa on telomeres and telomerase, as well as rates of apoptosis and senescence will likely strengthen the link between ecology and the evolution of cancer. Whether one is interested in finding better treatments, or intellectually driven to understand how multicellular life can exist: the search for a cancer proof organism is rewarding even if we never find one.

Acknowledgments

We are grateful for comments on the manuscript from Pat Monaghan and André Grüning.

References

Abegglen, L.M., Caulin, A.F., Chan, A., Lee, K., Robinson, R., Campbell, M.S., Kiso, W.K., Schmitt, D.L., Waddell, P.J., Bhaskara, S., Jensen, S.T., Maley, C.C., Schiffman, J.D., 2015. Potential mechanisms for cancer resistance in elephants and comparative cellular response to DNA damage in humans. JAMA 314, 1850–1860.

Alexandrov, L.B., Jones, P.H., Wedge, D.C., Sale, J.E., Campbell, P.J., Nik-Zainal, S., Stratton, M.R., 2015. Clock-like mutational processes in human somatic cells. Nature Genet. 47, 1402–1407.

Altrock, P.M., Liu, L.L., Michor, F., 2015. The mathematics of cancer: integrating quantitative models. Nature Rev. Cancer 15, 730–745.

Arnal, A., Ujvari, B., Crespi, B., Gatenby, R.A., Tissot, T., Vittecoq, M., Ewald, P.W., Casali, A., Ducasse, H., Jacqueline, C., Missé, D., Renaud, F., Roche, B., Thomas, F., 2015. Evolutionary perspective of cancer: myth, metaphors, and reality. Evol. Appl. 8, 541–544.

Arnal, A., Tissot, T., Ujvari, B., Nunney, L., Solary, E., Laplane, L., Bonhomme, F., Vittecoq, M., Tasiemski, A., Renaud, F., Pujol, P., Roche, B., Thomas, F., 2016. The guardians of inherited oncogenic vulnerabilities. Evolution 70, 1–6.

Baker, D.J., et al., 2016. Naturally occurring p16(Ink4a)-positive cells shorten healthy lifespan. Nature 530, 184–189.

Barfield, S., Aglyamova, G.V., Matz, M.V., 2016. Evolutionary origins of germline segregation in Metazoa: evidence for a germ stem cell lineage in the coral Orbicella faveolata (Cnidaria Anthozoa). Proc. R. Soc. B 283, 20152128.

Birkhead, T.R., Monaghan, P., 2009. Ingenious ideas: the history of behavioral ecology. In: Westneat, D.F., Fox, C. (Eds.), Evolutionary Behavioral Ecology.

Blanckenhorn, W.U., 2000. The evolution of body size: what keeps organisms small? Q. Rev. Biol. 75, 385–407.

Boettger, A., Alexandrova, O., 2007. Programmed cell death in Hydra. Semin. Cancer Biol. 17 (2), 134–146.

Bosch, T.C.G., David, C.N., 1986. Male and Female Stem-Cells and Sex Reversal in Hydra Polyps. Proc. Nat. Acad. Sci. 83 (24), 9478–9482.

Brown, J.S., Aktipis, C.A., 2015. Inclusive fitness effects can select for cancer suppression into old age. Phil. Trans. R. Soc. B 370, 20150160.

Browning, H.M., Gulland, F.M.D., Hammond, J.A., Colegrove, K.M., Hall, A.J., 2015. Common cancer in a wild animal: the California sea lion (Zalophus californianus) as an emerging model for carcinogenesis. Phil. Trans. R. Soc. B 370, 20140228.

Butler, R.J., Goswami, A., 2008. Body size evolution in Mesozoic birds: little evidence for Cope's rule. J. Evol. Biol. 21, 1673–1682.

Caulin, A.F., Maley, C.C., 2011. Peto's paradox: evolution's prescription for cancer prevention. Trends Ecol. Evol. 26, 175–182.

Caulin, A.F., Graham, T.A., Wang, L.-S., Maley, C.C., 2015. Solutions to Peto's paradox revealed by mathematical modelling and cross-species cancer gene analysis. Phil. Trans. R. Soc. B 370, 20140222.

Cunningham, J.J., Brown, J.S., Vincent, T.L., Gatenby, R.A., 2015. Divergent and convergent evolution in metastases suggest treatment strategies based on specific metastatic sites. Evol. Med. Public Health 76, 87.

Daan, S., Deerenberg, C., Dijkstra, C., 1996. Increased daily work precipitates natural death in the kestrel. J. Anim. Ecol. 65, 539–544.

Dang, C.V., 2015. A metabolic perspective of Peto's paradox and cancer. Phil. Trans. R. Soc. B 370, 20140223.

Dańko, M.J., Kozlowski, J., Schaible, R., 2015. Unraveling the non-senescence phenomenon in Hydra. J. Theor. Biol. 382, 137–149.

Dawkins, R., 2013. An Appetite for Wonder: The Making of a Scientist. Ecco, New York, 308 p.

Decker, B., et al., 2015. Comparison against 186 canid whole-genome sequences reveals survival strategies of an ancient clonally transmissible canine tumor. Genome Res. 25, 1646–1655.

Delaney, M.A., Ward, J.M., Walsh, T.F., Chinnadurai, S.K., Kerns, K., Kinsel, M.J., Treuting, P.M., 2016. Initial case reports of cancer in naked mole-rats (Heterocephalus glaber). Vet. Pathol. 53, 691–696.

Domazet-Lošo, T., Tautz, D., 2010. Phylostratigraphic tracking of cancer genes suggests a link to the emergence of multicellularity in metazoa. Biology 66, 1–10.

Domazet-Lošo, T., Klimovich, A., Anokhin, B., Anton-Erxleben, F., Hamm, M.J., Lange, C., Bosch, T.C.G., 2014. Naturally occurring tumours in the basal metazoan Hydra. Nature Comm. 5, 4222.

Ducasse, H., Ujvari, B., Solary, E., Vittecoq, M., Arnal, A., Bernex, F., Pirot, N., Misse, D., Bonhomme, F., Renaud, F., Thomas, F., Roche, B., 2015. Can Peto's paradox be used as the null hypothesis to identify the role of evolution in natural resistance to cancer? A Crit. Rev. BMC Cancer 15, 792.

Gardner, A., 2015. The genetical theory of multilevel selection. J. Evol. Biol. 28, 305–3019.

Gomes, N.M.V., Shay, J.W., Wright, W.E., 2010. Telomeres and telomerase. In: Wolf, N. (Ed.), The Comparative Biology of Aging. Springer, pp. 227–258.

Govindasamy, N., Murthy, S., Ghanekar, Y., 2014. Slow-cycling stem cells in hydra contribute to head regeneration. Biol. Open 3 (12), 1236–1244.

Grosberg, R.K., Strathmann, R.R., 1998. One cell, two cell, red cell, blue cell: the persistence of a unicellular stage in multicellular life histories. Trends Ecol. Evol. 13, 112–116.

Guy, M.K., Page, R.L., Jensen, W.A., Olson, P.N., Haworth, J.D., Searfoss, E.E., Brown, D.E., 2015. The golden retriever lifetime study: establishing an observational cohort study with translational relevance for human health. Phil. Trans. R. Soc. B 370, 20140230.

Haig, D., 2015. Maternal–fetal conflict, genomic imprinting and mammalian vulnerabilities to cancer. Phil. Trans. R. Soc. B 370, 20140178.

Haussmann, M.F., Winkler, D.W., O'Reilly, K.M., Huntington, C.E., Nisbet, I.C.T., Vleck, C.M., 2003. Telomeres shorten more slowly in long-lived birds and mammals than in short-lived ones. Proc. R. Soc. Lond. B 270, 1387–1392.

Hone, D.W.E., Benton, M.J., 2007. Cope's Rule in the Pterosauria, and differing perceptions of Cope's Rule at different taxonomic levels. J. Evol. Biol. 20, 1164–1170.

Huveneers, C., Klebe, S., Fox, A., Bruce, B., Robbins, R., Borucinska, J.D., Jones, R., Michael, M.Z., 2016. First histological examination of a neoplastic lesion from a free-swimming white shark, Carcharodon carcharias L. J. Fish Dis. 39, 1269–1273.

Kang, T.-W., Yevsa, T., Woller, N., Hoenicke, L., Wuestefeld, T., Dauch, D., Hohmeyer, A., Gereke, M., Rudalska, R., Potapova, A., Iken, M., Vucur, M., Weiss, S., Heikenwalder, M., Khan, S., Gil, J., Bruder, D., Manns, M., Schirmacher, P., Tacke, F., Ott, M., Tuedde, T., Longerich, T., Kubicka, S., Zender, L., 2012. Senescence surveillance of pre-malignant hepatocytes limits liver cancer development. Nature 479, 547–551.

Kingsolver, J.G., Pfennig, D.W., 2007. Individual-level selection as a cause of Cope's rule of phyletic size increase. Evolution 58, 1608–1612.

Kleiber, M., 1932. Body size and metabolism. Hilgardia 6, 315–349.

Kokko, H., Hochberg, M.E., 2015. Towards cancer-aware life-history modelling. Phil. Trans. R. Soc. B 370, 20140234.

Kolokotrones, T., Van Savage, E., Deeds, E.J., Fontana, W., 2010. Curvature in metabolic scaling. Nature 464, 753–756.

Lee, M.S.Y., Cau, A., Naish, D., Dyke, G.J., 2014. Sustained miniaturization and anatomical innovation in the dinosaurian ancestors of birds. Science 345, 562–566.

Lewis, Z., Price, T.A.R., Wedell, N., 2008. Sperm competition, immunity, selfish genes and cancer. Cell. Mol. Sci. 65, 3241–3254.

Lindstedt, S.L., Calder, III, W.A., 1981. Body size, physiological time, and longevity of homeothermic animals. Q. Rev. Biol. 56, 1–16.

Lynch, M., 2011. The lower bound to the evolution of mutation rates. Genome Biol. Evol. 3, 1107–1118.

Maciak, S., Michalak, P., 2015. Cell size and cancer: a new solution to Peto's paradox? Evol. Appl. 8, 2–8.

Maklakov, A.A., Immler, S., 2016. The expensive germline and the evolution of ageing. Curr. Biol. 26, R577–R586.

Martincorena, I., Campbell, P.J., 2015. Somatic mutation in cancer and normal cells. Science 349, 1483–1489.

Metzger, M.J., Reinisch, C., Sherry, J., Goff, S.P., 2015. Horizontal transmission of clonal cancer cells causes leukemia in soft-shell clams. Cell 161, 255–263.

Monaghan, P., 2014. Organismal stress, telomeres and life histories. J. Exp. Biol. 217, 57–66.

Murphy, M.E., 2013. The HSP70 family and cancer. Carcinogenesis 34, 1181–1188.

Nishimiya-Fujisawa, C., Kobayashi, S., 2012. Germline stem cells and sex determination in Hydra. Int. J. Dev. Biol. 56, 499–508.

Noble, R., Kaltz, O., Hochberg, M.E., 2015. Peto's paradox and human cancers. Phil. Trans. R. Soc. B 370, 20150104.

Nunney, L., 1999. Lineage selection and the evolution of multistage carcinogenesis. Proc. R. Soc. Lond. B 266, 493–498.

Nunney, L., 2013. The real war on cancer: the evolutionary dynamics of cancer suppression. Evol. Appl. 6, 11–19.

Nunney, L., Muir, B., 2015. Peto's paradox and the hallmarks of cancer: constructing an evolutionary framework for understanding the incidence of cancer. Phil. Trans. R. Soc. B 370, 20150161.

Peto, R., 2015. Quantitative implications of the approximate irrelevance of mammalian body size and lifespan to lifelong cancer risk. Phil. Trans. R. Soc. B 370, 20150198.

Ratcliff, W.C., Fankhauser, J.D., Rogers, D.W., Greig, D., Travisano, M., 2015. Origins of multicellular evolvability in snowflake yeast. Nature Comm. 6, 6102.

Rodriguez-Brenes, I.A., Wodarz, D., Komarova, N.L., 2015. Quantifying replicative senescence as a tumor suppressor pathway and a target for cancer therapy. Sci. Rep. 5, 17660.

Schaible, R., Scheuerlein, A., Danko, M.J., Gampe, J., Martinez, D.E., Vaupel, J.W., 2015. Constant mortality and fertility over age in Hydra. P. Natl. Acad. Sci. USA 112 (51), 15701–15706.

Smith, J., Daniel, R., 2012. Stem cells and aging: a chicken-or-the-egg issue? Aging Dis. 3 (3), 260–268.

Speijer, D., Lukes, J., Elias, M., 2015. Sex is a ubiquitous, ancient, and inherent attribute of eukaryotic life. Proc. Natl. Acad. Sci. USA 112, 8.

Sprouffske, K., Merlo, L.M.F., Gerrish, P.J., Maley, C.C., Sniegowski, P.D., 2012. Cancer in light of experimental evolution. Curr. Biol. 22, R762–R771.

Ujvari, B., Gatenby, R.A., Thomas, F., 2016. The evolutionary ecology of transmissible cancers. Infect. Genet. Evol. 39, 293–303.

Vogelstein, B., Lane, D., Levine, A.J., 2000. Surfing the p53 network. Nature 408, 307–310.

Weinstein, S.B., Kuris, A.M., 2016. Independent origins of parasitism in Animalia. Biol. Lett. 12, 20160324.

Weiss, R.A., Fassati, A., 2015. The clammy grid of parasitic tumors. Cell 161, 191–192.

Whitesell, L., Lindquist, S.L., 2005. HSP90 and the chaperoning of cancer. Nat. Rev. Cancer 5, 761–772.

23

Ecology, Evolution, and the Cancer Patient

Andrew F. Read

Center for Infectious Disease Dynamics, Pennsylvania State University,
University Park, PA, United States

Incredibly, experimental cancer treatments are no more successful today than they were in the middle of the last century. Patients are better off, of course: childhood cancers and breast cancers are much more curable than they used to be. That is because oncologists rapidly learn from successful clinical trials. But the chance that any particular experimental cancer treatment will work is no higher now than it was in the 1950s. We know this from extensive metaanalyses. Today, just as in the middle of the last century, patients in the control arm of randomized trials are as well off as those in the experimental arm. The average effect size is essentially zero, a sobering number that has not changed over 60 years (Djulbegovic et al., 2008, 2013; Kumar et al., 2005). In one important sense, this is good news: randomized control trials remain an ethically sound way to test the efficacy of new therapeutic interventions. But in another important sense, the observation is something of an affront. How is it that the spectacular achievements of molecular and cellular biology together with terrific advances in cell culture, animal models, and trial design have failed to improve our ability to identify novel treatments that help patients? It is as if something is missing.

Of course everyone in cancer research thinks they know what is missing: it is the thing they are working on. It is an inadequate knowledge of the epigenetic mechanisms, insufficient deep sequencing data, and a poor understanding of mechanisms of cell cycle control. Judging from the grants awarded by the US National Cancer Institute, perhaps the prevailing general view is the one given recently by *The Economist* magazine: "The main reason cancer has been such a hard problem to tackle is a lack of basic understanding of the underlying molecular mechanisms that drive it" (Anon, 2016a). Reading through the chapters in this book, it is clear that ignorance of molecular mechanisms is indeed important. But

every second chapter (almost literally) makes a strong case that ignorance of ecological and evolutionary mechanisms is just as important.

I agree. There has never been a more important time to study the ecology of neoplastic cells, and in particular to study them in what Michael Hochberg (in press) calls the disease ecosystem. We now know that the huge array of diseases we call cancer are all the result of evolutionary processes happening in clinically relevant time (Chapters 1, 10). Therapeutic breakthroughs have to involve finding new ways to control that evolution. Molecular mechanisms are the stuff that evolution chews on. But patient health depends on what evolution does with them—and that depends on the ecology. An analogy: if you want to fix traffic congestion, it is important to understand how a combustion engine works and the constraints on its performance. But alone, such knowledge will not produce a fix.

Evolution is one of the most potent life forces known. Whenever humans have tried to deliberately extinguish life, supposedly magic bullets such as antibiotics, antifungals, herbicides, pesticides, and rodenticides eventually lost efficacy in the face of evolution (Greene and Reid, 2012). "Use 'em and lose 'em" is the rule. The situation is even more sobering in cancer. Normal and neoplastic cells share a very recent common ancestor, and so there are few potential targets for magic bullets. Worse, cancers become more evolvable as malignancy progresses. This has several implications. Perhaps most importantly: prevention, prevention, prevention (Greaves and Maley, 2012). Second, treatment regimens have to be optimized to slow resistance emergence (Chapters 8, 10, 14). This, as I argue below, is a largely unstudied problem in applied population ecology. Importantly, conventional treatment aims, like minimizing tumor burden at the end of treatment, can exacerbate

Ecology and Evolution of Cancer. http://dx.doi.org/10.1016/B978-0-12-804310-3.00023-5

the resistance problem (Costa and Boldrini, 1997; Hansen et al., in press). And third, given the huge expense of novel therapeutics, investment has to be directed at evolution-proof targets. Note that "evolution-proofing" is like water-proofing. Ideally it is perfect, but substantially delaying water ingress is still a gain.

In other contexts, particularly agriculture and infectious diseases, there is a considerable theory and sometimes compelling evidence that partial or complete evolution-proofing is possible (Greene and Reid 2012; Consortium, 2013). A key message from those fields is that there is simply no understanding evolution without understanding the ecological context in which it is happening. It is tempting to think that with modern sequencing tools, cancer management is a question of reconstructing phylogenies, identifying driver mutations, and coming in hard with drugs targeted at cell lineages specific to an individual patient's tumor. But as with political history, it is easy to see how things could have been changed—after the event. More challenging is to predict and change the future before it happens. For cancer, that requires strategies that slow or prevent the emergence of molecular mechanisms which are not yet detectable.

That is possible. The process of a population evolving itself out of trouble is called evolutionary rescue (Gonzalez et al., 2013). We know that advanced malignancies are fantastically good at that: cancer cells have ferocious capacity to adapt to the insults oncologists throw at them. The rate at which adaptation can save a population from extinction depends primarily on the rate at which heritable variation arises (which can be fearsome in a tumor) and on natural selection and drift acting on that heritable variation. Natural selection and drift (demography) are ecological forces (Gonzalez et al., 2013; Uecker et al., 2014). Understanding the ecology is therefore an essential part of understanding cancer.

This is nowhere clearer than in the context of cancer chemotherapy. Every day, oncologists battle to keep their patients alive. When they lose that battle, as they will almost 600,000 times a year in the US alone (Anon, 2016b), it is largely because they could not tame resistance evolution. The fundamentals of that evolutionary process are essentially the same in all tumors. In the absence of drug treatment, the population size of resistant cells is tiny. Aggressive chemotherapy completely remodels the ecosystem experienced by those resistant cells. For instance, the therapy-sensitive cells are removed, enabling a vast amplification of resistance (Chapters 14, 19). That amplification process, or the ones that follow in subsequent rounds of the arm race between oncologists and tumors, is what kills the patient.

There are just two ways to prevent or delay therapeutic failure (Day and Read, 2016; Read et al., 2011). The first is to prevent resistance arising in the first place. This is what happens in modern HIV therapy; the right combination of drugs in fully compliant patients prevents resistant mutants. The second is to try to delay or prevent the emergence of resistance when it is present. Given that resistance to many chemotherapeutic agents is already present in a tumor when treatment begins, managing the emergence phase is often the only hope for the patient. This means managing the population dynamics of therapy-resistant cells. To do that rationally involves a more detailed understanding of the relevant ecology than we currently have. For instance, at the heart of the problem is a 'simple' trade-of involving therapy-sensitive neoplastic cells (Hansen et al., in press). Sensitive cells are a potential source of resistance, since they can acquire (epi)genetic changes that confer therapy-resistance (Chapters 5, 6, 10). But they also suppress populations of resistant cells. This must be one very important reason why resistance is so rare prior to treatment: sensitive cell lineages are keeping them in check. Thus two opposing forces—competitive suppression and resistance acquisition—together determine the fate of the patient. Several ingenious solutions for resistance management revolve around trying to tip these forces in the patients' favor (Baym et al., 2016; Maley et al., 2004; Willyard, 2016) (Chapters 10, 14, 21). For example, treatment regimens aimed at containing a tumor may prolong patient life longer than regimens which attempt to eliminate a tumor (Chapter 14). Critically, the ecology of cell–cell competition determines when containment can make things better, and when it makes the prognosis worse (Hansen et al., in press).

Yet we barely understand the nature of competition occurring within and between cancer cell lineages. It is clear that it can occur (Chapters 1, 4, 8, 10, 14, 19, 21). But is it scramble competition, with cell lineages proliferating in a resource rich environment at the edge of a tumor? Or are resources limiting, so that density dependent effects are important? In many cases, resistance mechanisms come with fitness costs (Chapter 14). Fitness costs are highly dependent on environmental context. Fitness costs can affect growth in an unconstrained environment, or they can affect competitive ability in density-dependent environment. The evolutionary consequences are very different. Quite possibly a variety of different competitive interactions are going on at once, even for the same type of genetic resistance mechanism in the same location. I know of no work looking quantitatively at competition between resistant and sensitive neoplastic cell lineages across resource gradients, the simplest and most fundamental ecological question. What nutrients or resources are involved? How does immunity modify competition? Where is any density-dependence coming from? How does therapy modify that? How best can we modify competition or fitness costs to enhance patient health?

Over recent decades, science has generated rich catalogs of the genetic events that can cause therapeutic resistance (Housman et al., 2014). But once resistance has arisen, ecological forces determine the fate of those resistance mechanisms and, among other things, how long a patient will live. Mathematical models can do a lot to capture the possible processes and study the potential impact of contrasting therapeutic options (Beerenwinkel et al., 2015; Bozic et al., 2013; Diaz et al., 2012; Foo and Michor, 2014). Indeed mathematical models are essential, not least because it is impossible to empirically investigate the wide variety of possible treatment regimens. But there is a frustrating lack of empirical cancer ecology from which such models can be parameterized. For instance, it is popular to model tumor growth as a Gompertz process (Foo and Michor, 2014). When is it Gompertzian and when is it Logistic, and in either case, what are the ecological mechanisms underlying whatever phenomenological model we do fit?

Solutions to global challenges in conservation biology, control of invasive species, and the management of resistance in agricultural all benefit from a thorough understanding of the ecological context (Alexander et al., 2014; Edward et al., 2009). Those problems are directly analogous to the problem of controlling resistance in the cancer ecosystem. General principles are usually clear; the solutions that flow from them are specific to the particular setting because details matter. One-size-fits-all rules seldom work (Day and Read, 2016; Hansen et al., in press). When we understand particular disease ecosystems in the way that card-carrying ecologists understand more traditional ecosystems, novel solutions to cancer will suggest themselves—and we will be able to make more effective use of the precious chemotherapeutic agents we already have.

In his small but powerful book *Ignorance, How it Drives Science*, Stuart Firestein, a neuroscientist, makes the interesting case that the discovery of voltage spikes in the brain and sensory organs in the early part of the 20th century was a mixed blessing (Firestein, 2012). Spike analysis occupied neuroscience for the better part of a century and generated a vast mountain of data and facts about spikes. But because of the focus on spikes, electrical signals more subtle than spikes and chemical processes that were not electrical were missed for decades. These nonspike processes may turn out to be as important as the spikes. I can't help, wonder if we will look back on this era of molecular oncology and wonder why for so long we missed the ecology.

Acknowledgment

I thank Elsa Hansen for clarifying my thinking.

References

Alexander, H.K., Martin, G., Martin, O.Y., Bonhoeffer, S., 2014. Evolutionary rescue: linking theory for conservation and medicine. Evol. Appl. 7, 1161–1179.

Anon, 2016a. Why cancer has not been cured. Available from: http://www.economist.com/blogs/economist-explains/2016/07/economist-explains-2: The Economist.

Anon, 2016b. Cancer Facts & Figures 2016. American Cancer Society, Atlanta.

Baym, M., Stone, L.K., Kishony, R., 2016. Multidrug evolutionary strategies to reverse antibiotic resistance. Science 351, aad3292.

Beerenwinkel, N., Schwarz, R.F., Gerstung, M., Markowetz, F., 2015. Cancer evolution: mathematical models and computational inference. Syst. Biol. 64, E1–E25.

Bozic, I., Reiter, J.G., Allen, B., Antal, T., Chatterjee, K., Shah, P., et al., 2013. Evolutionary dynamics of cancer in response to targeted combination therapy. elife 2, 15.

Consortium, R., 2013. Heterogeneity of selection and the evolution of resistance. Tr. Ecol. Evol. 28, 110–118.

Costa, M.I.S., Boldrini, J.L., 1997. Conflicting objectives in chemotherapy with drug resistance. Bull. Math. Biol. 59, 707–724.

Day, T., Read, A.F., 2016. Does high-dose antimicrobial chemotherapy prevent the evolution of resistance? PLoS Comput. Biol. 12, 20.

Diaz, L.A., Williams, R.T., Wu, J., Kinde, I., Hecht, J.R., Berlin, J., et al., 2012. The molecular evolution of acquired resistance to targeted EGFR blockade in colorectal cancers. Nature 486, 537–540.

Djulbegovic, B., Kumar, A., Soares, H.P., Hozo, I., Bepler, G., Clarke, M., et al., 2008. Treatment success in cancer. Arch. Intern. Med. 168, 632–642.

Djulbegovic, B., Kumar, A., Miladinovic, B., Reljic, T., Galeb, S., Mhaskar, A., et al., 2013. Treatment success in cancer: industry compared to publicly sponsored randomized controlled trials. PLoS One 8, 15.

Edward, B., Radcliffe, E.B., Hutchison, W.D., Cancelado, R. (Eds.), 2009. Integrated Pest Management: Concepts, Tactics, Strategies and Case Studies. Cambridge University Press, Cambridge.

Firestein, S., 2012. Ignorance. How it Drives Science. Oxford University Press, Oxford.

Foo, J., Michor, F., 2014. Evolution of acquired resistance to anti-cancer therapy. J. Theor. Biol. 355, 10–20.

Gonzalez, A., Ronce, O., Ferriere, R., Hochberg, M.E., 2013. Evolutionary rescue: an emerging focus at the intersection between ecology and evolution. Philos. Trans. R. Soc. B Biol. Sci. 368, 20120404.

Greaves, M., Maley, C.C., 2012. Clonal evolution in cancer. Nature 481, 306–313.

Greene, S.E., Reid, A., 2012. Moving Targets: Fighting the Evolution of Resistance in Infections, Pests and Cancer. American Academy of Microbiology, Washington.

Hansen, E., Woods, R.J., Read, A.F., in press. How to use a chemotherapeutic agent when resistance to it threatens the patient. PLoS Biol.

Hochberg, M.E., in press. Six wedges to curing disease. In: Dobson, A.P., Tilman, G.D., Holt, R.D. (Eds.), Unsolved Problems in Ecology, Princeton University Press, Princeton.

Housman, G., Byler, S., Heerboth, S., Lapinska, K., Longacre, M., Snyder, N., et al., 2014. Drug resistance in cancer: an overview. Cancers 6, 1769–1792.

Kumar, A., Soares, H., Wells, R., Clarke, M., Hozo, L., Bleyer, A., et al., 2005. Are experimental treatments for cancer in children superior to established treatments? Observational study of randomised controlled trials by the Children's Oncology Group. Br. Med. J. 331, 1295–1298B.

Maley, C.C., Reid, B.J., Forrest, S., 2004. Cancer prevention strategies that address the evolutionary dynamics of neoplastic cells: simulating benign cell boosters and selection for chemosensitivity. Cancer Epidemiol. Biomarkers Prev. 13, 1375–1384.

Read, A.F., Day, T., Huijben, S., 2011. The evolution of drug resistance and the curious orthodoxy of aggressive chemotherapy. Proc. Natl. Acad. Sci. USA 108, 10871–10877.

Uecker, H., Otto, S.P., Hermisson, J., 2014. Evolutionary rescue in structured populations. Am. Nat. 183, E17–E35.

Willyard, C., 2016. Cancer: an evolving threat. Nat. Genet. 532, 166–168.

Index

259

Printed in the United States
By Bookmasters